Categorical Data Analysis
Using
The *SAS* System

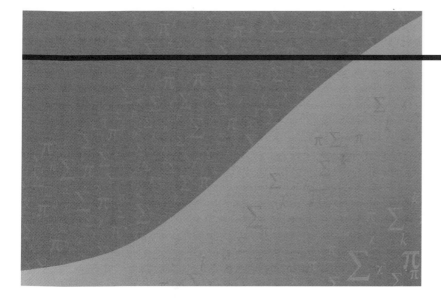

2nd Edition

Maura E. Stokes

Charles S. Davis

Gary G. Koch

The correct bibliographic citation for this manual is as follows: Stokes, Maura E., Charles S. Davis, and Gary G. Koch. 2000. *Categorical Data Analysis Using the SAS® System, Second Edition.* Cary, NC: SAS Institute Inc.

Categorical Data Analysis Using the SAS® System, Second Edition

Table of Contents

Preface to the Second Edition

This second edition contains several new topics and includes numerous updates to reflect Version 8 of the SAS System. Chapter 15, "Generalized Estimating Equations," is a new chapter that discusses the use of the GEE method, particularly as a tool for analyzing repeated measurements data. The book includes several comparisons of analyses using the GEE method, weighted least squares, and conditional logistic regression; the use of subject-specific models versus population-averaged models is discussed. Chapter 15 also describes the use of GEE methods for some univariate response situations.

Chapter 12, "Poisson Regression," is a new chapter on Poisson regression. Previously, this topic was described in the chapter on time-to-event categorical data. The methodology is illustrated with several examples.

Chapters on the analysis of tables now include much more material on the use of exact tests of association, particularly Chapter 2, "The 2×2 Table," and Chapter 5, "The $s \times r$ Table."

Exact logistic regression using the LOGISTIC procedure is discussed in Chapter 8, "Logistic Regression I: Dichotomous Response." Chapter 8 also describes the use of the CLASS statement in PROC LOGISTIC, and all of the examples in the various chapters using PROC LOGISTIC have been updated to take advantage of the new CLASS statement. Chapter 10, "Conditional Logistic Regression," has been largely revised to put more emphasis on the stratified data setting.

In addition, miscellaneous revisions and additions appear throughout the book.

Computing Details

Writing a book for software that is constantly changing is not straightforward. This second edition is targeted for Version 8 of the SAS System and takes advantage of many of the features of that release. The examples were executed with the 8.1 release on the HP UNIX platform, but most of the output can be reproduced using Version 8.0 with the following changes for Release 8.1:

- PROC LOGISTIC adds exact logistic regression.

- PROC GENMOD models, by default, the probability of the lowest ordered response variable levels. (The default has been changed from previous releases to make it consistent with other procedures.)

To make things a little more complicated, the authors used an output template for the LOGISTIC procedure that will become the default in Release 8.2. The main difference is that the label for the chi-square statistic in the parameter estimates table is "Wald Chi-Square" in Release 8.2 (which was the label used in Version 6).

Note that, because of limited space, not all of the output that is produced with the example SAS code is shown. Generally, the output pertinent to the discussion is displayed. An ODS SELECT statement is sometimes used in the example code to limit the tables produced.

For those users still running earlier versions of the SAS System, such as Release 6.09E on the mainframe and Release 6.12 on UNIX and PC platforms, the main additions to those releases with Version 8 are the CLASS statement in the LOGISTIC procedure, the inclusion of complete GEE facilities in the GENMOD procedure, and the availability of exact p-values for many of the tests produced by the FREQ procedure. The first example in Chapter 8 discusses how to use indicator variables, and the remaining logistic regression examples can be performed with indicator variables as well. Release 6.12 does contain a preliminary version of the GEE facility in PROC GENMOD; refer to the documentation for that release for more detail.

Some of the procedures such as PROC FREQ are printing more digits for various statistics and parameter estimates than they did in previous releases of the SAS System. This was done mainly to make the procedures more consistent with each other.

For More Information

The Website www.sas.com/catbook contains further information pertaining to topics in the book, including archives and errata. In the future, these Web pages will also provide information on using new features in SAS software for categorical data analysis, as well as contain examples and references on methodological advances.

Acknowledgments

The second edition proved to be a substantial undertaking. We are thankful for getting a lot of help along the way.

We would like to thank Ozkan Zengin for his assistance in bringing this book up to date in a number of ways, including adaptation to a new publishing system and running and checking all of the examples. Dan Spitzner provided careful proofing.

Numerous colleagues contributed to this book with their conversations, reviews, and suggestions, and we are very grateful for their time and effort. We thank Bob Derr, Diane Catellier, Gordon Johnston, Lisa LaVange, John Preisser, David Schlotzhauer, Todd Schwartz, and Donna Watts.

And, of course, we remain thankful to those persons who helped to launch the first edition with their sundry feedback. They include Sonia Davis, William Duckworth II, Suzanne Edwards, Stuart Gansky, Greg Goodwin, Wendy Greene, Duane Hayes, Allison Kinkead, Antonio Pedroso-de-Lima, Annette Sanders, Catherine Tangen, Lisa Tomasko, and Greg Weier.

We also thank our many readers who found the book useful and encouraged its continuing life in a second edition.

Virginia Clark edited this book.

Ginny Matsey designed the cover.

Tim Arnold provided documentation programming support.

Chapter 1

Introduction

Chapter Table of Contents

Chapter 1
Introduction

1.1 Overview

Data analysts often encounter response measures that are categorical in nature; their outcomes reflect categories of information rather than the usual interval scale. Frequently, categorical data are presented in tabular form, known as contingency tables. Categorical data analysis is concerned with the analysis of categorical response measures, regardless of whether any accompanying explanatory variables are also categorical or are continuous. This book discusses hypothesis testing strategies for the assessment of association in contingency tables and sets of contingency tables. It also discusses various modeling strategies available for describing the nature of the association between a categorical response measure and a set of explanatory variables.

An important consideration in determining the appropriate analysis of categorical variables is their scale of measurement. Section 1.2 describes the various scales and illustrates them with data sets used in later chapters. Another important consideration is the sampling framework that produced the data; it determines the possible analyses and the possible inferences. Section 1.3 describes the typical sampling frameworks and their ramifications. Section 1.4 introduces the various analysis strategies discussed in this book and describes how they relate to one another. It also discusses the target populations generally assumed for each type of analysis and what types of inferences you are able to make to them. Section 1.5 reviews how the SAS System handles contingency tables and other forms of categorical data. Finally, Section 1.6 provides a guide to the material in the book for various types of readers, including indications of the difficulty level of the chapters.

1.2 Scale of Measurement

The scale of measurement of a categorical response variable is a key element in choosing an appropriate analysis strategy. By taking advantage of the methodologies available for the particular scale of measurement, you can choose a well-targeted strategy. If you do not take the scale of measurement into account, you may choose an inappropriate strategy that could lead to erroneous conclusions. Recognizing the scale of measurement and using it properly are very important in categorical data analysis.

Categorical response variables can be

- dichotomous
- ordinal
- nominal
- discrete counts
- grouped survival times

Dichotomous responses are those that have two possible outcomes—most often they are yes and no. Did the subject develop the disease? Did the voter cast a ballot for the Democratic or Republican candidate? Did the student pass the exam? For example, the objective of a clinical trial for a new medication for colds is whether patients obtained relief from their pain-producing ailment. Consider Table 1.1, which is analyzed in Chapter 2, "The 2 × 2 Table."

Table 1.1.   Respiratory Outcomes

Treatment	Favorable	Unfavorable	Total
Placebo	16	48	64
Test	40	20	60

The placebo group contains 64 patients, and the test medication group contains 60 patients. The columns contain the information concerning the categorical response measure: 40 patients in the Test group had a favorable response to the medication, and 20 subjects did not. The outcome in this example is thus dichotomous, and the analysis investigates the relationship between the response and the treatment.

Frequently, categorical data responses represent more than two possible outcomes, and often these possible outcomes take on some inherent ordering. Such response variables have an *ordinal* scale of measurement. Did the new school curriculum produce little, some, or high enthusiasm among the students? Does the water exhibit low, medium, or high hardness? In the former case, the order of the response levels is clear, but there is no clue as to the relative distances between the levels. In the latter case, there is a possible distance between the levels: medium might have twice the hardness of low, and high might have three times the hardness of low. Sometimes the distance is even clearer: a 50% potency dose versus a 100% potency dose versus a 200% potency dose. All three cases are examples of ordinal data.

An example of an ordinal measure occurs in data displayed in Table 1.2, which is analyzed in Chapter 9, "Logistic Regression II: Polytomous Response." A clinical trial investigated a treatment for rheumatoid arthritis. Male and female patients were given either the active treatment or a placebo; the outcome measured was whether they showed marked, some, or no improvement at the end of the clinical trial. The analysis uses the proportional odds model to assess the relationship between the response variable and gender and treatment.

Table 1.2. Arthritis Data

Sex	Treatment	Improvement Marked	Some	None	Total
Female	Active	16	5	6	27
Female	Placebo	6	7	19	32
Male	Active	5	2	7	14
Male	Placebo	1	0	10	11

Note that categorical response variables can often be managed in different ways. You could combine the Marked and Some columns in Table 1.2 to produce a dichotomous outcome: No Improvement versus Improvement. Grouping categories is often done during an analysis if the resulting dichotomous response is also of interest.

If you have more than two outcome categories, and there is no inherent ordering to the categories, you have a *nominal* measurement scale. Which of four candidates did you vote for in the town council election? Do you prefer the beach, mountains, or lake for a vacation? There is no underlying scale for such outcomes and no apparent way in which to order them.

Consider Table 1.3, which is analyzed in Chapter 5, "The $s \times r$ Table." Residents in one town were asked their political party affiliation and their neighborhood. Researchers were interested in the association between political affiliation and neighborhood. Unlike ordinal response levels, the classifications Bayside, Highland, Longview, and Sheffeld lie on no conceivable underlying scale. However, you can still assess whether there is association in the table, which is done in Chapter 5.

Table 1.3. Distribution of Parties in Neighborhoods

Party	Neighborhood Bayside	Highland	Longview	Sheffeld
Democrat	221	160	360	140
Independent	200	291	160	311
Republican	208	106	316	97

Categorical response variables sometimes contain *discrete counts*. Instead of falling into categories that are labeled (yes, no) or (low, medium, high), the outcomes are numbers themselves. Was the litter size 1, 2, 3, 4, or 5 members? Did the house contain 1, 2, 3, or 4 air conditioners? While the usual strategy would be to analyze the mean count, the assumptions required for the standard linear model for continuous data are often not met with discrete counts that have small range; the counts are not distributed normally and may not have homogeneous variance.

For example, researchers examining respiratory disease in children visited children in different regions two times and determined whether they showed symptoms of respiratory illness. The response measure was whether the children exhibited symptoms in 0, 1, or 2 periods. Table 1.4 contains these data, which are analyzed in Chapter 13, "Weighted Least Squares."

Table 1.4. Colds in Children

Sex	Residence	Periods with Colds			Total
		0	1	2	
Female	Rural	45	64	71	180
Female	Urban	80	104	116	300
Male	Rural	84	124	82	290
Male	Urban	106	117	87	310

The table represents a cross-classification of gender, residence, and number of periods with colds. The analysis is concerned with modeling mean colds as a function of gender and residence.

Finally, another type of response variable in categorical data analysis is one that represents *survival times*. With survival data, you are tracking the number of patients with certain outcomes (possibly death) over time. Often, the times of the condition are grouped together so that the response variable represents the number of patients who fail during a specific time interval. Such data are called *grouped survival times*. For example, the data displayed in Table 1.5 are from Chapter 17, "Categorized Time-to-Event Data." A clinical condition is treated with an active drug for some patients and with a placebo for others. The response categories are whether there are recurrences, no recurrences, or whether the patients withdrew from the study. The entries correspond to the time intervals 0–1 years, 1–2 years, and 2–3 years, which make up the rows of the table.

Table 1.5. Life Table Format for Clinical Condition Data

Controls				
Interval	No Recurrences	Recurrences	Withdrawals	At Risk
0–1 Years	50	15	9	74
1–2 Years	30	13	7	50
2–3 Years	17	7	6	30
Active				
Interval	No Recurrences	Recurrences	Withdrawals	At Risk
0–1 Years	69	12	9	90
1–2 Years	59	7	3	69
2–3 Years	45	10	4	59

1.3   Sampling Frameworks

Categorical data arise from different sampling frameworks. The nature of the sampling framework determines the assumptions that can be made for the statistical analyses and in turn influences the type of analysis that can be applied. The sampling framework also determines the type of inference that is possible. Study populations are limited to target populations, those populations to which inferences can be made, by assumptions justified by the sampling framework.

Generally, data fall into one of three sampling frameworks: historical data, experimental data, and sample survey data. *Historical data* are observational data, which means that the study population has a geographic or circumstantial definition. These may include all the occurrences of an infectious disease in a multicounty area, the children attending a particular elementary school, or those persons appearing in court during a specified time period. Highway safety data concerning injuries in motor vehicles is another example of historical data.

Experimental data are drawn from studies that involve the random allocation of subjects to different treatments of one sort or another. Examples include studies where types of fertilizer are applied to agricultural plots and studies where subjects are administered different dosages of drug therapies. In the health sciences, experimental data may include patients randomly administered a placebo or treatment for their medical condition.

In *sample survey studies*, subjects are randomly chosen from a larger study population. Investigators may randomly choose students from their school IDs and survey them about social behavior; national health care studies may randomly sample Medicare users and investigate physician utilization patterns. In addition, some sampling designs may be a combination of sample survey and experimental data processes. Researchers may randomly select a study population and then randomly assign treatments to the resulting study subjects.

The major difference in the three sampling frameworks described in this section is the use of randomization to obtain them. Historical data involve no randomization, and so it is often difficult to assume that they are representative of a convenient population. Experimental data have good coverage of the possibilities of alternative treatments for the restricted protocol population, and sample survey data have very good coverage of the larger population from which they were selected.

Note that the unit of randomization can be a single subject or a cluster of subjects. In addition, randomization may be applied within subsets, called strata or blocks, with equal or unequal probabilities. In sample surveys, all of this can lead to more complicated designs, such as stratified random samples, or even multistage cluster random samples. In experimental design studies, such considerations lead to repeated measurements (or split-plot) studies.

1.4   Overview of Analysis Strategies

Categorical data analysis strategies can be classified into those that are concerned with hypothesis testing and those that are concerned with modeling. Many questions about a categorical data set can be answered by addressing a specific hypothesis concerning association. Such hypotheses are often investigated with randomization methods. In addition to making statements about association, you may also want to describe the nature of the association in the data set. Statistical modeling techniques using maximum likelihood estimation or weighted least squares estimation are employed to describe patterns of association or variation in terms of a parsimonious statistical model.

Most often the hypothesis of interest is whether association exists between the rows of a contingency table and its columns. The only assumption that is required is randomized allocation of subjects, either through the study design (experimental design) or through the hypothesis itself (necessary for historical data). In addition, particularly for the use of historical data, you often want to control for other explanatory variables that may have influenced the observed outcomes.

1.4.1 Randomization Methods

Table 1.1, the respiratory outcomes data, contains information obtained as part of a randomized allocation process. The hypothesis of interest is whether there is an association between treatment and outcome. For these data, the randomization is accomplished by the study design.

Table 1.6 contains data from a similar study. The main difference is that the study was conducted in two medical centers. The hypothesis of association is whether there is an association between treatment and outcome, controlling for any effect of center.

Table 1.6. Respiratory Improvement

Center	Treatment	Yes	No	Total
1	Test	29	16	45
1	Placebo	14	31	45
Total		43	47	90
2	Test	37	8	45
2	Placebo	24	21	45
Total		61	29	90

Chapter 2, "The 2 × 2 Table," is primarily concerned with the association in 2 × 2 tables; in addition, it discusses measures of association, that is, statistics designed to evaluate the strength of the association. Chapter 3, "Sets of 2 × 2 Tables," discusses the investigation of association in sets of 2 × 2 tables. When the table of interest has more than two rows and two columns, the analysis is further complicated by the consideration of scale of measurement. Chapter 4, "Sets of 2 × r and s × 2 Tables," considers the assessment of association in sets of tables where the rows (columns) have more than two levels.

Chapter 5 describes the assessment of association in the general $s \times r$ table, and Chapter 6, "Sets of $s \times r$ Tables," describes the assessment of association in sets of $s \times r$ tables. The investigation of association in tables and sets of tables is further discussed in Chapter 7, "Nonparametric Methods," which discusses traditional nonparametric tests that have counterparts among the strategies for analyzing contingency tables.

Another consideration in data analysis is whether you have enough data to support the asymptotic theory required for many tests. Often, you may have an overall table sample size that is too small or a number of zero or small cell counts that make the asymptotic assumptions questionable. Recently, exact methods have been developed for a number of association statistics that permit you to address the same hypotheses for these types of data. The above-mentioned chapters illustrate the use of exact methods for many situations.

1.4.2 Modeling Strategies

Often, you are interested in describing the variation of your response variable in your data with a statistical model. In the continuous data setting, you frequently fit a model to the expected mean response. However, with categorical outcomes, there are a variety of response functions that you can model. Depending on the response function that you choose, you can use weighted least squares or maximum likelihood methods to estimate the model parameters.

Perhaps the most common response function modeled for categorical data is the logit. If you have a dichotomous response and represent the proportion of those subjects with an event (versus no event) outcome as p, then the logit can be written

$$\log\left(\frac{p}{1-p}\right)$$

Logistic regression is a modeling strategy that relates the logit to a set of explanatory variables with a linear model. One of its benefits is that estimates of odds ratios, important measures of association, can be obtained from the parameter estimates. Maximum likelihood estimation is used to provide those estimates.

Chapter 8, "Logistic Regression I: Dichotomous Response," discusses logistic regression for a dichotomous outcome variable. Chapter 9, "Logistic Regression II: Polytomous Response," discusses logistic regression for the situation where there are more than two outcomes for the response variable. Logits called *generalized logits* can be analyzed when the outcomes are nominal. And logits called *cumulative logits* can be analyzed when the outcomes are ordinal. Chapter 10, "Conditional Logistic Regression," describes a specialized form of logistic regression that is appropriate when the data are highly stratified or arise from matched case-control studies. Chapter 8 and Chapter 10 describe the use of exact conditional logistic regression for those situations where you have limited or sparse data, and the asymptotic requirements for the usual maximum likelihood approach are not met.

In logistic regression, the objective is to predict a response outcome from a set of explanatory variables. However, sometimes you simply want to describe the structure of association in a set of variables for which there are no obvious outcome or predictor variables. This occurs frequently for sociological studies. The loglinear model is a traditional modeling strategy for categorical data and is appropriate for describing the association in such a set of variables. It is closely related to logistic regression, and the parameters in a loglinear model are also estimated with maximum likelihood estimation. Chapter 16, "Loglinear Models," discusses the loglinear model, including several typical applications.

Some application areas have features that led to the development of special statistical techniques. One of these areas for categorical data is bioassay analysis. Bioassay is the process of determining the potency or strength of a reagent or stimuli based on the response it elicits in biological organisms. Logistic regression is a technique often applied in bioassay analysis, where its parameters take on specific meaning. Chapter 11, "Quantal Bioassay Analysis," discusses the use of categorical data methods for quantal bioassay.

Poisson regression is a modeling strategy that is suitable for discrete counts, and it is discussed in Chapter 12, "Poisson Regression." Most often the log of the count is used as the response function so the model used is a loglinear one.

Besides the logit and log counts, other useful response functions that can be modeled include proportions, means, and measures of association. Weighted least squares estimation is a method of analyzing such response functions, based on large sample theory. These methods are appropriate when you have sufficient sample size and when you have a randomly selected sample, either directly through study design or indirectly via assumptions concerning the representativeness of the data. Not only can you model a variety of useful functions, but weighted least squares estimation also provides a useful framework for the analysis of repeated categorical measurements, particularly those limited to a small number of repeated values. Chapter 13, "Weighted Least Squares," addresses modeling categorical data with weighted least squares methods, and Chapter 14, "Modeling Repeated Measurements Data with WLS," discusses these techniques as applied to the analysis of repeated measurements data.

More recently, generalized estimating equations (GEE) has become a widely used method for the analysis of correlated responses, particularly for the analysis of categorical repeated measurements. The GEE method applies to a broad range of repeated measurements situations, such as those including time-dependent covariates and continuous explanatory variables, that weighted least squares doesn't handle. In addition, the GEE method is a useful technique for some univariate analyses such as modeling overdispersed Poisson counts and implementing the partial proportional odds model. Chapter 15, "Generalized Estimating Equations," discusses the GEE approach and illustrates its application with a number of examples.

Finally, another special application area for categorical data analysis is the analysis of grouped survival data. Chapter 17, "Categorized Time-to-Event Data," discusses some features of survival analysis that are pertinent to grouped survival data, including how to model them with the piecewise exponential model.

1.5   Working with Tables in the SAS System

This section discusses some considerations of managing tables with the SAS System. If you are already familiar with the FREQ procedure, you may want to skip this section.

Many times, categorical data are presented to the researcher in the form of tables, and other times, they are presented in the form of case record data. SAS procedures can handle either type of data. In addition, many categorical data have ordered categories, so that the order of the levels of the rows and columns takes on special meaning. There are numerous ways that you can specify a particular order to SAS procedures.

Consider the following SAS DATA step that inputs the data displayed in Table 1.1.

```
data respire;
   input treat $ outcome $ count;
   datalines;
placebo f 16
placebo u 48
test    f 40
test    u 20
;
proc freq;
   weight count;
   tables treat*outcome;
run;
```

The data set RESPIRE contains three variables: TREAT is a character variable containing values for treatment, OUTCOME is a character variable containing values for the outcome (f for favorable and u for unfavorable), and COUNT contains the number of observations that have the respective TREAT and OUTCOME values. Thus, COUNT effectively takes values corresponding to the cells of Table 1.1. The PROC FREQ statements request that a table be constructed using TREAT as the row variable and OUTCOME as the column variable. By default, PROC FREQ orders the values of the rows (columns) in alphanumeric order. The WEIGHT statement is necessary to tell the procedure that the data are count data, or frequency data; the variable listed in the WEIGHT statement contains the values of the count variable.

Output 1.1 contains the resulting frequency table.

Output 1.1 Frequency Table

```
                 Table of treat by outcome

        treat        outcome

        Frequency|
        Percent  |
        Row Pct  |
        Col Pct  |f       |u       | Total
        ---------+--------+--------+
        placebo  |    16  |    48  |    64
                 |  12.90 |  38.71 | 51.61
                 |  25.00 |  75.00 |
                 |  28.57 |  70.59 |
        ---------+--------+--------+
        test     |    40  |    20  |    60
                 |  32.26 |  16.13 | 48.39
                 |  66.67 |  33.33 |
                 |  71.43 |  29.41 |
        ---------+--------+--------+
        Total         56       68      124
                    45.16    54.84   100.00
```

Suppose that a different sample produced the numbers displayed in Table 1.7.

Table 1.7. Respiratory Outcomes

Treatment	Favorable	Unfavorable	Total
Placebo	5	10	15
Test	8	20	28

These data may be stored in case record form, which means that each individual is represented by a single observation. You can also use this type of input with the FREQ procedure. The only difference is that the WEIGHT statement is not required.

The following statements create a SAS data set for these data and invoke PROC FREQ for case record data. The @@ symbol in the INPUT statement means that the data lines contain multiple observations.

```
data respire;
   input treat $ outcome $ @@;
   datalines;
placebo f placebo f  placebo f
placebo f placebo f
placebo u placebo u  placebo u
placebo u placebo u  placebo u
placebo u placebo u  placebo u
placebo u
test     f test     f test     f
test     f test     f test     f
test     f test     f
test     u test     u test     u
test     u test     u test     u
test     u test     u test     u
test     u test     u test     u
test     u test     u test     u
test     u test     u
;
proc freq;
   tables treat*outcome;
run;
```

Output 1.2 displays the resulting frequency table.

Output 1.2   Frequency Table

```
                    Table of treat by outcome

            treat        outcome

            Frequency|
            Percent  |
            Row Pct  |
            Col Pct  |f       |u       |  Total
            ---------+--------+--------+
            placebo  |     5 |     10 |     15
                     | 11.63 | 23.26 |  34.88
                     | 33.33 | 66.67 |
                     | 38.46 | 33.33 |
            ---------+--------+--------+
            test     |     8 |     20 |     28
                     | 18.60 | 46.51 |  65.12
                     | 28.57 | 71.43 |
                     | 61.54 | 66.67 |
            ---------+--------+--------+
            Total          13      30      43
                        30.23   69.77  100.00
```

In this book, the data are generally presented in count form.

When ordinal data are considered, it becomes quite important to ensure that the levels of the rows and columns are sorted correctly. By default, the data are going to be sorted alphanumerically. If this isn't suitable, then you need to alter the default behavior.

Consider the data displayed in Table 1.2. IMPROVE is the outcome variable, and the values marked, some, and none are listed in decreasing order. Suppose that the data set ARTHRIT is created with the following statements.

```
data arthrit;
   length treat $7. sex $6. ;
   input sex $ treat $ improve $ count @@;
   datalines;
female active  marked 16 female active  some 5 female active  none  6
female placebo marked  6 female placebo some 7 female placebo none 19
male    active  marked  5 male    active  some 2 male    active  none  7
male    placebo marked  1 male    placebo some 0 male    placebo none 10
;
run;
```

If you invoked PROC FREQ for this data set and used the default sort order, the levels of the columns would be ordered marked, none, and some, which would be incorrect. One way to change this default sort order is to use the ORDER=DATA option in the PROC FREQ statement. This specifies that the sort order is the same order in which the values are encountered in the data set. Thus, since 'marked' comes first, it is first in the sort order. Since 'some' is the second value for IMPROVE encountered in the data set, then it is second in the sort order. And 'none' would be third in the sort order. This is the desired sort order. The following PROC FREQ statements produce a table displaying the sort order resulting from the ORDER=DATA option.

```
proc freq order=data;
   weight count;
   tables treat*improve;
run;
```

Output 1.3 displays the frequency table for the cross-classification of treatment and improvement for these data; the values for IMPROVE are in the correct order.

Output 1.3   Frequency Table from ORDER=DATA Option

```
                        Table of treat by improve

             treat      improve

             Frequency|
             Percent  |
             Row Pct  |
             Col Pct  |marked  |some    |none    | Total
             ---------+--------+--------+--------+
             active   |    21  |     7  |    13  |    41
                      | 25.00  |  8.33  | 15.48  | 48.81
                      | 51.22  | 17.07  | 31.71  |
                      | 75.00  | 50.00  | 30.95  |
             ---------+--------+--------+--------+
             placebo  |     7  |     7  |    29  |    43
                      |  8.33  |  8.33  | 34.52  | 51.19
                      | 16.28  | 16.28  | 67.44  |
                      | 25.00  | 50.00  | 69.05  |
             ---------+--------+--------+--------+
             Total         28       14       42       84
                        33.33    16.67    50.00   100.00
```

Other possible values for the ORDER= option include FORMATTED, which means sort by the formatted values. The ORDER= option is also available with the CATMOD, LOGISTIC, and GENMOD procedures. For information on the ORDER= option for the FREQ procedure, refer to the *SAS/STAT User's Guide, Version 8*. This option is used frequently in this book.

Often, you want to analyze sets of tables. For example, you may want to analyze the cross-classification of treatment and improvement for both males and females. You do this in PROC FREQ by using a three-way crossing of the variables SEX, TREAT, and IMPROVE.

```
proc freq order=data;
   weight count;
   tables sex*treat*improve / nocol nopct;
run;
```

The two rightmost variables in the TABLES statement determine the rows and columns of the table, respectively. Separate tables are produced for the unique combination of values of the other variables in the crossing. Since SEX has two levels, one table is produced for males and one table is produced for females. If there were four variables in this crossing, with the two variables on the left having two levels each, then four tables would be produced, one for each unique combination of the two leftmost variables in the TABLES statement.

Note also that the options NOCOL and NOPCT are included. These options suppress the printing of column percentages and cell percentages, respectively. Since generally you are interested in row percentages, these options are often specified in the code displayed in this book.

Output 1.4 contains the two tables produced with the preceding statements.

Output 1.4   Producing Sets of Tables

```
                   Table 1 of treat by improve
                     Controlling for sex=female

          treat       improve

          Frequency|
          Row Pct  |marked  |some    |none    |  Total
          ---------+--------+--------+--------+
          active   |    16 |     5 |     6 |     27
                   | 59.26 | 18.52 | 22.22 |
          ---------+--------+--------+--------+
          placebo  |     6 |     7 |    19 |     32
                   | 18.75 | 21.88 | 59.38 |
          ---------+--------+--------+--------+
          Total          22      12      25      59


                   Table 2 of treat by improve
                     Controlling for sex=male

          treat       improve

          Frequency|
          Row Pct  |marked  |some    |none    |  Total
          ---------+--------+--------+--------+
          active   |     5 |     2 |     7 |     14
                   | 35.71 | 14.29 | 50.00 |
          ---------+--------+--------+--------+
          placebo  |     1 |     0 |    10 |     11
                   |  9.09 |  0.00 | 90.91 |
          ---------+--------+--------+--------+
          Total           6       2      17      25
```

This section reviewed some of the basic table management necessary for using the FREQ procedure. Other related options are discussed in the appropriate chapters.

1.6   Using This Book

This book is intended for a variety of audiences, including novice readers with some statistical background (solid understanding of regression analysis), those readers with substantial statistical background, and those readers with background in categorical data analysis. Therefore, not all of this material will have the same importance to all readers. Some chapters include a good deal of tutorial material, while others have a good deal of advanced material. This book is not intended to be a comprehensive treatment of categorical data analysis, so some topics are mentioned briefly for completeness and some other topics are emphasized because they are not well documented.

The data used in this book come from a variety of sources and represent a wide breadth of application. However, due to the biostatistical background of all three authors, there is a certain inevitable weighting of biostatistical examples. Most of the data come from practice, and the original sources are cited when this is true; however, due to confidentiality concerns and pedagogical requirements, some of the data are altered or created. However, they still represent realistic situations.

Chapters 2–4 are intended to be accessible to all readers, as is most of Chapter 5. Chapter 6 is an integration of Mantel-Haenszel methods at a more advanced level, but scanning it is probably a good idea for any reader interested in the topic. In particular, the discussion about the analysis of repeated measurements data with extended Mantel-Haenszel methods is useful material for all readers comfortable with the Mantel-Haenszel technique.

Chapter 7 is a special interest chapter relating Mantel-Haenszel procedures to traditional nonparametric methods used for continuous data outcomes.

Chapters 8 and 9 on logistic regression are intended to be accessible to all readers, particularly Chapter 8. The last section of Chapter 8 describes the statistical methodology more completely for the advanced reader. Most of the material in Chapter 9 should be accessible to most readers. Chapter 10 is a specialized chapter that discusses conditional logistic regression and requires somewhat more statistical expertise. Chapter 11 discusses the use of logistic regression in analyzing bioassay data.

Chapter 12 describes Poisson regression and should be fairly accessible.

Chapter 13 discusses weighted least squares and is written at a somewhat higher statistical level than Chapters 8 and 9, but most readers should find this material useful, particularly the examples.

Chapters 14–17 discuss advanced topics and are necessarily written at a higher statistical level. Chapter 14 describes the analysis of repeated measurements data using weighted least squares and Chapter 15 discusses the use of generalized estimating equations. The opening sections both include a basic example that is intended to be accessible to a wide range of readers. Chapter 16 discusses loglinear model analysis, and Chapter 17 discusses the analysis of categorized time-to-event data.

All of the examples were executed with Release 8.1 of the SAS System with the few exceptions noted in the "Preface to the Second Edition." Software features upcoming in future releases are also mentioned.

Chapter 2

The 2 × 2 Table

Chapter Table of Contents

Chapter 2
The 2 × 2 Table

2.1 Introduction

The 2 × 2 contingency table is one of the most common ways to summarize categorical data. Categorizing patients by their favorable or unfavorable response to two different drugs, asking health survey participants whether they have regular physicians and regular dentists, and asking residents of two cities whether they desire more environmental regulations all result in data that can be summarized in a 2 × 2 table.

Generally, interest lies in whether there is an association between the row variable and the column variable that produce the table; sometimes there is further interest in describing the strength of that association. The data can arise from several different sampling frameworks, and the interpretation of the hypothesis of no association depends on the framework. Data in a 2 × 2 table can represent

- simple random samples from two groups that yield two independent binomial distributions for a binary response

  Asking residents from two cities whether they desire more environmental regulations is an example of this framework. This is a stratified random sampling setting, since the subjects from each city represent two independent random samples. Because interest lies in whether the proportion favoring regulation is the same for the two cities, the hypothesis of interest is the hypothesis of homogeneity. Is the distribution of the response the same in both groups?

- a simple random sample from one group that yields a single multinomial distribution for the cross-classification of two binary responses

  Taking a random sample of subjects and asking whether they see both a regular physician and a regular dentist is an example of this framework. The hypothesis of interest is one of independence. Are having a regular dentist and having a regular physician independent of each other?

- randomized assignment of patients to two equivalent treatments, resulting in the hypergeometric distribution

  This framework occurs when patients are randomly allocated to one of two drug treatments, and their response to that treatment is the binary outcome. Under the hypothesis that the effects of the two treatments are the same for each patient, a hypergeometric distribution applies to the response distributions for the two treatments. (A less frequent framework that produces data for the 2 × 2 table is the

Poisson distribution. Each count is considered to be the result of an independent Poisson process, and questions related to multiplicative effects in Poisson regression (discussed in Chapter 12) are addressed by testing the hypothesis of no association.)

Table 2.1 summarizes the information from a randomized clinical trial that compared two treatments (test, placebo) for a respiratory disorder.

Table 2.1.   Respiratory Outcomes

Treatment	Favorable	Unfavorable	Total
Placebo	16	48	64
Test	40	20	60

The question of interest is whether the rates of favorable response for test (67%) and placebo (25%) are the same. You can address this question by investigating whether there is a statistical association between treatment and outcome. The null hypothesis is stated

H_0: There is no association between treatment and outcome.

There are several ways of testing this hypothesis; many of the tests are based on the chi-square statistic. Section 2.2 discusses these methods. However, sometimes the counts in the table cells are too small to meet the sample size requirements necessary for the chi-square distribution to apply, and exact methods based on the hypergeometric distribution are used to test the hypothesis of no association. Exact methods are discussed in Section 2.3.

In addition to testing the hypothesis concerning the presence of association, you may be interested in describing the association or gauging its strength. Section 2.4 discusses the estimation of the difference in proportions from 2 × 2 tables. Section 2.5 discusses measures of association, which assess strength of association, and Section 2.6 discusses measures called sensitivity and specificity, which are useful when the two responses correspond to two different methods for determining whether a particular disorder is present. Finally, 2 × 2 tables often display data for matched pairs, and Section 2.7 discusses McNemar's Test for assessing association for matched pairs data.

2.2   Chi-Square Statistics

Table 2.2 displays the generic 2 × 2 table, including row and column marginal totals.

Table 2.2.   2 × 2 Contingency Table

Column	Row Levels		
Levels	1	2	Total
1	n_{11}	n_{12}	n_{1+}
2	n_{21}	n_{22}	n_{2+}
Total	n_{+1}	n_{+2}	n

Under the randomization framework that produced Table 2.1, the row marginal totals n_{1+} and n_{2+} are fixed since 60 patients were randomly allocated to one of the treatment groups and 64 to the other. The column marginal totals can be regarded as fixed under the null hypothesis of no treatment difference for each patient. Then, given that all of the marginal totals n_{1+}, n_{2+}, n_{+1}, and n_{+2} are fixed under the null hypothesis, the probability distribution from the randomized allocation of patients to treatment can be written

$$\Pr\{n_{ij}\} = \frac{n_{1+}! n_{2+}! n_{+1}! n_{+2}!}{n! n_{11}! n_{12}! n_{21}! n_{22}!}$$

which is the hypergeometric distribution. The expected value of n_{ij} is

$$E\{n_{ij} | H_0\} = \frac{n_{i+} n_{+j}}{n} = m_{ij}$$

and the variance is

$$V\{n_{ij} | H_0\} = \frac{n_{1+} n_{2+} n_{+1} n_{+2}}{n^2 (n-1)} = v_{ij}$$

For a sufficiently large sample, n_{11} approximately has a normal distribution, which implies that

$$Q = \frac{(n_{11} - m_{11})^2}{v_{11}}$$

approximately has a chi-square distribution with one degree of freedom. It is the ratio of a squared difference from the expected value versus its variance, and such quantities follow the chi-square distribution when the variable is distributed normally. Q is often called the randomization chi-square. It doesn't matter how the rows and columns are arranged, Q takes the same value since

$$|n_{11} - m_{11}| = |n_{ij} - m_{ij}| = \frac{|n_{11} n_{22} - n_{12} n_{21}|}{n}$$

A related statistic is the Pearson chi-square statistic. This statistic is written

$$Q_P = \sum_{i=1}^{2} \sum_{j=1}^{2} \frac{(n_{ij} - m_{ij})^2}{m_{ij}} = \frac{n}{(n-1)} Q$$

If the cell counts are sufficiently large, Q_P is distributed as chi-square with one degree of freedom. As n grows large, Q_P and Q converge. A useful rule for determining adequate sample size for both Q and Q_P is that the expected value m_{ij} should exceed 5 for all of the cells (and preferably 10). While Q is discussed here in the framework of a randomized allocation of patients to two groups, Q and Q_P are also appropriate for investigating the hypothesis of no association for all of the sampling frameworks described previously.

The following PROC FREQ statements produce a frequency table and the chi-square statistics for the data in Table 2.1. The data are supplied in frequency, or count, form. An observation is supplied for each configuration of the values of the variables TREAT and OUTCOME. The variable COUNT holds the total number of observations that have that particular configuration. The WEIGHT statement tells the FREQ procedure that the data are in frequency form and names the variable that contains the frequencies.

The CHISQ option in the TABLES statement produces chi-square statistics.

```
data respire;
   input treat $ outcome $ count;
   datalines;
placebo f 16
placebo u 48
test    f 40
test    u 20
;
proc freq;
   weight count;
   tables treat*outcome / chisq;
run;
```

Output 2.1 displays the data in a 2 × 2 table. With an overall sample size of 124, and all expected cell counts greater than 10, the sampling assumptions for the chi-square statistics are met. PROC FREQ prints out a warning message when more than 20% of the cells in a table have expected counts less than 5. (Note that you can specify the EXPECTED option in the TABLE statement to produce the expected cell counts along with the cell percentages.)

Output 2.1   Frequency Table

```
                Table of treat by outcome

        treat       outcome

        Frequency|
        Percent  |
        Row Pct  |
        Col Pct  |f        |u        |  Total
        ---------+--------+--------+
        placebo  |    16  |    48  |    64
                 | 12.90  | 38.71  | 51.61
                 | 25.00  | 75.00  |
                 | 28.57  | 70.59  |
        ---------+--------+--------+
        test     |    40  |    20  |    60
                 | 32.26  | 16.13  | 48.39
                 | 66.67  | 33.33  |
                 | 71.43  | 29.41  |
        ---------+--------+--------+
        Total          56       68      124
                    45.16    54.84   100.00
```

Output 2.2 contains the table with the chi-square statistics.

Output 2.2 Chi-Square Statistics

```
             Statistics for Table of treat by outcome

   Statistic                    DF       Value      Prob
   ------------------------------------------------------
   Chi-Square                    1      21.7087    <.0001
   Likelihood Ratio Chi-Square   1      22.3768    <.0001
   Continuity Adj. Chi-Square    1      20.0589    <.0001
   Mantel-Haenszel Chi-Square    1      21.5336    <.0001
   Phi Coefficient                     -0.4184
   Contingency Coefficient              0.3860
   Cramer's V                          -0.4184


                   Fisher's Exact Test
          ------------------------------------
          Cell (1,1) Frequency (F)         16
          Left-sided Pr <= F        2.838E-06
          Right-sided Pr >= F          1.0000

          Table Probability (P)     2.397E-06
          Two-sided Pr <= P         4.754E-06

                 Sample Size = 124
```

The randomization statistic Q is labeled "Mantel-Haenszel Chi-Square," and the Pearson chi-square Q_P is labeled "Chi-Square." Q has a value of 21.5336 and $p < 0.0001$; Q_P has a value of 21.7087 and $p < 0.0001$. Both of these statistics are clearly significant. There is a strong association between treatment and outcome such that the test treatment results in a more favorable response outcome than the placebo. The row percentages in Output 2.1 show that the test treatment resulted in 67% favorable response and the placebo treatment resulted in 25% favorable response.

Notice that the output also includes a statistic labeled "Likelihood Ratio Chi-Square." This statistic, often written Q_L, is asymptotically equivalent to Q and Q_P. The statistic Q_L is described later in chapters on modeling; it is not often used in the analysis of 2×2 tables. Some of the other statistics are discussed in the next section.

2.3 Exact Tests

Sometimes your data include small and zero cell counts. For example, consider the following data from a study on treatments for healing severe infections. A test treatment and a control are compared to determine whether the rates of favorable response are the same.

Table 2.3. Severe Infection Treatment Outcomes

Treatment	Favorable	Unfavorable	Total
Test	10	2	12
Control	2	4	6
Total	12	6	18

Obviously, the sample sizes requirements for the chi-square tests described in Section 2.2 are not met by these data. However, if you can consider the margins (12, 6, 12, 6) to be fixed, then you can assume that the data are distributed hypergeometrically and write

$$\Pr\{n_{ij}\} = \frac{n_{1+}! n_{2+}! n_{+1}! n_{+2}!}{n! n_{11}! n_{12}! n_{21}! n_{22}!}$$

The row margins may be fixed by the treatment allocation process; that is, subjects are randomly assigned to Test and Control. The column totals can be regarded as fixed by the null hypothesis; there are 12 patients with favorable response and 6 patients with unfavorable response, regardless of treatment. If the data are the result of a sample of convenience, you can still condition on marginal totals being fixed by addressing the null hypothesis that patients are interchangeable; that is, an individual patient is as likely to have a favorable response on Test as on Control.

Recall that a p-value is the probability of the observed data or more extreme data occurring under the null hypothesis. With Fisher's exact test, you determine the p-value for this table by summing the probabilities of the tables that are as likely or less likely, given the fixed margins. The following table includes all possible table configurations and their associated probabilities.

Table 2.4.   Table Probabilities

(1,1)	(1,2)	(2,1)	(2,2)	Probabilities
		Table Cell		
12	0	0	6	0.0001
11	1	1	5	0.0039
10	2	2	4	0.0533
9	3	3	3	0.2370
8	4	4	2	0.4000
7	5	5	1	0.2560
6	6	6	0	0.0498

To find the one-sided p-value, you sum the probabilities as small or smaller than those computed for the table observed, in the direction specified by the one-sided alternative. In this case, it would be those tables in which the Test treatment had the more favorable response, or

$$p = 0.0533 + 0.0039 + 0.0001 = 0.0573$$

To find the two-sided p-value, you sum all of the probabilities that are as small or smaller than that observed, or

$$p = 0.0533 + 0.0039 + 0.0001 + 0.0498 = 0.1071$$

Generally, you will be interested in the two-sided p-value. Note that when the row (or column) totals are nearly equal, the p-value for the two-sided Fisher's exact test is approximately twice the *p*-value for the one-sided Fisher's exact test for the better treatment. When the row (or column) totals are equal, the *p*-value for the two-sided Fisher's exact test is exactly twice the value of the *p*-value for the one-sided Fisher's exact test.

The following SAS code produces the 2×2 frequency table for Table 2.3. Specifying the CHISQ option also produces Fisher's exact test for a 2×2 table. In addition, the ORDER=DATA option specifies that PROC FREQ orders the levels of the rows (columns) in the same order in which the values are encountered in the data set.

```
data severe;
   input treat $ outcome $ count;
   datalines;
Test     f 10
Test     u 2
Control f 2
Control u 4
;
proc freq order=data;
   weight count;
   tables treat*outcome / chisq nocol;
run;
```

The NOCOL option suppresses the column percentages, as seen in Output 2.3.

Output 2.3   Frequency Table

```
                    Table of treat by outcome

            treat        outcome

            Frequency|
            Percent  |
            Row Pct  |f         |u        |  Total
            ---------+--------+--------+
            Test     |    10 |     2 |    12
                     | 55.56 | 11.11 | 66.67
                     | 83.33 | 16.67 |
            ---------+--------+--------+
            Control  |     2 |     4 |     6
                     | 11.11 | 22.22 | 33.33
                     | 33.33 | 66.67 |
            ---------+--------+--------+
            Total         12       6       18
                       66.67    33.33  100.00
```

Output 2.4 contains the chi-square statistics, including the exact test. Note that the sample size assumptions are not met for the chi-square tests: the warning beneath the table asserts that this is the case.

Output 2.4   Table Statistics

```
              Statistics for Table of treat by outcome

     Statistic                        DF       Value       Prob
     --------------------------------------------------------------
     Chi-Square                        1       4.5000      0.0339
     Likelihood Ratio Chi-Square       1       4.4629      0.0346
     Continuity Adj. Chi-Square        1       2.5313      0.1116
     Mantel-Haenszel Chi-Square        1       4.2500      0.0393
     Phi Coefficient                           0.5000
     Contingency Coefficient                   0.4472
     Cramer's V                                0.5000

        WARNING: 75% of the cells have expected counts less
                 than 5. Chi-Square may not be a valid test.


                        Fisher's Exact Test
              ----------------------------------
              Cell (1,1) Frequency (F)        10
              Left-sided Pr <= F          0.9961
              Right-sided Pr >= F         0.0573

              Table Probability (P)       0.0533
              Two-sided Pr <= P           0.1070

                     Sample Size = 18
```

Note that the SAS System produces both a left-tail and right-tail p-value for Fisher's exact test. The left-tail probability is the probability of all tables such that the (1,1) cell value is less than or equal to the one observed. The right-tail probability is the probability of all tables such that the (1,1) cell value is greater than or equal to the one observed. Thus, the one-sided p-value is the same as the right-tailed p-value in this case, since large values for the (1,1) cell correspond to better outcomes for Test treatment.

Both the two-sided p-value of 0.1070 and the one-sided p-value of 0.0573 are larger than the p-values associated with Q_P ($p = 0.0339$) and Q ($p = 0.0393$). Depending on your significance criterion, you may reach very different conclusions with these three test statistics. The sample size requirements for the chi-square distribution are not met with these data; hence the test statistics using this approximation are questionable. This example illustrates the usefulness of Fisher's exact test when the sample size requirements for the usual chi-square tests are not met.

The output also includes a statistic labeled the "Continuity Adj. Chi-Square"; this is the continuity-adjusted chi-square statistic suggested by Yates, which is intended to correct the Pearson chi-square statistic so that it more closely approximates Fisher's exact test. In this case, the correction produces a chi-square value of 2.5313 with $p = 0.1116$, which is certainly close to the two-sided Fisher's exact test value. However, many statisticians recommend that you should simply apply Fisher's exact test when the sample size requires it rather than try to approximate it. In particular, the continuity-corrected chi-square may be overly conservative for two-sided tests when the data are nonsymmetric, that is, the row (column) totals are very different, and the sample sizes are small.

Note that Fisher's exact test is always appropriate, even when the sample size is large.

2.3.1 Exact p-values for Chi-Square Statistics

For many years, the only practical way to assess association in 2×2 tables that had small or zero counts was with Fisher's exact test. This test is computationally quite easy for the 2×2 case. However, you can also obtain exact p-values for the statistics discussed in Section 2.2. This is possible due to the development of fast and efficient network algorithms that provide a distinct advantage over direct enumeration. Although such enumeration is reasonable for Fisher's exact test, it can prove prohibitive in other instances. Refer to Mehta, Patel, and Tsiatis (1984) for a description of these algorithms; Agresti (1992) provides a useful overview of the various algorithms for the computation of exact p-values.

In the case of Q, Q_P, and a closely related statistic, Q_L (likelihood ratio statistic), large values of the statistic imply a departure from the null hypothesis. The exact p-values for these statistics are the sum of the probabilities for the tables having a test statistic greater than or equal to the value of the observed test statistic.

The EXACT statement enables you to request exact p-values or confidence limits for many of the statistics produced by the FREQ procedure. Refer to the *SAS/STAT User's Guide, Version 8* for details about specification and the options that control computation time. Note that exact computations may take a considerable amount of memory and time for large problems.

For the Table 2.3 data, the following SAS statements produce the exact p-values for the chi-square tests of association. You include the keyword(s) for the statistics for which to compute exact p-values, CHISQ in this case.

```
proc freq order=data;
   weight count;
   tables treat*outcome / chisq nocol;
   exact chisq;
run;
```

First, the usual table for the CHISQ statistics is displayed.

Output 2.5   Statistics for Table of Treat by Outcome

```
              Statistics for Table of treat by outcome

      Statistic                      DF       Value      Prob
      -----------------------------------------------------------
      Chi-Square                      1       4.5000     0.0339
      Likelihood Ratio Chi-Square     1       4.4629     0.0346
      Continuity Adj. Chi-Square      1       2.5313     0.1116
      Mantel-Haenszel Chi-Square      1       4.2500     0.0393
      Phi Coefficient                         0.5000
      Contingency Coefficient                 0.4472
      Cramer's V                              0.5000

      WARNING: 75% of the cells have expected counts less than 5.
              (Asymptotic) Chi-Square may not be a valid test.
```

Then, individual tables for Q_P, Q_L, and Q are presented, including test values and both asymptotic and exact p-values.

Output 2.6   Pearson Chi-Square Test

```
              Statistics for Table of treat by outcome

              Pearson Chi-Square Test
      ---------------------------------------
      Chi-Square                      4.5000
      DF                                   1
      Asymptotic Pr >  ChiSq          0.0339
      Exact        Pr >= ChiSq        0.1070
```

Output 2.7   Likelihood Ratio Chi-Square Test

```
              Statistics for Table of treat by outcome

            Likelihood Ratio Chi-Square Test
      ---------------------------------------
      Chi-Square                      4.4629
      DF                                   1
      Asymptotic Pr >  ChiSq          0.0346
      Exact        Pr >= ChiSq        0.1070
```

Output 2.8   Mantel-Haenszel Chi-Square Test

```
              Statistics for Table of treat by outcome

            Mantel-Haenszel Chi-Square Test
      ---------------------------------------
      Chi-Square                      4.2500
      DF                                   1
      Asymptotic Pr >  ChiSq          0.0393
      Exact        Pr >= ChiSq        0.1070
```

$Q_P = 4.5$, with an exact p-value of 0.1070 (asymptotic $p = 0.0339$). $Q = 4.25$ with an exact p-value of 0.1070 (asymptotic $p = 0.0393$). Q_L is similar, with a value of 4.4629 and an exact p-value 0.1070 (asymptotic $p = 0.0346$). Thus, a researcher using the asymptotic p-values in this case may have found an inappropriate significance that is not there when exact p-values are considered. Note that Fisher's exact test provides an identical p-value of 0.1070, but this is not always the case.

Using the exact p-values for the association chi-square versus applying the Fisher exact test is a matter of preference. However, there may be some interpretation advantage in using the Fisher exact test since the comparison is to your actual table rather than to a test statistic based on the table.

2.4 Difference in Proportions

The previous sections have addressed the question of whether there is an association between the rows and columns of a 2×2 table. In addition, you may be interested in describing the association in the table. For example, once you have established that the proportions computed from a table are different, you may want to estimate their difference.

Consider the following table, which displays data from two independent groups:

Table 2.5. 2×2 Contingency Table

	Yes	No	Total	Proportion Yes
Group 1	n_{11}	n_{12}	n_{1+}	$p_1 = n_{11}/n_{1+}$
Group 2	n_{21}	n_{22}	n_{2+}	$p_2 = n_{21}/n_{2+}$
Total	n_{+1}	n_{+2}	n	

If the two groups are simple random samples from populations with corresponding probabilities Yes denoted as π_1 and π_2, you may be interested in estimating the difference between the proportions p_1 and p_2 with $d = p_1 - p_2$. You can show that the expected value is

$$E\{p_1 - p_2\} = \pi_1 - \pi_2$$

and the variance is

$$V\{p_1 - p_2\} = \frac{\pi_1(1 - \pi_1)}{n_{1+}} + \frac{\pi_2(1 - \pi_2)}{n_{2+}}$$

for which an unbiased estimate is

$$v_d = \frac{p_1(1 - p_1)}{n_{1+} - 1} + \frac{p_2(1 - p_2)}{n_{2+} - 1}$$

A $100(1 - \alpha)\%$ confidence interval for $(\pi_1 - \pi_2)$ is written

$$d \pm \left\{ z_{\alpha/2}\sqrt{v_d} + \frac{1}{2}\left\{ \frac{1}{n_{1+}} + \frac{1}{n_{2+}} \right\} \right\}$$

where $z_{\alpha/2}$ is the $100(1 - \alpha/2)$ percentile of the standard normal distribution; this confidence interval is based on Fleiss (1981, p. 29).

For example, consider Table 2.6, which reproduces the data analyzed in Section 2.2. In addition to determining that there is a statistical association between treatment and response, you may be interested in estimating the difference between the rates of favorable response for the test and placebo treatments, including a 95% confidence interval.

Table 2.6. Respiratory Outcomes

Treatment	Favorable	Unfavorable	Total	Favorable Proportion
Placebo	16	48	64	0.250
Test	40	20	60	0.667
Total	56	68	124	0.452

The difference is $d = 0.667 - 0.25 = 0.417$, and the confidence interval is written

$$0.417 \pm \left\{ (1.96) \left[\frac{0.667(1 - 0.667)}{60 - 1} + \frac{0.25(1 - 0.25)}{64 - 1} \right]^{1/2} + \frac{1}{2} \left(\frac{1}{60} + \frac{1}{64} \right) \right\}$$

$$= \ 0.417 \pm 0.177$$

$$= \ (0.240, 0.594)$$

A related measure of association is the Pearson correlation coefficient. This statistic is proportional to the difference of proportions. Since Q_P is also proportional to the squared difference in proportions, the Pearson correlation coefficient is also proportional to $\sqrt{Q_P}$.

The Pearson correlation coefficient can be written

$$r = \left\{ (n_{11} - \frac{n_{1+}n_{+1}}{n}) / \left[(n_{1+} - \frac{n_{1+}^2}{n})(n_{+1} - \frac{n_{+1}^2}{n}) \right]^{1/2} \right\}$$

$$= \left\{ (n_{11}n_{22} - n_{12}n_{21}) / [(n_{1+}n_{2+}n_{+1}n_{+2})]^{1/2} \right\}$$

$$= [n_{1+}n_{2+}/n_{+1}n_{+2}]^{1/2} d$$

$$= (Q_P/n)^{1/2}$$

For the data in Table 2.6, r is computed as

$$r = [(60)(64)/(56)(68)]^{1/2}(0.417) = 0.418$$

The FREQ procedure does produce the difference in proportions and a confidence interval, although the asymptotic confidence interval it produces requires a somewhat large sample size, say cell counts of at least 12. The confidence limits described above are appropriate for more moderate sample sizes, say cell counts of at least 8, and will likely be an option in a future PROC FREQ release.

You can request the difference of proportions with the RISKDIFF option in the TABLES statement. The following statements produce the difference along with the Pearson correlation coefficient, requested with the MEASURES option. Note that the table is input with the Test row first. This is so the first difference produced will be in agreement with that computed above, which is for Test versus Placebo.

The ODS SELECT statement is used to restrict the output produced to the RiskDiffCol1 table and the Measures table. You can use this statement, part of the Output Delivery System, to customize your output. The names of all the tables comprising the output for each SAS/STAT procedure are available in the "Details" section of each procedure chapter in *SAS/STAT User's Guide, Version 8*. Here, the RiskDiffCol1 table produces the difference for column 1 of the frequency table. There is also a table for the column 2 difference called RiskDiffCol1, which is not produced in this example.

```
ods select RiskDiffCol1 Measures;
data respire2;
   input treat $ outcome $ count @@;
   datalines;
test    f 40 test    u 20
placebo f 16 placebo u 48
;
proc freq order=data;
   weight count;
   tables treat*outcome / riskdiff measures;
run;
```

Output 2.9 contains the value for the Pearson correlation coefficient, which is rounded as 0.418, as calculated above.

Output 2.9   Pearson Correlation Coefficient

```
            Statistics for Table of treat by outcome

      Statistic                             Value       ASE
      ----------------------------------------------------
      Gamma                                 0.7143    0.0974
      Kendall's Tau-b                       0.4184    0.0816
      Stuart's Tau-c                        0.4162    0.0814

      Somers' D C|R                         0.4167    0.0814
      Somers' D R|C                         0.4202    0.0818

      Pearson Correlation                   0.4184    0.0816
      Spearman Correlation                  0.4184    0.0816

      Lambda Asymmetric C|R                 0.3571    0.1109
      Lambda Asymmetric R|C                 0.4000    0.0966
      Lambda Symmetric                      0.3793    0.0983

      Uncertainty Coefficient C|R           0.1311    0.0528
      Uncertainty Coefficient R|C           0.1303    0.0525
      Uncertainty Coefficient Symmetric     0.1307    0.0526
```

Output 2.10 contains the value for the difference of proportions for Test versus Placebo for the Favorable response, which is 0.4167 with confidence limits (0.2570, 0.5763). Note that these limits are a little narrower than those computed above; again, these limits may not provide adequate coverage for moderately small sample sizes. Note that this table also includes the proportions of column 1 response in both rows, along with the asymptotic and exact confidence limits. Although some methods for exact confidence limits for the difference in proportions are available, statistical research concerning their properties and the development of possibly better methods is still ongoing.

Output 2.10   Difference in Proportions

```
Statistics for Table of treat by outcome

                     Column 1 Risk Estimates

                                   (Asymptotic) 95%        (Exact) 95%
                 Risk      ASE     Confidence Limits     Confidence Limits
          ---------------------------------------------------------------------
Row 1          0.6667    0.0609    0.5474    0.7859      0.5331    0.7831
Row 2          0.2500    0.0541    0.1439    0.3561      0.1502    0.3740
Total          0.4516    0.0447    0.3640    0.5392      0.3621    0.5435

Difference     0.4167    0.0814    0.2570    0.5763

              Difference is (Row 1 - Row 2)
```

2.5   Odds Ratio and Relative Risk

Measures of association are used to assess the strength of an association. There are numerous measures of association available for the contingency table, some of which are described in Chapter 5, "The $s \times r$ Table." For the 2 × 2 table, one measure of association is the *odds ratio*, and a related measure of association is the *relative risk*.

Consider Table 2.5. The *odds ratio* compares the odds of the Yes proportion for Group 1 to the odds of the Yes proportion for Group 2. It is computed as

$$\text{OR} = \frac{p_1/(1-p_1)}{p_2/(1-p_2)} = \frac{n_{11}n_{22}}{n_{12}n_{21}}$$

The odds ratio ranges from 0 to infinity. When OR is 1, there is no association between the row variable and the column variable. When OR is greater than 1, Group 1 is more likely than Group 2 to have the yes response; when OR is less than 1, Group 1 is less likely than Group 2 to have the yes response.

Define the *logit* for general p as

$$\text{logit}(p) = \log\left\{\frac{p}{1-p}\right\}$$

If you take the log of the odds ratio,

$$f = \log\{OR\} = \log\left\{\frac{p_1(1-p_2)}{p_2(1-p_1)}\right\}$$
$$= \log\{p_1/(1-p_1)\} - \log\{p_2/(1-p_2)\}$$

you see that the odds ratio can be written in terms of the difference between two logits. The logit is the function that is modeled in logistic regression. As you will see in Chapter 8, "Logistic Regression I: Dichotomous Response," the odds ratio and logistic regression are closely connected.

The estimate of the variance of f is

$$v_f = \left\{\frac{1}{n_{11}} + \frac{1}{n_{12}} + \frac{1}{n_{21}} + \frac{1}{n_{22}}\right\}$$

so a $100(1-\alpha)\%$ confidence interval for OR can be written as

$$\exp(f \pm z_{\alpha/2}\sqrt{v_f})$$

The odds ratio is a useful measure of association regardless of how the data are collected. However, it has special meaning for retrospective studies because it can be used to estimate a quantity called *relative risk*, which is commonly used in epidemiological work. The relative risk is the risk of developing a particular condition (often a disease) for one group compared to another group. For data collected prospectively, the relative risk is written

$$RR = \frac{p_1}{p_2}$$

You can show that

$$RR = OR \times \frac{\{1 + (n_{21}/n_{22})\}}{\{1 + (n_{11}/n_{12})\}}$$

or that OR approximates RR when n_{11} and n_{21} are small relative to n_{12} and n_{22}, respectively. This is called the *rare outcome assumption*. Usually, the outcome of interest needs to occur less than 10% of the time for OR and RR to be similar. However, many times when the event under investigation is a relatively common occurrence, you are more interested in looking at the difference in proportions rather than at the odds ratio or the relative risk.

For cross-sectional data, the quantity p_1/p_2 is called the *prevalence ratio*; it does not indicate risk since the disease and risk factor are assessed at the same time, but it does give you an idea of the prevalence of a condition in one group compared to another.

It is important to realize that the odds ratio can always be used as a measure of association, and that relative risk and the odds ratio as an estimator of relative risk have meaning for certain types of studies and require certain assumptions.

Table 2.7 contains data from a study on how general daily stress affects one's opinion on a proposed new health policy. Since information on stress level and opinion were collected at the same time, the data are cross-sectional.

Table 2.7. Opinions on New Health Policy

Stress	Favorable	Unfavorable	Total
Low	48	12	60
High	96	94	190

To produce the odds ratio and other measures of association from PROC FREQ, you specify the MEASURES option in the TABLES statement. The ORDER=DATA option is used in the PROC FREQ statement to produce a table that looks the same as that displayed in Table 2.7. Without this option, the row corresponding to high stress would come first and the row corresponding to low stress would come last.

```
data stress;
   input stress $ outcome $ count;
   datalines;
low   f 48
low   u 12
high  f 96
high  u 94
;
proc freq order=data;
   weight count;
   tables stress*outcome / chisq measures nocol nopct;
run;
```

Output 2.11 contains the resulting frequency table. Since the NOCOL and NOPCT options are specified, only the row percentages are printed. 80% of the low stress group were favorable, while the high stress group was nearly evenly split between favorable and unfavorable.

Output 2.11   Frequency Table

```
                 Table of stress by outcome

         stress       outcome

         Frequency|
         Row Pct  |f        |u        |  Total
         ---------+--------+--------+
         low      |    48 |    12 |    60
                  | 80.00 | 20.00 |
         ---------+--------+--------+
         high     |    96 |    94 |   190
                  | 50.53 | 49.47 |
         ---------+--------+--------+
         Total        144      106      250
```

Output 2.12 displays the chi-square statistics. The statistics Q and Q_P indicate a strong association, with values of 16.1549 and 16.2198, respectively. Note how close the values for these statistics are for a sample size of 250.

Output 2.12   Chi-Square Statistics

```
              Statistics for Table of stress by outcome

        Statistic                    DF      Value      Prob
        -----------------------------------------------------
        Chi-Square                    1     16.2198    <.0001
        Likelihood Ratio Chi-Square   1     17.3520    <.0001
        Continuity Adj. Chi-Square    1     15.0354    0.0001
        Mantel-Haenszel Chi-Square    1     16.1549    <.0001
        Phi Coefficient                     0.2547
        Contingency Coefficient             0.2468
        Cramer's V                          0.2547
```

Output 2.13 contains the measures of association such as Kendall's tau-b, Pearson correlation, Spearman correlation, and uncertainty coefficients. See Chapter 5 for more information about some of these measures.

Output 2.13   Measures of Association

```
              Statistics for Table of stress by outcome

        Statistic                        Value       ASE
        -----------------------------------------------------
        Gamma                            0.5932     0.1147
        Kendall's Tau-b                  0.2547     0.0551
        Stuart's Tau-c                   0.2150     0.0489

        Somers' D C|R                    0.2947     0.0631
        Somers' D R|C                    0.2201     0.0499

        Pearson Correlation              0.2547     0.0551
        Spearman Correlation             0.2547     0.0551

        Lambda Asymmetric C|R            0.0000     0.0000
        Lambda Asymmetric R|C            0.0000     0.0000
        Lambda Symmetric                 0.0000     0.0000

        Uncertainty Coefficient C|R      0.0509     0.0231
        Uncertainty Coefficient R|C      0.0630     0.0282
        Uncertainty Coefficient Symmetric 0.0563    0.0253
```

Output 2.14 displays the odds ratio information.

Output 2.14   Odds Ratio

```
            Statistics for Table of stress by outcome


            Estimates of the Relative Risk (Row1/Row2)

    Type of Study                 Value      95% Confidence Limits
    ---------------------------------------------------------------
    Case-Control (Odds Ratio)     3.9167        1.9575      7.8366
    Cohort (Col1 Risk)            1.5833        1.3104      1.9131
    Cohort (Col2 Risk)            0.4043        0.2389      0.6841

                    Sample Size = 250
```

The odds ratio value is listed beside "Case-Control" in the section labeled "Estimates of the Relative Risk (Row1/Row2)." The estimated OR is 3.9167, which means that the odds of a favorable response are roughly four times higher for those with low stress than for those with high stress. The confidence intervals are labeled "Confidence Limits" and are 95% confidence intervals by default. To change them, use the ALPHA= option in the TABLES statement.

The values listed for "Cohort (Col1 Risk)" and "Cohort (Col2 Risk)" are the estimates of relative risk for a cohort (prospective) study. Since these data are cross-sectional, you cannot estimate relative risk. However, the value 1.5833 is the ratio of the prevalence of favorable opinions for the low stress group compared to the high stress group. (The value 0.4043 is the prevalence ratio of the unfavorable opinions of the low stress group compared to the high stress group.)

Table 2.8 contains data that concern respiratory illness. Two groups having the same symptoms of respiratory illness were selected via simple random sampling: one group was treated with a test treatment, and one group was treated with a placebo. This is an example of a cohort study since the comparison groups were chosen before the responses were measured. They are considered to come from independent binomial distributions.

Table 2.8.   Respiratory Improvement

Treatment	Yes	No	Total
Test	29	16	45
Placebo	14	31	45

In order to produce chi-square statistics, odds ratios, and relative risk measures for these data, the following statements are submitted. The ALL option has the same action as specifying both the CHISQ and the MEASURES options (and the CMH option, discussed in Chapter 3.

```
data respire;
   input treat $ outcome $ count;
   datalines;
test    yes  29
test    no   16
placebo yes  14
placebo no   31
;
proc freq order=data;
   weight count;
   tables treat*outcome / all nocol nopct;
run;
```

For these data, $Q = 9.9085$ and $Q_P = 10.0198$. Clearly, there is a strong association between treatment and improvement.

Output 2.15 Table Statistics

```
               Statistics for Table of treat by outcome

        Statistic                     DF      Value      Prob
        ---------------------------------------------------------

        Chi-Square                     1     10.0198    0.0015
        Likelihood Ratio Chi-Square    1     10.2162    0.0014
        Continuity Adj. Chi-Square     1      8.7284    0.0031
        Mantel-Haenszel Chi-Square     1      9.9085    0.0016
        Phi Coefficient                       0.3337
        Contingency Coefficient               0.3165
        Cramer's V                            0.3337


                     Fisher's Exact Test
              ------------------------------------
              Cell (1,1) Frequency (F)         29
              Left-sided Pr <= F           0.9997
              Right-sided Pr >= F          0.0015

              Table Probability (P)        0.0011
              Two-sided Pr <= P            0.0029
```

Output 2.16 displays the estimates of relative risk and the odds ratio (other measures of association produced by the ALL option are not displayed here). Two versions of the relative risk are supplied: one is the relative risk of the attribute corresponding to the first column, or the risk of improvement. The column 2 risk is the risk of no improvement. The relative risk for improvement is 2.0714, with a 95% confidence interval of (1.2742, 3.3675).

Note that if these data had been obtained retrospectively, the odds ratio couldn't be used as an estimate of the relative risk since the proportions with improvement are 0.36 and 0.69. The rare outcome assumption is not satisfied.

Output 2.16   Odds Ratio and Relative Risk

```
                   Statistics for Table of treat by outcome


                    Estimates of the Relative Risk (Row1/Row2)

       Type of Study                   Value       95% Confidence Limits
       --------------------------------------------------------------------
       Case-Control (Odds Ratio)       4.0134       1.6680        9.6564
       Cohort (Col1 Risk)              2.0714       1.2742        3.3675
       Cohort (Col2 Risk)              0.5161       0.3325        0.8011

                              Sample Size = 90
```

2.5.1   Exact Confidence Limits for the Odds Ratio

Section 2.3 discussed Fisher's exact test for assessing association in 2 × 2 tables that were too sparse for the usual asymptotic chi-square tests to apply. You may want to compute the odds ratio as a measure of association for these data, but the usual asymptotic confidence limits would not be appropriate because, again, the sparseness of the data violates the asymptotic assumptions.

You can obtain exact confidence limits for the odds ratio by using the FREQ procedure. The computation is based on work presented by Thomas (1971) and Gart (1971). These confidence limits are conservative; the coefficient is not exactly $1 - \alpha$, but it is at least $1 - \alpha$.

Consider the severe infection data in Table 2.3. To compute an odds ratio estimate for the odds of having a favorable outcome for the treatment group compared to the control group, you submit the following statements, including the EXACT statement with the OR keyword.

```
data severe;
   input treat $ outcome $ count;
   datalines;
Test     f 10
Test     u 2
Control f 2
Control u 4
;
proc freq order=data;
   weight count;
   tables treat*outcome / nocol;
   exact or;
run;
```

Output 2.17 displays the estimate of the odds ratio, which is 10. Test subjects have 10 times higher odds for the favorable response than the control subjects.

Output 2.17 Odds Ratio (Case-Control Study)

```
            Statistics for Table of treat by outcome


              Odds Ratio (Case-Control Study)
              ---------------------------------
              Odds Ratio                   10.0000

              Asymptotic Conf Limits
              95% Lower Conf Limit          1.0256
              95% Upper Conf Limit         97.5005

              Exact Conf Limits
              95% Lower Conf Limit          0.6896
              95% Upper Conf Limit        166.3562

                    Sample Size = 18
```

The exact confidence limits for the odds ratio are (0.6896, 166.3562), indicating low precision. Note that the exact confidence bands are much wider than the asymptotic ones.

2.6   Sensitivity and Specificity

Some other measures frequently calculated for 2×2 tables are *sensitivity* and *specificity*. These measures are of particular interest when you are determining the efficacy of screening tests for various disease outcomes. Sensitivity is the true proportion of positive results that a test elicits when performed on subjects known to have the disease; specificity is the true proportion of negative results that a test elicits when performed on subjects known to be disease free.

Often, a standard screening method is used to determine whether disease is present and compared to a new test method. Table 2.9 contains the results of a study investigating a new screening device for a skin disease. The distributions for positive and negative results for the test method are assumed to result from simple random samples from the corresponding populations of persons with disease present and those with disease absent.

Table 2.9.   Skin Disease Screening Test Results

Status	Test +	Test −	Total
Disease Present	52	8	60
Disease Absent	20	100	120

Sensitivity and specificity for these data are estimated by

$$\text{sensitivity} = (n_{11}/n_{1+}) \doteq \Pr(\text{Test} + |\text{disease present})$$

and

$$\text{specificity} = (n_{22}/n_{2+}) \doteq \Pr(\text{Test} - |\text{disease absent})$$

For these data, sensitivity = 52/60 = 0.867 and specificity =100/120 = 0.833.

You may know the underlying percentage of those with and without the disease in a population of interest. You may want to estimate the proportion of subjects with the disease among those who have a positive test. You can determine these proportions with the use of Bayes' theorem.

Suppose that the underlying prevalence of disease for an appropriate target population for these data is 15%. That is, 15% of the population have the disease and 85% do not. You can compute joint probabilities by multiplying the conditional probabilities by the marginal probabilities.

$$\Pr(T, D) = \Pr(T|D) \times \Pr(D)$$

Table 2.10.   How Test Should Perform in General Population

Status	Test +	Test −	Total
Disease Present	$0.867(.15) = 0.130$	$0.133(.15) = 0.020$	0.15
Disease Absent	$0.167(.85) = 0.142$	$0.833(.85) = 0.708$	0.85
Total	$0.130 + 0.142 = 0.272$	$0.020 + 0.708 = 0.728$	

The values in the row titled "Total" are Pr(Test +) and Pr(Test −), respectively. You can now determine the probability of those with the disease among those with a positive test:

$$\Pr(D|T) = \frac{\Pr(T, D)}{\Pr(T)}$$

Thus, Pr (disease|Test +) = 0.130/0.272 = 0.478 and Pr(no disease|Test −) = 0.708/0.728 = 0.972. Refer to Fleiss (1981, p. 4) for more detail, including the calculation of false negative and false positive rates.

2.7   McNemar's Test

The 2 × 2 table often contains information collected from *matched pairs*, experimental units for which two related responses are made. The sampling unit is no longer one individual but a pair of related individuals, which could be two locations on the same individual or two occasions for the same individual. For example, in case-control studies, cases are often matched to controls on the basis of demographic characteristics; interest lies in determining whether there is a difference between control exposure to a risk factor and case exposure to the same risk factor. Other examples of matched pairs are left eye and right eye measurements, and husband and wife voting preferences. Measurements at two different time points can also be considered a matched pair, such as before and after measurements.

Data from a study on matched pairs are represented in Table 2.11. The n_{11} in the (1,1) cell means that n_{11} pairs responded yes for both Response 1 and Response 2; the n_{21} in the (2,1) cell means that n_{21} pairs responded yes for Response 1 and no for Response 2.

Table 2.11. Matched Pairs Data

	Response 1		
Response 2	Yes	No	Total
Yes	n_{11}	n_{12}	n_{1+}
No	n_{21}	n_{22}	n_{2+}
Total	n_{+1}	n_{+2}	n

The question of interest for such data is whether the proportion of pairs responding yes for Response 1 is the same as the proportion of pairs responding yes for Response 2. This question cannot be addressed with the chi-square tests of association of previous sections, since the cell counts represent pairs instead of individuals.

The question is whether

$$p_1 = \frac{n_{+1}}{n}$$

and

$$p_2 = \frac{n_{1+}}{n}$$

are the same. McNemar (1947) developed a chi-square test based on the binomial distribution to address this situation. He shows that only the off-diagonal elements are important in determining whether there is a difference in these proportions. The test statistic is written

$$Q_M = \frac{(n_{12} - n_{21})^2}{(n_{12} + n_{21})}$$

and is approximately chi-square with one degree of freedom.

Table 2.12 displays data collected by political science researchers who polled husbands and wives on whether they approved of one of their U.S. senators. The cell counts represent the number of pairs of husbands and wives who fit the configurations indicated by the row and column levels.

Table 2.12. State Senator Approval Ratings

Husband	Wife Approval		
Approval	Yes	No	Total
Yes	20	5	25
No	10	10	20
Total	30	15	45

McNemar's test is easy to compute by hand.

$$Q_M = \frac{(5-10)^2}{(5+10)} = 1.67$$

Compared to a chi-square distribution with 1 df, this statistic is clearly nonsignificant.

The FREQ procedure computes McNemar's Test with the AGREE option in the TABLE statement (see Chapter 5 for other analyses available with the AGREE option for tables of other dimensions). The following SAS statements request McNemar's test. The ODS SELECT statement is used to restrict the output to that test.

```
data approval;
   input hus_resp $ wif_resp $ count;
   datalines;
yes yes 20
yes no   5
no yes  10
no no   10
;

ods select McNemarsTest;
proc freq order=data;
   weight count;
   tables hus_resp*wif_resp / agree;
run;
```

Output 2.18 displays the output that is produced. $Q_M = 1.67$, the same value as computed previously.

Output 2.18   McNemar's Test

```
           Statistics for Table of hus_resp by wif_resp

                         McNemar's Test
                 ----------------------
                 Statistic (S)     1.6667
                 DF                     1
                 Pr > S            0.1967
```

Note that exact p-values are also available for McNemar's test. You would include the statement

```
   exact mcnem;
```

in your PROC FREQ invocation. The computations work in a similar fashion to those for the chi-square tests of association; the exact p-value is the sum of the probabilities of those tables with a Q_M greater than or equal to the actual one.

Chapter 3
Sets of 2×2 Tables

Chapter Table of Contents

Chapter 3
Sets of 2 × 2 Tables

3.1 Introduction

The respiratory data displayed in Table 2.8 in the previous chapter are only a subset of the data collected in the clinical trial. The study included patients at two medical centers and produced the complete data shown in Table 3.1. These data comprise a set of two 2 × 2 tables.

Table 3.1. Respiratory Improvement

Center	Treatment	Yes	No	Total
1	Test	29	16	45
1	Placebo	14	31	45
Total		43	47	90
2	Test	37	8	45
2	Placebo	24	21	45
Total		61	29	90

Investigators were interested in whether there were overall differences in rates of improvement; however, they were concerned that the patient populations at the two centers were sufficiently different that center needed to be accounted for in the analysis. One strategy for examining the association between two variables while adjusting for the effects of others is *stratified analysis*.

In general, the strata may represent explanatory variables, or they may represent research sites or hospitals in a multicenter study. Each table corresponds to one stratum; the strata are determined by the levels of the explanatory variables (one for each unique combination of the levels of the explanatory variables). The idea is to evaluate the association between the row variable and the response variable, while *adjusting*, or *controlling*, for the effects of the stratification variables. In some cases, the stratification results from the study design, such as in the case of a multicenter clinical trial; in other cases, it may arise from a prespecified poststudy stratification performed to control for the effects of certain explanatory variables that are thought to be related to the response variable.

The analysis of sets of tables addresses the same questions as the analysis of a single table: is there an association between the row and column variables in the tables and what is the strength of that association? These questions are investigated with similar strategies involving chi-square statistics and measures of association such as the odds ratios; the key difference is that you are investigating overall association instead of the association in just one table.

3.2   Mantel-Haenszel Test

For the data in Table 3.1, interest lies in determining whether there is a difference in the favorable rates between Test and Placebo. Patients in both centers were randomized into two treatment groups, which induces independent hypergeometric distributions for the within-center frequencies under the hypothesis that treatments have equal effects for all patients. Thus, the distribution for the two tables is the product of these two hypergeometric distributions. You can induce the hypergeometric distribution via conditional distribution arguments when you have postrandomization stratification or when you have independent binomial distributions from simple random sampling.

Consider the following table as representative of q 2 × 2 tables, $h = 1, 2, \ldots, q$.

Table 3.2. hth 2 × 2 Contingency Table

	Yes	No	Total
Group 1	n_{h11}	n_{h12}	n_{h1+}
Group 2	n_{h21}	n_{h22}	n_{h2+}
Total	n_{h+1}	n_{h+2}	n_h

Under the null hypothesis of no treatment difference, the expected value of n_{h11} is

$$E\{n_{h11}|H_0\} = \frac{n_{h1+}n_{h+1}}{n_h} = m_{h11}$$

and its variance is

$$V\{n_{h11}|H_0\} = \frac{n_{h1+}n_{h2+}n_{h+1}n_{h+2}}{n_h^2(n_h - 1)} = v_{h11}$$

One method for assessing the overall association of group and response, adjusting for the stratification factor, is the Mantel-Haenszel (1959) statistic.

$$
\begin{aligned}
Q_{MH} &= \frac{\left\{\sum_{h=1}^{q} n_{h11} - \sum_{h=1}^{q} m_{h11}\right\}^2}{\sum_{h=1}^{q} v_{h11}} \\
&= \frac{\left\{\sum_{h=1}^{q} (n_{h1+}n_{h2+}/n_h)(p_{h11} - p_{h21})\right\}^2}{\sum_{h=1}^{q} v_{h11}}
\end{aligned}
$$

where $p_{hi1} = n_{hi1}/n_{hi+}$ is the proportion of subjects from the hth stratum and the ith group who have a favorable response. Q_{MH} approximately has the chi-square distribution with one degree of freedom when the combined row sample sizes ($\sum_{h=1}^{q} n_{hi+} = n_{+i+}$) are large, for example, greater than 30. This means that individual cell counts and table sample sizes may be small, so long as the overall row sample sizes are large. For the case of two tables, such as for Table 3.1, $q = 2$.

The Mantel-Haenszel strategy potentially removes the confounding influence of the explanatory variables that comprise the stratification and so can provide increased power for detecting association in a randomized study by comparing like subjects with like subjects. It can remove the bias that results in an observational study but possibly at the cost of decreased power. In some sense, the strategy is similar to adjustment for blocks in a two-way analysis of variance for randomized blocks; it is also like covariance adjustment for a categorical explanatory variable.

Q_{MH} is effective for detecting patterns of association across q strata when there is a strong tendency to expect the predominant majority of differences $\{p_{h11} - p_{h21}\}$ to have the same sign. For this reason, Q_{MH} is often called an *average partial association statistic*. Q_{MH} may fail to detect association when the differences are in opposite directions and are of similar magnitude. Q_{MH} as formulated here is directed at the n_{h11} cell; however, it is invariant to whatever cell is chosen. For a recent overview of Mantel-Haenszel methods, refer to Landis et al. (1998).

Mantel and Fleiss (1980) proposed a criterion for determining whether the chi-square approximation is appropriate for the distribution of the Mantel-Haenszel statistic for q strata:

$$\min\left\{\left[\sum_{h=1}^{q} m_{h11} - \sum_{h=1}^{q}(n_{h11})_L\right], \left[\sum_{h=1}^{q}(n_{h11})_U - \sum_{h=1}^{q} m_{h11}\right]\right\} > 5$$

where $(n_{h11})_L = \max(0, n_{h1+} - n_{h+2})$ and $(n_{h11})_U = \min(n_{h+1}, n_{h1+})$. The criterion specifies that the across-strata sum of expected values for a particular cell has a difference of at least 5 from both the minimum possible sum and the maximum possible sum of the observed values.

3.2.1 Respiratory Data Example

For the data in Table 3.1, there is interest in the association between treatment and respiratory outcome, after adjusting for the effects of the centers. The following DATA step puts all the respiratory data into the SAS data set RESPIRE.

```
data respire;
   input center treatment $ response $ count @@;
   datalines;
1 test      y 29 1 test     n 16
1 placebo y 14 1 placebo n 31
2 test      y 37 2 test     n  8
2 placebo y 24 2 placebo n 21
;
```

Producing a Mantel-Haenszel analysis from PROC FREQ requires the specification of multi-way tables. The triple crossing CENTER*TREATMENT*RESPONSE specifies that the data consists of sets of two-way tables. The two rightmost variables TREATMENT and RESPONSE determine the rows and columns of the tables, respectively, and the variables to the left (CENTER) determine the stratification scheme. There will be one table for each value of CENTER. If there are more variables to the left of the variables determining the rows and columns of the tables, there will be strata for each unique combination of values for those variables.

The CHISQ option specifies that chi-square statistics be printed for each table. The CMH option requests the Mantel-Haenszel statistics for the stratified analysis; these are also called summary statistics. The ORDER=DATA option specifies that PROC FREQ order the rows and columns according to the order in which the variable values are encountered in the input data.

```
proc freq order=data;
   weight count;
   tables center*treatment*response /
       nocol nopct chisq cmh;
run;
```

Output 3.1 and Output 3.2 display the frequency tables and chi-square statistics for each center. For Center 1, the favorable rate for test treatment is 64%, versus 31% for placebo. For Center 2, the favorable rate for test treatment is 82%, versus 53% for placebo. Q (the randomization statistic discussed in Chapter 2) for Center 1 is 9.908; Q for Center 2 is 8.503. With 1 df, both of these statistics are strongly significant.

Output 3.1   Table 1 Results

```
                 Table 1 of treatment by response
                    Controlling for center=1

            treatment        response

            Frequency|
            Row Pct  |y         |n        |  Total
            ---------+--------+--------+
            test     |     29 |     16 |     45
                     |  64.44 |  35.56 |
            ---------+--------+--------+
            placebo  |     14 |     31 |     45
                     |  31.11 |  68.89 |
            ---------+--------+--------+
            Total            43       47       90

    Statistic                     DF       Value      Prob
    ------------------------------------------------------------
    Chi-Square                     1      10.0198     0.0015
    Likelihood Ratio Chi-Square    1      10.2162     0.0014
    Continuity Adj. Chi-Square     1       8.7284     0.0031
    Mantel-Haenszel Chi-Square     1       9.9085     0.0016
    Phi Coefficient                        0.3337
    Contingency Coefficient                0.3165
    Cramer's V                             0.3337


                      Fisher's Exact Test
            ------------------------------------
            Cell (1,1) Frequency (F)          29
            Left-sided Pr <= F            0.9997
            Right-sided Pr >= F           0.0015

            Table Probability (P)        0.0011
            Two-sided Pr <= P            0.0029
```

Output 3.2 Table 2 Results

```
                Table 2 of treatment by response
                    Controlling for center=2

             treatment        response

             Frequency|
             Row Pct  |y       |n       |  Total
             ---------+--------+--------+
             test     |     37 |      8 |    45
                      |  82.22 |  17.78 |
             ---------+--------+--------+
             placebo  |     24 |     21 |    45
                      |  53.33 |  46.67 |
             ---------+--------+--------+
             Total          61       29       90

       Statistic                    DF      Value      Prob
       ------------------------------------------------------
       Chi-Square                    1      8.5981     0.0034
       Likelihood Ratio Chi-Square   1      8.8322     0.0030
       Continuity Adj. Chi-Square    1      7.3262     0.0068
       Mantel-Haenszel Chi-Square    1      8.5025     0.0035
       Phi Coefficient                      0.3091
       Contingency Coefficient              0.2953
       Cramer's V                           0.3091


                      Fisher's Exact Test
             -----------------------------------
             Cell (1,1) Frequency (F)       37
             Left-sided Pr <= F         0.9993
             Right-sided Pr >= F        0.0031

             Table Probability (P)      0.0025
             Two-sided Pr <= P          0.0063
```

Following the information for the individual tables, PROC FREQ prints out a section titled "Summary Statistics for treatment by response Controlling for center." This includes tables containing Mantel-Haenszel (MH) statistics, estimates of the common relative risk, and the Breslow-Day test for homogeneity of the odds ratio.

Output 3.3   Summary Statistics

```
                  Summary Statistics for treatment by response
                            Controlling for center

          Cochran-Mantel-Haenszel Statistics (Based on Table Scores)

        Statistic    Alternative Hypothesis    DF      Value      Prob
        --------------------------------------------------------------
            1        Nonzero Correlation        1     18.4106    <.0001
            2        Row Mean Scores Differ      1     18.4106    <.0001
            3        General Association         1     18.4106    <.0001


                Estimates of the Common Relative Risk (Row1/Row2)

        Type of Study    Method            Value    95% Confidence Limits
        ----------------------------------------------------------------
        Case-Control     Mantel-Haenszel   4.0288    2.1057      7.7084
           (Odds Ratio)  Logit             4.0286    2.1057      7.7072

        Cohort           Mantel-Haenszel   1.7368    1.3301      2.2680
           (Col1 Risk)   Logit             1.6760    1.2943      2.1703

        Cohort           Mantel-Haenszel   0.4615    0.3162      0.6737
           (Col2 Risk)   Logit             0.4738    0.3264      0.6877


                          Breslow-Day Test for
                     Homogeneity of the Odds Ratios
                     -----------------------------
                     Chi-Square            0.0002
                     DF                         1
                     Pr > ChiSq            0.9900


                      Total Sample Size = 180
```

To find the value of Q_{MH}, read the value for any of the statistics in the table labeled "Cochran-Mantel-Haenszel Statistics": "Nonzero Correlation," "Row Mean Scores Differ," or "General Association." These statistics pertain to the situation where you have sets of tables with two or more rows or columns: they are discussed in Chapter 6, "Sets of $s \times r$ Tables." However, they all reduce to the MH statistic when you have 2×2 tables and use the CMH option in its default mode (that is, no SCORE= option specified). Note that the General Association statistic is always appropriate regardless of the scores used.

Q_{MH} for these data is $Q_{MH} = 18.4106$, with 1 df. This is clearly significant. The associations in the individual tables reinforce each other so that the overall association is stronger than that seen in the individual tables. There is a strong association between treatment and response, adjusting for center. The test treatment had a significantly higher favorable response rate than placebo.

The information in the rest of the summary statistics output is discussed later in this chapter. Note that for these data, the Mantel-Fleiss criterion is satisfied:

$$\sum_{h=1}^{2} m_{h11} = 21.5 + 30.5 = 52$$

$$\sum_{h=1}^{2}(n_{h11})_L = 0 + 16 = 16$$

$$\sum_{h=1}^{2}(n_{h11})_U = 43 + 45 = 88$$

so that $(52 - 16) \geq 5$ and $(88 - 52) \geq 5$.

3.2.2 Health Policy Data

Another data set discussed in Chapter 2 was also a subset of the complete data. The health policy data displayed in Table 2.7 comes from a study that included interviews with subjects from both rural and urban geographic regions. Table 2.7 displays the information from the rural region, and Table 3.3 includes the complete data.

Table 3.3. Health Policy Opinion Data

Residence	Stress	Favorable	Unfavorable	Total
Urban	Low	48	12	60
Urban	High	96	94	190
	Total	144	106	250
Rural	Low	55	135	190
Rural	High	7	53	60
	Total	62	188	250

If you ignored region and pooled these two tables, you would obtain Table 3.4.

Table 3.4. Pooled Health Policy Opinion Data

Stress	Favorable	Unfavorable	Total
Low	103	147	250
High	103	147	250
Total	206	294	500

There is clearly no association in this table; the proportions for favorable opinion are the same for low stress and high stress. For this table, Q_P and Q take the value 0, and the odds ratio is exactly 1. These data illustrate the need to consider the sampling framework in any data analysis. If you note the row totals in Table 3.3, you see that high stress subjects were oversampled for the urban region, and the low stress subjects were oversampled for the rural region. This oversampling causes the pooled table to take its form, even though favorable response is more likely for low stress persons in both regions.

The fact that a marginal table (pooled over residence) may exhibit an association completely different from the partial tables (individual tables for urban and rural) is known as *Simpson's Paradox* (Simpson 1951, Yule 1903).

The following statements request a Mantel-Haenszel analysis for the health policy data.

```
data stress;
   input region $ stress $ outcome $ count @@;
   datalines;
urban low  f 48 urban  low   u   12
urban high f 96 urban  high  u   94
rural low  f 55 rural  low   u  135
rural high f  7 rural  high  u   53
;
proc freq order=data;
   weight count;
   tables region*stress*outcome / chisq cmh nocol nopct;
run;
```

Output 3.4 and Output 3.5 display the results for the individual tables. The urban region has a *Q* of 16.1549 for the association of stress level and health policy opinion; the *Q* for the rural region is 7.2724. The rate of favorable response is higher for the low stress group than for the high stress group in each region.

Output 3.4   Table 1 Results

```
                    Table 1 of stress by outcome
                    Controlling for region=urban

           stress       outcome

           Frequency|
           Row Pct  |f        |u        |  Total
           ---------+--------+--------+
           low      |     48 |     12 |     60
                    |  80.00 |  20.00 |
           ---------+--------+--------+
           high     |     96 |     94 |    190
                    |  50.53 |  49.47 |
           ---------+--------+--------+
           Total         144      106      250


      Statistic                       DF       Value      Prob
      -------------------------------------------------------------
      Chi-Square                       1     16.2198     <.0001
      Likelihood Ratio Chi-Square      1     17.3520     <.0001
      Continuity Adj. Chi-Square       1     15.0354     0.0001
      Mantel-Haenszel Chi-Square       1     16.1549     <.0001
      Phi Coefficient                         0.2547
      Contingency Coefficient                 0.2468
      Cramer's V                              0.2547


                    Fisher's Exact Test
           -----------------------------------
           Cell (1,1) Frequency (F)          48
           Left-sided Pr <= F            1.0000
           Right-sided Pr >= F        3.247E-05

           Table Probability (P)      2.472E-05
           Two-sided Pr <= P          4.546E-05
```

Output 3.5 Table 2 Results

```
                   Table 2 of stress by outcome
                   Controlling for region=rural

           stress      outcome

           Frequency|
           Row Pct  |f       |u       |  Total
           ---------+--------+--------+
           low      |    55  |   135  |   190
                    |  28.95 |  71.05 |
           ---------+--------+--------+
           high     |     7  |    53  |    60
                    |  11.67 |  88.33 |
           ---------+--------+--------+
           Total        62       188      250

   Statistic                     DF      Value      Prob
   ------------------------------------------------------
   Chi-Square                     1      7.3016    0.0069
   Likelihood Ratio Chi-Square    1      8.1976    0.0042
   Continuity Adj. Chi-Square     1      6.4044    0.0114
   Mantel-Haenszel Chi-Square     1      7.2724    0.0070
   Phi Coefficient                       0.1709
   Contingency Coefficient               0.1685
   Cramer's V                            0.1709


                    Fisher's Exact Test
           -----------------------------------
           Cell (1,1) Frequency (F)        55
           Left-sided Pr <= F          0.9988
           Right-sided Pr >= F         0.0041

           Table Probability (P)       0.0029
           Two-sided Pr <= P           0.0061
```

From Output 3.6 you can see that Q_{MH} has the value 23.050, which is strongly significant. Stress is highly associated with health policy opinion, adjusting for regional effects.

Output 3.6 Summary Statistics

```
              Summary Statistics for stress by outcome
                      Controlling for region

       Cochran-Mantel-Haenszel Statistics (Based on Table Scores)

       Statistic   Alternative Hypothesis    DF     Value     Prob
       -----------------------------------------------------------
           1       Nonzero Correlation        1    23.0502    <.0001
           2       Row Mean Scores Differ      1    23.0502    <.0001
           3       General Association        1    23.0502    <.0001
```

3.2.3   Soft Drink Example

The following data come from a study on soft drink tastes by a company interested in reactions to a new soft drink that was being targeted for both the United States and Great Britain. Investigators poststratified on gender because they thought it was potentially related to the response. After receiving a supply of the new soft drink and being given a week in which to try it, subjects were asked whether they would want to switch from their current soft drinks to this new soft drink.

Table 3.5.   Soft Drink Data

Gender	Country	Switch? Yes	No	Total
Male	American	29	6	35
Male	British	19	15	34
Total		48	21	69
Female	American	7	23	30
Female	British	24	29	53
Total		31	52	83

The following statements produce a Mantel-Haenszel analysis.

```
data soft;
   input gender $ country $ question $ count @@;
   datalines;
male    American  y 29 male    American  n  6
male    British   y 19 male    British   n 15
female American  y  7 female American  n 23
female British   y 24 female British   n 29
;
proc freq order=data;
   weight count;
   tables gender*country*question /
       chisq cmh nocol nopct;
run;
```

Output 3.7 and Output 3.8 display the table results for males and females.

Output 3.7   Summary Statistics for Males

```
                        Table 1 of country by question
                          Controlling for gender=male

                 country      question

                 Frequency|
                 Row Pct  |y        |n        |  Total
                 ---------+--------+--------+
                 American |    29 |       6 |    35
                          | 82.86 |  17.14 |
                 ---------+--------+--------+
                 British  |    19 |      15 |    34
                          | 55.88 |  44.12 |
                 ---------+--------+--------+
                 Total          48         21        69

      Statistic                       DF      Value      Prob
      --------------------------------------------------------
      Chi-Square                       1     5.9272     0.0149
      Likelihood Ratio Chi-Square      1     6.0690     0.0138
      Continuity Adj. Chi-Square       1     4.7216     0.0298
      Mantel-Haenszel Chi-Square       1     5.8413     0.0157
      Phi Coefficient                        0.2931
      Contingency Coefficient                0.2813
      Cramer's V                             0.2931


                        Fisher's Exact Test
             ----------------------------------
             Cell (1,1) Frequency (F)        29
             Left-sided  Pr <= F         0.9968
             Right-sided Pr >= F         0.0143

             Table Probability (P)       0.0112
             Two-sided Pr <= P           0.0194
```

Output 3.8   Summary Statistics for Females

```
                    Table 2 of country by question
                     Controlling for gender=female

            country       question

            Frequency|
            Row Pct  |y        |n        |  Total
            ---------+--------+--------+
            American |    7    |   23   |    30
                     |  23.33 |  76.67 |
            ---------+--------+--------+
            British  |   24    |   29   |    53
                     |  45.28 |  54.72 |
            ---------+--------+--------+
            Total          31       52       83

            Statistic                  DF      Value      Prob
            --------------------------------------------------
            Chi-Square                  1      3.9443    0.0470
            Likelihood Ratio Chi-Square 1      4.0934    0.0431
            Continuity Adj. Chi-Square  1      3.0620    0.0801
            Mantel-Haenszel Chi-Square  1      3.8968    0.0484
            Phi Coefficient                   -0.2180
            Contingency Coefficient            0.2130
            Cramer's V                        -0.2180


                       Fisher's Exact Test
            -----------------------------------
            Cell (1,1) Frequency (F)           7
            Left-sided Pr <= F            0.0385
            Right-sided Pr >= F           0.9881

            Table Probability (P)        0.0267
            Two-sided Pr <= P            0.0602
```

As indicated by Q for males (5.8413) and Q for females (3.8968), there is significant association in both tables between country and willingness to switch. However, look at Q_{MH} in the following output.

Output 3.9   Summary Statistics

```
                 Summary Statistics for country by question
                          Controlling for gender

         Cochran-Mantel-Haenszel Statistics (Based on Table Scores)

         Statistic   Alternative Hypothesis    DF     Value     Prob
         ----------------------------------------------------------------
             1       Nonzero Correlation        1     0.0243    0.8762
             2       Row Mean Scores Differ     1     0.0243    0.8762
             3       General Association        1     0.0243    0.8762
```

Q_{MH} takes the value 0.024, indicating that there is no association between country and willingness to switch, after adjusting for gender. However, if you examine the individual tables more closely, you see that the association is manifested in opposite directions. For males, Americans are overwhelmingly favorable, and the British are a little more favorable than unfavorable. For females, Americans are very opposed, while the British are mildly opposed.

Thus, for these data, Q_{MH} fails to detect an association because the association is of opposite directions with roughly the same magnitude. As discussed previously, Q_{MH} has power against the alternative hypothesis of consistent patterns of association; it has low power for detecting association in opposite directions. (However, regardless of these matters of power, the method always performs at the specified significance level [or less] under the null hypothesis, so it is always valid.)

Generally, this isn't a problem because if there is association, it is usually in the same direction across a set of tables, although often to varying degrees. However, you should always examine the individual tables, especially if your results are questionable, to determine if you have a situation in which the association is inconsistent and the Q_{MH} statistic is not very powerful.

3.3 Measures of Association

Section 2.5 discusses the odds ratio as a measure of association for the 2×2 table. You can compute average odds ratios for sets of 2×2 tables. For the hth stratum,

$$\text{OR}_h = \frac{p_{h1}/(1 - p_{h1})}{p_{h2}/(1 - p_{h2})} = \frac{n_{h11} n_{h22}}{n_{h12} n_{h21}}$$

so OR_h estimates ψ_h, the population odds ratio for the hth stratum. If the ψ_h are homogeneous, then you can compute the Mantel-Haenszel estimator for the common odds ratio ψ.

$$\hat{\psi}_{MH} = \sum_{h=1}^{q} \frac{n_{h11} n_{h22}}{n_h} \bigg/ \sum_{h=1}^{q} \frac{n_{h12} n_{h21}}{n_h}$$

The standard error for this estimator is based on work by Robins, Breslow, and Greenland (1986) in which they provide an estimated variance for log $\hat{\psi}_{MH}$. The $100\%(1 - \alpha)$ confidence interval for $\hat{\psi}_{MH}$ is

$$\left(\hat{\psi}_{MH} \cdot \exp(-z_{\alpha/2}\hat{\sigma}), \ \hat{\psi}_{MH} \cdot \exp(z_{\alpha/2}\hat{\sigma}) \right)$$

where

$$\begin{aligned} \hat{\sigma}^2 &= \hat{\text{var}}[\log(\hat{\psi}_{MH})] \\[2mm] &= \frac{\sum_h (n_{h11} + n_{h22})(n_{h11}\, n_{h22})/n_h^2}{2 \left(\sum_h n_{h11}\, n_{h22}/n_h \right)^2} \end{aligned}$$

$$+\frac{\sum_h [(n_{h11} + n_{h22})(n_{h12}\,n_{h21}) + (n_{h12} + n_{h21})(n_{h11}\,n_{h22})]/n_h^2}{2\left(\sum_h n_{h11}\,n_{h22}/n_h\right)\left(\sum_h n_{h12}\,n_{h21}/n_h\right)}$$

$$+\frac{\sum_h (n_{h12} + n_{h21})(n_{h12}\,n_{h21})/n_h^2}{2\left(\sum_h n_{h12}\,n_{h21}/n_h\right)^2}$$

Another estimator of ψ is the logit estimator. This is a weighted regression estimate with the form

$$\hat{\psi}_L = \exp\left\{\sum_{h=1}^{q} w_h f_h \Big/ \sum_{h=1}^{q} w_h\right\} = \exp\{\bar{f}\}$$

where $f_h = \log \mathrm{OR}_h$ and

$$w_h = \left\{\frac{1}{n_{h11}} + \frac{1}{n_{h12}} + \frac{1}{n_{h21}} + \frac{1}{n_{h22}}\right\}^{-1}$$

You can write a $100(1 - \alpha)\%$ confidence interval for $\hat{\psi}_L$ as

$$\exp\left\{\bar{f} \pm z_{\alpha/2}\left[\sum_{h=1}^{q} w_h\right]^{-1/2}\right\}$$

The logit estimator is also reasonable but requires adequate sample sizes (all $n_{hij} \geq 5$); it has problems with zero cells for the n_{hij}, in which case you should proceed cautiously. The Mantel-Haenszel estimator is not as sensitive to sample size.

Note that logistic regression provides a better strategy for estimating the common odds ratio and produces a confidence interval based on maximum likelihood methods. This is discussed in Chapter 8, "Logistic Regression I: Dichotomous Response." Currently, you need to perform an exact logistic regression to obtain exact confidence intervals for the odds ratio for sets of 2 × 2 tables. See Section 8.8 for an example.

3.3.1 Homogeneity of Odds Ratios

You are generally interested in whether the odds ratios in a set of tables are homogeneous. There are several test statistics that address the hypothesis of homogeneity, one of which is the Breslow-Day statistic.

Consider Table 3.6. The top table shows the expected counts m_{ij} for a 2 × 2 table, and the bottom table shows how you can write the expected counts for the rest of the cells if you know the (1,1) expected count m_{11}.

Table 3.6. Odds Ratios

	Yes	No	Total
Group 1	m_{11}	m_{12}	n_{1+}
Group 2	m_{21}	m_{22}	n_{2+}
Total	n_{+1}	n_{+2}	n
Group 1	m_{11}	$n_{1+} - m_{11}$	n_{1+}
Group 2	$n_{+1} - m_{11}$	$n - n_{1+} - n_{+1} + m_{11}$	$n - n_{1+}$
Total	n_{+1}	$n - n_{+1}$	n

If you assume that the odds ratio takes a certain value, $\psi = \psi_0$, then

$$\frac{m_{11}(n - n_{1+} - n_{+1} + m_{11})}{(n_{+1} - m_{11})(n_{1+} - m_{11})} = \psi_0$$

You can put this expression into the form of a quadratic equation and then solve for m_{11}; once you have m_{11}, you can solve for the other expected counts.

To compute the Breslow-Day statistic, you use ψ_{MH} as ψ_0 for each stratum and perform the preceding computations for the expected counts for each table; that is, you compute the m_{hij}. Then,

$$Q_{BD} = \sum_{h=1}^{q} \sum_{i=1}^{2} \sum_{j=1}^{2} \frac{(n_{hij} - m_{hij})^2}{m_{hij}}$$

Under the null hypothesis of homogeneity, Q_{BD} approximately has a chi-square distribution with $(q - 1)$ degrees of freedom. In addition, the cells in all of the tables must have expected cell counts greater than 5 (or at least 80% of them should). Note that a chi-square approximation for Q_{MH} requires only the total sample size to be large, but the chi-square approximation for Q_{BD} requires each table to have a large sample size. If the odds ratios are not homogeneous, then the overall odds ratio should be viewed cautiously; the within-strata odds ratios should be emphasized.

Note that the Mantel-Haenszel statistics do not require homogeneous odds ratios, so the Breslow-Day test should not be interpreted as an indicator of their validity. Refer to Breslow and Day (1980, p. 182) for more information.

3.3.2 Coronary Artery Disease Data Example

The following data are based on a study on coronary artery disease (Koch, Imrey, et al. 1985). The sample is one of convenience since the patients studied were people who came to a clinic and requested an evaluation.

Table 3.7. Coronary Artery Disease Data

Sex	ECG	Disease	No Disease	Total
Female	< 0.1 ST segment depression	4	11	15
Female	≥ 0.1 ST segment depression	8	10	18
Male	< 0.1 ST segment depression	9	9	18
Male	≥ 0.1 ST segment depression	21	6	27

Investigators were interested in whether electrocardiogram (ECG) measurement was associated with disease status. Gender was thought to be associated with disease status, so investigators poststratified the data into male and female groups. In addition, there was interest in examining the odds ratios.

The following statements produce the SAS data set CA and request a stratified analysis. The first TABLES statement requests chi-square tests for the association of gender and disease status. The second TABLES statement requests the stratified analysis, including the generation of odds ratios with the MEASURES option.

```
data ca;
    input gender $ ECG $ disease $ count;
    datalines;
female <0.1  yes   4
female <0.1  no    11
female >=0.1 yes   8
female >=0.1 no    10
male   <0.1  yes   9
male   <0.1  no    9
male   >=0.1 yes   21
male   >=0.1 no    6
;
proc freq;
    weight count;
    tables gender*disease / nocol nopct chisq;
    tables gender*ECG*disease / nocol nopct cmh chisq measures;
run;
```

Output 3.10 contains the table of GENDER by DISEASE. Q takes the value 6.9444 and Q_P takes the value 7.0346. Obviously there is a strong association between gender and disease status. Males are much more likely to have symptoms of coronary artery disease than females. The idea to control for gender in a stratified analysis is a good one.

Note that you are controlling for confounding in this example, which is different from the adjustment performed in previous examples. Confounding variables are those related to both the response and the factor under investigation. In previous examples, the stratification variable was part of the study design (center) or thought to be related to the response (gender in soft drink analysis). Adjusting for confounding is often required in epidemiological studies.

Output 3.10   GENDER × DISEASE

```
                    Table of gender by disease

              gender      disease

              Frequency|
              Row Pct  |no      |yes     |  Total
              ---------+--------+--------+
              female   |     21 |     12 |     33
                       |  63.64 |  36.36 |
              ---------+--------+--------+
              male     |     15 |     30 |     45
                       |  33.33 |  66.67 |
              ---------+--------+--------+
              Total          36       42       78

      Statistic                      DF     Value      Prob
      ------------------------------------------------------
      Chi-Square                      1     7.0346    0.0080
      Likelihood Ratio Chi-Square     1     7.1209    0.0076
      Continuity Adj. Chi-Square      1     5.8681    0.0154
      Mantel-Haenszel Chi-Square      1     6.9444    0.0084
      Phi Coefficient                       0.3003
      Contingency Coefficient               0.2876
      Cramer's V                            0.3003


                     Fisher's Exact Test
      -----------------------------------------------
      Cell (1,1) Frequency (F)              21
      Left-sided Pr <= F                0.9981
      Right-sided Pr >= F               0.0075

      Table Probability (P)             0.0056
      Two-sided Pr <= P                 0.0114
```

Output 3.11 and Output 3.12 display the individual tables results for ECG × disease status; included are the table of chi-square statistics generated by the CHISQ option and only the "Estimates of the Relative Risk" table part of the output generated by the MEASURES option.

Output 3.11   Results for Females

```
              Statistics for Table 1 of ECG by disease
                    Controlling for gender=female

        Statistic                    DF      Value      Prob
        ----------------------------------------------------------

        Chi-Square                    1      1.1175     0.2905
        Likelihood Ratio Chi-Square   1      1.1337     0.2870
        Continuity Adj. Chi-Square    1      0.4813     0.4879
        Mantel-Haenszel Chi-Square    1      1.0836     0.2979
        Phi Coefficient                      0.1840
        Contingency Coefficient              0.1810
        Cramer's V                           0.1840



                      Fisher's Exact Test
             ------------------------------------
             Cell (1,1) Frequency (F)          11
             Left-sided Pr <= F           0.9233
             Right-sided Pr >= F          0.2450

             Table Probability (P)        0.1683
             Two-sided Pr <= P            0.4688


            Estimates of the Relative Risk (Row1/Row2)

      Type of Study               Value      95% Confidence Limits
      ------------------------------------------------------------
      Case-Control (Odds Ratio)   2.2000     0.5036         9.6107
      Cohort (Col1 Risk)          1.3200     0.7897         2.2063
      Cohort (Col2 Risk)          0.6000     0.2240         1.6073
```

Q_{MH} is 1.084 for females, with a *p*-value of 0.2979. The odds ratio for the females is
OR = 2.2, with a 95% confidence interval that includes 1. Those females with higher ST
segment depression levels had 2.2 times the odds of CA disease than those with lower
levels.

Output 3.12   Results for Males

```
               Statistics for Table 2 of ECG by disease
                     Controlling for gender=male

     Statistic                       DF       Value       Prob
     ---------------------------------------------------------------
     Chi-Square                       1       3.7500      0.0528
     Likelihood Ratio Chi-Square      1       3.7288      0.0535
     Continuity Adj. Chi-Square       1       2.6042      0.1066
     Mantel-Haenszel Chi-Square       1       3.6667      0.0555
     Phi Coefficient                          0.2887
     Contingency Coefficient                  0.2774
     Cramer's V                               0.2887


                     Fisher's Exact Test
             ----------------------------------------
             Cell (1,1) Frequency (F)             9
             Left-sided  Pr <= F             0.9880
             Right-sided Pr >= F             0.0538

             Table Probability (P)          0.0417
             Two-sided Pr <= P              0.1049


          Estimates of the Relative Risk (Row1/Row2)

     Type of Study                Value       95% Confidence Limits
     ---------------------------------------------------------------
     Case-Control (Odds Ratio)    3.5000      0.9587       12.7775
     Cohort (Col1 Risk)           2.2500      0.9680        5.2298
     Cohort (Col2 Risk)           0.6429      0.3883        1.0642
```

Q_{MH} takes the value 3.667 for males, with a p-value of 0.056. The odds ratio for the males is OR = 3.5, with a 95% confidence interval that barely contains the value 1. Those men with higher ST segment depression levels had 3.5 times the odds of CA disease than those with lower levels.

Output 3.13 contains the Q_{MH} statistic, which takes the value 4.503 with a p-value of 0.0338. By combining the genders, the power has been increased so that the association detected by Q_{MH} is significant at the $\alpha = 0.05$ level of significance.

Output 3.13   Stratified Analysis

```
               Summary Statistics for ECG by disease
                       Controlling for gender

       Cochran-Mantel-Haenszel Statistics (Based on Table Scores)

     Statistic    Alternative Hypothesis    DF      Value       Prob
     ---------------------------------------------------------------
         1        Nonzero Correlation        1      4.5026     0.0338
         2        Row Mean Scores Differ      1      4.5026     0.0338
         3        General Association         1      4.5026     0.0338
```

Output 3.14 contains the estimates of the common odds ratios. $\hat{\psi}_{MH} = 2.847$ and $\hat{\psi}_L = 2.859$. The confidence intervals do not contain the value 1. On the average, those persons with higher ST segment depression levels had nearly three times the odds of CA disease than those with lower levels.

Output 3.14   Odds Ratios

```
                    Summary Statistics for ECG by disease
                          Controlling for gender

              Estimates of the Common Relative Risk (Row1/Row2)

    Type of Study     Method            Value     95% Confidence Limits
    ----------------------------------------------------------------------
    Case-Control      Mantel-Haenszel   2.8467     1.0765        7.5279
       (Odds Ratio)   Logit             2.8593     1.0807        7.5650

    Cohort            Mantel-Haenszel   1.6414     1.0410        2.5879
       (Col1 Risk)    Logit             1.5249     0.9833        2.3647

    Cohort            Mantel-Haenszel   0.6299     0.3980        0.9969
       (Col2 Risk)    Logit             0.6337     0.4046        0.9926


                       Breslow-Day Test for
                  Homogeneity of the Odds Ratios
                  ------------------------------
                  Chi-Square            0.2155
                  DF                         1
                  Pr > ChiSq            0.6425
```

Common measures of relative risk are also printed by the FREQ procedure. However, since these data do not come from a prospective study, these statistics are not relevant and should be ignored.

Finally, the Breslow-Day test is printed at the bottom and does not contradict the assumption of homogeneous odds ratios for these data. $Q_{B\,D} = 0.215$ with $p = 0.6425$.

Chapter 4

Sets of 2 × r and s × 2 Tables

Chapter Table of Contents

Chapter 4
Sets of 2 × r and s × 2 Tables

4.1 Introduction

While sets of 2 × 2 tables are very common, many sets of tables have other dimensions. This chapter focuses on sets of tables that also occur frequently: sets of $2 \times r$ tables in which the column variable is ordinally scaled and sets of $s \times 2$ tables in which the row variable is ordinally scaled. For $2 \times r$ tables, there is interest in investigating a response variable with multiple ordered outcomes for a combined set of strata. For example, you may be comparing a new treatment and a placebo on the extent of patient improvement that is rated as minimal, moderate, or substantial. For $s \times 2$ tables, there is interest in the trend of proportions across ordered groups for a combined set of strata. For example, you may be comparing the proportion of successful outcomes for different dosage levels of a new drug.

Extensions of the Mantel-Haenszel strategy address association in sets of tables with these characteristics. Section 4.2 addresses $2 \times r$ tables and Section 4.3 addresses $s \times 2$ tables. Each of these sections begins by discussing the assessment of association in a single table where the column (row) variable is ordinally scaled and the row (column) variable is dichotomous.

4.2 Sets of $2 \times r$ Tables

Consider the data from Koch and Edwards (1988) displayed in Table 4.1. The information comes from a randomized, double-blind clinical trial investigating a new treatment for rheumatoid arthritis. Investigators compared the new treatment with a placebo; the response measured was whether there was no, some, or marked improvement in the symptoms of rheumatoid arthritis.

Table 4.1. Rheumatoid Arthritis Data

| Gender | Treatment | Improvement | | | Total |
		None	Some	Marked	
Female	Test Drug	6	5	16	27
Female	Placebo	19	7	6	32
Total		25	12	22	59
Male	Test Drug	7	2	5	14
Male	Placebo	10	0	1	11
Total		17	2	6	25

These data comprise a set of two 2×3 tables. There is interest in the association between treatment and degree of improvement, adjusting for gender effects. Degree of improvement is an ordinal response, since none, some, and marked are gradations of improvement.

Mantel (1963) proposed an extension of the Mantel-Haenszel strategy for the analysis of $2 \times r$ tables when the response variable is ordinal. The extension involves computing mean scores for the responses and using the mean score differences across tables in the computation of a suitable test statistic, much like the difference in proportions across tables was the basis of the Mantel-Haenszel statistic.

4.2.1   The $2 \times r$ Table

Before discussing the strategies for assessing association in sets of $2 \times r$ tables, it is necessary to discuss the assessment of association in a single $2 \times r$ table that has an ordinal outcome. Consider Table 4.2 corresponding to patients pooled over gender for the rheumatoid arthritis data.

Table 4.2.   Combined Rheumatoid Arthritis Data

	Improvement			
Treatment	None	Some	Marked	Total
Test Drug	13	7	21	41
Placebo	29	7	7	43
Total	42	14	28	84

As discussed in Chapter 1, "Introduction," you want to use the information in the ordinal column variable in forming a test statistic. This involves assigning scores to the response levels, forming means, and then examining location shifts of the means across the levels of the row variable.

Define the mean for the Test Drug group as

$$\bar{f}_1 = \sum_{j=1}^{3} \frac{a_j n_{1j}}{n_{1+}}$$

where $\mathbf{a} = \{a_j\} = (a_1, a_2, a_3)$ are a set of scores reflecting the response levels. Then, if the null hypothesis H_0 is no location shifts,

$$E\{\bar{f}_1 | H_0\} = \sum_{j=1}^{3} \left(a_j \frac{n_{1+} n_{+j}}{n_{1+} n} \right) = \sum_{j=1}^{3} a_j \frac{n_{+j}}{n} = \mu_{\mathbf{a}}$$

It can be shown that

$$V\{\bar{f}_1 | H_0\} = \frac{n - n_{1+}}{n_{1+}(n-1)} \sum_{j=1}^{3} (a_j - \mu_{\mathbf{a}})^2 \left(\frac{n_{+j}}{n} \right)$$

$$= \frac{(n - n_{1+})v_{\mathbf{a}}}{n_{1+}(n-1)}$$

where $\mu_{\mathbf{a}}$ and $v_{\mathbf{a}}$ are the finite population mean and variance of scores $\mathbf{a}$ for the patients in the study. The quantity $\bar{f}_1$ approximately has a normal distribution by randomization central limit theory, so the quantity

$$Q_S = \frac{(\bar{f}_1 - \mu_{\mathbf{a}})^2}{\{(n - n_{1+})/[n_{1+}(n-1)]\}v_{\mathbf{a}}}$$

approximately has the chi-square distribution with one degree of freedom. Q_S is called the mean score statistic. By taking advantage of the ordinality of the response variable, Q_S can target the alternative hypothesis of location shifts to the hypothesis of no association with fewer degrees of freedom. While Q and Q_P are useful for detecting general types of association, they are not as effective as Q_S in detecting location shifts. Q_S is also a trend statistic for the tendency for the patients in one treatment group to have better scores than the patients in the other treatment group.

A very conservative sample size guideline is the guideline used for the Pearson chi-square statistic (that is, all expected values $n_{i+}n_{+j}/n = m_{ij}$ being greater than or equal to 5). However, one of the advantages of the mean score statistic is that it has less stringent sample size requirements. A more realistic but still conservative sample size guideline is to choose one or more cutpoints $j = (2, \ldots, (r-1))$, add the 1st through jth columns together and add the $(j+1)$th through rth columns together. If both of these sums are 5 or greater for each margin, then the sample size is adequate.

For example, for Table 4.2, choose $j = 2$. Adding the first and second columns together yields the sums 20 for the first row and 36 for the second; the remaining sums are just the third column cells (21 and 7, respectively). Thus, according to this criterion, the sample size is adequate.

The following PROC FREQ statements generate Q_S. Note the use of the ORDER=DATA option to ensure that the values for the variable RESPONSE are put in the correct order. If they are not, the resulting statistics do not account for the intended ordering. Ensuring the correct sort order is critical when you are using statistics that assume ordered values.

```
data arth;
   input gender $ treat $ response $ count @@;
   datalines;
female test    none 6  female test    some 5  female test    marked 16
female placebo none 19 female placebo some 7  female placebo marked 6
male    test    none 7  male    test    some 2  male    test    marked 5
male    placebo none 10 male    placebo some 0  male    placebo marked 1
;
proc freq data=arth order=data;
   weight count;
   tables treat*response / chisq nocol nopct;
run;
```

The results are contained in Output 4.1.

Output 4.1   Mean Score Statistic

```
                      Table of treat by response

         treat       response

         Frequency|
         Row Pct   |none     |some     |marked   |  Total
         ---------+--------+--------+--------+
         test      |     13 |      7 |     21 |     41
                   |  31.71 |  17.07 |  51.22 |
         ---------+--------+--------+--------+
         placebo   |     29 |      7 |      7 |     43
                   |  67.44 |  16.28 |  16.28 |
         ---------+--------+--------+--------+
         Total            42       14       28       84


                 Statistics for Table of treat by response

         Statistic                       DF       Value       Prob
         --------------------------------------------------------
         Chi-Square                        2     13.0550     0.0015
         Likelihood Ratio Chi-Square       2     13.5298     0.0012
         Mantel-Haenszel Chi-Square        1     12.8590     0.0003
         Phi Coefficient                          0.3942
         Contingency Coefficient                  0.3668
         Cramer's V                               0.3942

                         Sample Size = 84
```

For a $2 \times r$ table, the statistic labeled "Mantel-Haenszel Chi-Square" is Q_S. The scores (1, 2, 3) are used for the response levels none, some, and marked in Table 4.2. Q_S takes the value 12.8590, which is strongly significant. The test treatment performs better than the placebo treatment.

You can also produce Q_S by specifying the CMH option and generating the summary statistics, which will be for just one stratum. Q_S is the statistic labeled "Row Mean Scores Differ" in the resulting summary statistics table.

4.2.2   Extension to Q_{MH}

Assessing association for sets of $2 \times r$ tables where the response is ordinal also involves a strategy of computing means based on a scoring system and looking at shifts in location.

Consider the following table as representative of q $2 \times r$ tables, $h = 1, 2, \ldots, q$.

Table 4.3. hth 2 × r Contingency Table

	Level of Column Variable				
	1	2	...	r	Total
Group 1	n_{h11}	n_{h12}	...	n_{h1r}	n_{h1+}
Group 2	n_{h21}	n_{h22}	...	n_{h2r}	n_{h2+}
Total	n_{h+1}	n_{h+2}	...	n_{h+r}	n_h

For the rheumatoid arthritis data in Table 4.1, $r = 3$ and $q = 2$. Under the null hypothesis of no difference in treatment effects for each patient, the appropriate probability model is

$$
\Pr\{n_{hij}\} = \prod_{h=1}^{2} \frac{\prod_{i=1}^{2} n_{hi+}! \prod_{j=1}^{3} n_{h+j}!}{n_h! \prod_{i=1}^{2} \prod_{j=1}^{3} n_{hij}!}
$$

Here, n_{hij} represents the number of patients in the hth stratum who received the ith treatment and had the jth response.

Suppose $\{a_{hj}\}$ is a set of scores for the response levels in the hth stratum. Then you can compute the sum of strata scores for the 1st treatment, test, as

$$
f_{+1+} = \sum_{h=1}^{2} \sum_{j=1}^{3} a_{hj} n_{h1j} = \sum_{h=1}^{2} n_{h1+} \bar{f}_{h1}
$$

where

$$
\bar{f}_{h1} = \sum_{j=1}^{3} (a_{hj} n_{h1j} / n_{h1+})
$$

is the mean score for Group 1 in the hth stratum. Under the null hypothesis of no association, f_{+1+} has the expected value

$$
E\{f_{+1+} | H_0\} = \sum_{h=1}^{2} n_{h1+} \mu_h = \mu_*
$$

and variance

$$
V\{f_{+1+} | H_0\} = \sum_{h=1}^{2} \frac{n_{h1+}(n_h - n_{h1+})}{(n_h - 1)} v_h = v_*
$$

where $\mu_h = \sum_{j=1}^{3} (a_{hj} n_{h+j} / n_h)$ is the finite subpopulation mean and

$$
v_h = \sum_{j=1}^{3} (a_{hj} - \mu_h)^2 (n_{h+j} / n_h)
$$

is the variance of scores for the hth stratum.

If the across-strata sample sizes $n_{+i+} = \sum_{h=1}^{q} \sum_{j=1}^{r} n_{hij}$ are sufficiently large, then f_{+1+} approximately has a normal distribution, and so the quantity

$$
Q_{SMH} = \frac{(f_{+1+} - \mu_*)^2}{v_*}
$$

approximately has a chi-square distribution with one degree of freedom. Q_{SMH} is known as the extended Mantel-Haenszel mean score statistic; it is sometimes called the ANOVA statistic. You can show that Q_{SMH} is a linear function of the differences in the mean scores of the two treatments for the q strata.

$$Q_{SMH} = \frac{\left\{\sum_{h=1}^{q} n_{h1+}(\bar{f}_{h1} - \mu_h)\right\}^2}{\sum_{h=1}^{q} n_{h1+}n_{h2+}v_h/(n_h - 1)}$$

$$= \frac{\left\{\sum_{h=1}^{q} (n_{h1+}n_{h2+}/n_h)(\bar{f}_{h1} - \bar{f}_{h2})\right\}^2}{\sum_{h=1}^{q} (n_{h1+}n_{h2+}/n_h)^2 \bar{v}_h}$$

where the

$$\bar{v}_h = \left\{\frac{1}{n_{h1+}} + \frac{1}{n_{h2+}}\right\} \frac{n_h v_h}{n_h - 1}$$

are the variances of the mean score differences $\{\bar{f}_{h1} - \bar{f}_{h2}\}$ for the respective strata.

Q_{SMH} is effective for detecting consistent patterns of differences across the strata when the $(\bar{f}_{h1} - \bar{f}_{h2})$ predominantly have the same sign.

Besides the guideline that the across strata row totals (n_{+i+}) be sufficiently large, another guideline for sample size requirements for Q_{SMH} is to choose cutpoints and add columns together so that each stratum table is collapsed to a 2×2 table, similar to what is described in Section 4.2.1; the cutpoints don't have to be the same for each table. Then, you apply the Mantel-Fleiss criterion to these 2×2 tables (see Section 3.2).

4.2.3   Choosing Scores

Ordinal data analysis strategies do involve some choice on the part of the analyst, and that is the choice of scores to apply to the response levels. There are a variety of scoring systems to consider; the following are often used.

- *integer scores*

  Integer scores are defined as $a_j = j$ for $j = 1, 2, \ldots, r$. They are useful when the response levels are ordered categories that can be viewed as equally spaced and when the response levels correspond to discrete counts. They are also useful if you have equal interest in detecting group differences for any binary partition $\leq j$ versus $> j$ of outcomes for $j = 1, 2, \ldots, r$. Note that if you add the same number to a set of scores, or multiply a set of scores by the same number, both sets of scores produce the same test statistic because multiplication is cancelled by division by the same factor in the variance and addition is cancelled by subtraction of the same factor in the expected value. Thus, the integer scores $(1, 2, 3, \ldots)$ and $(0, 1, 2, \ldots)$ produce the same results.

- *standardized midranks*

    These scores are defined as

    $$a_j = \frac{2[\sum_{k=1}^{j} n_{+k}] - n_{+j} + 1}{2(n+1)}$$

    The $\{a_j\}$ are constrained to lie between 0 and 1. Their advantage over integer scores is that they require no scaling of the response levels other than that implied by their relative ordering. For sets of $2 \times r$ tables, they provide somewhat more power than actual midranks since they produce the van Elteren (1960) extension of the Wilcoxon rank sum test (refer to Lehmann 1975 for a discussion). Standardized midranks are also known as *modified ridit scores*.

- *logrank scores*

    $$a_j = 1 - \sum_{k=1}^{j} \left(\frac{n_{+k}}{\sum_{m=k}^{r} n_{+m}} \right)$$

    Logrank scores are useful when the distribution is thought to be L-shaped, and there is greater interest in treatment differences for response levels with higher values than with lower values.

    Other scores that are sometimes used are ridit and rank scores. For a single stratum, rank, ridit, and modified ridit scores produce the same result, which is the categorical counterpart of the Wilcoxon rank sum test. For stratified analyses, modified ridit scores produce van Elteren's extension of the Wilcoxon rank sum test, a property that makes them the preferred of these three types of scores. A possible shortcoming of rank scores, relative to ridit or modified ridit scores, is that their use tends to make the large strata overly influence the test statistic. See page 149 for additional discussion on choosing scores.

    You specify the choice of scores in the FREQ procedure by using the SCORES= option in the TABLES statement. If you don't specify SCORES=, then you get the default table scores. The column (row) numbers are the table scores for character data and the actual variable values are used as scores for numeric variables. Other SCORES= values are RANK, MODRIDIT, and RIDIT. If you are interested in using logrank scores, then you need to compute them in a DATA step and make them the values of the row and column variables you list in the TABLES statement.

4.2.1   Analyzing the Arthritis Data

Applying the extension of the Mantel-Haenszel strategy involves no new steps in the SAS System. You specify the CMH option in the TABLES statement of the FREQ procedure. Notice that the ORDER=DATA option is specified in the PROC statement to ensure that the levels of RESPONSE are sorted correctly. The columns will be ordered none, some, and marked; and the rows will be ordered test and placebo.

```
data arth;
   input gender $ treat $ response $ count @@;
   datalines;
female test    none 6  female test    some 5  female test    marked 16
female placebo none 19 female placebo some 7  female placebo marked 6
male    test    none 7  male    test    some 2  male    test    marked 5
male    placebo none 10 male    placebo some 0  male    placebo marked 1
;
proc freq data=arth order=data;
   weight count;
   tables gender*treat*response / cmh nocol nopct;
run;
```

Output 4.2   Tables by Gender

```
                   Table 1 of treat by response
                   Controlling for gender=female

         treat       response

         Frequency|
         Row Pct  |none    |some    |marked  | Total
         ---------+--------+--------+--------+
         test     |     6 |      5 |     16 |    27
                  |  22.22 |  18.52 |  59.26 |
         ---------+--------+--------+--------+
         placebo  |    19 |      7 |      6 |    32
                  |  59.38 |  21.88 |  18.75 |
         ---------+--------+--------+--------+
         Total         25       12       22       59


                   Table 2 of treat by response
                   Controlling for gender=male

         treat       response

         Frequency|
         Row Pct  |none    |some    |marked  | Total
         ---------+--------+--------+--------+
         test     |     7 |      2 |      5 |    14
                  |  50.00 |  14.29 |  35.71 |
         ---------+--------+--------+--------+
         placebo  |    10 |      0 |      1 |    11
                  |  90.91 |   0.00 |   9.09 |
         ---------+--------+--------+--------+
         Total         17        2        6       25
```

Output 4.2 displays the frequency tables for females and males. Output 4.3 displays the table of Mantel-Haenszel statistics. Note that the table heading includes "Table Scores" in parentheses. Q_{SMH} is the "Row Mean Scores Differ" statistic. It has the value 14.6319, with 1 df, and is clearly significant.

Note the small cell counts for several cells in the table for males. This is not a problem for Q_{SMH} since the adequacy of the sample sizes is determined by the across strata sample sizes n_{+i+}, which are $n_{+1+} = 41$ and $n_{+2+} = 43$ for these data.

Output 4.3   Mantel-Haenszel Results

```
                    Summary Statistics for treat by response
                            Controlling for gender

            Cochran-Mantel-Haenszel Statistics (Based on Table Scores)

            Statistic    Alternative Hypothesis    DF     Value     Prob
            ---------------------------------------------------------------
                1        Nonzero Correlation        1     14.6319   0.0001
                2        Row Mean Scores Differ      1     14.6319   0.0001
                3        General Association         2     14.6323   0.0007


                           Total Sample Size = 84
```

If you can't make the case that the response levels for degree of improvement are equally spaced, then modified ridit scores are an alternative strategy. The following PROC FREQ invocation requests that modified ridit scores be used in the computation of Q_{SMH} through the use of the SCORES=MODRIDIT option in the TABLES statement.

```
proc freq data=arth order=data;
   weight count;
   tables gender*treat*response/cmh scores=modridit nocol nopct;
run;
```

Output 4.4 contains the table of CMH statistics using modified ridit scores. Q_{SMH} takes the value 15.004 with 1 df, which is clearly significant. Note that the different scoring systems produced similar results. This is often the case.

Output 4.4   Mantel-Haenszel Results for Modified Ridit Scores

```
                    Summary Statistics for treat by response
                            Controlling for gender

            Cochran-Mantel-Haenszel Statistics (Modified Ridit Scores)

            Statistic    Alternative Hypothesis    DF     Value     Prob
            ---------------------------------------------------------------
                1        Nonzero Correlation        1     14.9918   0.0001
                2        Row Mean Scores Differ      1     15.0041   0.0001
                3        General Association         2     14.6323   0.0007


                           Total Sample Size = 84
```

4.2.5   Colds Example

The following data come from a study on the presence of colds in children in two regions (Stokes 1986). Researchers visited children several times and noted whether they had any symptoms of colds. The outcome measure is the number of periods in which a child exhibited cold symptoms.

Table 4.4.   Number of Periods with Colds by Gender and Residence

Gender	Residence	Periods With Colds			Total
		0	1	2	
Female	Urban	45	64	71	180
Female	Rural	80	104	116	300
Total		125	168	187	480
Male	Urban	84	124	82	290
Male	Rural	106	117	87	310
Total		190	141	169	600

These data consist of two 2 × 3 tables; there is interest in determining whether there is association between residence (urban or rural) and number of periods with colds (0, 1, or 2) while controlling for gender. The response levels for these data consist of small discrete counts, so number of colds can be considered an ordinal variable in which the levels are equally spaced. The usual ANOVA strategy for interval-scaled response variables is not appropriate since there is no reason to think that the number of periods with colds is normally distributed with homogeneous variance.

The following statements produce an extended Mantel-Haenszel analysis. The default table scores are used, which will be the actual scores of the variable PER_COLD (0, 1, 2).

```
data colds;
   input gender $ residence $ per_cold count @@;
   datalines;
female urban 0   45  female urban  1  64  female urban 2  71
female rural 0   80  female rural  1 104  female rural 2 116
male    urban 0   84  male    urban  1 124  male    urban 2  82
male    rural 0  106  male    rural  1 117  male    rural 2  87
;
proc freq data=colds order=data;
   weight count;
   tables gender*residence*per_cold / all nocol nopct;
run;
```

Output 4.5 and Output 4.6 contain the frequency tables for females and males and their associated chi-square statistics. There is no significant association between residence and number of periods with colds for females or males; $Q = 0.1059$ ($p = 0.7448$) for females and $Q = 0.7412$ ($p = 0.3893$) for males.

Output 4.5 Results for Females

```
                    Table 1 of residence by per_cold
                     Controlling for gender=female

             residence     per_cold

             Frequency|
             Row Pct  |       0|       1|       2| Total
             ---------+--------+--------+--------+
             urban    |     45 |     64 |     71 |    180
                      |  25.00 |  35.56 |  39.44 |
             ---------+--------+--------+--------+
             rural    |     80 |    104 |    116 |    300
                      |  26.67 |  34.67 |  38.67 |
             ---------+--------+--------+--------+
             Total         125      168      187     480
```

```
          Statistics for Table 1 of residence by per_cold
                     Controlling for gender=female

        Statistic                     DF       Value      Prob
        ------------------------------------------------------------
        Chi-Square                      2      0.1629    0.9218
        Likelihood Ratio Chi-Square     2      0.1634    0.9215
        Mantel-Haenszel Chi-Square      1      0.1059    0.7448
        Phi Coefficient                        0.0184
        Contingency Coefficient                0.0184
        Cramer's V                             0.0184
```

Output 4.6 Results for Males

```
                    Table 2 of residence by per_cold
                      Controlling for gender=male

             residence     per_cold

             Frequency|
             Row Pct  |       0|       1|       2| Total
             ---------+--------+--------+--------+
             urban    |     84 |    124 |     82 |    290
                      |  28.97 |  42.76 |  28.28 |
             ---------+--------+--------+--------+
             rural    |    106 |    117 |     87 |    310
                      |  34.19 |  37.74 |  28.06 |
             ---------+--------+--------+--------+
             Total         190      241      169     600
```

```
          Statistics for Table 2 of residence by per_cold
                      Controlling for gender=male

        Statistic                     DF       Value      Prob
        ------------------------------------------------------------
        Chi-Square                      2      2.2344    0.3272
        Likelihood Ratio Chi-Square     2      2.2376    0.3267
        Mantel-Haenszel Chi-Square      1      0.7412    0.3893
        Phi Coefficient                        0.0610
        Contingency Coefficient                0.0609
        Cramer's V                             0.0610
```

Output 4.7 contains the Mantel-Haenszel statistics. Q_{SMH} has the value 0.7379, with $p = 0.3903$. Even controlling for gender, there appears to be no association between residence and number of periods with colds for these data.

Output 4.7 Q_{SMH} Statistic

```
                Summary Statistics for residence by per_cold
                          Controlling for gender

          Cochran-Mantel-Haenszel Statistics (Based on Table Scores)

          Statistic    Alternative Hypothesis    DF    Value    Prob
          ------------------------------------------------------------
              1         Nonzero Correlation        1    0.7379   0.3903
              2         Row Mean Scores Differ      1    0.7379   0.3903
              3         General Association         2    1.9707   0.3733


                      Total Sample Size = 1080
```

4.3   Sets of $s \times 2$ Tables

The following data come from a study on adolescent usage of smokeless tobacco (Bauman, Koch, and Lentz 1989). Interest focused on factors that affected usage, such as perception of risk, father's usage of smokeless tobacco, and educational background. Table 4.5 contains two $s \times 2$ tables of risk perception (minimal, moderate, and substantial) and adolescent usage by father's usage. This time, the row variable is ordinally scaled. The question of interest is whether there is a discernible trend in the proportions of adolescent usage over the levels of risk perception. Does usage decline with higher risk perception?

Table 4.5.   Adolescent Smokeless Tobacco Usage

Father's Usage	Risk Perception	Adolescent Usage		Total
		No	Yes	
No	Minimal	59	25	84
No	Moderate	169	29	198
No	Substantial	196	9	205
Yes	Minimal	11	8	19
Yes	Moderate	33	11	44
Yes	Substantial	22	2	24

Since the response variable is dichotomous, both risk perception and adolescent usage can be considered ordinal variables. The strategy for assessing association when both row and column variables are ordinal involves assigning scores to the levels of both variables and evaluating their correlation.

4.3.1   The $s \times 2$ Table

Table 4.6 contains the data for those adolescents interviewed whose fathers did not use smokeless tobacco.

Table 4.6. Adolescent Smokeless Tobacco Usage When Fathers Did Not Use

Risk	Adolescent Usage		Total
Perception	No	Yes	
Minimal	59	25	84
Moderate	169	29	198
Substantial	196	9	205

Form the linear function

$$\bar{f} = \sum_{i=1}^{3} c_i \bar{f}_i \left(\frac{n_{i+}}{n} \right) = \sum_{i=1}^{3} \sum_{j=1}^{2} \frac{c_i a_j n_{ij}}{n}$$

where $\mathbf{c} = (c_1, c_2, c_3)$ represents scores for the groups and $\mathbf{a} = (a_1, a_2)$ represents scores for the columns (effectively 0, 1). Under H_0,

$$E\{\bar{f}|H_0\} = \sum_{i=1}^{3} c_i \left(\frac{n_{i+}}{n} \right) \sum_{j=1}^{2} a_j \left(\frac{n_{+j}}{n} \right) = \mu_{\mathbf{c}} \mu_{\mathbf{a}}$$

and

$$V\{\bar{f}|H_0\} = \left\{ \sum_{i=1}^{3} (c_i - \mu_{\mathbf{c}})^2 \left(\frac{n_{i+}}{n} \right) \sum_{j=1}^{2} \frac{(a_j - \mu_{\mathbf{a}})^2 (n_{+j}/n)}{(n-1)} \right\}$$

$$= \frac{v_{\mathbf{c}} v_{\mathbf{a}}}{(n-1)}$$

The quantity $\bar{f}$ has an approximate normal distribution for large samples, so for these situations

$$
\begin{aligned}
Q_{CS} &= \frac{(\bar{f} - E\{\bar{f}|H_0\})^2}{\text{Var}\{\bar{f}|H_0\}} \\
&= \frac{(n-1)[\sum_{i=1}^{3} \sum_{j=1}^{2} (c_i - \mu_{\mathbf{c}})(a_j - \mu_{\mathbf{a}}) n_{ij})]^2}{[\sum_{i=1}^{2} (c_i - \mu_{\mathbf{c}})^2 n_{i+}][\sum_{j=1}^{2} (a_i - \mu_{\mathbf{a}})^2 n_{+j}]} \\
&= (n-1) r_{\mathbf{ac}}^2
\end{aligned}
$$

where $r_{\mathbf{ac}}$ is the Pearson correlation coefficient. Thus, Q_{CS} is known as the correlation statistic. It is approximately chi-square with one degree of freedom. This test is comparable to the Cochran-Armitage trend test (Cochran 1954, Armitage 1955), which tests for trends in binomial proportions across the levels of an ordinal covariate. In fact, multiplying Q_{CS} by $n/(n-1)$ yields the same value as the z^2 of the Cochran-Armitage test.

4.3.2   Correlation Statistic

Mantel (1963) also proposed a statistic for the association of two variables that were
ordinal for a combined set of strata, based on assigning scores $\{\mathbf{a}\}$ and $\{\mathbf{c}\}$ to the columns
and rows of the tables.

$$
\begin{aligned}
Q_{CSMH} &= \frac{\left\{\sum_{h=1}^{q} n_h(\bar{f}_h - E\{\bar{f}_h | H_0\})\right\}^2}{\sum_{h=1}^{q} n_h^2 \mathrm{var}\{f_h | H_0\}} \\
&= \frac{\left\{\sum_{h=1}^{q} n_h (v_{h\mathbf{c}} v_{h\mathbf{a}})^{1/2} r_{\mathbf{ca},h}\right\}^2}{\sum_{h=1}^{q} [n_h^2 v_{h\mathbf{c}} v_{h\mathbf{a}}/(n_h - 1)]}
\end{aligned}
$$

Q_{CSMH} is called the extended Mantel-Haenszel correlation statistic. It approximately
follows the chi-square distribution with one degree of freedom when the combined strata
sample sizes are sufficiently large, that is,

$$
\sum_{h=1}^{q} n_h \geq 40
$$

4.3.3   Analysis of Smokeless Tobacco Data

The following SAS statements request that Mantel-Haenszel correlation statistics be
computed for the smokeless tobacco data. Two TABLES statements are included to specify
analyses using both integer scores and modified ridit scores. The Cochran-Armitage trend
test is also requested directly with the TREND option in the TABLES statement. You can
include as many TABLES statements in a PROC FREQ invocation as you like.

```
data tobacco;
   length risk $11. ;
   input f_usage $ risk $ usage $ count @@;
   datalines;
no minimal       no   59 no  minimal     yes 25
no moderate      no  169 no  moderate    yes 29
no substantial   no  196 no  substantial yes  9
yes minimal      no   11 yes minimal     yes  8
yes moderate     no   33 yes moderate    yes 11
yes substantial no   22 yes substantial yes  2
;
proc freq;
   weight count;
   tables f_usage*risk*usage /cmh chisq measures trend;
   tables f_usage*risk*usage /cmh scores=modridit;
run;
```

Output 4.8 contains the statistics for the table of risk perception by adolescent usage when
there is no father's usage. Note that $Q_{CS} = 34.2843$, with 1 df, signifying a strong
correlation between risk perception and smokeless tobacco usage.

Output 4.8   Results for No Father's Usage

```
            Statistics for Table 1 of risk by usage
                   Controlling for f_usage=no

Statistic                      DF       Value       Prob
--------------------------------------------------------
Chi-Square                      2      34.9217     <.0001
Likelihood Ratio Chi-Square     2      34.0684     <.0001
Mantel-Haenszel Chi-Square      1      34.2843     <.0001
Phi Coefficient                         0.2678
Contingency Coefficient                 0.2587
Cramer's V                              0.2678
```

Output 4.9 contains the Cochran-Armitage trend test table. The test statistic, Z, is 5.8613 and is highly significant. There is an increasing trend in binomial proportions as you go from minimal to substantial risk perception.

Output 4.9   Cochran-Armitage Trend Test

```
            Statistics for Table 1 of risk by usage
                   Controlling for f_usage=no

              Cochran-Armitage Trend Test
              ---------------------------
              Statistic (Z)        5.8613
              One-sided Pr >  Z    <.0001
              Two-sided Pr > |Z|   <.0001
```

Output 4.10 contains the same association test results for those whose fathers used smokeless tobacco, and Output 4.11 contains the corresponding trend text.

Output 4.10   Results for Father's Usage

```
            Statistics for Table 2 of risk by usage
                   Controlling for f_usage=yes

Statistic                      DF       Value       Prob
--------------------------------------------------------
Chi-Square                      2       6.6413     0.0361
Likelihood Ratio Chi-Square     2       7.0461     0.0295
Mantel-Haenszel Chi-Square      1       6.5644     0.0104
Phi Coefficient                         0.2763
Contingency Coefficient                 0.2663
Cramer's V                              0.2763
```

Output 4.11   Cochran-Armitage Trend Test

```
            Statistics for Table 2 of risk by usage
                   Controlling for f_usage=yes

              Cochran-Armitage Trend Test
              ---------------------------
              Statistic (Z)        2.5770
              One-sided Pr >  Z    0.0050
              Two-sided Pr > |Z|   0.0100
```

There is still a correlation between risk perception and adolescent usage, although it is not as strong. The Cochran-Armitage Z statistic has the value 2.5770 and a two-sided p-value of 0.0100. Note that exact p-values are available for the trend test for sparse data.

Output 4.12 contains the results for the combined tables. Q_{CSMH} is Statistic 1 in the table, labeled the "Nonzero Correlation" statistic. It takes the value 40.6639 for integer scores, and it takes the value 39.3048 for modified ridit scores. Both results are similar, with strongly significant statistics; often, different sets of scores produce essentially the same results.

Output 4.12   Results for Combined Tables

```
                   Summary Statistics for risk by usage
                         Controlling for f_usage

        Cochran-Mantel-Haenszel Statistics (Based on Table Scores)

        Statistic    Alternative Hypothesis    DF      Value      Prob
        ------------------------------------------------------------------
            1        Nonzero Correlation        1      40.6639    <.0001
            2        Row Mean Scores Differ      2      41.0577    <.0001
            3        General Association         2      41.0577    <.0001


                        Total Sample Size = 574




                   Summary Statistics for risk by usage
                         Controlling for f_usage

        Cochran-Mantel-Haenszel Statistics (Modified Ridit Scores)

        Statistic    Alternative Hypothesis    DF      Value      Prob
        ------------------------------------------------------------------
            1        Nonzero Correlation        1      39.3048    <.0001
            2        Row Mean Scores Differ      2      41.0826    <.0001
            3        General Association         2      41.0577    <.0001


                        Total Sample Size = 574
```

4.3.4   Pain Data Analysis

Clinical trials not only investigate measures of efficacy, or how well a drug works for its designed purpose, but also address the matter of adverse effects, or whether the drug has harmful side effects. Table 4.7 contains data from a study concerned with measuring the adverse effects of a pain relief treatment that was given at five different dosages, including placebo, to patients with one of two diagnoses. Investigators were interested in whether there was a trend in the proportions with adverse effects.

Table 4.7. Adverse Effects for Pain Treatment

	Diagnosis			
	I		II	
	Adverse Effects		Adverse Effects	
Treatment	No	Yes	No	Yes
Placebo	26	6	26	6
Dosage1	26	7	12	20
Dosage2	23	9	13	20
Dosage3	18	14	1	31
Dosage4	9	23	1	31

The following SAS statements request a Q_{CSMH} statistic from PROC FREQ, using both integer scores and modified ridit scores. First, a TABLES statement requesting the table of treatment by response pooled over the two diagnoses is requested. Note the use of the ORDER=DATA option in the PROC statement. If this option was omitted, the levels of TREATMENT would be ordered incorrectly, with placebo being placed last instead of first.

```
data pain;
   input diagnosis $ treatment $ response $ count @@;
   datalines;
I placebo  no 26 I  placebo yes  6
I dosage1  no 26 I  dosage1 yes  7
I dosage2  no 23 I  dosage2 yes  9
I dosage3  no 18 I  dosage3 yes 14
I dosage4  no  9 I  dosage4 yes 23
II placebo no 26 II placebo yes  6
II dosage1 no 12 II dosage1 yes 20
II dosage2 no 13 II dosage2 yes 20
II dosage3 no  1 II dosage3 yes 31
II dosage4 no  1 II dosage4 yes 31
;
proc freq order=data;
   weight count;
   tables treatment*response / chisq;
   tables diagnosis*treatment*response / chisq cmh;
   tables diagnosis*treatment*response / scores=modridit cmh;
run;
```

Q_{CS} for the combined table is strongly significant, with a value of 65.4730 and 1 df.

Output 4.13   Results for Combined Diagnoses

```
                  Table of treatment by response

            treatment       response

            Frequency|
            Percent  |
            Row Pct  |
            Col Pct  |no      |yes     |  Total
            ---------+--------+--------+
            placebo  |     52 |     12 |     64
                     |  16.15 |   3.73 |  19.88
                     |  81.25 |  18.75 |
                     |  33.55 |   7.19 |
            ---------+--------+--------+
            dosage1  |     38 |     27 |     65
                     |  11.80 |   8.39 |  20.19
                     |  58.46 |  41.54 |
                     |  24.52 |  16.17 |
            ---------+--------+--------+
            dosage2  |     36 |     29 |     65
                     |  11.18 |   9.01 |  20.19
                     |  55.38 |  44.62 |
                     |  23.23 |  17.37 |
            ---------+--------+--------+
            dosage3  |     19 |     45 |     64
                     |   5.90 |  13.98 |  19.88
                     |  29.69 |  70.31 |
                     |  12.26 |  26.95 |
            ---------+--------+--------+
            dosage4  |     10 |     54 |     64
                     |   3.11 |  16.77 |  19.88
                     |  15.63 |  84.38 |
                     |   6.45 |  32.34 |
            ---------+--------+--------+
            Total         155      167      322
                        48.14    51.86   100.00



          Statistics for Table of treatment by response

          Statistic                    DF      Value      Prob
          ------------------------------------------------------

          Chi-Square                    4     68.0752    <.0001
          Likelihood Ratio Chi-Square   4     73.2533    <.0001
          Mantel-Haenszel Chi-Square    1     65.4730    <.0001
          Phi Coefficient                      0.4598
          Contingency Coefficient              0.4178
          Cramer's V                           0.4598
```

Output 4.14 contains the statistics for the individual tables. Q_{CS} takes the value 22.8188 for Diagnosis I and the value 52.3306 for Diagnosis II.

Output 4.14   Results for Separate Diagnoses

```
          Statistics for Table 1 of treatment by response
                   Controlling for diagnosis=I

     Statistic                    DF       Value      Prob
     ---------------------------------------------------------
     Chi-Square                    4       26.6025    <.0001
     Likelihood Ratio Chi-Square   4       26.6689    <.0001
     Mantel-Haenszel Chi-Square    1       22.8188    <.0001
     Phi Coefficient                       0.4065
     Contingency Coefficient               0.3766
     Cramer's V                            0.4065

                    Sample Size = 161



          Statistics for Table 2 of treatment by response
                   Controlling for diagnosis=II

     Statistic                    DF       Value      Prob
     ---------------------------------------------------------
     Chi-Square                    4       60.5073    <.0001
     Likelihood Ratio Chi-Square   4       68.7446    <.0001
     Mantel-Haenszel Chi-Square    1       52.3306    <.0001
     Phi Coefficient                       0.6130
     Contingency Coefficient               0.5226
     Cramer's V                            0.6130

                    Sample Size = 161
```

Output 4.15 contains the stratified analysis results. Integer scores produce a Q_{CSMH} of 71.7263, and modified ridit scores produce a Q_{CSMH} of 71.6471. These statistics are clearly significant. The proportion of patients with adverse effects is correlated with level of dosage; higher dosages produce more reports of adverse effects.

Output 4.15 Combined Results

```
            Summary Statistics for treatment by response
                       Controlling for diagnosis

          Cochran-Mantel-Haenszel Statistics (Based on Table Scores)

          Statistic    Alternative Hypothesis    DF      Value      Prob
          ---------------------------------------------------------------
              1         Nonzero Correlation        1     71.7263    <.0001
              2         Row Mean Scores Differ      4     74.5307    <.0001
              3         General Association         4     74.5307    <.0001


                        Total Sample Size = 322




            Summary Statistics for treatment by response
                       Controlling for diagnosis

          Cochran-Mantel-Haenszel Statistics (Modified Ridit Scores)

          Statistic    Alternative Hypothesis    DF      Value      Prob
          ---------------------------------------------------------------
              1         Nonzero Correlation        1     71.6471    <.0001
              2         Row Mean Scores Differ      4     74.5307    <.0001
              3         General Association         4     74.5307    <.0001


                        Total Sample Size = 322
```

4.4   Relationships Between Sets of Tables

Suppose you transposed the rows and columns of Table 4.7. You would obtain the following:

Table 4.8.   Adverse Effects for Pain Treatment

Diagnosis	Adverse Effects	Placebo	Dosage1	Dosage2	Dosage3	Dosage4
I	No	26	26	23	18	9
I	Yes	6	7	9	14	23
II	No	26	12	13	1	1
II	Yes	6	20	20	31	31

Furthermore, suppose you analyzed these tables as two $2 \times r$ tables, making the response variable the row variable and the grouping variable the column variable.

```
proc freq order=data;
   weight count;
   tables diagnosis*response*treatment / cmh;
   tables diagnosis*treatment*response / cmh;
run;
```

Look at the resulting table of Mantel-Haenszel statistics for DIAGNOSIS by RESPONSE by TREATMENT and compare it to the reprinted table of DIAGNOSIS by TREATMENT by RESPONSE.

Output 4.16 Combined Results

```
                 Summary Statistics for response by treatment
                          Controlling for diagnosis

          Cochran-Mantel-Haenszel Statistics (Based on Table Scores)

          Statistic    Alternative Hypothesis    DF      Value     Prob
          --------------------------------------------------------------
             1         Nonzero Correlation        1      71.7263   <.0001
             2         Row Mean Scores Differ      1      71.7263   <.0001
             3         General Association         4      74.5307   <.0001


                          Total Sample Size = 322




                 Summary Statistics for treatment by response
                          Controlling for diagnosis

          Cochran-Mantel-Haenszel Statistics (Based on Table Scores)

          Statistic    Alternative Hypothesis    DF      Value     Prob
          --------------------------------------------------------------
             1         Nonzero Correlation        1      71.7263   <.0001
             2         Row Mean Scores Differ      4      74.5307   <.0001
             3         General Association         4      74.5307   <.0001


                          Total Sample Size = 322
```

Q_{SMH} and Q_{CSMH} are identical here. One degree of freedom is needed to compare the mean differences across two groups, in the case of Q_{SMH}, and one degree of freedom is needed to assess correlation, in the case of Q_{CSMH}.

In Chapter 6, "Sets of $s \times r$ Tables," the Mantel-Haenszel statistic is extended to sets of $s \times r$ tables. The mean score statistic for the case of more than two groups has $(s - 1)$ degrees of freedom, since you are comparing mean differences across s groups. Thus, Q_S for the $2 \times r$ table is a special case of the more general mean score statistic and has $(s - 1) = (2 - 1) = 1$ degree of freedom. When $s = 2$, Q_{SMH} and Q_{CSMH} take the same value with table scores and can be used interchangeably. Thus, transposing the Table 4.7 data and computing these statistics produced identical mean score and correlation statistics, since the transposed data produced a mean score statistic with one degree of freedom.

Similarly, when $s = 2$, Q_S and Q_{CS} take the same value. This is why, in Section 4.2.1, you are able to use the Mantel-Haenszel statistic produced by the CHISQ option of PROC FREQ. That statistic is actually Q_{CS}, but for $2 \times r$ tables it is also the mean score statistic.

Table 4.9 summarizes the Mantel-Haenszel statistics for the tables discussed in this chapter; it also lists the labels associated with these statistics in PROC FREQ output.

Table 4.9.   Summary of Extended Mantel-Haenszel Statistics

Table Dimensions	Statistic	DF	Corresponding PROC FREQ MH Label
2×2	Q_{MH}	1	Nonzero Correlation Row Mean Scores Differ General Association
$2 \times r$	Q_{SMH}	1	Nonzero Correlation Row Mean Scores Differ
$s \times 2$	Q_{CSMH}	1	Nonzero Correlation

Chapter 5
The s × r Table

Chapter Table of Contents

Chapter 5
The s × r Table

5.1 Introduction

Previous chapters address the concepts of association and measures of association in 2×2 tables and association in $2 \times r$ and $s \times 2$ tables. This chapter extends these concepts to the general $s \times r$ table. The main difference from these earlier chapters is that scale of measurement is always a consideration; the statistics you choose depend on whether the rows and columns of the table are nominally or ordinally scaled. This is true for investigating whether association exists and for summarizing the degree of association. Section 5.2 addresses tests for association, and Section 5.4 addresses measures of association.

Often, subjects or experimental units are observed by two or more researchers, and the question of interest is how closely their evaluations agree. Such studies are called *observer agreement* studies. The columns of the resulting table are the classifications of one observer, and the rows are the classifications of the other observer. Subjects are cross-classified into table cells according to their observed profiles. Observer agreement is discussed in Section 5.5. Sometimes you are interested in ordered alternatives to the hypothesis of no association. Section 5.6 discusses the Jonckheere-Terpstra test for ordered differences.

Exact p-values are now available for many tests of association and measures of association. The Fisher exact test for the $s \times r$ table and exact p-values for several chi-square statistics are discussed in Section 5.3. Exact p-values are also discussed for measures of association, observer agreement, and the Jonckheere-Terpstra test in those respective sections. Note that the exact p-value computations for the actual test statistics such as chi-square statistics, which take only non-negative values, are based on the sum of the exact probabilities for those tables where the test statistic is greater than or equal to the one you observe. The tables you consider are those with the same margins as the table you observe.

For those tests where you may consider one-sided or two-sided alternative hypotheses, such as for the kappa coefficient, the computation is a bit more involved. For one-sided tests, the FREQ procedure computes the right-sided p-value when the observed value of the test statistic is greater than its expected value, and it computes the left-sided p-value when the test statistic is less than or equal to its expected value. In each case, the p-value is the sum of the probabilities for those tables having a more extreme test statistic than the observed one. The two-sided p-value is computed as the sum of the one-sided p-value and the area in the other tail of the distribution for the statistic that is at least as far from the expected value. Refer to Agresti (1992) for a review of the strategies for exact p-value computations for table statistics.

5.2   Association

5.2.1   Tests for General Association

Table 5.1 contains data from a study concerning the distribution of party affiliation in a city suburb. The interest was whether there was an association between registered political party and neighborhood.

Table 5.1.   Distribution of Parties in Neighborhoods

Party	Bayside	Highland	Longview	Sheffeld
	Neighborhood			
Democrat	221	160	360	140
Independent	200	291	160	311
Republican	208	106	316	97

For these data, both row and column variables are nominally scaled; there is no inherent ordering of the response values for either neighborhood or political party. Thus, the alternative to the null hypothesis of no association is general association, defined as heterogeneous patterns of distribution of the response (column) levels across the row levels. The following table represents the general $s \times r$ table.

Table 5.2.   s × r Contingency Table

Group	1	2	...	r	Total
	Response Variable Categories				
1	n_{11}	n_{12}	...	n_{1r}	n_{1+}
2	n_{21}	n_{22}	...	n_{2r}	n_{2+}
⋮	⋮	⋮		⋮	⋮
s	n_{s1}	n_{s2}	...	n_{sr}	n_{s+}
Total	n_{+1}	n_{+2}	...	n_{+r}	n

One test statistic for the hypothesis of no general association is the Pearson chi-square. This statistic is defined the same as for the 2×2 table, except that the summation for i is from 1 to s, and the summation for j is from 1 to r.

$$Q_P = \sum_{i=1}^{s} \sum_{j=1}^{r} \frac{(n_{ij} - m_{ij})^2}{m_{ij}}$$

where

$$m_{ij} = E\{n_{ij}|H_0\} = \frac{n_{i+}n_{+j}}{n}$$

is the expected value of the frequencies in the ith row and jth column.

If the sample size is sufficiently large, that is, all expected cell counts $m_{ij} \geq 5$, then Q_P approximately has the chi-square distribution with $(s-1)(r-1)$ degrees of freedom. In the case of the 2×2 table, $r = 2$ and $s = 2$ so that Q_P has 1 df.

Just as for 2×2 tables, the randomization statistic Q can be written

$$Q = \frac{n-1}{n} Q_P$$

and it also has an approximate chi-square distribution with $(s-1)(r-1)$ degrees of freedom under the null hypothesis.

For more detail, recall from Chapter 2, "The 2×2 Table," that the derivation of Q depends on the assumption of fixed marginal totals such that the table frequencies have a hypergeometric distribution. For the $s \times r$ table, the distribution is multivariate hypergeometric under the null hypothesis of no association.

You can write the probability distribution as

$$Pr\{n_{ij}\} = \frac{\prod_{i=1}^{s} n_{i+}! \prod_{j=1}^{r} n_{+j}!}{n! \prod_{i=1}^{s} \prod_{j=1}^{r} n_{ij}!}$$

The covariance structure under H_0 is

$$\text{Cov}\{n_{ij}, n_{i'j'} | H_0\} = \frac{m_{ij}(n\delta_{ii'} - n_{i'+})(n\delta_{jj'} - n_{+j'})}{n(n-1)}$$

where $\delta_{kk'} = 1$ if $k = k'$ and $\delta_{kk'} = 0$ if $k \neq k'$.

Q is computed from the quadratic form

$$Q = (\mathbf{n} - \mathbf{m})' \mathbf{A}' (\mathbf{A}\mathbf{V}\mathbf{A}')^{-1} \mathbf{A}(\mathbf{n} - \mathbf{m})$$

where $\mathbf{n} = (n_{11}, n_{12}, \ldots, n_{1r}, \ldots, n_{s1}, \ldots, n_{sr})'$ is the compound vector of observed frequencies, $\mathbf{m}$ is the corresponding vector of expected frequencies, $\mathbf{V}$ is the covariance matrix, and $\mathbf{A}$ is a matrix of coefficients defined such that $\mathbf{A}\mathbf{V}\mathbf{A}'$ is nonsingular. The symbol $\otimes$ denotes the left-hand Kronecker product (the matrix on the left of the $\otimes$ multiplies each element in the matrix on the right).

The usual choice for $\mathbf{A}$ for testing general association is

$$\mathbf{A} = \left[\mathbf{I}_{(r-1)}, \mathbf{0}_{(r-1)} \right] \otimes \left[\mathbf{I}_{(s-1)}, \mathbf{0}_{(s-1)} \right]$$

where $\mathbf{I}_{(u-1)}$ is the $(u-1) \times (u-1)$ identity matrix and $\mathbf{0}_{(u-1)}$ is a $(u-1)$ vector of 0s.

For example, for a 2×3 table,

$$\mathbf{A} = \begin{bmatrix} 1 & 0 & 0 & 0 & 0 & 0 \\ 0 & 1 & 0 & 0 & 0 & 0 \end{bmatrix}$$

Generating Q_P and Q requires no new PROC FREQ features. The CHISQ option in the TABLES statement produces Q_P, and the CMH option produces Q. The following statements produce these statistics for the neighborhood data.

```
data neighbor;
   length party $ 11 neighborhood $ 10;
   input party $ neighborhood $ count @@;
   datalines;
democrat    longview   360 democrat    bayside  221
democrat    sheffeld   140 democrat    highland 160
republican  longview   316 republican  bayside  208
republican  sheffeld    97 republican  highland 106
independent longview   160 independent bayside  200
independent sheffeld   311 independent highland 291
;
proc freq ;
   weight count;
   tables party*neighborhood / chisq cmh nocol nopct;
run;
```

Output 5.1 contains the frequency table.

Output 5.1 Frequency Table

```
                   Table of party by neighborhood

      party           neighborhood

      Frequency    |
      Row Pct      |bayside |highland|longview|sheffeld|   Total
      ------------+--------+--------+--------+--------+
      democrat     |   221 |   160 |   360 |   140 |    881
                   | 25.09 | 18.16 | 40.86 | 15.89 |
      ------------+--------+--------+--------+--------+
      independent  |   200 |   291 |   160 |   311 |    962
                   | 20.79 | 30.25 | 16.63 | 32.33 |
      ------------+--------+--------+--------+--------+
      republican   |   208 |   106 |   316 |    97 |    727
                   | 28.61 | 14.58 | 43.47 | 13.34 |
      ------------+--------+--------+--------+--------+
      Total           629     557     836     548     2570
```

Output 5.2 displays the table statistics. $Q_P = 273.9188$ with 6 df, $p < 0.0001$.

Output 5.2 Pearson Chi-Square

```
            Statistics for Table of party by neighborhood

     Statistic                     DF       Value       Prob
     ------------------------------------------------------------
     Chi-Square                     6      273.9188     <.0001
     Likelihood Ratio Chi-Square    6      282.3266     <.0001
     Mantel-Haenszel Chi-Square     1        0.8124      0.3674
     Phi Coefficient                         0.3265
     Contingency Coefficient                 0.3104
     Cramer's V                              0.2308

                      Sample Size = 2570
```

Output 5.3 contains the MH statistics. PROC FREQ computes Q as the extended Mantel-Haenszel statistic for one stratum. Q is the "General Association" statistic, with a value of 273.8122 and 6 df. Notice how close the values of Q and Q_P are for these data; this is expected since the sample size is large (2570).

Output 5.3 Randomization Q

```
            Summary Statistics for party by neighborhood

     Cochran-Mantel-Haenszel Statistics (Based on Table Scores)

     Statistic    Alternative Hypothesis    DF      Value      Prob
     ------------------------------------------------------------------
         1        Nonzero Correlation        1       0.8124    0.3674
         2        Row Mean Scores Differ      2      13.8938    0.0010
         3        General Association         6     273.8122    <.0001

                      Total Sample Size = 2570
```

Political party and neighborhood are statistically associated. If you study the column percentages, you can see that the neighborhoods that have relatively high numbers of Democrats (Bayside, Longview) also have high numbers of Republicans. The neighborhoods that have relatively high numbers of Independents, Highland and Sheffeld, also have low numbers of both Democrats and Republicans.

5.2.2 Mean Score Test

The following data come from a study on headache pain relief. A new treatment was compared with the standard treatment and a placebo. Researchers measured the number of hours of substantial relief from headache pain.

Table 5.3. Pain Study Data

	Hours of Relief				
Treatment	0	1	2	3	4
Placebo	6	9	6	3	1
Standard	1	4	6	6	8
Test	2	5	6	8	6

Clearly, number of hours of relief is an ordinally scaled response measure. While Q and Q_P are good strategies for detecting general association, they aren't as good as other strategies when the response variable is ordinally scaled and the alternative to no association is location shifts. Section 4.2.1 discusses the mean score test for a $2 \times r$ table. Scores are assigned to the levels of the response variable, and row mean scores are computed. The statistic Q_S is then derived. Q_S also applies to $s \times r$ tables, in which case it has $(s-1)$ degrees of freedom since you are comparing mean scores across s groups.

For more detail, the statistic Q_S is derived from the same general quadratic form as Q discussed in Section 5.2. You choose **A** so that it assigns scores to the response levels and then compares the resulting linear functions of scores for $(s-1)$ groups to their expected values. **A** is the $(s-1) \times sr$ matrix

$$
\mathbf{A} = \begin{bmatrix} \mathbf{a}' & \underline{0}' & \cdots & \underline{0}' & \underline{0}' \\ \underline{0}' & \mathbf{a}' & \cdots & \underline{0}' & \underline{0}' \\ \vdots & \vdots & & & \vdots \\ \underline{0}' & \underline{0}' & \cdots & \mathbf{a}' & \underline{0}' \end{bmatrix}
$$

For example, if the actual values were used as scores for the columns in Table 5.3, then $\mathbf{a}' = (0\ 1\ 2\ 3\ 4)$.

It is interesting to note that Q_S can be written in a one-way analysis of variance form

$$
Q_S = \frac{(n-1) \sum_{i=1}^{s} n_{i+}(\bar{f}_i - \mu_{\mathbf{a}})^2}{n v_{\mathbf{a}}}
$$

where, as discussed in Section 4.2.1,

$$
\bar{f}_i = \sum_{j=1}^{r} \frac{a_j n_{ij}}{n_{i+}}
$$

and $\mu_{\mathbf{a}}$ is its expected value

$$
\mu_{\mathbf{a}} = E\{\bar{f}_i | H_0\} = \sum_{j=1}^{r} \frac{a_j n_{+j}}{n}
$$

$$
v_{\mathbf{a}} = \sum_{j=1}^{r} (a_j - \mu_{\mathbf{a}})^2 \left(\frac{n_{+j}}{n}\right)
$$

See Section 4.2.3 for choices of scoring systems. For the pain data, integer scores make sense. The following statements request the mean score test Q_S for the pain data.

```
data pain;
   input treatment $ hours count @@;
   datalines;
placebo  0 6 placebo  1  9 placebo   2 6 placebo  3 3 placebo   4 1
standard 0 1 standard 1  4 standard  2 6 standard 3 6 standard  4 8
test     0 2 test     1  5 test      2 6 test     3 8 test       4 6
;
proc freq;
   weight count;
   tables treatment*hours/ cmh nocol nopct;
run;
```

Output 5.4 contains the frequency table produced by PROC FREQ.

Output 5.4 Frequency Table

```
                    Table of treatment by hours

       treatment     hours

       Frequency|
       Row Pct  |      0|      1|      2|      3|      4|  Total
       ---------+-------+-------+-------+-------+-------+
       placebo  |    6  |    9  |    6  |    3  |    1  |    25
                | 24.00 | 36.00 | 24.00 | 12.00 |  4.00 |
       ---------+-------+-------+-------+-------+-------+
       standard |    1  |    4  |    6  |    6  |    8  |    25
                |  4.00 | 16.00 | 24.00 | 24.00 | 32.00 |
       ---------+-------+-------+-------+-------+-------+
       test     |    2  |    5  |    6  |    8  |    6  |    27
                |  7.41 | 18.52 | 22.22 | 29.63 | 22.22 |
       ---------+-------+-------+-------+-------+-------+
       Total         9      18      18      17      15      77
```

Output 5.5 displays the summary statistics.

Output 5.5 Mean Score Statistic

```
              Summary Statistics for treatment by hours

         Cochran-Mantel-Haenszel Statistics (Based on Table Scores)

       Statistic    Alternative Hypothesis    DF      Value     Prob
       ---------------------------------------------------------------
           1        Nonzero Correlation        1      8.0668    0.0045
           2        Row Mean Scores Differ     2     13.7346    0.0010
           3        General Association        8     14.4030    0.0718


                     Total Sample Size = 77
```

Q_S is the "Row Mean Scores Differ" statistic. $Q_S = 13.7346$, with 2 df, and is clearly significant. Note that Q for these data takes the value 14.403, which has a p-value of

0.0718 with 8 df. In fact, there are a number of cells whose expected values are less than or equal to 5, so the chi-square approximation for the test for general association may not even be valid. However, since the row totals of the table are all greater than 20, and each row has counts ≥ 5 for both outcomes ≤ 1 and ≥ 2, there is sufficient sample size for Q_S. This is an example of where taking advantage of the ordinality of the data not only is the more appropriate approach, it may be the only possible Mantel-Haenszel strategy due to sample size constraints.

5.2.3 Correlation Test

Sometimes, both the row variable and the column variable are ordinally scaled. This is common when you are studying responses that are evaluated on an ordinal scale and what is being compared are different dosage levels, which are also ordinally scaled. Consider the data in Table 5.4. A water treatment company is studying water additives and investigating how they affect clothes washing. The treatments studied were no treatment (plain water), the standard treatment, and a double dose of the standard treatment, called super. Washability was measured as low, medium, and high.

Table 5.4. Washability Data

Treatment	Washability			Total
	Low	Medium	High	
Water	27	14	5	46
Standard	10	17	26	53
Super	5	12	50	67

As discussed in Section 4.2.1, the appropriate statistic to investigate association for this situation is one that takes advantage of the ordinality of both the row variable and the column variable and tests the null hypothesis of no association against the alternative of linear association. In Chapter 4, "Sets of $2 \times r$ and $s \times 2$ Tables," the test statistic Q_{CS} was developed for the $s \times 2$ table and was shown to have one degree of freedom. A similar strategy applies to the $s \times r$ table. You assign scores both to the levels of the response variable and to the levels of the grouping variable to obtain Q_{CS}, which is approximately chi-square with one degree of freedom. Thus, whether the table is 2×2, $s \times 2$, or $s \times r$, Q_{CS} always has one degree of freedom. (See Section 4.4 for a related discussion.)

For more detail, this statistic is also derived from the general quadratic form

$$Q = (\mathbf{n} - \mathbf{m})'\mathbf{A}'(\mathbf{AVA}')^{-1}\mathbf{A}(\mathbf{n} - \mathbf{m})$$

You obtain Q_{CS} by choosing $\mathbf{A}$ to be

$$\mathbf{A} = \begin{bmatrix} \mathbf{a}' \otimes \mathbf{c}' \end{bmatrix} = [a_1 c_1, \ldots, a_r c_1, \ldots, a_r c_s]$$

where $\mathbf{a}' = (a_1, a_2, \ldots, a_r)$ are scores for the response levels and $\mathbf{c}' = (c_1, c_2, \ldots, c_s)$ are scores for the levels of the grouping variable. $\mathbf{A}$ has dimension $1 \times sr$.

The following PROC FREQ invocation produces the correlation statistic for the washability data. It is of interest to use both integer scores and modified ridit scores and compare the results. The following statements request both integer scores (the default) and modified ridit scores. The ORDER= option maintains the desired order of the levels of the rows and columns; it is the same as the order in which the variable values are encountered in the DATA step. The NOPRINT option suppresses the printing of the individual tables.

```
data wash;
   input treatment $ washability $ count @@;
   datalines;
water low 27 water medium 14 water high 5
standard low 10 standard medium 17 standard high 26
super low 5 super medium 12 super high 50
;
proc freq order=data;
   weight count;
   tables treatment*washability / chisq cmh nocol nopct;
   tables treatment*washability / scores=modridit cmh
                                  noprint nocol nopct;
run;
```

Output 5.6 displays the frequency table.

Output 5.6   Frequency Table

```
                   Table of treatment by washability

            treatment       washability

            Frequency|
            Row Pct  |low     |medium  |high    |   Total
            ---------+--------+--------+--------+
            water    |     27 |     14 |      5 |      46
                     |  58.70 |  30.43 |  10.87 |
            ---------+--------+--------+--------+
            standard |     10 |     17 |     26 |      53
                     |  18.87 |  32.08 |  49.06 |
            ---------+--------+--------+--------+
            super    |      5 |     12 |     50 |      67
                     |   7.46 |  17.91 |  74.63 |
            ---------+--------+--------+--------+
            Total          42       43       81      166
```

The CHISQ option always produces the correlation statistic Q_{CS}. Compare its value, $Q_{CS} = 50.6016$ (displayed in Output 5.7), with the statistic displayed under "Nonzero Correlation" in Output 5.8. These statistics are the same. Thus, you don't need to specify CMH to obtain Q_{CS} for a single table. For a 2×2 table, Q_{CS} is equivalent to Q and Q_S; for a $2 \times r$ table, Q_{CS} is equivalent to Q_S.

Output 5.7   Statistics for Table

```
         Statistics for Table of treatment by washability

         Statistic                  DF       Value      Prob
         -------------------------------------------------------
         Chi-Square                  4       55.0879   <.0001
         Likelihood Ratio Chi-Square 4       58.0366   <.0001
         Mantel-Haenszel Chi-Square  1       50.6016   <.0001
         Phi Coefficient                      0.5761
         Contingency Coefficient              0.4992
         Cramer's V                           0.4073

                       Sample Size = 166
```

Output 5.8 Q_{CS} for Integer Scores

```
          Summary Statistics for treatment by washability

       Cochran-Mantel-Haenszel Statistics (Based on Table Scores)

      Statistic   Alternative Hypothesis    DF       Value      Prob
      -----------------------------------------------------------------
          1       Nonzero Correlation        1      50.6016   <.0001
          2       Row Mean Scores Differ     2      52.7786   <.0001
          3       General Association        4      54.7560   <.0001

                        Total Sample Size = 166
```

Q_{CS} is clearly significant. Washability increases with the degree of additive to the water. Output 5.9 displays Q_{CS} for the modified ridit scores. It has the value 49.541, which is clearly significant.

Output 5.9 Q_{CS} for Modified Ridit Scores

```
          Summary Statistics for treatment by washability

       Cochran-Mantel-Haenszel Statistics (Modified Ridit Scores)

      Statistic   Alternative Hypothesis    DF       Value      Prob
      -----------------------------------------------------------------
          1       Nonzero Correlation        1      49.5410   <.0001
          2       Row Mean Scores Differ     2      52.5148   <.0001
          3       General Association        4      54.7560   <.0001

                        Total Sample Size = 166
```

5.3   Exact Tests for Association

5.3.1   General Association

In some cases, there is not sufficient sample size for the chi-square statistics discussed earlier in this chapter to be valid (several $m_{ij} \leq 5$). An alternative strategy for these situations is the Fisher exact test for $s \times r$ tables. This method follows the same principles as Fisher's exact test for the 2×2 table, except that the probabilities that are summed are taken from the multivariate hypergeometric distribution. Mehta and Patel (1983) describe a network algorithm for obtaining exact p-values that works much faster and more efficiently than direct enumeration; Baglivo, Olivier, and Pagano (1988), Cox and Plackett (1980), and Pagano and Halvorsen (1981) have also done work in this area. Besides Fisher's exact test, which produces the exact p-value for the table, exact p-values are available for general association tests such as Q, Q_P, and Q_L.

Consider Table 5.5. A marketing research firm organized a focus group to consider issues of new car marketing. Members of the group included those persons who had purchased a car from a local dealer in the last month. Researchers were interested in whether there was an association between the type of car bought and the manner in which group members found out about the car in the media. Cars were classified as sedans, sporty, and utility. The types of media included television, magazines, newspapers, and radio.

Table 5.5. Car Marketing Data

Type of Car	Advertising Source				Total
	TV	Magazine	Newspaper	Radio	
Sedan	4	0	0	2	6
Sporty	0	3	3	4	10
Utility	5	5	2	2	14

It is clear that the data do not meet the requirements for the usual tests of association via the Pearson chi-square or the randomization chi-square. There are a number of zero cells and a number of other cells whose expected values are less than 5. Under these circumstances, the exact test for no association is an appropriate strategy.

The following SAS statements produce the exact test for the car marketing data. Recall that Fisher's exact test is produced automatically for 2×2 tables with the CHISQ option; to generate the exact test for $s \times r$ tables, you need to specify the EXACT option in the TABLES statement. This generates the usual statistics produced with the CHISQ option and the exact test. Since the ORDER= option isn't specified, the columns of the resulting table will be ordered alphabetically. No ordering is assumed for this test, so this does not matter.

```
data market;
   length AdSource $ 9. ;
   input car $ AdSource $ count @@;
   datalines;
sporty  paper 3 sporty  radio 4 sporty  tv 0 sporty  magazine 3
sedan   paper 0 sedan   radio 2 sedan   tv 4 sedan   magazine 0
utility paper 2 utility radio 2 utility tv 5 utility magazine 5
;
run;
proc freq;
   weight count;
   table car*AdSource / norow nocol nopct;
   exact fisher pchi lrchi;
run;
```

Output 5.10 contains the frequency table.

Output 5.10   Car Marketing Frequency Table

```
                        Table of car by AdSource

        car         AdSource

        Frequency|magazine|paper   |radio   |tv      | Total
        ---------+--------+--------+--------+--------+
        sedan    |     0  |     0  |     2  |     4  |     6
        ---------+--------+--------+--------+--------+
        sporty   |     3  |     3  |     4  |     0  |    10
        ---------+--------+--------+--------+--------+
        utility  |     5  |     2  |     2  |     5  |    14
        ---------+--------+--------+--------+--------+
        Total          8        5        8        9       30
```

Output 5.11 displays the Fisher exact p-value for the table, as well as the asymptotic and exact results for Q_P and Q_L. For these data, the exact p-value for the table is $p = 0.0473$. Note that Q_P is 11.5984 with 6 df, $p = 0.0716$, and Q_L has the value 16.3095 with $p = 0.0122$. Q_P tends to be more conservative and Q_L tends to be more liberal than the exact test. Note that since the alternative hypothesis is general association, there is no analogy to the left-tail or right-tail as there is for Fisher's exact test for 2×2 tables, when the alternative can be directional association.

The exact p-value for Q_P is 0.0664 and the exact p-value for Q_L is 0.0272. Thus, with the exact computations, Q_P became somewhat stronger and Q_L became somewhat weaker but still significant at the $\alpha = 0.05$ level. For this table, you would typically use the Fisher exact p-value as your indication of the strength of the association and consider the association to be significant at the 0.05 level of significance.

While you can't directly produce an exact p-value for the general association Q (that is, the test produced by PROC FREQ with the CMH option for the one stratum case), the exact distribution for Q is identical to the exact distribution for Q_P so the exact p-value for Q_P is the same as the exact p-value for Q. This is because

$$Q = \frac{(n-1)Q_P}{n}$$

for general association.

Output 5.11 Exact Test Results

```
             Statistics for Table of car by AdSource


                    Pearson Chi-Square Test
           ------------------------------------
           Chi-Square                    11.5984
           DF                                  6
           Asymptotic Pr >   ChiSq         0.0716
           Exact       Pr >= ChiSq         0.0664


                Likelihood Ratio Chi-Square Test
           ------------------------------------
           Chi-Square                    16.3095
           DF                                  6
           Asymptotic Pr >   ChiSq         0.0122
           Exact       Pr >= ChiSq         0.0272


                    Fisher's Exact Test
           ------------------------------------
           Table Probability (P)       2.545E-05
           Pr <= P                        0.0473

                    Sample Size = 30
```

Notes on Exact Computations

Even though the network algorithms used to produce these exact tests are very fast compared to direct enumeration, exact methods are computationally intensive. The memory requirements and CPU time requirements can be quite high. As the sample size becomes larger, the test is likely to become computationally infeasible. For most situations when the sample size is moderately large, asymptotic methods are valid. An exception would be data that have marked sparseness in the row and column marginal totals. The exact test is mainly useful when significance is suggested by the approximate results of Q_P and Q_L. Also, in these situations, the computations are not overly lengthy. Computations are lengthy when the p-value is somewhere around 0.5, and in this situation, the exact p-value is usually not needed.

When the SAS System is performing exact computations, it prints a message to the log stating that you can press the system interrupt key if you want to terminate them. In addition, you can specify the MAXTIME= option in the EXACT statement to request, in seconds, a length of time after which the procedure is to stop exact computations.

There are some data for which computing the exact p-values is going to be very memory and time intensive and yet the asymptotic tests are not quite justifiable. You can request Monte Carlo estimation for these situations by specifying MC as an EXACT statement option. PROC FREQ uses Monte Carlo methods to estimate the exact p-value and give a confidence interval for the estimate. Refer to Agresti, Wackerly, and Boyett (1979) for more detail. With PROC FREQ, you can specify the number of samples (the default is 10,000) and the random number seed. For example, you can request Monte Carlo estimation for the MHCHI option with the statement

```
exact mhchi / mc;
```

As mentioned before, exact computations should be used only when the data require them and then used judiciously. Refer to the *SAS/STAT User's Guide, Version 8* for more information about the exact computational algorithms used in the FREQ procedure.

5.3.2 Test of Correlation

Section 1.2 discusses Mantel-Haenszel test statistics for the evaluation of general association, location shifts, and correlation. While exact p-values are not yet available in the SAS System for the MH tests of location shifts, you can obtain exact p-values for the correlation test for the case when both the rows and columns of your table are ordinally scaled.

Consider the following data from a study on a new drug for a skin disorder. Subjects were randomly assigned to one of four dosage levels and, after a suitable period of time, the affected skin area was examined and classified on a five-point scale ranging from 0 for terrible to 4 for excellent.

Table 5.6. Skin Disorder Data

Dose in Mg	Response Poor	Fair	Good	Excellent
25	1	1	1	0
50	1	2	1	1
75	0	0	2	2
100	0	0	7	0

Since both the rows and columns can be considered to be on an ordinal scale, the type of association involved is linear and the correlation Mantel-Haenszel statistic is suitable. However, note that there are several zero cells, many other cells with counts of 1 or 2, and a total sample size of 19. This is on the border of too small for the asymptotic MH test, which requires an overall sample size of at least 20. In addition, if you collapse this table into various 2×2 tables, many of the resulting cell counts are less than 5; for the asymptotic MH correlation test you generally want any cell count of a collapsed 2×2 table to be 5 or larger.

However, you can compute an exact p-value, which is the sum of the exact p-values associated with the tables where the test statistic is larger than the one you observe.

The following DATA step creates SAS data set DISORDER.

```
data disorder;
   input dose outcome count @@;
   datalines;
 25 0 1   25 1 1   25 2 1   25 3 0
 50 0 1   50 1 2   50 2 1   50 3 1
 75 0 0   75 1 0   75 2 2   75 3 2
100 0 0  100 1 0  100 2 7  100 3 0
 ;
```

Specifying the EXACT statement with the MHCHI keyword produces both the asymptotic and the exact MH test.

```
proc freq;
   weight count;
   tables dose*outcome / nocol norow nopct;
   exact mhchi;
run;
```

Output 5.12 displays the frequency table for the skin disorder data.

Output 5.12   Skin Disorder Data

```
                     Table of dose by outcome

        dose        outcome

        Frequency|      0|      1|      2|      3| Total
        ---------+--------+--------+--------+--------+
            25 |   1 |   1 |   1 |   0 |     3
        ---------+--------+--------+--------+--------+
            50 |   1 |   2 |   1 |   1 |     5
        ---------+--------+--------+--------+--------+
            75 |   0 |   0 |   2 |   2 |     4
        ---------+--------+--------+--------+--------+
           100 |   0 |   0 |   7 |   0 |     7
        ---------+--------+--------+--------+--------+
        Total        2        3       11        3      19
```

Output 5.13 displays Q_{CS} and both the asymptotic and exact p-values.

Output 5.13   Exact Results for Correlation MH

```
                Statistics for Table of dose by outcome


              Mantel-Haenszel Chi-Square Test
              -----------------------------------
              Chi-Square                 3.9314
              DF                              1
              Asymptotic Pr >  ChiSq     0.0474
              Exact      Pr >= ChiSq     0.0488

                   Sample Size = 19
```

With a Q_{CS} value of 3.9314 and 1 df, the chi-square approximation provides a significant p-value of 0.0474. The exact p-value is a little higher, p=0.0488, but is still significant at $\alpha = 0.05$. These data clearly have an association that is detected with a linear correlation statistic.

5.4   Measures of Association

Analysts are sometimes interested in assessing the strength of association in the $s \times r$ table. Although there is no counterpart to the odds ratios in 2×2 tables, there are several measures of association available, and, as you might expect, their choice depends on the scale of measurement.

5.4.1   Ordinal Measures of Association

If the data in the table have an interval scale or have scores that are equally spaced, then the Pearson correlation coefficient is an appropriate measure of association, and one that is familiar to most readers.

If the data do not lie on an obvious scale, but are ordinal in nature, then there are other measures of association that apply. The Spearman rank correlation coefficient is produced by substituting ranks as variable values for the Pearson correlation coefficient. Other measures are based on the classification of all possible pairs of subjects in the table as *concordant* or *discordant* pairs. If a pair is concordant, then the subject ranking higher on the row variable also ranks higher on the column variable. If a pair is discordant, then the subject ranking higher on the row variable also ranks lower on the column variable. The pair can also be tied on the row and column variables.

The gamma, Kendall's tau-*b*, Stuart's tau-*c*, and Somer's D statistics are all based on concordant and discordant pairs; that is, they use the relative ordering on the levels of the variables to determine whether association is negative, positive, or present at all. For example, gamma is estimated by

$$\hat{\gamma} = \frac{(C - D)}{(C + D)}$$

where C is the total number of concordant pairs and D is the total number of discordant pairs.

These measures, like the Pearson correlation coefficient, take values between -1 and 1. They differ mainly in their strategies for adjusting for ties and sample size. Somer's D depends on which variable is considered to be explanatory (the grouping variable—adjustments for ties are made only on it). Somer's D, Stuart's tau-*c*, and Kendall's tau-*b* generally express less strength of association than gamma.

Asymptotic standard errors are available for these measures. Although the measure of association is always valid, these standard errors are only valid if the sample size is large. Very conservative guidelines are the usual requirements for the Pearson chi-square that the expected cell counts are 5 or greater. A more realistic guideline is to collapse the $s \times r$ table to a 2×2 table by choosing cutpoints and then adding the appropriate rows and columns. Think of this as a line under one row and beside one column; the 2×2 table is the result of summing the cells in the resulting quadrants. The sample size is adequate if each of the cells of this 2×2 table is 5 or greater.

If the sample size is adequate, then the measure of association is approximately normally distributed and you can form the confidence intervals of interest. For example,

$$\text{measure} \pm 1.96 \times \text{ASE}$$

forms the bounds of a 95% confidence interval. Refer to the *SAS/STAT User's Guide, Version 8* for more information on these ordinal measures of association.

Measures of association are produced in the PROC FREQ output by specifying MEASURES as an option in the TABLES statement. In addition, you can request

confidence limits by specifying the CL option. The following statements produce
measures of association for the washability data listed in Table 5.4. Using
SCORES=RANK in the second TABLES statement requests that rank scores are to be
used in calculating Pearson's correlation coefficient.

```
data wash;
   input treatment $ washability $ count @@;
   datalines;
water     low 27 water     medium 14 water     high  5
standard low 10 standard medium 17 standard high 26
super     low  5 super     medium 12 super     high 50
;
proc freq order=data;
   weight count;
   tables treatment*washability / measures noprint nocol nopct cl;
   tables treatment*washability / measures scores=rank noprint cl;
run;
```

Output 5.14 contains the table produced by the first PROC FREQ invocation. All of the
measures of ordinal association indicate a positive association. Note also that the Somer's
D statistics, Kendall's tau-*b*, and Stuart's tau-*c* all have smaller values than gamma.
Somer's D statistic has two forms: Somer's D C|R means that the column variable is
considered the dependent, or response, variable, and Somer's D R|C means that the row
variable is considered the response variable.

Output 5.14   Measures of Association

```
          Statistics for Table of treatment by washability

                                                          95%
Statistic                             Value    ASE    Confidence Limits
----------------------------------------------------------------------
Gamma                                 0.6974  0.0636  0.5728    0.8221
Kendall's Tau-b                       0.4969  0.0553  0.3885    0.6053
Stuart's Tau-c                        0.4803  0.0545  0.3734    0.5872

Somers' D C|R                         0.4864  0.0542  0.3802    0.5926
Somers' D R|C                         0.5077  0.0572  0.3956    0.6197

Pearson Correlation                   0.5538  0.0590  0.4382    0.6693
Spearman Correlation                  0.5479  0.0596  0.4311    0.6648

Lambda Asymmetric C|R                 0.2588  0.0573  0.1465    0.3711
Lambda Asymmetric R|C                 0.2727  0.0673  0.1409    0.4046
Lambda Symmetric                      0.2663  0.0559  0.1567    0.3759

Uncertainty Coefficient C|R           0.1668  0.0389  0.0906    0.2431
Uncertainty Coefficient R|C           0.1609  0.0372  0.0880    0.2339
Uncertainty Coefficient Symmetric     0.1638  0.0380  0.0893    0.2383

                    Sample Size = 166
```

Output 5.15 contains the output produced by the second PROC FREQ invocation. The only difference is that rank scores were used in the calculation of Pearson's correlation coefficient. When rank scores are used, Pearson's correlation coefficient is equivalent to Spearman's correlation, as illustrated in the output. (However, the asymptotic standard errors are not equivalent.)

Output 5.15   Rank Scores for Pearson's Correlation

```
            Statistics for Table of treatment by washability

                                                        95%
Statistic                         Value     ASE    Confidence Limits
-------------------------------------------------------------------
Gamma                            0.6974   0.0636   0.5728     0.8221
Kendall's Tau-b                  0.4969   0.0553   0.3885     0.6053
Stuart's Tau-c                   0.4803   0.0545   0.3734     0.5872

Somers' D C|R                    0.4864   0.0542   0.3802     0.5926
Somers' D R|C                    0.5077   0.0572   0.3956     0.6197

Pearson Correlation (Rank Scores) 0.5479  0.0591   0.4322     0.6637
Spearman Correlation             0.5479   0.0596   0.4311     0.6648

Lambda Asymmetric C|R            0.2588   0.0573   0.1465     0.3711
Lambda Asymmetric R|C            0.2727   0.0673   0.1409     0.4046
Lambda Symmetric                 0.2663   0.0559   0.1567     0.3759

Uncertainty Coefficient C|R      0.1668   0.0389   0.0906     0.2431
Uncertainty Coefficient R|C      0.1609   0.0372   0.0880     0.2339
Uncertainty Coefficient Symmetric 0.1638  0.0380   0.0893     0.2383

                    Sample Size = 166
```

5.4.2   Exact Tests for Ordinal Measures of Association

In addition to estimating measures of association, you can also test whether a particular measure is equal to zero. In the case of the correlation coefficients, you can produce exact p-values for this test. Thus, you have access to exact methods in the evaluation of the correlation coefficients.

Table 5.7 displays data that a recreation supervisor collected from her girls's soccer league coaches. Hearing complaints about too-intense parental involvement, she surveyed each coach to see whether they considered the parental interference to be of low, medium, or high intensity for the three different grade leagues. Interference was considered to be parents questioning their child's playing time or position, questioning referee calls during the games, or yelling very specific instructions to the children on the team. She was interested in whether interference was associated with league grade. Since both grade level and interference level lie on an ordinal scale, the Spearman rank correlation coefficient is an appropriate statistic to consider.

Table 5.7.   Soccer Coach Interviews

Grades	Parental Interference		
	Low	Medium	High
1–2	3	1	0
3–4	3	2	1
5–6	1	3	2

Since there were four teams in the 1st and 2nd grade league, and six teams each in the 3rd and 4th and 5th and 6th grade leagues, the counts are necessarily small. If you apply various cutpoints to produce collapsed 2×2 table as suggested previously to determine if the asymptotic confidence intervals for the measures of association would be valid, you determine that no cutpoints exist so that each component cell is ≥ 5. However, you can apply exact methods to get an exact p-value for the hypothesis that the Spearman rank correlation coefficient is equal to zero.

The following DATA step inputs the soccer data into SAS data set SOCCER:

```
data soccer;
   input grades $ degree $ count @@;
   datalines;
1-2 low 3 1-2 medium 1 1-2 high 0
3-4 low 3 3-4 medium 2 3-4 high 1
5-6 low 1 5-6 medium 3 5-6 high 2
;
run;
```

In order to produce the exact p-value for the Spearman's rank test, you specify an EXACT statement that includes the keyword SCORR. The ORDER=DATA option is specified in the PROC statement to ensure that the columns and rows maintain the correct ordering.

```
proc freq order=data;
   weight count;
   tables grades*degree / nocol nopct norow;
   exact scorr;
run;
```

Output 5.16 displays the frequency table for the soccer data.

Output 5.16   Soccer Frequency Table

```
                    Table of grades by degree

        grades        degree

        Frequency|low      |medium  |high    |  Total
        ---------+--------+--------+--------+
        1-2      |    3 |      1 |      0 |      4
        ---------+--------+--------+--------+
        3-4      |    3 |      2 |      1 |      6
        ---------+--------+--------+--------+
        5-6      |    1 |      3 |      2 |      6
        ---------+--------+--------+--------+
        Total         7        6        3       16
```

Output 5.17 contains the Spearman correlation coefficient, which has the value 0.4878, indicating the possibility of modest correlation.

Output 5.17 Spearman Correlation Coefficient

```
             Statistics for Table of grades by degree


                 Spearman Correlation Coefficient
                 --------------------------------
                 Correlation (r)            0.4878
                 ASE                        0.1843
                 95% Lower Conf Limit       0.1265
                 95% Upper Conf Limit       0.8491

                      Sample Size = 16
```

The results for the hypothesis test that the correlation is equal to zero is listed in
Output 5.18. The exact two-sided p-value is 0.0637, a borderline result, although you
would reject the hypothesis at a strict $\alpha = 0.05$ level of confidence. Note that the
asymptotic test results in a two-sided p-value of 0.0092. However, with these counts, you
could not justify the use of the asymptotic test.

Output 5.18 Hypothesis Test for Spearman's Rank Test

```
             Statistics for Table of grades by degree

                 Test of H0: Correlation = 0

                 ASE under H0               0.1872
                 Z                          2.6055
                 One-sided Pr >  Z          0.0046
                 Two-sided Pr > |Z|         0.0092

                 Exact Test
                 One-sided Pr >=  r         0.0354
                 Two-sided Pr >= |r|        0.0637

                      Sample Size = 16
```

Note that you can also test whether the asymptotic statistics produced by the MEASURES
option are equal to zero. You request such tests with the TEST statement in the FREQ
procedure; note that you can also test hypotheses concerning the kappa statistics discussed
in Section 5.5.

Refer to *SAS/STAT User's Guide: Version 8* for more information.

5.4.3 Nominal Measures of Association

Measures of association when one or both variables are nominally scaled are more difficult
to define, since you can't think of association in these circumstances as negative or positive
in any sense. However, indices of association in the nominal case have been constructed,
and most are based on mimicking R-squared in some fashion. One such measure is the
uncertainty coefficient, and another is the lambda coefficient. More information about
these statistics can be obtained in the *SAS/STAT User's Guide, Version 8*, including the
appropriate references. Agresti (1990) also discusses some of these measures.

The following PROC FREQ invocation produces nominal measures of association for the neighborhood data.

```
data neighbor;
   length party $ 11 neighborhood $ 10;
   input party $ neighborhood $ count @@;
   datalines;
democrat      longview   360 democrat      bayside  221
democrat      sheffeld   140 democrat      highland 160
republican    longview   316 republican    bayside  208
republican    sheffeld    97 republican    highland 106
independent   longview   160 independent   bayside  200
independent   sheffeld   311 independent   highland 291
;
proc freq ;
   weight count;
   tables party*neighborhood / chisq measures nocol nopct;
run;
```

Output 5.19 displays the resulting table.

Output 5.19   Nominal Measures of Association

```
          Statistics for Table of party by neighborhood


   Statistic                                Value       ASE
   -----------------------------------------------------------
   Gamma                                   -0.0183    0.0226
   Kendall's Tau-b                         -0.0130    0.0161
   Stuart's Tau-c                          -0.0137    0.0169

   Somers' D C|R                           -0.0138    0.0170
   Somers' D R|C                           -0.0123    0.0152

   Pearson Correlation                     -0.0178    0.0190
   Spearman Correlation                    -0.0150    0.0189

   Lambda Asymmetric C|R                    0.0871    0.0120
   Lambda Asymmetric R|C                    0.1374    0.0177
   Lambda Symmetric                         0.1113    0.0119

   Uncertainty Coefficient C|R              0.0401    0.0046
   Uncertainty Coefficient R|C              0.0503    0.0058
   Uncertainty Coefficient Symmetric        0.0446    0.0051

                 Sample Size = 2570
```

You should ignore the ordinal measures of association here since the data are not ordinally scaled. There are three versions of both the lambda coefficient and the uncertainty coefficient: column variable as the response variable, row variable as the response variable, and a symmetric version. Obviously, this makes a difference in the resulting statistic.

5.5   Observer Agreement

5.5.1   Computing the Kappa Statistic

For many years, researchers in medicine, epidemiology, psychiatry, and psychological measurement and testing have been aware of the importance of observer error as a major source of measurement error. In many cases, different observers, or even the same observer at a different time, may examine an x-ray or perform a physical examination and reach different conclusions. It is important to evaluate observer agreement, both to understand the possible contributions to measurement error and as part of the evaluation of testing new instruments and procedures.

Often, the data collected as part of an observer agreement study form a contingency table, where the column levels represent the ratings of one observer and the row levels represent the ratings of another observer. Each cell represents one possible profile of the observers' ratings. The cells on the diagonal represent the cases where the observers agree.

Consider Table 5.8. These data come from a study concerning the diagnostic classification of multiple sclerosis patients. Patients from Winnipeg and New Orleans were classified into one of four diagnostic classes by both a Winnipeg neurologist and a New Orleans neurologist. Table 5.8 contains the data for the Winnipeg patients (Landis and Koch 1977).

Table 5.8.   Ratings of Neurologists

New Orleans Neurologist	Winnipeg Neurologist			
	1	2	3	4
1	38	5	0	1
2	33	11	3	0
3	10	14	5	6
4	3	7	3	10

Certainly one way to assess the association between these two raters is to compute the usual measures of association. However, while measures of association can reflect the strength of the predictable relationship between two raters or observers, they don't target how well they agree. Agreement can be considered a special case of association—to what degree do different observers classify a particular subject into the identical category? All measures of agreement target the diagonal cells of a contingency table in their computations, and some measures take into consideration how far away from the diagonal elements other cells fall.

Suppose π_{ij} is the probability of a subject being classified in the ith category by the first observer and the jth category by the second observer. Then

$$\Pi_o = \sum \pi_{ii}$$

is the probability that the observers agree. If the ratings are independent, then the probability of agreement is

$$\Pi_e = \sum \pi_{i+} \pi_{+i}$$

So, $\Pi_o - \Pi_e$ is the amount of agreement beyond that expected by chance. The *kappa coefficient* (Cohen 1960) is defined as

$$\kappa = \frac{\Pi_o - \Pi_e}{1 - \Pi_e}$$

Since $\Pi_o = 1$ when there is perfect agreement (all non-diagonal elements are zero), κ equals 1 when there is perfect agreement, and κ equals 0 when the agreement equals that expected by chance. The closer the value is to 1, the more agreement there is in the table. It is possible to obtain negative values, but that rarely occurs. Note that κ is analogous to the intraclass correlation coefficient obtained from ANOVA models for quantitative measurements; it can be used as a measure of reliability of multiple determinations on the same subject (Fleiss and Cohen 1973, Fleiss 1975).

You may be interested in distinguishing degrees of agreement in a table, particularly if the categories are ordered in some way. For example, you may want to take into account those disagreements that are just one category away. A weighted form of the kappa statistic allows you to assign weights, or scores, to the various categories so that you can incorporate such considerations into the construction of the test statistic.

Weighted κ is written

$$\kappa_w = \frac{\sum\sum w_{ij}\pi_{ij} - \sum\sum w_{ij}\pi_{i+}\pi_{+j}}{1 - \sum\sum_{ij} w_{ij}\pi_{i+}\pi_{+j}}$$

where w_{ij} represents weights with values between 0 and 1. One possible set of weights is

$$w_{ij} = 1 - \frac{|\mathrm{score}(i) - \mathrm{score}(j)|}{\mathrm{score}(dim) - \mathrm{score}(1)}$$

where score(i) is the score for the ith row, score(j) is the score for the jth column, and *dim* is the dimension of an $s \times s$ table. This scoring system puts more weight on those cells closest to the diagonal. These weights are known as Cicchetti-Allison weights (Cicchetti and Allision 1969) and are the default weights for the weighted kappa statistic in PROC FREQ. Fleiss-Cohen weights are also available (Fleiss and Cohen 1973).

The following SAS statements generate kappa statistics for the Winnipeg data. To produce measures of agreement, you specify AGREE in the TABLES statement.

```
data classify;
   input no_rater w_rater count @@;
   datalines;
1 1 38 1 2  5 1 3 0 1 4  1
2 1 33 2 2 11 2 3 3 2 4  0
3 1 10 3 2 14 3 3 5 3 4  6
4 1  3 4 2  7 4 3 3 4 4 10
;
proc freq;
   weight count;
   tables no_rater*w_rater / agree norow nocol nopct;
run;
```

Output 5.20 contains the table.

Output 5.20   Winnipeg Data

```
                        Table of no_rater by w_rater

        no_rater        w_rater

        Frequency|         1|        2|        3|        4|  Total
        ---------+--------+--------+--------+--------+
               1 |     38 |      5 |      0 |      1 |     44
        ---------+--------+--------+--------+--------+
               2 |     33 |     11 |      3 |      0 |     47
        ---------+--------+--------+--------+--------+
               3 |     10 |     14 |      5 |      6 |     35
        ---------+--------+--------+--------+--------+
               4 |      3 |      7 |      3 |     10 |     23
        ---------+--------+--------+--------+--------+
        Total          84       37       11       17     149
```

Output 5.21 displays the measures of association.

Output 5.21   Kappa Statistics

```
              Statistics for Table of no_rater by w_rater

                        Test of Symmetry
                 -----------------------
                 Statistic (S)      46.7492
                 DF                        6
                 Pr > S               <.0001


                        Kappa Statistics

        Statistic          Value       ASE      95% Confidence Limits
        -----------------------------------------------------------
        Simple Kappa       0.2079    0.0505       0.1091      0.3068
        Weighted Kappa     0.3797    0.0517       0.2785      0.4810

                        Sample Size = 149
```

$\hat{\kappa}$ has the value 0.2079. This is indicative of slight agreement. Values of 0.4 or above are considered to indicate moderate agreement, and values of 0.8 or higher indicate excellent agreement. The asymptotic standard error is also printed, as well as confidence bounds. Since the confidence bounds do not contain the value 0, you can reject the hypothesis that κ is 0 for these data (no agreement) at the $\alpha = 0.05$ level of significance.

Using the default scores, $\hat{\kappa}_w$ takes the value 0.3797. This means that if you consider disagreement close to the diagonals less heavily than disagreement further away from the diagonals, you get higher agreement. $\hat{\kappa}$ treats all off-diagonal cells the same. When $\hat{\kappa}_w$ is high, for example, ≥ 0.6 for moderate sample size, it may be preferable to produce confidence bounds on a transformed scale like logarithms or the Fisher z transformation and then exponentiate to compute the limits.

The test of symmetry is Bowker's test of symmetry (Bowker 1948). The null hypothesis of this test is that the square table is symmetric, or that the cell probabilities p_{ij} and p_{ji} are equal. When you have a 2×2 table, the test is the same as McNemar's test.

5.5.2 Exact p-values for the Kappa Statistic

Exact p-values are also available for the kappa statistic. Consider the data in Table 5.9. Elderly residents of a midwestern community enrolled in a pilot program that provided resources to seniors and also sought to identify those persons requiring additional living assistance. Researchers tested a tool for in-home evaluation of a resident's agility. The test rated the ease with which basic tasks could be performed and provided an overall rating of ability on a four-point scale ranging from 1 for poor to 4 for excellent. Two raters evaluated the same 24 people.

Table 5.9.   Ratings of Social Workers

Rater One	Rater Two			
	1	2	3	4
1	4	0	1	0
2	0	2	6	1
3	1	0	2	1
4	0	2	1	3

The table includes numerous 0 and 1 counts, too many for the asymptotic requirements to be fulfilled. However, rater agreement can still be assessed with the use of exact p-values for the test of null agreement with kappa statistics.

The following PROC FREQ statements produce the desired exact results. You specify the keyword KAPPA in the EXACT statement to generate a table with both the asymptotic and exact results.

```
data pilot;
   input rater1 rater2 count @@;
   datalines;
1 1 4 1 2 0 1 3 1 1 4 0
2 1 0 2 2 2 2 3 6 2 4 1
3 1 1 3 2 0 3 3 2 3 4 1
4 1 0 4 2 2 4 3 1 4 4 3
;
proc freq;
   weight count;
   tables rater1*rater2 /norow nocol nopct;
   exact kappa;
run;
```

Output 5.22 contains the table of ratings.

Output 5.22 Pilot Data

```
                      Table of rater1 by rater2

        rater1      rater2

        Frequency|        1|        2|        3|        4|  Total
        ---------+--------+--------+--------+--------+
               1 |     4 |     0 |     1 |     0 |     5
        ---------+--------+--------+--------+--------+
               2 |     0 |     2 |     6 |     1 |     9
        ---------+--------+--------+--------+--------+
               3 |     1 |     0 |     2 |     1 |     4
        ---------+--------+--------+--------+--------+
               4 |     0 |     2 |     1 |     3 |     6
        ---------+--------+--------+--------+--------+
        Total           5        4       10        5       24
```

Output 5.23 contains the estimate of the kappa coefficient, which is 0.2989 with 95%
confidence limits of (0.0469, 0.5509). The table "Test of HO: Kappa=0" presents both the
asymptotic test for the hypothesis that a constructed Z statistic is equal to zero as well as
the exact p-value for that test. The exact p-value is 0.0088, both one-sided and two-sided.
There is some agreement between raters.

Output 5.23 Exact Results for Kappa Test

```
             Statistics for Table of rater1 by rater2

                    Simple Kappa Coefficient
                ---------------------------------
                Kappa (K)                 0.2989
                ASE                       0.1286
                95% Lower Conf Limit      0.0469
                95% Upper Conf Limit      0.5509

                     Test of HO: Kappa = 0

                ASE under H0              0.1066
                Z                         2.8032
                One-sided Pr >   Z        0.0025
                Two-sided Pr > |Z|        0.0051

                Exact Test
                One-sided Pr >=  K        0.0088
                Two-sided Pr >= |K|       0.0088

                      Sample Size = 24
```

5.6   Test for Ordered Differences

Sometimes you have a contingency table in which the columns represent an ordinal
outcome and the rows are either nominal or ordinal. One test of interest is whether there
are location shifts in the mean response; this is evaluated with the mean score test as
discussed in Section 5.2.2. However, you may also be interested in testing against an

ordered alternative; that is, are the mean scores strictly increasing (or decreasing) across the levels of the row variable? The Jonckheere-Terpstra test is designed to test the null hypothesis that the distribution of the ordered responses is the same across the various rows of the table. This test detects whether there are differences in

$$d_1 \leq d_2 \leq \ldots \leq d_s \text{ or } d_s \geq d_{s-1} \geq \ldots \geq d_1$$

where d_i represents the ith group effect.

The Jonckheere-Terpstra test is a nonparametric test that is based on sums of Mann-Whitney test statistics; the asymptotic p-values are produced by using the normal approximation for the distribution of the standardized test statistic. Refer to the *SAS/STAT User's Guide, Version 8* for more computational detail, and refer to Pirie (1983) and Hollander and Wolfe (1973) for more information on the Jonckheere-Terpstra test.

Table 5.10 displays the dumping syndrome data, which have appeared frequently in the categorical data analysis literature, beginning with Grizzle, Starmer, and Koch (1969). Investigators conducted a randomized clinical trial in four hospitals, where patients were assigned to one of four surgical procedures for the treatment of severe duodenal ulcers. The treatments include:

v + d: vagotomy and drainage

v + a: vagotomy and antrectomy (removal of 25% of gastric tissue)

v + h: vagatomy and hemigastrectomy (removal of 50% of gastric tissue)

gre: gastric resection (removal of 75% of gastric tissue)

The response measured was the severity (none, slight, moderate) of the dumping syndrome, which is expected to increase directly with the proportion of gastric tissue removed. This response, an adverse effect of surgery, can be considered ordinally scaled, as can operation. Investigators wanted to determine if type of operation was associated with severity of dumping syndrome, after adjusting for hospital. This analysis is performed in the next chapter.

Table 5.10. Dumping Syndrome Data

| Hospital | Operation | Severity of Symptoms | | | Total |
		None	Slight	Moderate	
1	v + d	23	7	2	32
1	v + a	23	10	5	38
1	v + h	20	13	5	38
1	gre	24	10	6	40
2	v + d	18	6	1	25
2	v + a	18	6	2	26
2	v + h	13	13	2	28
2	gre	9	15	2	26
3	v + d	8	6	3	17
3	v + a	12	4	4	20
3	v + h	11	6	2	19
3	gre	7	7	4	18
4	v + d	12	9	1	22
4	v + a	15	3	2	20
4	v + h	14	8	3	25
4	gre	13	6	4	23

Ignoring hospital, there is interest in determining whether the responses are ordered the same across the operations. The Jonckheere-Terpstra test is appropriate here. The following SAS statements input these data.

```
data operate;
   input hospital trt $ severity $ wt @@;
   datalines;
1 v+d none 23    1 v+d slight  7    1 v+d moderate 2
1 v+a none 23    1 v+a slight 10    1 v+a moderate 5
1 v+h none 20    1 v+h slight 13    1 v+h moderate 5
1 gre none 24    1 gre slight 10    1 gre moderate 6
2 v+d none 18    2 v+d slight  6    2 v+d moderate 1
2 v+a none 18    2 v+a slight  6    2 v+a moderate 2
2 v+h none 13    2 v+h slight 13    2 v+h moderate 2
2 gre none  9    2 gre slight 15    2 gre moderate 2
3 v+d none  8    3 v+d slight  6    3 v+d moderate 3
3 v+a none 12    3 v+a slight  4    3 v+a moderate 4
3 v+h none 11    3 v+h slight  6    3 v+h moderate 2
3 gre none  7    3 gre slight  7    3 gre moderate 4
4 v+d none 12    4 v+d slight  9    4 v+d moderate 1
4 v+a none 15    4 v+a slight  3    4 v+a moderate 2
4 v+h none 14    4 v+h slight  8    4 v+h moderate 3
4 gre none 13    4 gre slight  6    4 gre moderate 4
;
```

The following PROC FREQ statements request the Jonckheere-Terpstra test by specifying the JT option in the TABLES statement. Note that the order of the table columns is very important for such a test; in this PROC FREQ invocation, the ORDER=DATA option in the PROC statement produces the desired order.

```
proc freq order=data;
   weight wt;
   tables trt*severity / norow nocol nopct jt;
run;
```

Output 5.24 contains the contingency table of treatment by severity.

Output 5.24 Dumping Syndrome Data

```
                      Table of trt by severity

          trt        severity

          Frequency|none    |slight |moderate|  Total
          ---------+--------+--------+--------+
          v+d      |    61 |    28 |     7 |      96
          ---------+--------+--------+--------+
          v+a      |    68 |    23 |    13 |     104
          ---------+--------+--------+--------+
          v+h      |    58 |    40 |    12 |     110
          ---------+--------+--------+--------+
          gre      |    53 |    38 |    16 |     107
          ---------+--------+--------+--------+
          Total         240     129      48     417
```

Output 5.25 displays the results.

Output 5.25 Jonckheere-Terpstra

```
              Statistics for Table of trt by severity

                   Jonckheere-Terpstra Test
              ------------------------------
              Statistic               35697.0000
              Z                           2.5712
              One-sided Pr >  Z           0.0051
              Two-sided Pr > |Z|          0.0101

                   Sample Size = 417
```

The value of the actual Jonckheere-Terpstra statistic is 35697. The corresponding Z-statistic has the value 2.5712 with a two-sided p-value of 0.0101. At a 0.05 α level, you would conclude that there are significant differences among groups in their respective ordering for the ordered response variable represented by the columns of the table.

Note that an exact version of the Jonckheere-Terpstra test is available. You simply specify the option JT in an EXACT statement.

Chapter 6

Sets of s × r Tables

Chapter Table of Contents

Chapter 6
Sets of s × r Tables

6.1 Introduction

Previous chapters address stratified analysis as the assessment of association in sets of 2×2 tables, $2 \times r$ tables where the response variable, represented in the table columns, is ordinally scaled, and $s \times 2$ tables where the groups for the row variable are ordinally scaled. Such analyses are special cases of the analysis of sets of $s \times r$ tables, which includes the cases where the row and column variables are both nominally scaled, the row variable is nominally scaled and the column variable is ordinally scaled, and the row variable and the column variable are both ordinally scaled. The Mantel-Haenszel procedure can be extended to handle these situations. It provides statistics that detect general association, mean score differences, and linear correlation as alternatives to the null hypothesis of no association; the choice of statistic depends on the scale of the row and column variables.

The general idea of stratified analyses is that you control for the effects of factors that are part of the research design, such as medical centers or hospitals in a randomized clinical trial, or factors that represent a prespecified poststudy stratification to adjust for explanatory variables that are thought to be related to the response variable. This is a common strategy for retrospective and observational studies. As mentioned in previous chapters, the Mantel-Haenszel procedure potentially removes the confounding influence of the explanatory variables that comprise the stratification and provides a gain of power for detecting association by comparing like subjects. In some sense, the strategy is similar to adjustment for blocks in a two-way analysis of variance for randomized blocks; it is also similar to covariance adjustment for a categorical explanatory variable.

Historically, the principle of combining information across strata was identified by Cochran (1954): this was in the context of combining differences of proportions from binomial distributions. Mantel and Haenszel (1959) refined the procedure to apply to hypergeometric distributions and produced a statistic to which central limit theory was more applicable for the combined strata. Thus, only the overall sample size needed to be reasonably large. The Mantel-Haenszel statistic proved more useful than Cochran's method. (Cochran's influence is the reason why the FREQ procedure output is labeled "Cochran-Mantel-Haenszel Statistics"; current literature tends to use the terms "extended Mantel-Haenszel statistics" and "Mantel-Haenszel statistics.")

Mantel (1963) discussed extensions to the MH strategy, including strategies for sets of $2 \times r$ tables, sets of $s \times 2$ tables, and the correlation statistic for $s \times r$ tables. The method was further elaborated by Landis, Heyman, and Koch (1978) to encompass the family of Mantel-Haenszel statistics, which included the statistics for general association, nonparametric ANOVA (mean score), the correlation statistic, and other special cases. Kuritz, Landis, and Koch (1988) present a useful overview of the Mantel-Haenszel strategy, and so do Landis et al. (1998).

The Mantel-Haenszel procedure requires minimal assumptions. The methods it encompasses are based on randomization considerations; the only assumptions required are the randomization of the subjects to levels of the row variable. This can be done explicitly, such as for randomized clinical trials; implicitly, via hypothesis; or conditionally, such as for retrospective studies or observational data. The minimal assumptions often allow you to perform hypothesis tests on data that do not meet the more rigorous assumptions concerning random sampling or underlying distributions that are required for statistical modeling. However, the conclusions of the analysis may be restricted to the study population at hand, versus inference to a larger population. Most often, a complete analysis includes the applications of these minimal assumption methods to perform hypothesis tests and then statistical modeling to describe more completely the variation in the data.

Another advantage of the Mantel-Haenszel procedure is the fact that sample size requirements are based on total frequencies, or quantities summed across tables, rather than on individual cell sizes. This is partly because the Mantel-Haenszel methods are targeted at detecting average effects across strata; they are often called methods of assessing average partial association.

Section 6.2 discusses the formulation of the Mantel-Haenszel statistics in matrix terminology. Section 6.3 illustrates the use of the Mantel-Haenszel strategy for several applications. Finally, Section 6.4 includes the advanced topic of the use of the Mantel-Haenszel procedure in repeated measurements analysis.

6.2   General Mantel-Haenszel Methodology

Table 6.1 represents the generic $s \times r$ table in a set of q $s \times r$ tables.

Table 6.1. hth $s \times r$ Contingency Table

Group	Response Variable Categories				Total
	1	2	...	r	
1	n_{h11}	n_{h12}	...	n_{h1r}	n_{h1+}
2	n_{h21}	n_{h22}	...	n_{h2r}	n_{h2+}
$\vdots$	$\vdots$	$\vdots$		$\vdots$	$\vdots$
s	n_{hs1}	n_{hs2}	...	n_{hsr}	n_{hs+}
Total	n_{h+1}	n_{h+2}	...	n_{h+r}	n_h

Under the assumption that the marginal totals n_{hi+} and n_{h+j} are fixed, the overall null hypothesis of no partial association can be stated as follows:

> For each of the levels of the stratification variable $h = 1, 2, \ldots, q$, the response variable is distributed at random with respect to the groups (row variable levels).

Suppose $\mathbf{n}_h' = (n_{h11}, n_{h12}, \ldots, n_{h1r}, \ldots, n_{hs1}, \ldots, n_{hsr})$, where n_{hij} is the number of subjects in the hth stratum in the ith group in the jth response category. The probability distribution for the vector $\mathbf{n}_h$ under H_0 can be written

$$Pr\{\mathbf{n}_h | H_0\} = \frac{\prod_{i=1}^{s} n_{hi+}! \prod_{j=1}^{r} n_{h+j}!}{n_h! \prod_{i=1}^{s} \prod_{j=1}^{r} n_{hij}!}$$

For the hth stratum, suppose that $p_{hi+} = n_{hi+}/n_h$ denotes the marginal proportion of subjects belonging to the ith group, and suppose that $p_{h+j} = n_{h+j}/n_h$ denotes the marginal proportion of subjects classified as belonging to the jth response category. These proportions can be denoted in vector notation as

$$\mathbf{p}_{h*+}' = (p_{h1+}, \ldots, p_{hs+})$$

$$\mathbf{p}_{h+*}' = (p_{h+1}, \ldots, p_{h+r})$$

Then,

$$E\{n_{hij} | H_0\} = m_{hij} = n_h p_{hi+} p_{h+j}$$

and the expected value of $\mathbf{n}_h$ can be written

$$E\{\mathbf{n}_h | H_0\} = \mathbf{m}_h = n_h \left[\mathbf{p}_{h+*} \otimes \mathbf{p}_{h*+} \right]$$

where $\otimes$ denotes the left-hand Kronecker product (the matrix on the left of the $\otimes$ multiplies each element of the matrix on the right).

The variance of $\mathbf{n}_h$ under H_0 is

$$V_h = \text{Var}\{\mathbf{n}_h | H_0\} = \frac{n_h^2}{(n_h - 1)} \left\{ [\mathbf{D}_{\mathbf{p}_{h+*}} - \mathbf{p}_{h+*}\mathbf{p}_{h+*}'] \otimes [\mathbf{D}_{\mathbf{p}_{h*+}} - \mathbf{p}_{h*+}\mathbf{p}_{h*+}'] \right\}$$

where $\mathbf{D}_{\mathbf{p}_{h+*}}$ and $\mathbf{D}_{\mathbf{p}_{h*+}}$ are diagonal matrices with elements of the vectors $\mathbf{p}_{h+*}$ and $\mathbf{p}_{h*+}$ as the main diagonals.

The general form of the extended Mantel-Haenszel statistic for $s \times r$ tables is

$$Q_{EMH} = \left\{ \sum_{h=1}^{q} (\mathbf{n}_h - \mathbf{m}_h)' \mathbf{A}_h' \right\} \left\{ \sum_{h=1}^{q} \mathbf{A}_h \mathbf{V}_h \mathbf{A}_h' \right\}^{-1} \left\{ \sum_{h=1}^{q} \mathbf{A}_h (\mathbf{n}_h - \mathbf{m}_h) \right\}$$

where $\mathbf{A}_h$ is a matrix that specifies the linear functions of the $\{\mathbf{n}_h - \mathbf{m}_h\}$ at which the test statistic is directed. Choices of the $\{\mathbf{A}_h\}$ provide stratification-adjusted counterparts to the randomization chi-square statistic Q, the mean score statistic Q_S, and the correlation statistic Q_{CS} that are discussed in Chapter 5, "The $s \times r$ Table."

6.2.1 General Association Statistic

When both the row and column variables are nominally scaled, the alternative hypothesis of interest is that of general association, where the pattern of distribution of the response levels across the row levels is heterogeneous. This is the most general alternative hypothesis and is always valid, no matter how the row and column variables are scaled.

In this case,

$$\mathbf{A}_h = \left\{ [\mathbf{I}_{(r-1)}, \mathbf{0}_{(r-1)}] \otimes [\mathbf{I}_{(s-1)}, \mathbf{0}_{(s-1)}] \right\}$$

which, applied to $(\mathbf{n}_h - \mathbf{m}_h)$, produces the differences between the observed and expected frequencies under H_0 for the $(s-1)(r-1)$ cells of the table after eliminating the last row and column. This results in Q_{GMH}, which is approximately chi-square with $(s-1)(r-1)$ degrees of freedom. Q_{GMH} is often called the test of general association.

6.2.2 Mean Score Statistic

When the response levels are ordinally scaled, you can assign scores to them to compute row mean scores. In this case, the alternative hypothesis to the null hypothesis of no association is that there are location shifts for these mean scores across the levels of the row variables.

Here,

$$\mathbf{A}_h = \mathbf{a}_h' \otimes [\mathbf{I}_{(s-1)}, \mathbf{0}_{(s-1)}]$$

where $\{\mathbf{a}_h\} = (a_{h1}, a_{h2}, \ldots, a_{hr})$ specifies scores for the jth response level in the hth stratum, from which the means

$$\bar{y}_{hi} = \sum_{j=1}^{r} (a_{hj} n_{hij}/n_{hi+})$$

are created for comparisons of the s populations across the strata.

This produces the extended Mantel-Haenszel Q_{SMH}, which is approximately chi-square with $(s-1)$ degrees of freedom under H_0. Q_{SMH} is called the mean score statistic and is the general form of the Q_{SMH} statistic for $2 \times r$ tables discussed in Chapter 4, "Sets of $2 \times r$ and $s \times 2$ Tables," where $(s-1) = 1$. If marginal rank or ridit scores are used, with midranks assigned for ties, Q_{SMH} is equivalent to an extension of the Kruskal-Wallis ANOVA test on ranks to account for strata and the Friedman ANOVA test on ranks to account for more than one subject per group within strata. See Chapter 7, "Nonparametric Methods," for further discussion on nonparametric tests that are special cases of Mantel-Haenszel strategies.

6.2.3 Correlation Statistic

When both the response variable (columns) and the row variable (or groups) are ordinally scaled, you can assign scores to both the response levels and the row levels in the hth stratum. The alternative hypothesis to no association in this situation is a linear trend on the mean scores across the levels of the row variable. In this case,

$$\mathbf{A}_h = [\mathbf{a}_h' \otimes \mathbf{c}_h']$$

where the $\{\mathbf{a}_h\}$ are defined as before and the $\{\mathbf{c}_h\} = (c_{h1}, c_{h2}, \ldots, c_{hs})$ specify a set of scores for the ith level of the row variable in the hth stratum. This produces the differences between the observed and expected sum of products of the row and column scores with the frequencies n_{hij}, so that the resulting test statistic is directed at detecting correlation.

This test statistic is Q_{CSMH}, which is approximately chi-square with one degree of freedom under H_0. It is the general form of Q_{CSMH} discussed in Chapter 4 for stratified $s \times 2$ tables where the row variable is ordinally scaled. It has increased power relative to Q_{GMH} or Q_{SMH} for linear association alternatives to the null hypothesis of no association.

6.2.4 Summary

Table 6.2 summarizes the various types of extended Mantel-Haenszel statistics.

Table 6.2.   Extended Mantel-Haenszel Statistics

MH Statistic	Alternative Hypothesis	SAS Output Label	Degrees of Freedom	Scale Requirements	Nonparametric Equivalents
Q_{GMH}	general association	General Association	$(s-1) \times (r-1)$	none	
Q_{SMH}	mean score location shifts	Row Means Scores Differ	$(s-1)$	column variable ordinal	Kruskal-Wallis
Q_{CSMH}	linear association	Nonzero Correlation	1	row and column variable ordinal	Spearman correlation

6.3   Mantel-Haenszel Applications

The Mantel-Haenszel strategy has applications in many different settings, including a number of different sampling frameworks. Chapter 3, "Sets of 2×2 Tables," demonstrates the use of this strategy for analyzing sets of 2×2 tables, and Chapter 4 demonstrates the use of the strategy for sets of $2 \times r$ and $s \times 2$ tables. If you haven't read these chapters, you should review them since they contain many general remarks on the application of Mantel-Haenszel methods. Section 6.3 illustrates the use of these methods for sets of $s \times r$ tables, including examples from clinical trials, observational studies, and prospective studies.

6.3.1 Dumping Syndrome Data

The previous chapter described the dumping syndrome data, which are displayed in Table 5.10. The response measured was the severity (none, slight, moderate) of the dumping syndrome, which is expected to increase directly with the proportion of gastric tissue removed. This response, an adverse effect of surgery, can be considered ordinally scaled, as can operation. Investigators wanted to determine if type of operation was associated with severity of dumping syndrome, after adjusting for hospital.

Since both the row and column variables are ordinally scaled, you can use the correlation statistic Q_{CSMH} to assess the null hypothesis of no association against the alternative that type of operation and severity of response are linearly associated.

The following SAS statements input the data into the SAS data set OPERATE and request the MH analysis. Note the use of the option ORDER=DATA, as well as the request for both integer scores (the default table scores) and standardized midrank scores (SCORES=MODRIDIT).

```
data operate;
   input hospital trt $ severity $ wt @@;
   datalines;
1 v+d none 23    1 v+d slight  7    1 v+d moderate 2
1 v+a none 23    1 v+a slight 10    1 v+a moderate 5
1 v+h none 20    1 v+h slight 13    1 v+h moderate 5
1 gre none 24    1 gre slight 10    1 gre moderate 6
2 v+d none 18    2 v+d slight  6    2 v+d moderate 1
2 v+a none 18    2 v+a slight  6    2 v+a moderate 2
2 v+h none 13    2 v+h slight 13    2 v+h moderate 2
2 gre none  9    2 gre slight 15    2 gre moderate 2
3 v+d none  8    3 v+d slight  6    3 v+d moderate 3
3 v+a none 12    3 v+a slight  4    3 v+a moderate 4
3 v+h none 11    3 v+h slight  6    3 v+h moderate 2
3 gre none  7    3 gre slight  7    3 gre moderate 4
4 v+d none 12    4 v+d slight  9    4 v+d moderate 1
4 v+a none 15    4 v+a slight  3    4 v+a moderate 2
4 v+h none 14    4 v+h slight  8    4 v+h moderate 3
4 gre none 13    4 gre slight  6    4 gre moderate 4
;
proc freq order=data;
   weight wt;
   tables hospital*trt*severity / cmh;
   tables hospital*trt*severity / cmh scores=modridit;
run;
```

Output 6.1 contains the results for the extended Mantel-Haenszel analysis using integer scores. Q_{CSMH} takes the value 6.3404, which is significant at the $\alpha = 0.05$ level; note that the statistics for general association, Q_{GMH}, and mean score differences, Q_{SMH}, are not significant at the $\alpha = 0.05$ level of significance. This is an example of the utility of taking advantage of the correlation statistic when it is appropriate; its greater power against the alternative hypothesis of linear association has detected significant evidence against the null hypothesis.

Output 6.1   Table Scores

```
                 Summary Statistics for trt by severity
                        Controlling for hospital

       Cochran-Mantel-Haenszel Statistics (Based on Table Scores)

       Statistic    Alternative Hypothesis    DF      Value      Prob
       -------------------------------------------------------------------
           1         Nonzero Correlation        1      6.3404    0.0118
           2         Row Mean Scores Differ      3      6.5901    0.0862
           3         General Association         6     10.5983    0.1016


                      Total Sample Size = 417
```

Output 6.2 contains the results for the standardized midrank scores. $Q_{CSMH} = 6.9266$, with $p = 0.0085$. As with the integer scores, the other statistics do not detect as much evidence against the null hypothesis of no association. Since the response variable levels are subjective and undoubtedly not equally spaced, the analysis of standardized midrank scores may provide the most appropriate test.

Output 6.2   Standardized Midrank Scores

```
                 Summary Statistics for trt by severity
                        Controlling for hospital

       Cochran-Mantel-Haenszel Statistics (Modified Ridit Scores)

       Statistic    Alternative Hypothesis    DF      Value      Prob
       -------------------------------------------------------------------
           1         Nonzero Correlation        1      6.9266    0.0085
           2         Row Mean Scores Differ      3      7.6370    0.0541
           3         General Association         6     10.5983    0.1016


                      Total Sample Size = 417
```

This analysis shows that, adjusting for hospital, there is a clear monotonic association between degree of gastric tissue removal and severity of dumping syndrome. The greater the degree of gastric tissue removal, the worse the dumping syndrome.

6.3.2   Shoulder Harness Data

The following data were collected in a study of shoulder harness usage in observations for a sample of North Carolina cars (Hochberg, Stutts, and Reinfurt 1977).

Table 6.3.   Shoulder Harness Data

Area	Location	Larger Cars No	Larger Cars Yes	Medium No	Medium Yes	Smaller Cars No	Smaller Cars Yes	Total
Coast	Urban	174	69	134	56	150	54	637
Coast	Rural	52	14	31	14	25	17	153
Piedmont	Urban	127	62	94	63	112	93	551
Piedmont	Rural	35	29	32	30	46	34	206
Mountains	Urban	111	26	120	47	145	68	517
Mountains	Rural	62	31	44	32	85	43	297

For these data, researchers were interested in whether there was an association between the size of car and shoulder harness usage, after controlling for geographic area and location. First, there is interest in looking at the pooled table of car size × usage. Then, a Mantel-Haenszel analysis is requested for a stratification consisting of the combinations of levels of area and location, resulting in six strata. Finally, Mantel-Haenszel analyses are requested for the association of size with usage stratified on area and location, singly. Standardized midrank scores are specified.

The following SAS statements request these analyses. Note that the NOPRINT option is specified in the last two TABLES statements to suppress the printing of tables.

```
data shoulder;
    input area $ location $ size $ usage $ count @@;
    datalines;
coast       urban large  no 174 coast       urban large  yes 69
coast       urban medium no 134 coast       urban medium yes 56
coast       urban small  no 150 coast       urban small  yes 54
coast       rural large  no  52 coast       rural large  yes 14
coast       rural medium no  31 coast       rural medium yes 14
coast       rural small  no  25 coast       rural small  yes 17
piedmont urban large  no 127 piedmont urban large  yes 62
piedmont urban medium no  94 piedmont urban medium yes 63
piedmont urban small  no 112 piedmont urban small  yes 93
piedmont rural large  no  35 piedmont rural large  yes 29
piedmont rural medium no  32 piedmont rural medium yes 30
piedmont rural small  no  46 piedmont rural small  yes 34
mountain urban large  no 111 mountain urban large  yes 26
mountain urban medium no 120 mountain urban medium yes 47
mountain urban small  no 145 mountain urban small  yes 68
mountain rural large  no  62 mountain rural large  yes 31
mountain rural medium no  44 mountain rural medium yes 32
mountain rural small  no  85 mountain rural small  yes 43
;
proc freq;
    weight count;
    tables size*usage / chisq;
    tables area*location*size*usage / cmh scores=modridit;
    tables area*size*usage / noprint cmh scores=modridit;
    tables location*size*usage / noprint cmh scores=modridit;
run;
```

Output 6.3 displays the pooled frequency table. The "Mantel-Haenszel Chi-Square," Q_{CS}, is valid for these data since SIZE is ordinally scaled, and the response is dichotomous; it indicates that there is a strong association between size of car and shoulder harness usage ($Q_{CS} = 7.2050$). By looking at the row percentages in the table cells, you can see that drivers of small and medium sized cars exhibit a greater tendency to use shoulder harnesses than the drivers of large cars.

Output 6.3   Pooled Table

```
                      Table of size by usage

           size        usage

           Frequency|
           Percent  |
           Row Pct  |
           Col Pct  |no       |yes      |  Total
           ---------+--------+--------+
           large    |    561 |    231 |    792
                    |  23.76 |   9.78 |  33.55
                    |  70.83 |  29.17 |
                    |  35.53 |  29.54 |
           ---------+--------+--------+
           medium   |    455 |    242 |    697
                    |  19.27 |  10.25 |  29.52
                    |  65.28 |  34.72 |
                    |  28.82 |  30.95 |
           ---------+--------+--------+
           small    |    563 |    309 |    872
                    |  23.85 |  13.09 |  36.93
                    |  64.56 |  35.44 |
                    |  35.66 |  39.51 |
           ---------+--------+--------+
           Total        1579      782     2361
                        66.88    33.12   100.00


           Statistics for Table of size by usage

     Statistic                      DF      Value      Prob
     ------------------------------------------------------
     Chi-Square                      2      8.5049     0.0142
     Likelihood Ratio Chi-Square     2      8.5980     0.0136
     Mantel-Haenszel Chi-Square      1      7.2050     0.0073
     Phi Coefficient                        0.0600
     Contingency Coefficient                0.0599
     Cramer's V                             0.0600

                  Sample Size = 2361
```

This association holds when you control for area and location. Output 6.4 contains the frequency table for rural locations in the coast region (the other tables are not reproduced here).

Output 6.4   Table for AREA=coast and LOCATION=rural

```
                    Table 1 of size by usage
              Controlling for area=coast location=rural

          size        usage

          Frequency|
          Percent  |
          Row Pct  |
          Col Pct  |no       |yes      |   Total
          ---------+--------+--------+
          large    |     52 |     14 |      66
                   |  33.99 |   9.15 |   43.14
                   |  78.79 |  21.21 |
                   |  48.15 |  31.11 |
          ---------+--------+--------+
          medium   |     31 |     14 |      45
                   |  20.26 |   9.15 |   29.41
                   |  68.89 |  31.11 |
                   |  28.70 |  31.11 |
          ---------+--------+--------+
          small    |     25 |     17 |      42
                   |  16.34 |  11.11 |   27.45
                   |  59.52 |  40.48 |
                   |  23.15 |  37.78 |
          ---------+--------+--------+
          Total          108       45        153
                       70.59    29.41     100.00
```

Output 6.5 displays the Mantel-Haenszel results for the stratified analysis where the strata are all combinations of area and location. $Q_{CSMH} = 6.6398$, which is strongly significant. Controlling for area and location, shoulder harness usage is clearly associated with size of car.

Output 6.5   Stratified by Area and Location

```
                Summary Statistics for size by usage
                  Controlling for area and location

      Cochran-Mantel-Haenszel Statistics (Modified Ridit Scores)

      Statistic    Alternative Hypothesis    DF      Value      Prob
      ------------------------------------------------------------------
          1        Nonzero Correlation        1      6.6398    0.0100
          2        Row Mean Scores Differ     2      8.4226    0.0148
          3        General Association        2      8.4258    0.0148


                     Total Sample Size = 2361
```

Output 6.6 and Output 6.7 contain the Mantel-Haenszel results for the association of size and shoulder harness usage controlling for area and location singly. $Q_{CSMH} = 6.5097$ and 7.0702, respectively. Controlling only for area or location, the significant association between shoulder harness and size of car remains evident. Q_{GMH} and Q_{SMH} are significant too, for the preceding analyses, but most of the information is contained in the correlation statistic Q_{CSMH}.

However, you should use caution in interpreting the mean score statistic for modified ridit scores when the outcome is a dichotomous response. Ordinarily, you would want the values 0 and 1 to be maintained in such an analysis; by using modified ridit scores you are effectively assigning different values from 0 and 1 to the columns, and these scores will be different in the different strata.

Also, the fact that Q_{GMH} and Q_{SMH} have very close or the same values in Output 6.5, Output 6.6, and Output 6.7 is an artifact. However, these statistics are identical for sets of $s \times 2$ tables when integer scores are used.

Output 6.6   Stratified By Area

```
              Summary Statistics for size by usage
                     Controlling for area

      Cochran-Mantel-Haenszel Statistics (Modified Ridit Scores)

      Statistic    Alternative Hypothesis    DF    Value    Prob
      ---------------------------------------------------------------
          1         Nonzero Correlation        1    6.5097   0.0107
          2         Row Mean Scores Differ      2    8.1203   0.0172
          3         General Association         2    8.1203   0.0172


                    Total Sample Size = 2361
```

Output 6.7   Stratified By Location

```
              Summary Statistics for size by usage
                   Controlling for location

      Cochran-Mantel-Haenszel Statistics (Modified Ridit Scores)

      Statistic    Alternative Hypothesis    DF    Value    Prob
      ---------------------------------------------------------------
          1         Nonzero Correlation        1    7.0702   0.0078
          2         Row Mean Scores Differ      2    8.5794   0.0137
          3         General Association         2    8.5789   0.0137


                    Total Sample Size = 2361
```

6.3.3   Learning Preference Data

In this study, educational researchers compared three different approaches to mathematics instruction for third graders. During the year, students were rotated through three different styles: a self-instructional mode that was largely based on computer use, a team approach in which students solved problems in groups of four students, and a traditional class approach. Researchers were interested in how other school programs influenced the effectiveness of the styles, as well as how they influenced the students' perceptions of the different styles. Table 6.4 displays data that reflect the students' preferences of styles, cross-classified by the school program they are in: Regular, which is a regular school schedule, and After, which supplements the regular school day with an afternoon school program involving the same classmates. The study included three different schools.

Table 6.4.   School Program Data

School	Program	Learning Style Preference		
		Self	Team	Class
1	Regular	10	17	26
1	After	5	12	50
2	Regular	21	17	26
2	After	16	12	36
3	Regular	15	15	16
3	After	12	12	20

The question of interest is whether students' learning style preference is associated with their school day program, after adjusting for any effects of individual school. There may be some ordinality to the response measure, in the sense of increasing group participation, but that doesn't stand up when you try to distinguish the team approach from the classroom approach. Thus, the appropriate extended Mantel-Haenszel statistic for the stratified analysis of these data is the test for general association. Since $(s-1)(r-1)$ for these data is equal to 2, Q_{GMH} has two degrees of freedom.

The following SAS statements request the appropriate analysis.

```
data school;
   input school program $ style $ count @@;
   datalines;
1 regular   self 10  1 regular   team 17 1 regular class   26
1 after     self 5   1 after     team 12 1 after   class   50
2 regular   self 21  2 regular   team 17 2 regular class   26
2 after     self 16  2 after     team 12 2 after   class   36
3 regular   self 15  3 regular   team 15 3 regular class   16
3 after     self 12  3 after     team 12 3 after   class   20
;
proc freq;
   weight count;
   tables school*program*style / cmh chisq measures;
run;
```

Output 6.8 contains the results for the stratified analysis. Q_{GMH} has a value of 10.9577, with 2 df, $p = 0.0042$. School program and learning style preference are strongly associated. Note that for these data, the general association statistic is most appropriate. The other statistics printed in this table are not applicable since the scale of the row and column variables of these tables do not justify their use. Note that since the ORDER=DATA option is not specified, the columns and rows of the tables are arranged alphabetically. This has no bearing on the general association statistic. However, if you wanted to order the rows and columns of the table as displayed in Table 6.4, then you would use ORDER=DATA.

Output 6.8 Stratified Analysis

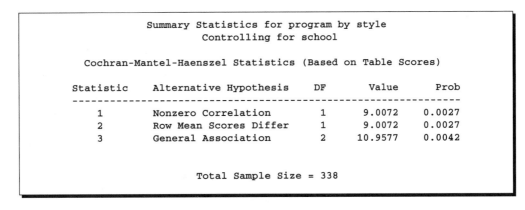

```
                 Summary Statistics for program by style
                         Controlling for school

          Cochran-Mantel-Haenszel Statistics (Based on Table Scores)

          Statistic    Alternative Hypothesis    DF      Value      Prob
          ---------------------------------------------------------------
              1         Nonzero Correlation        1     9.0072     0.0027
              2         Row Mean Scores Differ      1     9.0072     0.0027
              3         General Association         2    10.9577     0.0042


                          Total Sample Size = 338
```

Output 6.9, Output 6.10, and Output 6.11 contain the results for the individual tables. Note that most of the association seems to be occurring in School 1, judging by Q_P.

Output 6.9 Results for School 1

```
                      Table 1 of program by style
                         Controlling for school=1

               program       style

               Frequency|
               Percent  |
               Row Pct  |
               Col Pct  |class   |self    |team    |   Total
               ---------+--------+--------+--------+
               after    |     50 |      5 |     12 |      67
                        |  41.67 |   4.17 |  10.00 |   55.83
                        |  74.63 |   7.46 |  17.91 |
                        |  65.79 |  33.33 |  41.38 |
               ---------+--------+--------+--------+
               regular  |     26 |     10 |     17 |      53
                        |  21.67 |   8.33 |  14.17 |   44.17
                        |  49.06 |  18.87 |  32.08 |
                        |  34.21 |  66.67 |  58.62 |
               ---------+--------+--------+--------+
               Total          76       15       29      120
                            63.33    12.50    24.17   100.00




                   Statistics for Table 1 of program by style
                           Controlling for school=1

          Statistic                         DF      Value      Prob
          ----------------------------------------------------------
          Chi-Square                         2     8.5913     0.0136
          Likelihood Ratio Chi-Square        2     8.6385     0.0133
          Mantel-Haenszel Chi-Square         1     6.4209     0.0113
          Phi Coefficient                          0.2676
          Contingency Coefficient                  0.2585
          Cramer's V                               0.2676
```

Output 6.10   Results for School 2

```
                      Table 2 of program by style
                        Controlling for school=2

          program      style

          Frequency|
          Percent  |
          Row Pct  |
          Col Pct  |class  |self   |team   |  Total
          ---------+-------+-------+-------+
          after    |    36 |    16 |    12 |     64
                   | 28.13 | 12.50 |  9.38 |  50.00
                   | 56.25 | 25.00 | 18.75 |
                   | 58.06 | 43.24 | 41.38 |
          ---------+-------+-------+-------+
          regular  |    26 |    21 |    17 |     64
                   | 20.31 | 16.41 | 13.28 |  50.00
                   | 40.63 | 32.81 | 26.56 |
                   | 41.94 | 56.76 | 58.62 |
          ---------+-------+-------+-------+
          Total          62      37      29      128
                      48.44   28.91   22.66   100.00




               Statistics for Table 2 of program by style
                        Controlling for school=2

          Statistic                    DF      Value      Prob
          ------------------------------------------------------
          Chi-Square                    2     3.1506     0.2069
          Likelihood Ratio Chi-Square   2     3.1641     0.2056
          Mantel-Haenszel Chi-Square    1     2.7062     0.1000
          Phi Coefficient                      0.1569
          Contingency Coefficient              0.1550
          Cramer's V                           0.1569
```

Output 6.11   Results for School 3

```
                    Table 3 of program by style
                      Controlling for school=3

            program      style

            Frequency|
            Percent  |
            Row Pct  |
            Col Pct  |class   |self    |team    |  Total
            ---------+--------+--------+--------+
            after    |    20  |    12  |    12  |    44
                     | 22.22  | 13.33  | 13.33  | 48.89
                     | 45.45  | 27.27  | 27.27  |
                     | 55.56  | 44.44  | 44.44  |
            ---------+--------+--------+--------+
            regular  |    16  |    15  |    15  |    46
                     | 17.78  | 16.67  | 16.67  | 51.11
                     | 34.78  | 32.61  | 32.61  |
                     | 44.44  | 55.56  | 55.56  |
            ---------+--------+--------+--------+
            Total          36       27       27        90
                        40.00    30.00    30.00    100.00



                Statistics for Table 3 of program by style
                      Controlling for school=3

        Statistic                    DF       Value      Prob
        ------------------------------------------------------
        Chi-Square                    2      1.0672     0.5865
        Likelihood Ratio Chi-Square   2      1.0690     0.5860
        Mantel-Haenszel Chi-Square    1      0.8259     0.3635
        Phi Coefficient                      0.1089
        Contingency Coefficient              0.1083
        Cramer's V                           0.1089
```

6.4   Advanced Topic: Application to Repeated Measures

6.4.1   Introduction

The Mantel-Haenszel strategy has a useful application to the analysis of repeated measurements data. Such data occur when measurements are obtained over time; when responses from experimental units are measured under multiple conditions, such as multiple teeth in the same subject; and when multiple measurements are obtained from the same experimental unit, such as from two or more observers. Using repeated measurements enables comparisons among different times or conditions to avoid being obscured by subject-to-subject variability.

By specifying the appropriate tables for the data, you construct a setting in which Mantel-Haenszel methods can address the hypothesis of no association between a repeated measurement factor, such as time or condition, and a response variable, adjusting for the effect of subject. This type of analysis may be sufficient, or there may also be interest in statistical modeling of the repeated measurements data, which is discussed in Chapter 14, "Modeling Repeated Measurements Data with WLS," and Chapter 15, "Generalized Estimating Equations."

Consider the general situation in which t measurements of a univariate response variable Y are obtained from each of n experimental units. One common application is to longitudinal studies, in which repeated measurements are obtained at t time points for each subject. In other applications, the responses from each experimental unit are measured under multiple conditions rather than at multiple time points. In some settings in which repeated measures data are obtained, the independent experimental units are not individual subjects. For example, in a matched case-control study, the experimental units are matched sets and responses are obtained from the individual members of each set. In a toxicological study, the experimental units may be litters; responses are then obtained from the multiple newborns in each litter. In a genetic study, experimental units may be defined by families; responses are then obtained from the members of each family.

Although interest will focus primarily on the situation in which Y is categorical, the response may be either continuous or categorical. Let y_{ij} denote the response for subject i at time (or condition) j. The resulting data are commonly displayed in an $n \times t$ data matrix, as shown in Table 6.5.

Table 6.5.   One-Sample Repeated Measures Data

Subject	Time Point				
	1	...	j	...	t
1	y_{11}	...	y_{1j}	...	y_{1t}
$\vdots$	$\vdots$	$\ddots$	$\vdots$	$\ddots$	$\vdots$
i	y_{i1}	...	y_{ij}	...	y_{it}
$\vdots$	$\vdots$	$\ddots$	$\vdots$	$\ddots$	$\vdots$
n	y_{n1}	...	y_{nj}	...	y_{nt}

Alternatively, suppose c denotes the number of distinct values of Y and suppose indicator variables

$$n_{ijk} = \begin{cases} 1 & \text{if subject } i \text{ is classified in response category } k \text{ at time } j \\ 0 & \text{otherwise} \end{cases}$$

for $i = 1, \ldots, n$; $j = 1, \ldots, t$; and $k = 1, \ldots, c$. In this case, the data from subject i can be displayed in a $t \times c$ contingency table, as shown in Table 6.6. Thus, the data from a one-sample repeated measures study can be viewed as a set of n independent two-way contingency tables, where each table has t rows and c columns.

Table 6.6.   Contingency Table Layout for Subject i

Time Point	Response Category		Total
	1	...	c	
1	n_{i11}	...	n_{i1c}	n_{i1+}
$\vdots$	$\vdots$	$\ddots$	$\vdots$	$\vdots$
t	n_{it1}	...	n_{itc}	n_{it+}
Total	n_{i+1}	...	n_{i+c}	n_i

If the response variable Y is categorical with a limited number of possible values, the number of columns in each table, c, will be relatively small. On the other hand, if Y is a continuous variable, the number of distinct values of Y may be very large. The most extreme case results when each of the n subjects has a unique response at each time. In this situation, c is equal to nt and every column marginal total n_{i+k} is equal to zero or one.

When the data are complete, the total sample size for each of the n tables is $n_i = t$ and every row marginal total n_{ij+} is equal to 1. In this case, each row of Table 6.6 has exactly one n_{ijk} value equal to 1 and the remaining values are equal to 0. This situation occurs when the outcome variable is measured once at every time point for each subject.

However, if a particular subject has a missing response at one or more time points, the corresponding row of the subject's table will have each n_{ijk} value equal to 0 and the marginal total n_{ij+} will consequently equal 0. In this case, the total sample size n_i equals t minus the number of missing observations.

Based on the framework displayed in Table 6.6, Mantel-Haenszel statistics can be used to test the null hypothesis of no association between the row dimension (time) and the column dimension (response), adjusted for subject. Under the assumption that the marginal totals $\{n_{ij+}\}$ and $\{n_{i+k}\}$ of each table are fixed, the null hypothesis is that, for each subject, the response variable Y is distributed at random with respect to the t time points. As discussed in Landis et al. (1988), this null hypothesis is precisely the interchangeability hypothesis of Madansky (1963). Interchangeability states that all permutations of responses across conditions within a subject are equally likely. In turn, the hypothesis of interchangeability implies *marginal homogeneity* in the distribution of Y across the t time points; that is, the marginal distribution of Y is the same at each of the t time points.

Although the interchangeability hypothesis is a somewhat stronger condition than marginal homogeneity, the general association statistic Q_{GMH}, mean score statistic Q_{SMH}, and correlation statistic Q_{CSMH} are directed at alternatives that correspond to various types of departures from marginal homogeneity. The following examples demonstrate the use of MH statistics in testing marginal homogeneity for repeated measures.

6.4.2 Dichotomous Response: Two Time Points (McNemar's Test)

A running shoe company produces a new model of running shoe that includes a harder material for the insert that corrects for overpronation. However, the company is concerned that the material will induce heel tenderness as a result of some loss of cushioning on the strike of each step. It conducted a study on 87 runners who used the new shoe for a month. Researchers asked the participants whether they experienced occasional heel tenderness before and after they used the new shoe.

The data was collected as one observation per time period, that is, two measurements were collected for each subject, and they included the time period (before or after) and whether heel tenderness was experienced (yes or no). Table 6.7 contains the contingency table summarizing these data.

Table 6.7.   Heel Tenderness for Runners

Before	After No	Yes	Total
No	48	15	63
Yes	5	19	24

However, you can think of the measurements for each subject as being one of four 2×2 tables, corresponding to the four cells of Table 6.7. These tables are displayed in Table 6.8 through Table 6.11. Each subject's set of responses can be represented by one of these tables.

Table 6.8.   (No, No) Configuration Table (48)

Time	Heel Tenderness No	Yes	Total
Before	1	0	1
After	1	0	1

Table 6.9.   (No, Yes) Configuration Table (15)

Time	Heel Tenderness No	Yes	Total
Before	1	0	1
After	0	1	1

Table 6.10.   (Yes, No) Configuration Table (5)

Time	Heel Tenderness No	Yes	Total
Before	0	1	1
After	1	0	1

Table 6.11.   (Yes, Yes) Configuration Table (19)

Time	Heel Tenderness No	Yes	Total
Before	0	1	1
After	0	1	1

You can determine whether there is an association between the response and time for before and after responses by performing a stratified analysis where each subject constitutes a stratum. There are 87 tables altogether: 48 with the (no, no) configuration, 15 with the (no, yes) configuration, 5 with the (yes, no) configuration, and 19 with the (yes, yes) configuration.

If you study Table 6.7, you can see that these data effectively have the matched pairs framework that was discussed in Section 2.7. In fact, the Mantel-Haenszel statistic for the described analysis is equivalent to McNemar's test. The following analysis demonstrates the Mantel-Haenszel approach to analyzing these repeated measurements data. The same strategy is followed when the tables involved have dimensions greater than 2×2 and there is no alternative strategy such as McNemar's test.

The following SAS statements input the running shoes data. The data are in case record form: one observation per time point per subject. Thus, there are 174 observations altogether.

```
data pump;
   input subject time $ response $ @@;
   datalines;
 1 before no    1 after no    2 before no    2 after no
 3 before no    3 after no    4 before no    4 after no
 5 before no    5 after no    6 before no    6 after no
 7 before no    7 after no    8 before no    8 after no
 9 before no    9 after no   10 before no   10 after no
11 before no   11 after no   12 before no   12 after no
13 before no   13 after no   14 before no   14 after no
15 before no   15 after no   16 before no   16 after no
17 before no   17 after no   18 before no   18 after no
19 before no   19 after no   20 before no   20 after no
21 before no   21 after no   22 before no   22 after no
23 before no   23 after no   24 before no   24 after no
25 before no   25 after no   26 before no   26 after no
27 before no   27 after no   28 before no   28 after no
29 before no   29 after no   30 before no   30 after no
31 before no   31 after no   32 before no   32 after no
33 before no   33 after no   34 before no   34 after no
35 before no   35 after no   36 before no   36 after no
37 before no   37 after no   38 before no   38 after no
39 before no   39 after no   40 before no   40 after no
41 before no   41 after no   42 before no   42 after no
43 before no   43 after no   44 before no   44 after no
45 before no   45 after no   46 before no   46 after no
47 before no   47 after no   48 before no   48 after no
49 before no   49 after yes  50 before no   50 after yes
51 before no   51 after yes  52 before no   52 after yes
53 before no   53 after yes  54 before no   54 after yes
55 before no   55 after yes  56 before no   56 after yes
57 before no   57 after yes  58 before no   58 after yes
59 before no   59 after yes  60 before no   60 after yes
61 before no   61 after yes  62 before no   62 after yes
63 before no   63 after yes  64 before yes  64 after no
65 before yes  65 after no   66 before yes  66 after no
67 before yes  67 after no   68 before yes  68 after no
69 before yes  69 after yes  70 before yes  70 after yes
71 before yes  71 after yes  72 before yes  72 after yes
73 before yes  73 after yes  74 before yes  74 after yes
75 before yes  75 after yes  76 before yes  76 after yes
77 before yes  77 after yes  78 before yes  78 after yes
```

```
79 before yes 79 after yes 80 before yes 80 after yes
81 before yes 81 after yes 82 before yes 82 after yes
83 before yes 83 after yes 84 before yes 84 after yes
85 before yes 85 after yes 86 before yes 86 after yes
87 before yes 87 after yes
;
```

The next statements request the Mantel-Haenszel analysis. Since the data are in case record form, no WEIGHT statement is required. Since 87 tables are to be computed, the NOPRINT option is specified so that the tables are not printed.

```
proc freq;
    tables subject*time*response/ noprint cmh out=freqtab;
run;
```

Output 6.12 contains the Mantel-Haenszel results. Q_{MH} has the value 5.0000 with $p=0.0253$. This is clearly significant. Runners reported more heel tenderness with the new running shoes than with their old running shoes.

Output 6.12   Mantel-Haenszel Results

```
            Summary Statistics for time by response
                    Controlling for subject

    Cochran-Mantel-Haenszel Statistics (Based on Table Scores)

    Statistic    Alternative Hypothesis    DF    Value    Prob
    ----------------------------------------------------------------
        1        Nonzero Correlation        1    5.0000   0.0253
        2        Row Mean Scores Differ     1    5.0000   0.0253
        3        General Association        1    5.0000   0.0253


                 Total Sample Size = 174
```

Note that the CMH option always produces the "Estimates of Relative Risk" table and the "Breslow-Day Test for Homogeneity" for sets of 2×2 tables. However, Q_{BD} is not valid here.

Another way of obtaining these results for sets of 2×2 tables is to input the original 2×2 table and specify the AGREE option to obtain McNemar's test (available in Release 6.10 of the SAS System).

```
data shoes;
    input before $ after $ count ;
    datalines;
yes yes 19
yes no  5
no yes  15
no no   48
;
```

```
proc freq;
   weight count;
   tables before*after / agree;
run;
```

Output 6.13 contains the resulting frequency table and McNemar's test. $Q_M = 5.0000$, the same value as was obtained for Q_{MH}.

Output 6.13   Frequency Table and McNemar's Test

```
                    Table of before by after

            before      after

            Frequency|
            Percent  |
            Row Pct  |
            Col Pct  |no        |yes      |  Total
            ---------+--------+--------+
            no         |      48 |      15 |      63
                       |   55.17 |   17.24 |   72.41
                       |   76.19 |   23.81 |
                       |   90.57 |   44.12 |
            ---------+--------+--------+
            yes        |       5 |      19 |      24
                       |    5.75 |   21.84 |   27.59
                       |   20.83 |   79.17 |
                       |    9.43 |   55.88 |
            ---------+--------+--------+
            Total              53        34        87
                            60.92     39.08    100.00


            Statistics for Table of before by after

                    McNemar's Test
            ----------------------
            Statistic (S)       5.0000
            DF                       1
            Pr > S              0.0253

                Sample Size = 87
```

Recall that McNemar's test did not make use of the diagonal cells, that is, the (no, no) and (yes, yes) cells. Thus, if you repeated the Mantel-Haenszel analysis and eliminated the tables corresponding to the (no, no) and (yes, yes) configurations, you would obtain identical results.

6.4.3   Dichotomous Response: Three Repeated Measurements

Grizzle, Starmer, and Koch (1969) analyze data in which 46 patients were each treated with three drugs (A, B, and C). The response to each drug was recorded as favorable (F) or unfavorable (U). Table 6.12 summarizes the eight possible combinations of favorable or unfavorable response for the three drugs and the number of patients with each response pattern.

Table 6.12.   Drug Response Data

Drug			Frequency
A	B	C	
F	F	F	6
F	F	U	16
F	U	F	2
F	U	U	4
U	F	F	2
U	F	U	4
U	U	F	6
U	U	U	6

The objective of the analysis is to determine whether the three drugs have similar probabilities for favorable response. Thus, the null hypothesis is interchangeability (that is, no association between drug and response for each patient), which implies equality of the marginal probabilities of a favorable response for the three drugs across patients. This hypothesis can be tested using the general association statistic Q_{GMH}. The data in Table 6.12 must first be restructured so that there are forty-six 3×2 contingency tables, one for each of the 46 patients. For example, Table 6.13 shows the underlying table for a patient who responded favorably to drugs A and C and unfavorably to drug B.

Table 6.13.   Sample Contingency Table for a Single Patient

Drug	Response		Total
	F	U	
A	1	0	1
B	0	1	1
C	1	0	1
Total	2	1	3

To apply the Mantel-Haenszel strategy to this data, you have to create a SAS data set that contains $46 \times 3 = 138$ observations (one observation per measurement) and three variables representing patient, drug, and measurement, respectively. If the data are supplied in frequency count form, they must be rearranged. The following SAS statements read the data in frequency form, as displayed in Table 6.12, and rearrange them into the form displayed in Table 6.13. Thus, three observations are created for each patient, one for each drug. Each of the observations in data set DRUG2 contains an arbitrary patient identifier (numbered from 1 to 46), the drug code (A, B, or C), and the response (F or U).

Finally, the FREQ procedure computes the MH statistics that assess the association of drug and response, adjusting for patient. The NOPRINT option of the TABLES statement suppresses the printing of the 46 individual contingency tables. You almost always use this option when analyzing repeated measures data using MH methods.

```
data drug;
   input druga $ drugb $ drugc $ count;
   datalines;
F F F  6
F F U 16
F U F  2
F U U  4
U F F  2
U F U  4
U U F  6
U U U  6
;
data drug2; set drug;
   keep patient drug response;
   retain patient 0;
   do i=1 to count;
   patient=patient+1;
   drug='A';  response=druga;  output;
   drug='B';  response=drugb;  output;
   drug='C';  response=drugc;  output;
   end;
proc freq;
   tables patient*drug*response / noprint cmh;
run;
```

Output 6.14 displays the results from PROC FREQ. Since the response is dichotomous, the general association and mean score statistics both have 2 df. With table scores, their values are identical. Since the repeated measures factor (drug) is not ordered, the correlation statistic does not apply.

Output 6.14 Test of Marginal Homogeneity

```
                    Summary Statistics for drug by response
                            Controlling for patient

        Cochran-Mantel-Haenszel Statistics (Based on Table Scores)

        Statistic    Alternative Hypothesis    DF      Value      Prob
        ----------------------------------------------------------------
            1         Nonzero Correlation        1      6.3529    0.0117
            2         Row Mean Scores Differ      2      8.4706    0.0145
            3         General Association         2      8.4706    0.0145


                        Total Sample Size = 138
```

The value of Q_{GMH} is 8.4706. With reference to the approximate chi-square distribution with 2 df, there is a clearly significant association between drug and response. This test is the same as Cochran's Q statistic (Cochran 1950). In order to summarize the nature of the association, it is helpful to report the estimated marginal probabilities of a favorable response for drugs A, B, and C. These can be computed from Table 6.12 and are equal to 28/46 = 0.61, 28/46 = 0.61, and 16/46 = 0.35, respectively. It is evident that the marginal proportion for drug C differs considerably from that of drugs A and B. Drugs A and B have a much greater probability of favorable response than Drug C.

6.4.4   Ordinal Response

The same Mantel-Haenszel strategy is appropriate when the repeated measurements response variable is ordinally scaled. In this case, the statistic of interest is Q_{SMH}, the mean score statistic.

Macknin, Mathew, and Medendorp (1990) studied the efficacy of steam inhalation in the treatment of common cold symptoms. Thirty patients with colds of recent onset (symptoms of nasal drainage, nasal congestion, and sneezing for three days or less) received two 20-minute steam inhalation treatments. On four successive days, these patients self-assessed the severity of nasal drainage on a four-point ordinal scale (0=no symptoms, 1=mild symptoms, 2=moderate symptoms, 3=severe symptoms). Table 6.14 displays the resulting data.

Table 6.14.   Nasal Drainage Data

Patient ID	Study Day 1	Study Day 2	Study Day 3	Study Day 4	Patient ID	Study Day 1	Study Day 2	Study Day 3	Study Day 4
1	1	1	2	2	16	2	1	1	1
2	0	0	0	0	17	1	1	1	1
3	1	1	1	1	18	2	2	2	2
4	1	1	1	1	19	3	1	1	1
5	0	2	2	0	20	1	1	2	1
6	2	0	0	0	21	2	1	1	2
7	2	2	1	2	22	2	2	2	2
8	1	1	1	0	23	1	1	1	1
9	3	2	1	1	24	2	2	3	1
10	2	2	2	3	25	2	0	0	0
11	1	0	1	1	26	1	1	1	1
12	2	3	2	2	27	0	1	1	0
13	1	3	2	1	28	1	1	1	1
14	2	1	1	1	29	1	1	1	0
15	2	3	3	3	30	3	3	3	3

The objective of the study was to determine if nasal drainage becomes less severe following steam inhalation treatment. Thus, the relevant null hypothesis is that the distribution of the symptom severity scores is the same on each of the four study days for each patient. Since there are only four possible values of the response variable, the assumptions for the usual parametric methods are not directly applicable. In addition, the sample size is too small to justify analysis of the full 4^4 contingency table obtained by the joint cross-classification of the four-level response variable on four days. Thus, randomization model MH methods seem appropriate.

Although the general association statistic Q_{GMH} may be considered for this example, its use of 9 df would have low power to detect departures from marginal homogeneity in a sample of only 30 patients. Since the response is ordinal, the mean score statistic Q_{SMH}, with 3 df, can be used to compare the average symptom scores across the four days. The adequacy of the sample size to support the use of this statistic may also be questionable. Alternatively, since the repeated measures factor (study day) is also ordinal, you could test for a linear trend over study day for symptom severity using the correlation statistic Q_{CSMH}.

Both Q_{SMH} and Q_{CSMH} require that scores be assigned to the values of the repeated measures and response variables. Since study day is quantitative, it is natural to use the scores 1–4 for this variable. If it is reasonable to assume that the symptom severity ratings are equally spaced, the actual scores 0–3 can be used. You could also assign scores that incorporate unequal spacing between the four levels of symptom severity.

Another possibility is to use rank scores for the symptom severity ratings. In PROC FREQ, the SCORES=RANK option of the TABLES statement uses rank scores for both the row and column variables. However, since each patient contributes exactly one observation on each of the four days, the rank scores for study day are also 1, 2, 3, and 4. Thus, this option only affects the scoring of the symptom severity levels. The SCORES=RIDIT and SCORES=MODRIDIT options compute rank scores that are standardized by a function of the stratum-specific sample size. Since the sample sizes in the 30 underlying 4×4 contingency tables are all equal to 4, the results from the SCORES=RANK, SCORES=RIDIT, and SCORES=MODRIDIT options would be identical.

The following SAS statements read in the data in case record form with responses for all days on the same record and rearrange it so that there are four observations per patient. PROC FREQ computes the MH statistics, first using equally spaced table scores and then using rank scores.

```
data cold;
    keep id day drainage;
    input id day1-day4;
    day=1; drainage=day1; output;
    day=2; drainage=day2; output;
    day=3; drainage=day3; output;
    day=4; drainage=day4; output;
    datalines;
 1 1 1 2 2
 2 0 0 0 0
 3 1 1 1 1
 4 1 1 1 1
 5 0 2 2 0
 6 2 0 0 0
 7 2 2 1 2
 8 1 1 1 0
 9 3 2 1 1
10 2 2 2 3
11 1 0 1 1
12 2 3 2 2
13 1 3 2 1
14 2 1 1 1
15 2 3 3 3
16 2 1 1 1
17 1 1 1 1
18 2 2 2 2
19 3 1 1 1
20 1 1 2 1
```

```
21 2 1 1 2
22 2 2 2 2
23 1 1 1 1
24 2 2 3 1
25 2 0 0 0
26 1 1 1 1
27 0 1 1 0
28 1 1 1 1
29 1 1 1 0
30 3 3 3 3
;
proc freq;
   tables id*day*drainage / cmh noprint;
   tables id*day*drainage / cmh noprint scores=rank;
run;
```

Output 6.15 displays the MH statistics based on table scores, and Output 6.16 displays the corresponding results using rank scores. Using the default table scores, the test statistic that the mean symptom severity scores are the same at all four days is not statistically significant ($Q_{SMH} = 4.9355$, $p = 0.1766$). However, there is a statistically significant trend between study day and nasal drainage severity (Q_{CSMH}=4.3548, $p = 0.0369$). The observed mean scores at days 1–4 are 1.50, 1.37, 1.37, and 1.17; thus, symptom severity is decreasing over time.

Output 6.15   MH Tests Using Table Scores

```
              Summary Statistics for day by drainage
                         Controlling for id

      Cochran-Mantel-Haenszel Statistics (Based on Table Scores)

    Statistic    Alternative Hypothesis    DF      Value      Prob
    ---------------------------------------------------------------
        1        Nonzero Correlation        1      4.3548     0.0369
        2        Row Mean Scores Differ     3      4.9355     0.1766
        3        General Association        9     10.1267     0.3403


                    Total Sample Size = 120
```

Output 6.16   MH Tests Using Rank Scores

```
              Summary Statistics for day by drainage
                         Controlling for id

      Cochran-Mantel-Haenszel Statistics (Based on Rank Scores)

    Statistic    Alternative Hypothesis    DF      Value      Prob
    ---------------------------------------------------------------
        1        Nonzero Correlation        1      2.6825     0.1015
        2        Row Mean Scores Differ     3      3.3504     0.3407
        3        General Association        9     10.1267     0.3403


                    Total Sample Size = 120
```

In this example, the use of rank scores leads to a less clear conclusion regarding the statistical significance of the correlation statistic ($Q_{CSMH} = 2.6825$, $p = 0.1015$). Some authors recommend the routine use of rank scores in preference to the arbitrary assignment of scores (for example, Fleiss 1986, pp. 83–84). However, as demonstrated by Graubard and Korn (1987), rank scores can be a poor choice when the column margin is far from uniformly distributed. This occurs because rank scores also assign a spacing between the levels of the categories. This spacing is generally not known by the analyst and may not be as powerful as other spacings for certain patterns of differences among distributions. Graubard and Korn (1987) recommend that you specify the scores whenever possible. If the choice of scores is not apparent, they recommend integer (or equally spaced) scores.

When there is no natural set of scores, Agresti (1990, p. 294) recommends that the data be analyzed using several reasonably assigned sets of scores to determine whether substantive conclusions depend on the choice of scores. This type of sensitivity analysis seems especially appropriate in this example, since the results assuming equally spaced scores differ from those obtained using rank scores. For example, the scores 0, 1, 3, 5 assume that the moderate category is equally spaced between the mild and severe categories, while none and mild are less far apart. Another possibility would be 0, 1, 2, 4; this choice places severe symptoms further from the other three categories. These alternative scoring specifications are easily implemented by redefining the values of the DRAINAGE variable in the DATA step and then using the default table scores, which are just the input numeric values for drainage.

Note that since the general association statistic does not use scores, the value of Q_{GMH} is the same in both analyses.

6.4.5 Ordinal Response with Missing Data

Researchers at the C. S. Mott Children's Hospital in Ann Arbor, Michigan, investigated the effect of pulse duration on the development of acute electrical injury during transesophageal atrial pacing in animals. In brief, this procedure involves placing a pacemaker in the esophagus. Each of the 14 animals available for experimentation then received atrial pacing at pulse durations of 2, 4, 6, 8, and 10 milliseconds (ms), with each pulse delivered at a separate site in the esophagus for 30 minutes. The response variable, lesion severity, was classified according to depth of injury by histologic examination using an ordinal staging scale from 0 to 5 (0=no lesion, 5=acute inflammation of extraesophageal fascia). Table 6.15 displays the resulting data (missing observations are denoted by –). Landis et al. (1988) previously analyzed the data from the first 11 animals.

Table 6.15. Lesion Severity Data

ID	\multicolumn{5}{c}{Pulse Duration (ms)}				
	2	4	6	8	10
6	0	0	5	0	3
7	0	3	3	4	5
8	0	3	4	3	2
9	2	2	3	0	4
10	0	0	4	4	3
12	0	0	0	4	4
13	0	4	4	4	0
15	0	4	0	0	0
16	0	3	0	1	1
17	–	–	0	1	0
19	0	0	1	1	0
20	–	0	0	2	2
21	0	0	2	3	3
22	–	0	0	3	0

The investigators were primarily interested in determining the extent to which increasing the pulse duration from 2 to 10 ms tends to increase the severity of the lesion. In an experiment in which five repeated measurements of a six-category ordinal response are obtained from only 14 experimental units, the choice of statistical methodology is limited. The study is further complicated by the fact that 3 of the 14 animals have incomplete data.

The general association statistic Q_{GMH} has 20 df in this case ($s = 5, r = 6$). In addition to the fact that Q_{GMH} will not have a chi-square distribution when the sample size is so small relative to the degrees of freedom, there will be very low power to detect general departures from the null hypothesis of interchangeability. Although the alternative of location shift for mean responses across the five pulse durations can be addressed using the mean score statistic Q_{SMH}, this statistic does not take into account the ordering of the pulse durations. The 1 df correlation statistic Q_{CSMH} specifically focuses on the narrow alternative of a monotone relationship between lesion severity and pulse duration. This test addresses the objective of the investigators and is also best justified given the small sample size.

The following SAS statements read in the data in the format shown in Table 6.15, rearrange them so that each subject has five observations, one for each pulse duration, and request the MH statistics using table scores and all three types of rank scores. Since Q_{GMH} is unaffected by the choice of scores, the CMH2 option is used in all but the first TABLES statement. This option specifies that only the Mantel-Haenszel statistics Q_{SMH} and Q_{CSMH} be computed. (The CMH1 option specifies that only the correlation statistic Q_{CSMH} be computed.)

```
data animals;
   keep id pulse severity;
   input id sev2 sev4 sev6 sev8 sev10;
   pulse=2;  severity=sev2;  output;
   pulse=4;  severity=sev4;  output;
   pulse=6;  severity=sev6;  output;
   pulse=8;  severity=sev8;  output;
   pulse=10; severity=sev10; output;
   datalines;
 6 0 0 5 0 3
 7 0 3 3 4 5
 8 0 3 4 3 2
 9 2 2 3 0 4
10 0 0 4 4 3
12 0 0 0 4 4
13 0 4 4 4 0
15 0 4 0 0 0
16 0 3 0 1 1
17 . . 0 1 0
19 0 0 1 1 0
20 . 0 0 2 2
21 0 0 2 3 3
22 . 0 0 3 0
;
proc freq;
   tables id*pulse*severity / noprint cmh;
   tables id*pulse*severity / noprint cmh2 scores=rank;
   tables id*pulse*severity / noprint cmh2 scores=ridit;
   tables id*pulse*severity / noprint cmh2 scores=modridit;
run;
```

Output 6.17 displays the results using the default table scores. In this case, lesion severity is scored using the integers $0, \ldots, 5$ in computing both the mean score statistic Q_{SMH} and the correlation statistic Q_{CSMH}. In addition, pulse duration is scored as 2, 4, 6, 8, or 10 in computing Q_{CSMH}. The correlation statistic Q_{CSMH} shows a highly significant monotone association (trend) between pulse duration and lesion severity; the results from the mean score statistic are also statistically significant.

Output 6.17   MH Tests Using Table Scores

```
           Summary Statistics for pulse by severity
                     Controlling for id

     Cochran-Mantel-Haenszel Statistics (Based on Table Scores)

     Statistic   Alternative Hypothesis   DF    Value    Prob
     --------------------------------------------------------------
         1       Nonzero Correlation       1    8.8042   0.0030
         2       Row Mean Scores Differ     4   12.3474   0.0149
         3       General Association       20   22.8460   0.2964


                  Effective Sample Size = 66
                  Frequency Missing = 4
```

Output 6.18, Output 6.19, and Output 6.20 display the corresponding results using rank, ridit, and modified ridit scores, respectively. In this example, the values of the mean score and correlation statistics differ slightly among the three types of rank statistics. This is due to the fact that the sample sizes are no longer the same across the 14 tables (due to the occurrence of missing data).

Output 6.18 MH Tests Using Rank Scores

```
                 Summary Statistics for pulse by severity
                            Controlling for id

        Cochran-Mantel-Haenszel Statistics (Based on Rank Scores)

    Statistic    Alternative Hypothesis     DF      Value      Prob
    ---------------------------------------------------------------
        1        Nonzero Correlation         1      9.9765    0.0016
        2        Row Mean Scores Differ      4     13.6796    0.0084


                     Effective Sample Size = 66
                     Frequency Missing = 4
```

Output 6.19 MH Tests Using Ridit Scores

```
                 Summary Statistics for pulse by severity
                            Controlling for id

        Cochran-Mantel-Haenszel Statistics (Based on Ridit Scores)

    Statistic    Alternative Hypothesis     DF      Value      Prob
    ---------------------------------------------------------------
        1        Nonzero Correlation         1     10.0335    0.0015
        2        Row Mean Scores Differ      4     14.2628    0.0065


                     Effective Sample Size = 66
                     Frequency Missing = 4
```

Output 6.20 MH Tests Using Modified Ridit Scores

```
                 Summary Statistics for pulse by severity
                            Controlling for id

        Cochran-Mantel-Haenszel Statistics (Modified Ridit Scores)

    Statistic    Alternative Hypothesis     DF      Value      Prob
    ---------------------------------------------------------------
        1        Nonzero Correlation         1     10.1102    0.0015
        2        Row Mean Scores Differ      4     14.1328    0.0069


                     Effective Sample Size = 66
                     Frequency Missing = 4
```

As shown in Table 6.15, 3 of the 14 animals had incomplete data. Table 6.16 through Table 6.18 display the underlying contingency tables for these strata (ID numbers 17, 20, and 22). Although each of these three tables has one or more rows with a marginal total of zero, the remaining rows provide useful information concerning the association between pulse duration and lesion severity.

Table 6.16. Contingency Table for ID 17

Pulse Duration	Lesion Severity						Total
	0	1	2	3	4	5	
2	0	0	0	0	0	0	0
4	0	0	0	0	0	0	0
6	1	0	0	0	0	0	1
8	0	1	0	0	0	0	1
10	1	0	0	0	0	0	1
Total	2	1	0	0	0	0	3

Table 6.17. Contingency Table for ID 20

Pulse Duration	Lesion Severity						Total
	0	1	2	3	4	5	
2	0	0	0	0	0	0	0
4	1	0	0	0	0	0	1
6	1	0	0	0	0	0	1
8	0	0	1	0	0	0	1
10	0	0	1	0	0	0	1
Total	2	0	2	0	0	0	4

Table 6.18. Contingency Table for ID 22

Pulse Duration	Lesion Severity						Total
	0	1	2	3	4	5	
2	0	0	0	0	0	0	0
4	1	0	0	0	0	0	1
6	1	0	0	0	0	0	1
8	0	0	0	1	0	0	1
10	1	0	0	0	0	0	1
Total	3	0	0	1	0	0	4

The following statements exclude these three animals from the analysis and compute the test statistics for the subset of complete cases. In this case, all three types of rank scores produce the same results; thus, only the SCORES=RANK option is used. The WHERE clause is used to delete those observations with ID equal to 17, 20, or 22.

```
proc freq data=animals;
   where id notin(17,20,22);
   tables id*pulse*severity / noprint cmh;
   tables id*pulse*severity / noprint cmh scores=rank;
run;
```

Output 6.21 and Output 6.22 display the results from the analysis of complete cases. The value of each of the test statistics is somewhat smaller than the corresponding value computed using all available data. Thus, the partial data from the incomplete cases strengthen the evidence in favor of the existence of a significant trend between pulse duration and lesion severity.

Output 6.21   MH Tests Using Table Scores: Complete Cases Only

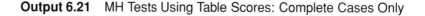

```
                    Summary Statistics for pulse by severity
                             Controlling for id

          Cochran-Mantel-Haenszel Statistics (Based on Table Scores)

          Statistic    Alternative Hypothesis    DF      Value      Prob
          ---------------------------------------------------------------
              1         Nonzero Correlation        1      7.5610    0.0060
              2         Row Mean Scores Differ      4     11.5930    0.0206
              3         General Association        20     21.2489    0.3826


                          Total Sample Size = 55
```

Output 6.22   MH Tests Using Rank Scores: Complete Cases Only

```
                    Summary Statistics for pulse by severity
                             Controlling for id

          Cochran-Mantel-Haenszel Statistics (Based on Rank Scores)

          Statistic    Alternative Hypothesis    DF      Value      Prob
          ---------------------------------------------------------------
              1         Nonzero Correlation        1      8.5099    0.0035
              2         Row Mean Scores Differ      4     12.2637    0.0155
              3         General Association        20     21.2489    0.3826


                          Total Sample Size = 55
```

6.4.6   Continuous Response

Table 6.19 displays artificial data collected for the purpose of determining if pH level alters action potential characteristics following administration of a drug (Harrell 1989). The response variable of interest (Vmax) was measured at up to four pH levels for each of 25 patients. While at least two measurements were obtained from each patient, only three patients provided data at all four pH levels.

Table 6.19. Action Potential Data

Patient	pH Level 6.5	6.9	7.4	7.9	Patient	pH Level 6.5	6.9	7.4	7.9
1		284	310	326	14	204	234	268	
2			261	292	15			258	267
3		213	224	240	16		193	224	235
4		222	235	247	17	185	222	252	263
5			270	286	18		238	301	300
6			210	218	19		198	240	
7		216	234	237	20		235	255	
8		236	273	283	21		216	238	
9	220	249	270	281	22		197	212	219
10	166	218	244		23		234	238	
11	227	258	282	286	24			295	281
12	216		284		25			261	272
13			257	284					

Although the response is a continuous measurement, MH statistics can still be used to determine if the average Vmax differs among the four pH values (Q_{SMH}) and if there is a trend between Vmax and pH (Q_{CSMH}). This approach offers the advantage of not requiring any assumptions concerning the distribution of Vmax. In addition, the MH methodology accommodates the varying numbers of observations per patient (under the assumption that missing values are missing completely at random and the test statistic is specified with either table scores or ranks).

The following SAS statements read in the data in the format shown in Table 6.19, restructure the data set for PROC FREQ, and compute the MH mean score and correlation statistics. The CMH2 option is used since it is not possible (or sensible) to compute the general association statistic Q_{GMH}. Since both pH and Vmax are quantitative variables, the default table scores are used. In addition, the trend is also assessed using modified ridit scores.

```
data ph_vmax;
   keep subject ph vmax;
   input subject vmax1-vmax4;
   ph=6.5;  vmax=vmax1;  output;
   ph=6.9;  vmax=vmax2;  output;
   ph=7.4;  vmax=vmax3;  output;
   ph=7.9;  vmax=vmax4;  output;
   datalines;
1  .  284 310 326
2  .   .  261 292
3  .  213 224 240
4  .  222 235 247
5  .   .  270 286
6  .   .  210 218
7  .  216 234 237
8  .  236 273 283
9 220 249 270 281
```

```
10 166 218 244   .
11 227 258 282 286
12 216   .  284   .
13   .   .  257 284
14 204 234 268   .
15   .   .  258 267
16   .  193 224 235
17 185 222 252 263
18   .  238 301 300
19   .  198 240   .
20   .  235 255   .
21   .  216 238   .
22   .  197 212 219
23   .  234 238   .
24   .   .  295 281
25   .   .  261 272
;
proc freq;
   tables subject*ph*vmax / noprint cmh2;
   tables subject*ph*vmax / noprint cmh2 scores=modridit;
run;
```

Output 6.23 shows that the mean Vmax differs significantly among the four pH levels
($Q_{SMH} = 27.7431$, 3 df, $p < 0.0001$). In addition, there is a highly significant linear trend
between pH and Vmax ($Q_{CSMH} = 27.3891$, 1 df, $p < 0.0001$).

Output 6.23   MH Mean Score and Correlation Tests: Table Scores

```
                 Summary Statistics for ph by vmax
                      Controlling for subject

        Cochran-Mantel-Haenszel Statistics (Based on Table Scores)

     Statistic    Alternative Hypothesis    DF      Value      Prob
     ---------------------------------------------------------------
         1        Nonzero Correlation        1     27.3891    <.0001
         2        Row Mean Scores Differ      3     27.7431    <.0001


                    Effective Sample Size = 66
                      Frequency Missing = 34

                 WARNING: 34% of the data are missing.
```

The mean score and correlation statistics are even more significant when modified ridit
scores are used (Output 6.24). Note that Vmax tends to progressively increase with pH for
almost all patients (patients 18 and 24 are the exception).

Output 6.24 MH Mean Score and Correlation Tests: Modified Ridit Scores

```
                    Summary Statistics for ph by vmax
                         Controlling for subject

          Cochran-Mantel-Haenszel Statistics (Modified Ridit Scores)

          Statistic     Alternative Hypothesis     DF     Value     Prob
          -------------------------------------------------------------
              1          Nonzero Correlation         1    35.3818   <.0001
              2          Row Mean Scores Differ       3    34.7945   <.0001


                         Effective Sample Size = 66
                           Frequency Missing = 34

                    WARNING: 34% of the data are missing.
```

In this example, the column variable of each table was continuous and the row variable, although quantitative, had only four possible values. Thus, both Q_{SMH} and Q_{CSMH} could be used. The MH approach to the analysis of one-sample repeated measures can also be very useful when the row and column variables are both continuous. In this case, only Q_{CSMH} can be used. This can be specified by using the CMH1 option in the TABLES statement.

The methodology is also applicable when there are multiple groups (samples). However, the observations are viewed as a single group when comparing conditions within subjects.

Chapter 7

Nonparametric Methods

Chapter Table of Contents

Chapter 7
Nonparametric Methods

7.1 Introduction

Parametric methods of statistical inference require you to assume that your data come from some underlying distribution whose general form is known, such as the normal, binomial, Poisson, or Weibull distribution. Statistical methods for estimation and hypothesis testing are then based on these assumptions. The focus is on estimating parameters and testing hypotheses about them.

In contrast, nonparametric statistical methods make few assumptions about the underlying distribution from which the data are sampled. One of their main advantages is that inference is not focused on specific population parameters, and it is thus possible to test hypotheses that are more general than statements about parameters. For example, they allow you to test whether two distributions are the same without having to test hypotheses concerning population parameters. Nonparametric procedures can also be used when the underlying distribution is unknown or when parametric assumptions are not valid.

The main disadvantage is that a nonparametric test is generally less powerful than the corresponding parametric test when the assumptions are satisfied. However, for many of the commonly used nonparametric methods, the decrease in power is not large.

Most of this book concentrates on the analysis of categorical response variables measured on nominal or ordinal scales. This chapter focuses on the analysis of continuous response variables with the use of nonparametric statistical methods. The reason for considering these methods is that many of the commonly used nonparametric tests, such as the Wilcoxon-Mann-Whitney, Kruskal-Wallis, Spearman correlation, Friedman, and Durbin tests, can be computed using Mantel-Haenszel procedures. While previous chapters have shown how to use Mantel-Haenszel procedures to analyze two-way tables and sets of two-way tables, this chapter shows how to use the same procedures to perform nonparametric analyses of continuous response variables.

7.2 Wilcoxon-Mann-Whitney Test

The Wilcoxon-Mann-Whitney test is a nonparametric test of the null hypothesis that the distribution of an ordinally scaled response variable is the same in two independently sampled populations. It is sensitive to the alternative hypothesis that there is a location difference between the two populations. This test was first proposed for the case of two samples of equal size by Wilcoxon (1945). Mann and Whitney (1947) introduced an equivalent statistic, and they were the first to consider unequal sample sizes and to furnish tables suitable for use with small samples. (Such tables are found in the appendices of standard statistics textbooks.)

The Wilcoxon-Mann-Whitney test can be used whenever the two-sample t-test is appropriate. One approach to comparing the nonparametric Wilcoxon-Mann-Whitney test to the parametric two-sample t-test is based on the concept of asymptotic relative efficiency, as developed by Pitman (1948) in a series of unpublished lecture notes. In brief, the efficiency of a test T_2 relative to a test T_1 is the ratio n_1/n_2 of the sample sizes needed to obtain the same power for the two tests. For normal distributions with a shift in the mean, the asymptotic efficiency of the Wilcoxon-Mann-Whitney test relative to the two-sample t-test is 0.955. Thus, a small price is paid for using the nonparametric test, in return for greater applicability. If the underlying populations are not normally distributed (for example, they have asymmetric distributions), the power of the Wilcoxon-Mann-Whitney test can be much higher than that of the two-sample t-test. In fact, the asymptotic relative efficiency can be as high as infinity.

When the sample sizes in the two groups are small, tables of the exact distribution of the test statistic should be used. Alternatively, you can carry out exact tests of significance for small sample sizes by using the EXACT statement in the NPAR1WAY procedure. If there are at least 10 observations per group, the p-value can be approximated using the asymptotic normal distribution of the test criteria. The asymptotic test is simply the Mantel-Haenszel mean score statistic for the special case of one stratum when rank scores are used.

Table 7.1 displays data from a study of the relationship between sodium chloride preference and hypertension (Schechter, Horwitz, and Henkin 1973). Two groups of subjects, 12 normal and 10 hypertensive, were isolated for a week and compared with respect to their average daily Na^+ intakes.

Table 7.1.   Sodium Chloride Preference Data

Normal Group		Hypertensive Group	
Subject	NA^+	Subject	NA^+
1	10.2	1	92.8
2	2.2	2	54.8
3	0.0	3	51.6
4	2.6	4	61.7
5	0.0	5	250.8
6	43.1	6	84.5
7	45.8	7	34.7
8	63.6	8	62.2
9	1.8	9	11.0
10	0.0	10	39.1
11	3.7		
12	0.0		

The following statements create a SAS data set containing the data of Table 7.1 and compare the two groups using the Wilcoxon-Mann-Whitney test. The statistic is computed using the Mantel-Haenszel mean score test based on rank scores.

```
data sodium;
   input group $ subject intake;
   datalines;
Normal    1  10.2
Normal    2   2.2
Normal    3   0.0
Normal    4   2.6
Normal    5   0.0
Normal    6  43.1
Normal    7  45.8
Normal    8  63.6
Normal    9   1.8
Normal   10   0.0
Normal   11   3.7
Normal   12   0.0
Hyperten  1  92.8
Hyperten  2  54.8
Hyperten  3  51.6
Hyperten  4  61.7
Hyperten  5 250.8
Hyperten  6  84.5
Hyperten  7  34.7
Hyperten  8  62.2
Hyperten  9  11.0
Hyperten 10  39.1
;
proc freq;
   tables group*intake / noprint cmh2 scores=rank;
run;
```

The CMH2 option in the TABLES statement specifies that only the Mantel-Haenszel correlation and mean score statistics are to be computed, since the general association statistic is not useful in this example (for example, its use is for strictly categorical data). As shown in Output 7.1, the mean score statistic indicates that there is a significant difference between normal and hypertensive subjects (chi-square=9.6589, 1 df, $p = 0.0019$). Since there are only two groups, the correlation and mean score statistics are identical.

Output 7.1 Wilcoxon-Mann-Whitney Test Using the MH Mean Score Statistic

```
           Summary Statistics for group by intake

      Cochran-Mantel-Haenszel Statistics (Based on Rank Scores)

   Statistic   Alternative Hypothesis    DF     Value     Prob
   ------------------------------------------------------------
       1        Nonzero Correlation       1     9.6589    0.0019
       2        Row Mean Scores Differ     1     9.6589    0.0019


              Total Sample Size = 22
```

The Wilcoxon-Mann-Whitney test is also equivalent to the extended Mantel-Haenszel correlation statistic in the tests of no association for a two-way contingency table (with

$s \times 2$ or $2 \times r$ structure), provided that rank scores are specified. The following SAS statements produce the results shown in Output 7.2.

```
proc freq;
    tables group*intake / noprint chisq scores=rank;
run;
```

The chi-square test statistic labeled "MH Chi-Square (Rank Scores)" is also equal to 9.6589. Note that chi-square and other contingency table results for this specification should be ignored because of insufficient cell sizes.

Output 7.2 Wilcoxon-Mann-Whitney Test Using the MH Chi-Square Statistic

```
             Statistics for Table of group by intake

  Statistic                      DF      Value      Prob
  -------------------------------------------------------
  Chi-Square                     18     22.0000     0.2320
  Likelihood Ratio Chi-Square    18     30.3164     0.0345
  MH Chi-Square (Rank Scores)     1      9.6589     0.0019
  Phi Coefficient                        1.0000
  Contingency Coefficient                0.7071
  Cramer's V                             1.0000

  WARNING: 100% of the cells have expected counts less
           than 5. Chi-Square may not be a valid test.

                  Sample Size = 22
```

You can also use the NPAR1WAY procedure to compute the Wilcoxon-Mann-Whitney statistic. You specify the WILCOXON option in the PROC statement, list GROUP in the CLASS statement, and list INTAKE in the VAR statement. Output 7.3 displays the results of the following statements.

```
proc npar1way wilcoxon;
    class group;
    var intake;
run;
```

Output 7.3   Wilcoxon-Mann-Whitney Test Using PROC NPAR1WAY

```
              Wilcoxon Scores (Rank Sums) for Variable intake
                      Classified by Variable group

                          Sum of      Expected      Std Dev         Mean
        group      N      Scores      Under H0      Under H0        Score
        ------------------------------------------------------------------
        Normal    12       91.0        138.0       15.122873     7.583333
        Hyperten  10      162.0        115.0       15.122873    16.200000

                     Average scores were used for ties.


                       Wilcoxon Two-Sample Test

             Statistic                   162.0000

             Normal Approximation
             Z                             3.0748
             One-Sided Pr >  Z             0.0011
             Two-Sided Pr > |Z|            0.0021

             t Approximation
             One-Sided Pr >  Z             0.0029
             Two-Sided Pr > |Z|            0.0057

          Z includes a continuity correction of 0.5.


                      Kruskal-Wallis Test

             Chi-Square                    9.6589
             DF                                 1
             Pr > Chi-Square               0.0019
```

The NPAR1WAY procedure computes a continuity-corrected Wilcoxon-Mann-Whitney test that yields slightly different results from PROC FREQ. The normal approximation statistic is $Z = 3.0748$ with a two-sided p-value of 0.0021. Output 7.3 also displays a chi-square statistic labeled "Kruskal-Wallis Test." The value of this statistic is identical to that resulting from the tests computed by the FREQ procedure.

7.3   Kruskal-Wallis Test

The Kruskal-Wallis (1952) test is a generalization of the two-sample Wilcoxon-Mann-Whitney test to three or more groups. It is a nonparametric test of the null hypothesis that the distribution of a response variable is the same in multiple independently sampled populations. The test requires an ordinally scaled response variable and is sensitive to the alternative hypothesis that there is a location difference among the populations. The Kruskal-Wallis test can be used whenever a one-way analysis of variance (ANOVA) model is appropriate.

When the sample sizes in the groups are small, tables of the exact distribution of the test statistic should be used. Alternatively, you can carry out exact tests of significance for small sample sizes (see page 162). If there are at least five observations per group, the p-value can be approximated using the asymptotic chi-square distribution with $s - 1$ degrees of freedom, where s is the number of groups. The approximate test is simply the

Mantel-Haenszel mean score statistic for the special case of one stratum when rank scores are used.

Table 7.2 displays data from a study of antecubital vein cortisol levels at time of delivery in pregnant women (Cawson et al. 1974). The investigators wanted to determine if median cortisol levels differed among three groups of women, all of whom had delivery between 38 and 42 weeks gestation. The data were obtained before the onset of labor at elective Caesarean section (Group I), at emergency Caesarean section during induced labor (Group II), or at the time of vaginal or Caesarean delivery in women in whom spontaneous labor occurred (Group III).

Table 7.2. Antecubital Vein Cortisol Levels at Time of Delivery

Group I		Group II		Group III	
Patient	Level	Patient	Level	Patient	Level
1	262	1	465	1	343
2	307	2	501	2	772
3	211	3	455	3	207
4	323	4	355	4	1048
5	454	5	468	5	838
6	339	6	362	6	687
7	304				
8	154				
9	287				
10	356				

The following statements create a SAS data set containing the data of Table 7.2 and request the Mantel-Haenszel mean score statistic comparing the mean rank scores in the three groups of subjects.

```
data cortisol;
   input group $ subject cortisol;
   datalines;
I    1   262
I    2   307
I    3   211
I    4   323
I    5   454
I    6   339
I    7   304
I    8   154
I    9   287
I   10   356
II   1   465
II   2   501
II   3   455
II   4   355
II   5   468
II   6   362
III  1   343
```

```
III  2   772
III  3   207
III  4  1048
III  5   838
III  6   687
;


proc freq;
    tables group*cortisol / noprint cmh2 scores=rank;
run;
```

The Kruskal-Wallis statistic, labeled "Row Mean Scores Differ" in Output 7.4, is equal to 9.2316 with 2 df, corresponding to a p-value of 0.0099. Thus, the cortisol level distributions differ among the three groups of patients. Since there are more than two groups, the Mantel-Haenszel correlation statistic, labeled "Nonzero Correlation," does not produce the same results as the Kruskal-Wallis test. The correlation statistic uses rank scores to test the null hypothesis that there is no association between group and cortisol level, versus the alternative hypothesis of a monotone association between the two variables. Thus, this statistic is only valid if the three groups are ordered (which might be realistic for this example in terms of the timing for cortisol level determination).

Output 7.4   Kruskal-Wallis Test Using PROC FREQ

```
            Summary Statistics for group by cortisol

     Cochran-Mantel-Haenszel Statistics (Based on Rank Scores)

   Statistic    Alternative Hypothesis    DF    Value    Prob
   ------------------------------------------------------------
       1        Nonzero Correlation        1    8.2857   0.0040
       2        Row Mean Scores Differ     2    9.2316   0.0099


                 Total Sample Size = 22
```

The Kruskal-Wallis test can also be computed using the WILCOXON option of the NPAR1WAY procedure.

```
proc npar1way wilcoxon;
    class group;
    var cortisol;
run;
```

The Kruskal-Wallis test displayed in Output 7.5 is identical to the value shown in Output 7.4. The NPAR1WAY procedure gives additional results showing that the mean rank scores in groups II and III are nearly equivalent and are substantially greater than the mean rank score in group I.

Output 7.5   Kruskal-Wallis Test Using PROC NPAR1WAY

```
                Wilcoxon Scores (Rank Sums) for Variable cortisol
                          Classified by Variable group

                            Sum of      Expected      Std Dev         Mean
         group      N       Scores      Under H0      Under H0        Score
         -------------------------------------------------------------------
         I          10      69.0        115.0         15.165751       6.900000
         II         6       90.0        69.0          13.564660       15.000000
         III        6       94.0        69.0          13.564660       15.666667


                            Kruskal-Wallis Test

                       Chi-Square            9.2316
                       DF                         2
                       Pr > Chi-Square       0.0099
```

7.4   Friedman's Chi-Square Test

Friedman's test (1937) is a nonparametric method for analyzing a randomized complete block design. This type of study design is applicable when interest is focused on one particular factor, but there are other factors whose effects you want to control. The experimental units are first divided into blocks (groups) in such a way that units within a block are relatively homogeneous. The size of each block is equal to the number of treatments or conditions under study. The treatments are then assigned at random to the experimental units within each block so that each treatment is given once and only once per block. The basic design principle is to partition the experimental units in such a way that background variability between blocks is maximized so that the variability within blocks is minimized.

The standard parametric ANOVA methods for analyzing randomized complete block designs require the assumption that the experimental errors are normally distributed. The Friedman test, which does not require this assumption, depends only on the ranks of the observations within each block and is sometimes called the two-way analysis of variance by ranks.

For small randomized complete block designs, the exact distribution of the Friedman test statistic should be used; for example, Odeh et al. (1977) tabulate the critical values of the Friedman test for up to six blocks and up to six treatments. Alternatively, you can carry out exact tests of significance for small sample sizes (see page 162). As the number of blocks increases, the distribution of the Friedman statistic approaches that of a chi-square random variable with $s - 1$ degrees of freedom, where s is the number of treatments. The approximate test is simply the Mantel-Haenszel mean score statistic for the special case of rank scores and one subject per treatment group in each block.

Table 7.3 displays data from an experiment designed to determine if five electrode types performed similarly (Berry 1987). In this study, all five types were applied to the arms of 16 subjects and the resistance was measured. Each subject is a block in this example, and all five treatments are applied once and only once per block.

Table 7.3. Electrical Resistance Data

Subject	Electrode Type				
	1	2	3	4	5
1	500	400	98	200	250
2	660	600	600	75	310
3	250	370	220	250	220
4	72	140	240	33	54
5	135	300	450	430	70
6	27	84	135	190	180
7	100	50	82	73	78
8	105	180	32	58	32
9	90	180	220	34	64
10	200	290	320	280	135
11	15	45	75	88	80
12	160	200	300	300	220
13	250	400	50	50	92
14	170	310	230	20	150
15	66	1000	1050	280	220
16	107	48	26	45	51

The following statements read in one record per subject and create a SAS data set containing one observation per electrode per subject. The Mantel-Haenszel mean score statistic is then computed using rank scores, where the 16 subjects define 16 strata.

```
data electrod;
   input subject resist1-resist5;
   type=1;   resist=resist1;   output;
   type=2;   resist=resist2;   output;
   type=3;   resist=resist3;   output;
   type=4;   resist=resist4;   output;
   type=5;   resist=resist5;   output;
   datalines;
 1   500   400    98   200   250
 2   660   600   600    75   310
 3   250   370   220   250   220
 4    72   140   240    33    54
 5   135   300   450   430    70
 6    27    84   135   190   180
 7   100    50    82    73    78
 8   105   180    32    58    32
 9    90   180   220    34    64
10   200   290   320   280   135
11    15    45    75    88    80
12   160   200   300   300   220
13   250   400    50    50    92
14   170   310   230    20   150
15    66  1000  1050   280   220
16   107    48    26    45    51
;
```

```
proc freq;
    tables subject*type*resist / noprint cmh2 scores=rank;
run;
```

Output 7.6 displays the results. The value of the test statistic is 5.4522 with 4 df. The
p-value of 0.2440 indicates that there is little evidence of a statistically significant
difference among the five types of electrodes.

Output 7.6   Friedman Test

```
                    Summary Statistics for type by resist
                         Controlling for subject

          Cochran-Mantel-Haenszel Statistics (Based on Rank Scores)

         Statistic    Alternative Hypothesis    DF    Value    Prob
         -------------------------------------------------------------
             1        Nonzero Correlation        1    2.7745   0.0958
             2        Row Mean Scores Differ      4    5.4522   0.2440


                         Total Sample Size = 80
```

In experimental situations with more than one subject per group in each block, PROC
FREQ can be used to compute generalizations of the Friedman test. The general principle
is that the strata are defined by the blocks, and the treatments or groups define the rows of
each table.

7.5   Aligned Ranks Test for Randomized Complete Blocks

When the number of blocks or treatments is small, the Friedman test has relatively low
power. This results from the fact that the test statistic is based on ranking the observations
within each block, which provides comparisons only of the within-block responses. Thus,
direct comparison of responses in different blocks is not meaningful, due to variation
between blocks. If the blocks are small, there are too few comparisons to permit an
effective overall comparison of the treatments. As an example, the Friedman test reduces
to the sign test if there are only two treatments. This disadvantage becomes less serious as
the number of treatments increases or as the number of subjects per block increases for a
fixed number s of treatments.

An alternative to the Friedman test is to use *aligned ranks*. The basic idea is to make the
blocks more comparable by subtracting from each observation within a block some
estimate of the location of the block, such as the average or median of the observations.
The resulting differences are called *aligned observations*. Instead of separately ranking the
observations within each block, you rank the complete set of aligned observations relative
to each other. Thus, the ranking scheme is the same as that used in computing the
Kruskal-Wallis statistic. The resulting ranks are called aligned ranks.

The aligned rank test was introduced by Hodges and Lehmann (1962). Koch and
Sen (1968) considered four cases of interest in the analysis of randomized complete block
experiments and independently proposed the aligned rank procedure for their Case IV.

Apart from the fact that one set of aligned ranks is used instead of separate within-block ranks, the computation of the aligned rank statistic is the same as for the Friedman test.

The exact distribution of the test statistic is cumbersome to compute. In addition, tables are not feasible since the distribution depends on the way the aligned ranks are distributed over the blocks. However, the null distribution of the test statistic is approximately chi-square with $s - 1$ degrees of freedom, where s is the number of treatments (or block size when there is one observation per treatment in each block). Tardif (1980, 1981, 1985) studied the asymptotic efficiency and other aspects of aligned rank tests in randomized block designs.

In Section 7.4, Friedman's test was used to analyze data from an experiment designed to determine if five electrode types performed similarly (Table 7.3). Using the SAS data set created in Section 7.4, the following statements compute the aligned rank statistic.

```
proc standard mean=0;
   by subject;
   var resist;
proc rank;
   var resist;
proc freq;
   tables subject*type*resist / noprint cmh2;
run;
```

The STANDARD procedure standardizes the observations within each block (subject) to have mean zero. Thus, the subject-specific sample mean is subtracted from each response. The RANK procedure computes a single set of rankings for the combined aligned observations. Using the resulting aligned ranks as scores, the FREQ procedure computes the aligned rank statistic.

Output 7.7 displays the results. The test statistic is equal to 13.6003 with 4 df. With reference to the chi-square distribution with four degrees of freedom, there is a clearly significant difference among the five electrode types. Recall that the Friedman test (Output 7.6) was not statistically significant. Thus, this example illustrates the potentially greater power of the aligned ranks test.

Output 7.7   Aligned Ranks Test

```
                 Summary Statistics for type by resist
                       Controlling for subject

       Cochran-Mantel-Haenszel Statistics (Based on Table Scores)

       Statistic    Alternative Hypothesis    DF      Value      Prob
       ----------------------------------------------------------------
            1        Nonzero Correlation        1     4.9775     0.0257
            2        Row Mean Scores Differ      4    13.6003     0.0087


                      Total Sample Size = 80
```

7.6 Durbin's Test for Balanced Incomplete Blocks

In the randomized complete block design, every treatment is applied in every block. However, it is sometimes impractical or impossible for all of the treatments to be applied to each block, especially when the number of treatments is large and the block size is limited. Experimental designs in which not all treatments are applied to each block are called incomplete block designs. If the design is balanced so that every block contains c experimental units, every treatment appears in u blocks, and every treatment appears with every other treatment an equal number of times, the design is then called a balanced incomplete block design.

Durbin (1951) presented a rank test that can be used to test the null hypothesis of no differences among treatments in a balanced incomplete block design. This test can be computed using the stratified Mantel-Haenszel mean score statistic based on rank scores. The Durbin test reduces to the Friedman test if the number of treatments equals the number of experimental units per block.

As an example, Table 7.4 displays data taken from a mirror drawing experiment conducted in 20 psychiatric patients (Ogilvie 1965). The subject's task was to trace along a straight line, seen in a mirror, with his or her hand hidden. The straight lines were oriented at five different angles to the median plane of the subject (0°, 22.5°, 45°, 67.5°, and 90°), and the outcome variable was the time (in seconds) taken to complete the task. Ideally, every subject should draw lines at each angle. However, the effect of practice on performance could be considerable. In addition, it was difficult to maintain the subject's interest and cooperation in the experiment for more than a brief period. Consequently, a balanced incomplete block design was used. Each of the 20 subjects completed the experiment at two of the five angles; thus, each angle was studied eight times.

Table 7.4.   Drawing Times (Seconds) from a Mirror Tracing Experiment

Subject	Angle (Degrees) 0	22.5	45	67.5	90	Subject	Angle (Degrees) 0	22.5	45	67.5	90
1	7	15				11			17	9	
2	20		72			12			100		15
3	8			26		13	16			32	
4	33				36	14		19		32	
5	7	16				15			36	39	
6		68	67			16				44	54
7		33		64		17	16				38
8		34			12	18		17			12
9	10		96			19			37		11
10		29	59			20				56	6

The following statements read in one record per subject. The four input variables for each subject contain the two angles used and the drawing times at these angles. The resulting SAS data set contains two observations per subject and three variables per observation: subject identifier, angle, and drawing time. The PROC FREQ statements compute the extended Mantel-Haenszel mean score statistic, with one stratum for each subject. The rows of each table are defined by the five angles studied, and rank scores are used for the response variable (drawing time).

```
data tracing;
   keep subject angle time;
   input subject angle1 angle2 time1 time2;
   angle=angle1;  time=time1;  output;
   angle=angle2;  time=time2;  output;
   datalines;
 1  0.0 22.5    7  15
 2  0.0 45.0   20  72
 3  0.0 67.5    8  26
 4  0.0 90.0   33  36
 5 22.5  0.0   16   7
 6 22.5 45.0   68  67
 7 22.5 67.5   33  64
 8 22.5 90.0   34  12
 9 45.0  0.0   96  10
10 45.0 22.5   59  29
11 45.0 67.5   17   9
12 45.0 90.0  100  15
13 67.5  0.0   32  16
14 67.5 22.5   32  19
15 67.5 45.0   39  36
16 67.5 90.0   44  54
17 90.0  0.0   38  16
18 90.0 22.5   12  17
19 90.0 45.0   11  37
20 90.0 67.5    6  56
;
proc freq;
   tables subject*angle*time / noprint cmh2 scores=rank;
run;
```

Output 7.8 displays the results of Durbin's test. The chi-square statistic is 10.4000 with 4 df (since there are five groups). The p-value of 0.0342 indicates that there is a significant difference among the drawing time distributions at the five angles.

Output 7.8   Results of Durbin's Test

```
                    Summary Statistics for angle by time
                           Controlling for subject

          Cochran-Mantel-Haenszel Statistics (Based on Rank Scores)

          Statistic    Alternative Hypothesis    DF    Value    Prob
          ----------------------------------------------------------
              1        Nonzero Correlation        1    1.8000   0.1797
              2        Row Mean Scores Differ     4   10.4000   0.0342


                        Total Sample Size = 40
```

While Durbin's test uses within-block ranks, you could perform a similar test using aligned ranks, as described in Section 7.5. In addition, Benard and van Elteren (1953) generalized the Durbin test to the case where some experimental units may contain several observations per treatment. This generalization can also be computed using the CMH option of PROC FREQ.

7.7   Rank Analysis of Covariance

The analysis of covariance (ANCOVA) is a standard statistical methodology that combines the features of analysis of variance (ANOVA) and linear regression to determine if there is a difference in some response variable between two or more groups. The basic idea is to augment the ANOVA model containing the group effects with one or more additional categorical or quantitative variables that are related to the response variable. These additional variables *covary* with the response and so are called covariables or covariates.

One of the main uses of ANCOVA is to increase precision in randomized experiments by using the relationship between the response variable and the covariates to reduce the error variability in comparing treatment groups. In this setting, ANCOVA often results in more powerful tests, shorter confidence intervals, and a reduction in the sample size required to establish differences among treatment groups. ANCOVA is also useful in adjusting for sources of bias in observational studies.

The validity of classical parametric ANCOVA depends on several assumptions, including normality of error terms, equality of error variances for different treatments, equality of slopes for the different treatment regression lines, and linearity of regression. For situations in which these assumptions may not be satisfied, Quade (1967) proposed the use of rank analysis of covariance. This technique can be combined with the randomization model framework of extended Mantel-Haenszel statistics to carry out nonparametric comparisons between treatment groups, after adjusting for the effects of one or more covariates. The methodology, which has been described by Koch et al. (1982, 1990), can easily be implemented using the SAS System.

Table 7.5 displays exercise data from treadmill testing of healthy males and females (Bruce, Kusumi, and Hosmer 1973; Fisher and van Belle 1993). The purpose of the analysis is to determine if men and women use the same amount of oxygen. The outcome, VO_2MAX, is computed by determining the volume of oxygen used per minute per kilogram of body weight. Since the effort expended to go further on the treadmill increases with the duration of time on the treadmill, there should be some relationship between VO_2MAX and duration on the treadmill; thus, this variable is used as a covariate.

Table 7.5. Exercise Data for Healthy Males and Females

Males						Females					
ID	Durat.	VO$_2$	ID	Durat.	VO$_2$	ID	Durat.	VO$_2$	ID	Durat.	VO$_2$
1	706	41.5	23	582	35.8	1	660	38.1	23	461	30.5
2	732	45.9	24	503	29.1	2	628	38.4	24	540	25.9
3	930	54.5	25	747	47.2	3	637	41.7	25	588	32.7
4	900	60.3	26	600	30.0	4	575	33.5	26	498	26.9
5	903	60.5	27	491	34.1	5	590	28.6	27	483	24.6
6	976	64.6	28	694	38.1	6	600	23.9	28	554	28.8
7	819	47.4	29	586	28.7	7	562	29.6	29	521	25.9
8	922	57.0	30	612	37.1	8	495	27.3	30	436	24.4
9	600	40.2	31	610	34.5	9	540	33.2	31	398	26.3
10	540	35.2	32	539	34.4	10	470	26.6	32	366	23.2
11	560	33.8	33	559	35.1	11	408	23.6	33	439	24.6
12	637	38.8	34	653	40.9	12	387	23.1	34	549	28.8
13	593	38.9	35	733	45.4	13	564	36.6	35	360	19.6
14	719	49.5	36	596	36.9	14	603	35.8	36	566	31.4
15	615	37.1	37	580	41.6	15	420	28.0	37	407	26.6
16	589	32.2	38	550	22.7	16	573	33.8	38	602	30.6
17	478	31.3	39	497	31.9	17	602	33.6	39	488	27.5
18	620	33.8	40	605	42.5	18	430	21.0	40	526	30.9
19	710	43.7	41	552	37.4	19	508	31.2	41	524	33.9
20	600	41.7	42	640	48.2	20	565	31.2	42	562	32.3
21	660	41.0	43	500	33.6	21	464	23.7	43	496	26.9
22	644	45.9	44	603	45.0	22	495	24.5			

The following statements create a SAS data set containing the sex, subject ID, duration of exercise (seconds), and VO$_2$MAX values for each subject.

```
data exercise;
   input sex $ case duration vo2max @@;
   datalines;
M  1 706 41.5    M  2 732 45.9    M  3 930 54.5    M  4 900 60.3
M  5 903 60.5    M  6 976 64.6    M  7 819 47.4    M  8 922 57.0
M  9 600 40.2    M 10 540 35.2    M 11 560 33.8    M 12 637 38.8
M 13 593 38.9    M 14 719 49.5    M 15 615 37.1    M 16 589 32.2
M 17 478 31.3    M 18 620 33.8    M 19 710 43.7    M 20 600 41.7
M 21 660 41.0    M 22 644 45.9    M 23 582 35.8    M 24 503 29.1
M 25 747 47.2    M 26 600 30.0    M 27 491 34.1    M 28 694 38.1
M 29 586 28.7    M 30 612 37.1    M 31 610 34.5    M 32 539 34.4
M 33 559 35.1    M 34 653 40.9    M 35 733 45.4    M 36 596 36.9
M 37 580 41.6    M 38 550 22.7    M 39 497 31.9    M 40 605 42.5
M 41 552 37.4    M 42 640 48.2    M 43 500 33.6    M 44 603 45.0
F  1 660 38.1    F  2 628 38.4    F  3 637 41.7    F  4 575 33.5
F  5 590 28.6    F  6 600 23.9    F  7 562 29.6    F  8 495 27.3
F  9 540 33.2    F 10 470 26.6    F 11 408 23.6    F 12 387 23.1
F 13 564 36.6    F 14 603 35.8    F 15 420 28.0    F 16 573 33.8
F 17 602 33.6    F 18 430 21.0    F 19 508 31.2    F 20 565 31.2
F 21 464 23.7    F 22 495 24.5    F 23 461 30.5    F 24 540 25.9
```

```
F 25 588 32.7     F 26 498 26.9     F 27 483 24.6     F 28 554 28.8
F 29 521 25.9     F 30 436 24.4     F 31 398 26.3     F 32 366 23.2
F 33 439 24.6     F 34 549 28.8     F 35 360 19.6     F 36 566 31.4
F 37 407 26.6     F 38 602 30.6     F 39 488 27.5     F 40 526 30.9
F 41 524 33.9     F 42 562 32.3     F 43 496 26.9
;
run;
```

The first step of the analysis is to compute the ranks of the response variable and covariate in the combined group of males and females. You do this using PROC RANK, as follows:

```
proc rank out=ranks;
   var duration vo2max;
run;
```

The next step is to calculate the residuals from the linear regression of the VO_2MAX ranks on the duration ranks using PROC REG. The residuals are saved in an output data set.

```
proc reg noprint;
   model vo2max=duration;
   output out=residual r=resid;
run;
```

Finally, the Mantel-Haenszel mean score statistic is used to compare the mean values of the residuals in males and females using TABLE scores.

```
proc freq;
   tables sex*resid / noprint cmh2;
run;
```

Output 7.9 displays the results, which indicate a clearly significant difference between males and females (chi-square=11.7626, 1 df, $p = 0.0006$).

Output 7.9   Rank Analysis of Covariance Results

```
                   Summary Statistics for sex by resid

          Cochran-Mantel-Haenszel Statistics (Based on Table Scores)

     Statistic    Alternative Hypothesis      DF      Value      Prob
     ---------------------------------------------------------------------
         1        Nonzero Correlation          1     11.7626    0.0006
         2        Row Mean Scores Differ       1     11.7626    0.0006


                     Total Sample Size = 87
```

The methodology can also be modified for the situation in which there are multiple strata. Table 7.6 displays data from an experiment to evaluate the effectiveness of topically applied stannous fluoride and acid phosphate fluoride in reducing the incidence of dental caries, as compared with a placebo treatment of distilled water (Cartwright, Lindahl, and Bawden 1968; Quade 1982). These data are from 69 female children from three centers who completed the two-year study. The stannous fluoride, acid phosphate fluoride, and distilled water treatment groups are denoted by SF, APF, and W. The columns labeled B and A represent the number of decayed, missing, or filled teeth (DMFT) before and after the study, respectively. In this example, the response to be compared among the three groups is the number of DMFT after treatment; the number of DMFT before treatment is used as a covariate. In addition, the analysis is stratified by center.

Table 7.6. Dental Caries Data

Center 1				Center 2				Center 3							
ID	Grp	B	A	ID	Grp	B	A	ID	Grp	B	A	ID	Grp	B	A
1	W	7	11	1	W	10	14	1	W	2	4	18	APF	10	12
2	W	20	24	2	W	13	17	2	W	13	18	19	APF	7	11
3	W	21	25	3	W	3	4	3	W	9	12	20	APF	13	12
4	W	1	2	4	W	4	7	4	W	15	18	21	APF	5	8
5	W	3	7	5	W	4	9	5	W	13	17	22	APF	1	3
6	W	20	23	6	SF	15	18	6	W	2	5	23	APF	8	9
7	W	9	13	7	SF	6	8	7	W	9	12	24	APF	4	5
8	W	2	4	8	SF	4	6	8	SF	4	6	25	APF	4	7
9	SF	11	13	9	SF	18	19	9	SF	10	14	26	APF	14	14
10	SF	15	18	10	SF	11	12	10	SF	7	11	27	APF	8	10
11	APF	7	10	11	SF	9	9	11	SF	14	15	28	APF	3	5
12	APF	17	17	12	SF	4	7	12	SF	7	10	29	APF	11	12
13	APF	9	11	13	SF	5	7	13	SF	3	6	30	APF	16	18
14	APF	1	5	14	SF	11	14	14	SF	9	12	31	APF	8	8
15	APF	3	7	15	SF	4	6	15	SF	8	10	32	APF	0	1
				16	APF	4	4	16	SF	19	19	33	APF	3	4
				17	APF	7	7	17	SF	10	13				
				18	APF	0	4								
				19	APF	3	3								
				20	APF	0	1								
				21	APF	8	8								

The following SAS statements read in the variables CENTER, ID, GROUP, BEFORE, and AFTER, whose values are displayed in Table 7.6.

```
data caries;
   input center id group $ before after @@;
   datalines;
1  1 W    7 11    1  2 W   20 24    1  3 W   21 25    1  4 W     1  2
1  5 W    3  7    1  6 W   20 23    1  7 W    9 13    1  8 W     2  4
1  9 SF  11 13    1 10 SF  15 18    1 11 APF  7 10    1 12 APF  17 17
1 13 APF  9 11    1 14 APF  1  5    1 15 APF  3  7    2  1 W    10 14
2  2 W   13 17    2  3 W    3  4    2  4 W    4  7    2  5 W     4  9
```

```
 2  6 SF  15 18      2  7 SF   6  8      2  8 SF   4  6      2  9 SF  18 19
 2 10 SF  11 12      2 11 SF   9  9      2 12 SF   4  7      2 13 SF   5  7
 2 14 SF  11 14      2 15 SF   4  6      2 16 APF  4  4      2 17 APF  7  7
 2 18 APF  0  4      2 19 APF  3  3      2 20 APF  0  1      2 21 APF  8  8
 3  1 W    2  4      3  2 W   13 18      3  3 W    9 12      3  4 W   15 18
 3  5 W   13 17      3  6 W    2  5      3  7 W    9 12      3  8 SF   4  6
 3  9 SF  10 14      3 10 SF   7 11      3 11 SF  14 15      3 12 SF   7 10
 3 13 SF   3  6      3 14 SF   9 12      3 15 SF   8 10      3 16 SF  19 19
 3 17 SF  10 13      3 18 APF 10 12      3 19 APF  7 11      3 20 APF 13 12
 3 21 APF  5  8      3 22 APF  1  3      3 23 APF  8  9      3 24 APF  4  5
 3 25 APF  4  7      3 26 APF 14 14      3 27 APF  8 10      3 28 APF  3  5
 3 29 APF 11 12      3 30 APF 16 18      3 31 APF  8  8      3 32 APF  0  1
 3 33 APF  3  4
;
run;
```

The next statements produce standardized ranks for the covariate BEFORE and the response variable AFTER in each of the three centers. Standardized ranks are used to adjust for the fact that the number of patients differs among centers.

```
proc rank nplus1 ties=mean out=ranks;
   by center;
      var before after;
run;
```

The NPLUS1 option of the RANK procedure requests fractional ranks using the denominator $n + 1$, where n is the center-specific sample size. The TIES=MEAN option requests that tied values receive the mean of the corresponding ranks (midranks). Since TIES=MEAN is the default for PROC RANK, this option was not specified in the previous example. However, when fractional ranks are requested using either the FRACTION (denominator is n) or NPLUS1 (denominator is $n + 1$) options, the TIES=HIGH option is the default. Thus, you must specify both the NPLUS1 and TIES=MEAN options.

PROC REG is then used to fit separate linear regression models for the three centers. In each model, the standardized ranks of the AFTER and BEFORE variables are used as the dependent and independent variables, respectively. The following statements request these models and output the corresponding residuals into an output data set named RESIDUAL.

```
proc reg noprint;
   by center;
   model after=before;
   output out=residual r=resid;
run;
```

Finally, the stratified mean score test, using the values of the residuals as scores, compares the three groups.

```
proc freq;
   tables center*group*resid / noprint cmh2;
run;
```

Output 7.10 displays the results. The difference among the three treatment groups, after adjusting for the baseline number of DMFT and center, is clearly significant (row mean score chi-square=17.5929, 2 df, $p = 0.0002$).

Output 7.10 Results of Stratified Rank Analysis of Covariance

```
                 Summary Statistics for group by resid
                       Controlling for center

         Cochran-Mantel-Haenszel Statistics (Based on Table Scores)

         Statistic   Alternative Hypothesis    DF     Value     Prob
         ----------------------------------------------------------------
             1       Nonzero Correlation        1    17.1716   <.0001
             2       Row Mean Scores Differ      2    17.5929   0.0002


                       Total Sample Size = 69
```

The analyses described in this section are generally limited to randomized clinical trials, since the covariables should have similar distributions in the groups being compared. In the dental caries example, patients were randomly assigned to one of the three treatment groups, and Cartwright, Lindahl, and Bawden (1968) reported that the groups were comparable with respect to the number of DMFT at baseline, as well as with respect to other baseline variables. Although the patients in the APF group appear to have fewer DMFT at baseline than the patients in the SF and W groups (the corresponding medians were 7, 9, and 9, respectively), there is insufficient evidence to conclude that the distributions are significantly different (Kruskal-Wallis chi-square=4.4 with 2 df, $p = 0.11$); thus, rank analysis of covariance methods are appropriate.

In contrast, the exercise data (Table 7.5) were obtained from a nonrandomized experiment comparing men and women, and the distributions of the covariate, duration of time on the treadmill, differ significantly in the two samples. In particular, the average durations in males and females are 647.4 and 514.9 seconds, respectively ($p < 0.001$ from the two-sample t-test). Therefore, although the analysis presented in this section is a useful illustration of the rank analysis of covariance methodology, its results should be interpreted cautiously.

Chapter 8

Logistic Regression I: Dichotomous Response

Chapter Table of Contents

Chapter 8
Logistic Regression I: Dichotomous Response

8.1   Introduction

The previous chapters discussed the investigation of statistical association, primarily by testing the hypothesis of no association between a set of groups and outcomes for a response with adjustment for a set of strata. Recall that Mantel-Haenszel strategies produced tests for specific alternatives to no association: general association, location shifts for means, and linear trends. This chapter shifts the focus to statistical models, methods aimed at describing the nature of the association in terms of a parsimonious number of parameters. Besides describing the variation in the data, statistical modeling allows you to address questions about association in terms of hypotheses concerning model parameters.

If certain realistic sampling assumptions are plausible, a statistical model can be used to make inferences from a study population to a larger target population. If you are analyzing a clinical trial that assigned its subjects to a randomized protocol, then you can generalize your results to the population from which the subjects were selected and possibly to a more general target population. If you are analyzing observational data, and you can argue that your study subjects are conceptually representative of some larger target population, then you may make inferences to that target population.

Logistic regression is a form of statistical modeling that is often appropriate for categorical outcome variables. It describes the relationship between a categorical response variable and a set of explanatory variables. The response variable is usually dichotomous, but it may be polytomous, that is, have more than two response levels. These multiple-level response variables can be nominally or ordinally scaled. This chapter addresses logistic regression when the response is dichotomous; typically the two outcomes are yes and no. Logistic regression with more than two response variable levels is covered in Chapter 9, "Logistic Regression II: Polytomous Response." Another kind of logistic regression is called conditional logistic regression and is often used for stratified data. Chapter 10, "Conditional Logistic Regression," describes this methodology.

Chapter 8 and Chapter 9 focus on asymptotic methods that require a certain sample size in order for model fit and effect assessment tests to be valid. However, sometimes your data are so sparse or have such small cell counts that these methods are not valid. This chapter also discusses exact logistic regression, which is an alternative strategy for these situations.

The explanatory variables in logistic regression can be categorical or continuous.

Sometimes the term "logistic regression" is restricted to analyses that include continuous explanatory variables, and the term "logistic analysis" is used for those situations where all the explanatory variables are categorical. In this book, logistic regression refers to both cases. Logistic regression has applications in fields such as epidemiology, medical research, banking, market research, and social research. As you will see, one of its advantages is that model interpretation is possible through odds ratios, which are functions of model parameters.

Several procedures in the SAS System can be used to perform logistic regression, including the LOGISTIC procedure, the CATMOD procedure, and the GENMOD procedure. The LOGISTIC procedure is designed primarily for logistic regression analysis, and it provides useful information such as odds ratio estimates and model diagnostics. The CATMOD procedure is a general procedure designed to fit models to functions of categorical response variables. PROC GENMOD is a procedure for analyzing generalized linear models, of which logistic regression is a simple case. In this chapter, attention is focused on the use of the LOGISTIC procedure to perform logistic regression.

8.2 Dichotomous Explanatory Variables

8.2.1 Logistic Model

Table 8.1 displays the coronary artery disease data that were analyzed in Chapter 3, "Sets of 2×2 Tables." Recall that the study population consists of people who visited a clinic on a walk-in basis and required a catheterization. The response, presence of coronary artery disease (CA), is dichotomous, as are the explanatory variables, sex and ECG. These data were analyzed in Section 3.3.2 with Mantel-Haenszel methods; also, odds ratios and the common odds ratio were computed. Recall that ECG was clearly associated with disease status, adjusted for gender.

Table 8.1. Coronary Artery Disease Data

Sex	ECG	Disease	No Disease	Total
Female	< 0.1 ST segment depression	4	11	15
Female	≥ 0.1 ST segment depression	8	10	18
Male	< 0.1 ST segment depression	9	9	18
Male	≥ 0.1 ST segment depression	21	6	27

Assume that these data arise from a stratified simple random sample so that presence of coronary artery disease is distributed binomially for each sex $\times$ ECG combination, that is, for each row of Table 8.1. These rows are called groups or subpopulations. You can then write a model for the probability, or the likelihood, of these data. The sex by ECG by disease status classification has the product binomial distribution

$$\Pr\{n_{hij}\} = \prod_{h=1}^{2} \prod_{i=1}^{2} \frac{n_{hi+}!}{n_{hi1}! n_{hi2}!} \theta_{hi}^{n_{hi1}} (1 - \theta_{hi})^{n_{hi2}}$$

The quantity θ_{hi} is the probability that a person of the hth sex with an ith ECG status has coronary artery disease, and n_{hi1} and n_{hi2} are the numbers of persons of the hth sex and *i*th ECG with and without coronary artery disease, respectively ($h = 1$ for females, $h = 2$ for males; $i = 1$ for ECG < 0.1, $i = 2$ for ECG ≥ 0.1; $j = 1$ for disease, $j = 2$ for no disease, and $n_{hi+} = (n_{hi1} + n_{hi2})$). You can apply the logistic model to describe the variation among the $\{\theta_{hi}\}$:

$$\theta_{hi} = \frac{1}{1 + \exp\{-(\alpha + \sum_{k=1}^{t} \beta_k x_{hik})\}}$$

Another form of this equation that is often used is

$$\theta_{hi} = \frac{\exp\{\alpha + \sum_{k=1}^{t} \beta_k x_{hik}\}}{1 + \exp\{\alpha + \sum_{k=1}^{t} \beta_k x_{hik}\}}$$

The quantity α is the intercept parameter; the $\{x_{hik}\}$ are the t explanatory variables for the *h*th sex and *i*th ECG; $k = 1, \ldots, t$; and the $\{\beta_k\}$ are the t regression parameters.

The matrix form of this equation is

$$\theta_{hi} = \frac{\exp(\alpha + \mathbf{x}'_{hi}\boldsymbol{\beta})}{1 + \exp(\alpha + \mathbf{x}'_{hi}\boldsymbol{\beta})}$$

where the quantity $\boldsymbol{\beta}$ is a vector of t regression parameters, and $\mathbf{x}_{hi}$ is a vector of explanatory variables corresponding to the *h*i*th group.

You can show that the odds of CA disease for the *h*i*th group is

$$\frac{\theta_{hi}}{1 - \theta_{hi}} = \exp\{\alpha + \sum_{k=1}^{t} \beta_k x_{hik}\}$$

By taking natural logarithms on both sides, you obtain a linear model for the logit:

$$\log\left\{\frac{\theta_{hi}}{1 - \theta_{hi}}\right\} = \alpha + \sum_{k=1}^{t} \beta_k x_{hik}$$

The logit is the log of an odds, so this model is for the log odds of coronary artery disease versus no coronary artery disease for the *h*i*th group. The log odds for the *h*i*th group can be written as the sum of an intercept and a linear combination of explanatory variable values multiplied by the appropriate parameter values. This result allows you to obtain the model-predicted odds ratios for variation in the x_{hik} by exponentiating model parameter estimates for the β_k, as explained below.

Besides taking the familiar linear form, the logistic model has the useful property that all possible values of $(\alpha + \mathbf{x}'_{hi}\boldsymbol{\beta})$ in $(-\infty, \infty)$ map into $(0, 1)$ for θ_{hi}. Thus, predicted probabilities produced by this model are constrained to lie between 0 and 1. This model produces no negative predicted probabilities and no predicted probabilities greater than 1. Maximum likelihood methods are generally used to estimate α and β. PROC LOGISTIC uses the Fisher scoring method, which is equivalent to model fitting with iteratively weighted least squares. PROC CATMOD and PROC GENMOD use Newton-Raphson algorithms. When the overall sample size $n = \sum_h \sum_i n_{hi}$ is sufficiently large, the resulting estimates for α and β have a multivariate normal distribution for which a consistent estimate of the corresponding covariance matrix is conveniently available. On this basis, confidence intervals and test statistics are straightforward to construct for inferences concerning α and β. See Appendix A in this chapter for more methodological detail.

8.2.2   Model Fitting

A useful first model for the coronary disease data is one that includes main effects for sex and ECG. Since these effects are dichotomous, there are three parameters in this model, including the intercept.

You can write this main effects model as

$$
\begin{bmatrix} \text{logit}(\theta_{11}) \\ \text{logit}(\theta_{12}) \\ \text{logit}(\theta_{21}) \\ \text{logit}(\theta_{22}) \end{bmatrix} = \begin{bmatrix} \alpha & & & \\ \alpha & & + & \beta_2 \\ \alpha & + & \beta_1 & \\ \alpha & + & \beta_1 & + & \beta_2 \end{bmatrix} = \begin{bmatrix} 1 & 0 & 0 \\ 1 & 0 & 1 \\ 1 & 1 & 0 \\ 1 & 1 & 1 \end{bmatrix} \begin{bmatrix} \alpha \\ \beta_1 \\ \beta_2 \end{bmatrix}
$$

This type of parameterization is often called *incremental effects* parameterization. It has a model matrix (also called a design matrix) composed of 0s and 1s. The quantity α is the log odds of coronary artery disease for females with an ECG of less than 0.1. Since females with ST segment depression less than 0.1 are described by the intercept, this group is known as the *reference cell* in this parameterization. The parameter β_1 is the increment in log odds for males, and β_2 is the increment in log odds for having an ECG of at least 0.1. Table 8.2 displays the probabilities and odds predicted by this model.

Table 8.2.   Model-Predicted Probabilities and Odds

Sex	ECG	Pr{CA Disease}=θ_{hi}	Odds of CA Disease
Females	< 0.1	$e^{\alpha}/(1 + e^{\alpha})$	e^{α}
Females	≥ 0.1	$e^{\alpha+\beta_2}/(1 + e^{\alpha+\beta_2})$	$e^{\alpha+\beta_2}$
Males	< 0.1	$e^{\alpha+\beta_1}/(1 + e^{\alpha+\beta_1})$	$e^{\alpha+\beta_1}$
Males	≥ 0.1	$e^{\alpha+\beta_1+\beta_2}/(1 + e^{\alpha+\beta_1+\beta_2})$	$e^{\alpha+\beta_1+\beta_2}$

You can calculate the odds ratio for males versus females by forming the ratio of male odds of CA disease to female odds of CA disease for either low or high ECG (see Chapter 2,"The 2 × 2 Table," for a discussion of odds ratios):

$$\frac{e^{\alpha+\beta_1}}{e^{\alpha}} = e^{\beta_1} \quad \text{or} \quad \frac{e^{\alpha+\beta_1+\beta_2}}{e^{\alpha+\beta_2}} = e^{\beta_1}$$

Similarly, the odds ratio for high ECG versus low ECG is determined by forming the corresponding ratio of the odds of CA disease for either sex:

$$\frac{e^{\alpha+\beta_1+\beta_2}}{e^{\alpha+\beta_1}} = e^{\beta_2} \quad \text{or} \quad \frac{e^{\alpha+\beta_2}}{e^{\alpha}} = e^{\beta_2}$$

Thus, you can obtain odds ratios as functions of the model parameters in logistic regression. With incremental effects parameterization for a main effects model, you simply exponentiate the parameter estimates. However, unlike the odds ratios you calculate from individual 2 × 2 tables, these odds ratios have been adjusted for all other explanatory variables in the model.

8.2.3 Goodness of Fit

Once you have applied the model, you need to assess how well it fits the data, or how close the model-predicted values are to the corresponding observed values. Test statistics that assess fit in this manner are known as *goodness-of-fit statistics*. They address the differences between observed and predicted values, or their ratio, in some appropriate manner. Departures of the predicted proportions from the observed proportions should be essentially random. The test statistics have approximate chi-square distributions when the $\{n_{hij}\}$ are sufficiently large. If they are larger than a tolerable value, then you have an oversimplified model and you need to identify some other factors to better explain the variation in the data.

Two traditional goodness-of-fit tests are the Pearson chi-square, Q_P, and the likelihood ratio chi-square, Q_L, also known as the *deviance*.

$$Q_P = \sum_{h=1}^{2}\sum_{i=1}^{2}\sum_{j=1}^{2}(n_{hij} - m_{hij})^2/m_{hij}$$

$$Q_L = \sum_{h=1}^{2}\sum_{i=1}^{2}\sum_{j=1}^{2}2n_{hij}\log\left(\frac{n_{hij}}{m_{hij}}\right)$$

where the m_{hij} are the model-predicted counts defined as

$$m_{hij} = \begin{cases} n_{hi+}\hat{\theta}_{hi} & \text{for j=1} \\ n_{hi+}(1 - \hat{\theta}_{hi}) & \text{for j=2} \end{cases}$$

The quantity $\hat{\theta}_{hi}$ is the estimate of θ_{hi} using the estimates of α and the β_k. If the model fits, both Q_P and Q_L are approximately distributed as chi-square with degrees of freedom equal to the number of rows in the table minus the number of parameters. For the main effects model being discussed, there are four rows in the table (four groups) and three parameters, including the intercept, and so Q_P and Q_L have $4 - 3 = 1$ degree of freedom. Sample size guidelines for these statistics to be approximately chi-square include

- each of the groups has at least 10 subjects ($n_{hi+} \geq 10$)
- 80% of the predicted counts (m_{hij}) are at least 5
- all other expected counts are greater than 2, with essentially no 0 counts

When the above guidelines do not apply, there is usually a tendency for the chi-square approximation to Q_P and Q_L to overstate lack of fit, and so tolerably small values for them are robustly interpretable as supporting goodness of fit. For a more rigorous evaluation of goodness of fit when the $\{n_{hij}\}$ are not large enough to justify chi-square approximations for Q_L and Q_P, exact methods for logistic regression are available (see Section 8.8).

8.2.4   Using PROC LOGISTIC

The LOGISTIC procedure was designed specifically to fit logistic regression models. You specify the response variable and the explanatory variables in a MODEL statement, and it fits the model via maximum likelihood estimation. PROC LOGISTIC produces the parameter estimates, their standard errors, and statistics to assess model fit. In addition, it also provides several model selection methods, puts predicted values and other statistics into output data sets, and includes a number of options for controlling the model-fitting process.

The following SAS code creates the data set CORONARY.

```
data coronary;
   input sex ecg ca count @@;
   datalines;
0 0 0  11 0 0 1  4
0 1 0  10 0 1 1  8
1 0 0   9 1 0 1  9
1 1 0   6 1 1 1 21
;
run;
```

The variable CA is the response variable, and SEX and ECG are the explanatory variables. The variable SEX takes the value 0 for females and 1 for males, and ECG takes the value 0 for lower ST segment depression and 1 for higher ST segment depression. Thus, these variables provide the values for the model matrix. Such coding is known as *indicator-coding* or *dummy-coding*.

The variable CA takes the value 1 if CA disease is present and is 0 otherwise. By default, PROC LOGISTIC orders the response variable values alphanumerically so that, for these data, it bases its model on the probability of the smallest value, Pr{CA=0}, which is Pr{no coronary artery disease}. This means that it models the log odds of {no coronary artery disease}. If you want to change the basis of the model to be Pr{CA=1}, which is

Pr{coronary artery disease}, you have to alter this default behavior. Data analysts usually want their models to be based on the probability of the event (disease, success), which is often coded as 1.

The DESCENDING option in the PROC LOGISTIC statement requests that the response value ordering be reversed. For these data, this means that PROC LOGISTIC will model Pr{coronary artery disease}. For a dichotomous response variable, the effect of reversing the order of the response values is to change the sign of the parameter estimates. Thus, if your estimates for the parameters have opposite signs from another logistic regression run, you have modeled opposite forms for the dichotomous response variable.

The next group of SAS statements invokes PROC LOGISTIC. Note the use of the DESCENDING option. Since the data are in frequency, or count, form, you need to indicate that to PROC LOGISTIC. This is done with the FREQ statement, which is similar in use to the WEIGHT statement in PROC FREQ. (Note that a WEIGHT statement is available with the LOGISTIC procedure; however, it is used somewhat differently.) The main effects model is specified in the MODEL statement, which also includes the options SCALE=NONE and AGGREGATE. The SCALE option produces goodness-of-fit statistics; the AGGREGATE option requests that PROC LOGISTIC treat each unique combination of the explanatory variable values as a distinct group in computing the goodness-of-fit statistics.

```
proc logistic descending;
    freq count;
    model ca=sex ecg / scale=none aggregate;
run;
```

Output 8.1 displays the resulting "Response Profile" table. The response variable values are listed according to their PROC LOGISTIC *ordered values*. The DESCENDING option has made CA=1 the first ordered value (1) and CA=0 the second ordered value (2). Thus, the model is based on Pr{coronary artery disease}. It is always important to check the "Response Profile" table to ensure that PROC LOGISTIC is ordering response variable values the way you want. You can also use the ORDER= option in the PROC LOGISTIC statement to establish a different set of ordered values, for example, by creating formats for the levels of the response variable and using ORDER=FORMATTED.

Output 8.1   Response Profile

```
                    Response Profile

        Ordered                        Total
         Value          ca           Frequency

           1             1               42
           2             0               36
```

Output 8.2 contains the goodness-of-fit statistics. Q_P has the value 0.2155, and Q_L has the value 0.2141. Compared to a chi-square distribution with 1 df, these values suggest that the model fits the data adequately. The note that the number of unique profiles is 4 means that these statistics are computed based on the 4 groups that are the rows of Table 8.1, the result of the AGGREGATE option.

Output 8.2   Goodness-of-Fit Statistics

```
            Deviance and Pearson Goodness-of-Fit Statistics

        Criterion        DF        Value     Value/DF     Pr > ChiSq

        Deviance          1        0.2141     0.2141        0.6436
        Pearson           1        0.2155     0.2155        0.6425

                    Number of unique profiles: 4
```

Output 8.3 lists various criteria for assessing model fit through the quality of the explanatory capacity of the model; for $-2 \log L$ and the score statistic, this is done by testing whether the explanatory variables are jointly significant relative to the chi-square distribution. AIC and SC serve a similar purpose while adjusting for the number of explanatory variables in the model. All of these statistics are analogous to the overall F test for the model parameters in a linear regression setting. Refer to the *SAS/STAT User's Guide, Version 8*, for more information on these statistics.

Output 8.3   Testing Joint Significance of the Explanatory Variables

```
                      Model Fit Statistics

                                              Intercept
                                Intercept        and
                 Criterion        Only        Covariates

                 AIC            109.669         101.900
                 SC            112.026         108.970
                 -2 Log L      107.669          95.900
```

8.2.5   Interpretation of Main Effects Model

With the satisfactory goodness of fit, it is appropriate to examine the parameter estimates from the model. Note that these results apply only to the population consisting of those persons who visited this medical clinic and required catheterization. The "Analysis of Maximum Likelihood Estimates" table in Output 8.4 lists the estimated model parameters, their standard errors, Wald chi-square tests, and p-values. A Wald test is a statistic that takes the form of the squared value ratio for the estimate to its standard error; it follows an approximate chi-square distribution when the sample size is sufficiently large. Wald statistics are easy to compute and are based on normal theory; however, their statistical properties are somewhat less optimal than those of the likelihood ratio statistics for small samples. Moreover, when there is concern for the statistical properties of results from small samples, exact methods can be helpful; see Section 8.8.

Output 8.4 Main Effects Model: ANOVA Table

```
              Analysis of Maximum Likelihood Estimates

                                  Standard        Wald
    Parameter    DF   Estimate     Error      Chi-Square   Pr > ChiSq

    Intercept     1    -1.1747     0.4854       5.8571       0.0155
    sex           1     1.2770     0.4980       6.5750       0.0103
    ecg           1     1.0545     0.4980       4.4844       0.0342
```

The variable SEX is significant compared to a significance level of 0.05, with a Wald statistic (usually denoted Q_W) of 6.5750. The variable ECG is also significant, with $Q_W = 4.4844$.

The model equation can be written as follows:

$$\text{logit}(\theta_{hi}) = -1.1747 + 1.2770 \text{ SEX} + 1.0545 \text{ ECG}$$

Table 8.3 lists the parameter interpretations, and Table 8.4 displays the predicted logits and odds of coronary disease.

Table 8.3. Interpretation of Parameters

Parameter	Estimate	Standard Error	Interpretation
α	-1.1747	0.485	log odds of coronary disease for females with ECG < 0.1
β_1	1.2770	0.498	increment to log odds for males
β_2	1.0545	0.498	increment to log odds for high ECG

Table 8.4. Model-Predicted Logits and Odds of CA Disease

Sex	ECG	Logit	Odds of Coronary Artery Disease
Female	< 0.1	$\hat{\alpha} = -1.1747$	$e^{\hat{\alpha}} = e^{-1.1747} = 0.3089$
Female	≥ 0.1	$\hat{\alpha} + \hat{\beta}_2 = -0.1202$	$e^{\hat{\alpha}+\hat{\beta}_2} = e^{-0.1202} = 0.8867$
Male	< 0.1	$\hat{\alpha} + \hat{\beta}_1 = 0.1023$	$e^{\hat{\alpha}+\hat{\beta}_1} = e^{0.1023} = 1.1077$
Male	≥ 0.1	$\hat{\alpha} + \hat{\beta}_1 + \hat{\beta}_2 = 1.1568$	$e^{\hat{\alpha}+\hat{\beta}_1+\hat{\beta}_2} = e^{1.1568} = 3.1797$

The odds ratio for males compared to females is the ratio of the predicted odds of CA disease for males versus females, which, on page 187, was shown to be

$$e^{\hat{\beta}_1} = e^{1.2770} = 3.586$$

Men in the study have three times higher odds for coronary artery disease than women in the study. The odds ratio for ECG ≥ 0.1 versus ECG < 0.1 is the ratio of the predicted odds of CA disease for high ECG versus low ECG, which was shown to be

$$e^{\hat{\beta}_2} = e^{1.0545} = 2.871$$

Those persons with ECG ≥ 0.1 have nearly three times the odds of coronary artery disease as those with ECG < 0.1. This quantity is very similar to the common odds ratio estimates computed by PROC FREQ and displayed in Section 3.3.2 ($\hat{\psi}_{MH} = 2.847$ and $\hat{\psi}_L = 2.859$).

Output 8.5 contains the adjusted odds ratios and their 95% Wald confidence limits. The point estimates have the values calculated above. Neither of the confidence limits includes the value 1 in agreement with the statistical significance of each factor relative to the hypothesis of no association.

Predicted values are easily produced. The OUTPUT statement specifies that predicted values for the first ordered value (CA=1) be put into the variable PROB and output into the SAS data set PREDICT along with the variables from the input data set. You can print these values with the PRINT procedure.

```
proc logistic descending;
   freq count;
   model ca=sex ecg;
   output out=predict pred=prob;
run;
proc print data=predict;
run;
```

Output 8.5 Confidence Limits for Odds Ratios

	Odds Ratio Estimates		
Effect	Point Estimate	95% Wald Confidence Limits	
sex	3.586	1.351	9.516
ecg	2.871	1.082	7.618

The data set PREDICT contains model-predicted values for each observation in the input data set. The created variable named PROB contains these predicted values; the created variable _LEVEL_ tells you that they are the predicted values for the first ordered value, or Pr{coronary artery disease}. Observations 7 and 8 display the predicted value 0.76075 for males with high ECG.

Output 8.6   Predicted Values Output Data Set

Obs	sex	ecg	ca	count	_LEVEL_	prob
1	0	0	0	11	1	0.23601
2	0	0	1	4	1	0.23601
3	0	1	0	10	1	0.46999
4	0	1	1	8	1	0.46999
5	1	0	0	9	1	0.52555
6	1	0	1	9	1	0.52555
7	1	1	0	6	1	0.76075
8	1	1	1	21	1	0.76075

In conclusion, the main effects model is satisfactory. Being male and having ECG ≥ 0.1 are risk indicators for the presence of coronary artery disease for these data. If you can make the argument that this convenience sample is representative of a target group of coronary artery disease patients, possibly those persons who visit clinics on a walk-in basis, then these results may also apply to that population.

8.2.6   Alternative Methods of Assessing Goodness of Fit

There are other strategies available for assessing goodness of fit; these are based on fitting an appropriate expanded model and then evaluating whether the contribution of the additional terms is nonsignificant. If so, you then conclude that the original model has an adequate fit. You can compute likelihood ratio tests for the significance of the additional terms by taking the difference in the log likelihood for both models ($-2 \log L$ in the "Model Fit Statistics" table); this difference has an approximate chi-square distribution with degrees of freedom equal to the difference in the number of parameters in the models. You can also examine the Wald statistic for the additional parameters in order to assess goodness of fit.

For these data, the expanded model would be the one that contains the main effects for sex and ECG and their interaction. The desired likelihood ratio statistic tests the significance of the interaction term and thus serves as a goodness-of-fit test for the main effects model.

You can write this model as

$$
\begin{bmatrix} \text{logit}(\theta_{11}) \\ \text{logit}(\theta_{12}) \\ \text{logit}(\theta_{21}) \\ \text{logit}(\theta_{22}) \end{bmatrix} = \begin{bmatrix} \alpha \\ \alpha & + & & \beta_2 \\ \alpha & + & \beta_1 \\ \alpha & + & \beta_1 & + & \beta_2 & + & \beta_3 \end{bmatrix} = \begin{bmatrix} 1 & 0 & 0 & 0 \\ 1 & 0 & 1 & 0 \\ 1 & 1 & 0 & 0 \\ 1 & 1 & 1 & 1 \end{bmatrix} \begin{bmatrix} \alpha \\ \beta_1 \\ \beta_2 \\ \beta_3 \end{bmatrix}
$$

The model matrix column corresponding to β_3, the interaction term, is constructed by multiplying the columns for β_1 and β_2 together. Note that this model is a *saturated* model, since there are as many parameters as there are logit functions being modeled.

The following SAS code fits this model. Since PROC LOGISTIC now includes a complete model-building facility, you simply cross SEX and ECG in the MODEL statement to specify their interaction. The interaction term in the resulting model matrix has the value 1 if both SEX and ECG are 1; otherwise, it is 0.

```
ods select FitStatistics ParameterEstimates;
proc logistic descending;
    freq count;
    model ca=sex ecg sex*ecg;
run;
```

The resulting tables titled "Model Fit Statistics" and "Analysis of Maximum Likelihood Estimates" follow.

Output 8.7   Results for Saturated Model

```
                        Model Fit Statistics

                                             Intercept
                                 Intercept       and
                   Criterion       Only      Covariates

                   AIC           109.669       103.686
                   SC            112.026       113.112
                   -2 Log L      107.669        95.686


             Analysis of Maximum Likelihood Estimates

                                      Standard        Wald
     Parameter    DF    Estimate       Error      Chi-Square    Pr > ChiSq

     Intercept     1     -1.0116       0.5839       3.0018        0.0832
     sex           1      1.0116       0.7504       1.8172        0.1776
     ecg           1      0.7885       0.7523       1.0985        0.2946
     sex*ecg       1      0.4643       1.0012       0.2151        0.6428
```

The value for -2(log likelihood) is 95.686 for the saturated model; this is the value for -2 Log L listed under "Intercept and Covariates." The value for the main effects model is 95.900 (see Output 8.3), yielding a difference of 0.214. This difference is the likelihood ratio test value, with 1 df (4 parameters for the expanded model $-$ 3 parameters for the main effects model). Compared with a chi-square distribution with 1 df, the non-significance of this statistic supports the adequacy of the main effects model. Note that you can always compute a likelihood ratio test in this manner for the contribution of a particular model term or a set of model terms.

This likelihood ratio test value is the same as the deviance reported for the main effects model in Output 8.2. This is because the deviance statistic is effectively comparing the model for which it is computed with a saturated model.

Note that the p-value for the Wald statistic is 0.2151 for the interaction listed in the "Analysis of Maximum Likelihood Estimates" table. Both the likelihood ratio statistic and the Wald statistic are evaluating the same hypothesis: whether or not the interaction explains any of the variation among the different log odds beyond that explained by the main effects. They support goodness of fit of the main effects model by indicating nonsignificance of the interaction between sex and ECG. The Wald statistic and the likelihood ratio statistic are essentially equivalent for large samples.

8.2.7 Overdispersion

Sometimes a logistic model is considered reasonable, but the goodness-of-fit statistics indicate that too much variation remains (usually the deviance or deviance/df is examined). This condition is known as *overdispersion*, and it occurs when the data do not follow a binomial distribution well; the condition is also known as heterogeneity.

You can model the overdispersion by adjusting, or scaling, the covariance matrix to account for it. This involves the additional estimation of a dispersion parameter, often called a scaling parameter. PROC LOGISTIC allows you to specify a scaling parameter through the use of the SCALE= option; this explains why the SCALE=NONE option is used to generate the goodness-of-fit statistics, including the deviance, when no scale adjustment is desired. McCullagh and Nelder (1989) and Collett (1991) discuss overdispersion comprehensively. The SAS Institute publication *SAS/STAT User's Guide, Version 8*, describes these options in detail. Another method for addressing overdispersion is discussed in Section 15.12 in the context of methods involving generalized estimating equations.

8.3 Using the CLASS Statement

In the previous example, PROC LOGISTIC used the values of the explanatory variables to construct the model matrix. These values were already coded as 0s and 1s. However, often your SAS data set contains a response variable or explanatory variables that have character values. Or, your model may involve a number of explanatory variables and/or interaction terms so that constructing and managing the required terms as variables becomes a chore. The LOGISTIC procedure handles character-valued response variables by creating ordered values based on the alphabetical order of the response variable values. In addition, PROC LOGISTIC now includes a CLASS statement and allows GLM-like model specification so that constructing variables to be model terms is no longer necessary. This next example illustrates how the procedure handles character-valued response variables and how the CLASS statement simplifies the use of classification variables in your model.

8.3.1 Analysis of Sentencing Data

Table 8.5 displays data based on a study on prison sentencing for persons convicted of a burglary or larceny. Investigators collected information on whether there was a prior arrest record and whether the crime was a nonresidential burglary, residential burglary, or something else—usually some sort of larceny. Here, type of crime is divided into nonresidential burglary versus all others. Sentence was recorded as to whether the offender was sent to prison.

Table 8.5. Sentencing Data

Type	Prior Arrest	Prison	No Prison	Total
Nonresidential	Some	42	109	151
Nonresidential	None	17	75	92
Other	Some	33	175	208
Other	None	53	359	412

Assume that these data arise from a stratified simple random sample so that sentence is distributed binomially for each offense type × prior arrest record combination, that is, for each row of Table 8.5. The type of offense by prior arrest status by sentence classification has the product binomial distribution.

$$\Pr\{n_{hij}\} = \prod_{h=1}^{2} \prod_{i=1}^{2} \frac{n_{hi+}!}{n_{hi1}!n_{hi2}!} \theta_{hi}^{n_{hi1}} (1 - \theta_{hi})^{n_{hi2}}$$

The quantity θ_{hi} is the probability that a person arrested for a crime of type h with an ith prior arrest record receives a prison sentence, and n_{hi1} and n_{hi2} are the number of persons of the hth type and ith prior record who did and did not receive prison sentences, respectively (h=1 for nonresidential, h=2 for other; i=1 for prior arrest, i=2 for no arrest).

Similar to the previous example, a useful preliminary model for the sentencing data is one that includes main effects for type of offense and prior arrest record. There are three parameters in this model. The parameter α is the intercept, β_1 is the increment in log odds for committing a nonresidential burglary, and β_2 is the increment in log odds for having a prior arrest record. The probabilities and odds predicted by this model have identical structure to those presented in Table 8.2, replacing the first column with the values Nonresidential and Other and replacing the second column with the values Some and None. The model matrix is identical to the one displayed on page 186.

The following DATA step creates the SAS data set SENTENCE.

```
data sentence;
   input type $ prior $ sentence $ count @@;
   datalines;
nrb    some   y   42 nrb    some   n 109
nrb    none   y   17 nrb    none   n  75
other  some   y   33 other  some   n 175
other  none   y   53 other  none   n 359
;
run;
```

The variable SENTENCE is the response variable, and TYPE and PRIOR are the explanatory variables. Note that SENTENCE is character valued, with values 'y' for prison sentence and 'n' for no prison sentence. PROC LOGISTIC orders these values alphabetically by default so that it bases its model on the probability of the value 'n', or Pr{no prison sentence}. If you want to change the basis of the model to be Pr{prison sentence}, you have to alter this default behavior.

The following group of SAS statements invoke PROC LOGISTIC. Note that since the desired model is based on Pr{prison sentence}, the DESCENDING option is specified to request that 'y' be the first ordered value.

```
proc logistic descending;
   class type prior(ref=first) / param=ref;
   freq count;
   model sentence = type prior / scale=none aggregate;
run;
```

You list your classification variables in the CLASS statement. If you desire the incremental effects parameterization, you specify the option PARAM=REF after a '/'. The procedure provides a number of other parameterizations as well, including the effect (deviation from the mean) parameterization used in PROC CATMOD and the less than full rank parameterization used in PROC GLM and PROC GENMOD. The incremental effects parameterization is a full rank parameterization.

By default, PROC LOGISTIC uses the last ordered value of the explanatory variable as the reference level and assigns it the value 0. If you want another value to be the reference level, you specify it with the REF= option after the slash or after each individual variable, enclosed in parentheses. Here, REF=FIRST indicates that the 'none' level of PRIOR is the reference. You could also specify this value directly with REF='none'. Since 'other' is the last alphanumerical value in TYPE, it becomes the reference value for that effect, which is desired.

The "Response Profile" table indicates that SENTENCE='y' corresponds to the first ordered value. Thus, the model is based on Pr{prison sentence}.

Output 8.8 Response Profiles

```
                        Response Profile

           Ordered                       Total
           Value        sentence       Frequency

                1         y                   145
                2         n                   718
```

The "Class Level Information" table informs you how the model matrix is constructed. The design variables are the values associated with the explanatory variable levels. Since you want PRIOR='some' and TYPE='nrb' to be the incremental effects, the design variables take the value 1 for those levels.

Output 8.9 Class Level Information

```
                  Class Level Information

                                        Design
                                       Variables

          Class        Value              1

          type         nrb                1
                       other              0

          prior        none               0
                       some               1
```

The goodness-of-fit statistics $Q_L = 0.5076$ and $Q_P = 0.5025$ indicate an adequate model fit. Note that if these statistics have values that are dissimilar, it is an indication that sample sizes in the groups are not large enough to support their use as goodness-of-fit statistics.

Output 8.10   Goodness of Fit

```
          Deviance and Pearson Goodness-of-Fit Statistics

     Criterion        DF         Value     Value/DF    Pr > ChiSq

     Deviance          1        0.5076      0.5076       0.4762
     Pearson           1        0.5025      0.5025       0.4784

                   Number of unique profiles: 4
```

Since there are CLASS variables in the model, PROC LOGISTIC prints out the "TYPE III Analysis of Effects" table. These are Wald tests for the effects. Since both TYPE and PRIOR have 1 df, these tests are the same as for the parameter estimates in Output 8.12

Output 8.11   TYPE III Analysis of Effects

```
                   Type III Analysis of Effects

                                   Wald
              Effect      DF    Chi-Square    Pr > ChiSq

              type         1       9.0509       0.0026
              prior        1       3.3127       0.0687
```

The variable TYPE is clearly significant, with $Q_W = 9.0509$. The variable PRIOR nearly approaches significance, with $Q_W = 3.3127$ and $p = 0.0687$. While some analysts might delete any effects that do not meet their designated 0.05 significance level, it is sometimes reasonable to keep modestly suggestive effects in the model to avoid potential bias for estimates of the other effects. In fact, for main effects models where presumably each explanatory variable chosen has some potential basis for its inclusion, many analysts keep all effects in the model, regardless of their significance. The model still appropriately describes the data, and it is easier to compare with other researchers' models where those nonsignificant effects may prove to be more important.

Output 8.12   Main Effects Model

```
              Analysis of Maximum Likelihood Estimates

                                   Standard      Wald
     Parameter         DF   Estimate    Error  Chi-Square   Pr > ChiSq

     Intercept          1    -1.9523   0.1384    199.0994      <.0001
     type      nrb      1     0.5920   0.1968      9.0509      0.0026
     prior     some     1     0.3469   0.1906      3.3127      0.0687
```

However, you may want to consider removing modest or clearly nonsignificant effects if some of them are redundant; that is, they are reflecting essentially the same factor. This can induce collinearity, and sometimes the association of explanatory variables with each other may mask the true effect. The additional model terms lead to poorer quality of the individual parameter estimates since they will be less precise (higher standard errors). In this case, PRIOR is kept in the model.

The model equation can be written as follows:

$$\text{logit}(\theta_{hi}) = -1.9523 + 0.5920 \text{ TYPE } + 0.3469 \text{ PRIOR}$$

The "Analysis of Maximum Likelihood Estimates" table for this model is displayed in Output 8.12. The estimates of the βs are printed as well as standard errors and significance tests. Output 8.13 displays the odds ratio estimates and confidence limits. The odds ratios are 1.808 ($e^{0.5920}$) for type of offense and 1.415 ($e^{0.3469}$) for prior arrest record. Thus, those persons committing a nonresidential burglary have nearly twice the odds of receiving prison sentences as those committing another offense. Those with a prior arrest record are somewhat more likely to receive a prison sentence than those with no prior record.

Output 8.13   Odds Ratio Estimates and Confidence Limits

```
                        Odds Ratio Estimates

                            Point           95% Wald
           Effect        Estimate      Confidence Limits

           type  nrb vs other     1.808     1.229     2.658
           prior some vs none     1.415     0.974     2.056
```

8.3.2   Requesting Goodness-of-Fit Statistics for Single Main Effect Model

Suppose that you did decide to fit the model with a single main effect, TYPE, and you wanted to generate the appropriate goodness-of-fit statistics for that model. Using the SCALE=NONE and AGGREGATE options would not work for this model, since the AGGREGATE option creates groups on which to base the goodness-of-fit statistic according to the values of the explanatory variables. Since there is just one dichotomous explanatory variable remaining in the model, only two groups would be created. To produce the groups consistent with the sampling framework, you need to specify AGGREGATE=(TYPE PRIOR), where the list of variables inside the parentheses are those whose unique values determine the rows of Table 8.5.

The following statements request the main effects model. The ODS SELECT statement restricts the output to the goodness-of-fit information.

```
ods select GoodnessOfFit;
proc logistic descending;
   class type prior (ref=first) / param=ref;
   freq count;
   model sentence = type / scale=none aggregate=(type prior);
run;
```

Output 8.14 includes the goodness-of-fit statistics. Note the SAS message that there are 4 unique covariate profiles; this tells you that the correct groups were formed and that the statistics are based on the intended subpopulations.

Output 8.14   Single Effect Model

```
              Deviance and Pearson Goodness-of-Fit Statistics

       Criterion        DF        Value      Value/DF     Pr > ChiSq

       Deviance          2        3.8086       1.9043        0.1489
       Pearson           2        3.7527       1.8763        0.1532

                    Number of unique profiles: 4
```

Since $Q_L = 3.8086$ and $Q_P = 3.7527$, both with 2 df and p-values of about 0.15, this single main effect model has a satisfactory fit.

8.3.3   Deviation from the Mean Parameterization

The preceding example used incremental effects parameterization, also called reference cell parameterization. However, that is not the default parameterization for the LOGISTIC procedure. If you do not specify the PARAM= option, you would obtain *deviation from the mean* parameterization, also known as *effect* parameterization. You can specify this explicitly in PROC LOGISTIC with the option PARAM=EFFECT in the CLASS statement. Note that this is also the default parameterization used in the CATMOD procedure.

In this parameterization, also a full rank parameterization like the incremental effects parameterization, the effects are differential rather than incremental. This model is written as follows:

$$
\begin{bmatrix} \text{logit}(\theta_{11}) \\ \text{logit}(\theta_{12}) \\ \text{logit}(\theta_{21}) \\ \text{logit}(\theta_{22}) \end{bmatrix} = \begin{bmatrix} \alpha & + & \beta_1 & + & \beta_2 \\ \alpha & + & \beta_1 & - & \beta_2 \\ \alpha & - & \beta_1 & + & \beta_2 \\ \alpha & - & \beta_1 & - & \beta_2 \end{bmatrix} = \begin{bmatrix} 1 & 1 & 1 \\ 1 & 1 & -1 \\ 1 & -1 & 1 \\ 1 & -1 & -1 \end{bmatrix} \begin{bmatrix} \alpha \\ \beta_1 \\ \beta_2 \end{bmatrix}
$$

Here, α is the average log odds (across the four populations) of a prison sentence, β_1 is the average differential change in log odds for whether a nonresidential burglary was committed, and β_2 is the differential change in log odds for having a prior arrest record. β_1 is an added amount for a nonresidential burglary and a subtracted amount for other burglary. β_2 is an added amount for a prior arrest record and a subtracted amount for no previous arrest record. The formulas for the model-predicted probabilities and odds for this parameterization are listed in Table 8.6.

Table 8.6. Model-Predicted Probabilities and Odds

Type	Prior Arrest	Pr{Prison}	Odds of Prison
Nonresidential	Some	$e^{\alpha+\beta_1+\beta_2}/(1+e^{\alpha+\beta_1+\beta_2})$	$e^{\alpha+\beta_1+\beta_2}$
Nonresidential	None	$e^{\alpha+\beta_1-\beta_2}/(1+e^{\alpha+\beta_1-\beta_2})$	$e^{\alpha+\beta_1-\beta_2}$
Other	Some	$e^{\alpha-\beta_1+\beta_2}/(1+e^{\alpha-\beta_1+\beta_2})$	$e^{\alpha-\beta_1+\beta_2}$
Other	None	$e^{\alpha-\beta_1-\beta_2}/(1+e^{\alpha-\beta_1-\beta_2})$	$e^{\alpha-\beta_1-\beta_2}$

The odds of a prison sentence for nonresidential burglary (nrb) versus other is obtained by forming the ratio of the odds for nrb versus other for either prior arrest level. Using some prior arrest, this is computed as

$$\frac{e^{\alpha+\beta_1+\beta_2}}{e^{\alpha-\beta_1+\beta_2}} = e^{2\beta_1}$$

The odds of a prison sentence for some arrest record versus none is obtained by forming the ratio of the odds for some prior arrest versus no prior arrest for either level of burglary type. Using nrb, this is computed as

$$\frac{e^{\alpha+\beta_1+\beta_2}}{e^{\alpha+\beta_1-\beta_2}} = e^{2\beta_2}$$

Thus, with this parameterization for a two-level explanatory variable, you need to exponentiate twice the parameter estimates to calculate the odds ratios, instead of simply exponentiating them, as was true for the reference cell model. However, this is taken care of by the LOGISTIC procedure.

The following SAS statements request an analysis of the sentencing data with the differential effects parameterization.

```
ods select ClassLevelInfo GoodnessOfFit
          ParameterEstimates OddsRatios;
proc logistic data=sentence descending;
   class type prior(ref='none');
   freq count;
   model sentence = type prior / scale=none aggregate;
run;
```

Since not all of the output from the LOGISTIC procedure is desired, the ODS SELECT statement is used to request that only specific tables be generated. Since no PARAM= option is specified, the differential effects parameterization is used.

The "Class Level Information" table details the way in which the parameterization is constructed. The values of the CLASS variables are ordered alphanumerically, and the first ordered value gets the value 1 and the second gets the value -1, as illustrated in the case of variable TYPE. Since REF='none' was specified for variable PRIOR, the -1 is assigned to 'none' as the reference level, and the 1 is assigned to the value 'some'.

Output 8.15   Class Level Information

```
               Class Level Information

                                      Design
                                     Variables

          Class     Value               1

          type      nrb                 1
                    other              -1

          prior     none               -1
                    some                1
```

Next, the goodness-of-fit statistics Q_P and Q_L have the values 0.5025 and 0.5076 respectively, the same as in the analysis with the incremental effects parameterization. In both cases, the test is assessing the same effect. A geometric way of looking at this is to say that the sets of explanatory variables for the two parameterizations span the same space, and so their estimated parameters produce the same predicted values.

Output 8.16   Goodness of Fit

```
       Deviance and Pearson Goodness-of-Fit Statistics

    Criterion        DF       Value     Value/DF     Pr > ChiSq

    Deviance          1       0.5076     0.5076         0.4762
    Pearson           1       0.5025     0.5025         0.4784

              Number of unique profiles: 4
```

Output 8.17 displays the "Analysis of Maximum Likelihood Estimates" table.

Output 8.17   Analysis of Maximum Likelihood Estimates

```
           Analysis of Maximum Likelihood Estimates

                                       Standard      Wald
    Parameter          DF    Estimate     Error   Chi-Square   Pr > ChiSq

    Intercept           1     -1.4828    0.0951    243.2458      <.0001
    type      nrb       1      0.2960    0.0984      9.0509      0.0026
    prior     some      1      0.1735    0.0953      3.3127      0.0687
```

However, the parameter estimates are very different. This is because they represent very different quantities. The intercept is now the average log odds (across the four populations) of a prison sentence and the other parameters are the differential changes in the log odds for prior arrest and type of offense.

Output 8.18 displays the "Odds Ratio Estimates" table.

Output 8.18   Odds Ratio Estimates

Odds Ratio Estimates			
Effect	Point Estimate	95% Wald Confidence Limits	
type nrb vs other	1.808	1.229	2.658
prior some vs none	1.415	0.974	2.056

The estimate for the odds ratio for a prison sentence comparing nonresidential burglary to other is $e^{2\beta_1} = 1.808$, which is the exponentiation of 2×0.2960; thus, PROC LOGISTIC has computed the odds ratio correctly. Similarly, the odds ratio for a prison sentence comparing prior arrest record to no arrest record is 1.415. The confidence limits for these point estimates are (1.229, 2.658) and (0.974, 2.056), respectively.

8.4   Qualitative Explanatory Variables

The previous examples have been concerned with analyses of dichotomous outcomes when the explanatory variables were also dichotomous. However, explanatory variables can be nominal (qualitative) with three or more levels, ordinal, or continuous. Logistic regression allows for any combination of these types of explanatory variables. This section is concerned with handling explanatory variables that are qualitative and contain three or more levels.

The following data come from a study on urinary tract infections (Koch, Imrey, et al. 1985). Investigators applied three treatments to patients who had either a complicated or uncomplicated diagnosis of urinary tract infection. Since complicated cases of urinary tract infections are difficult to cure, investigators were interested in whether the pattern of treatment differences are the same across diagnoses: did the diagnosis status of the patients affect the relative effectiveness of the three treatments? This is the same as determining whether there is a treatment × diagnosis interaction. Diagnosis is a dichotomous explanatory variable and treatment is a nominal explanatory variable consisting of levels for treatments A, B, and C. Table 8.7 displays the data.

Table 8.7. Urinary Tract Infection Data

Diagnosis	Treatment	Cured	Not Cured	Proportion Cured
Complicated	A	78	28	0.736
Complicated	B	101	11	0.902
Complicated	C	68	46	0.596
Uncomplicated	A	40	5	0.889
Uncomplicated	B	54	5	0.915
Uncomplicated	C	34	6	0.850

These data can be assumed to arise from a stratified simple random sample so that the response (cured or not cured) is distributed binomially for each diagnosis × treatment combination, that is, for each row of Table 8.7. The diagnosis by treatment classification has the product binomial distribution.

$$\Pr\{n_{hij}\} = \prod_{h=1}^{2} \prod_{i=1}^{3} \frac{n_{hi+}!}{n_{hi1}! n_{hi2}!} \theta_{hi}^{n_{hi1}} (1 - \theta_{hi})^{n_{hi2}}$$

The quantity θ_{hi} is the probability that a person with the hth diagnosis receiving the ith treatment is cured, and n_{hi1} and n_{hi2} are the numbers of patients of the hth diagnosis and ith treatment who were and were not cured, respectively ($h = 1$ for complicated, $h = 2$ for uncomplicated; $i = 1$ for treatment A, $i = 2$ for treatment B, $i = 3$ for treatment C). You can then apply the logistic model to describe the variation among the $\{\theta_{hi}\}$. This is the same likelihood function as in the previous example except that i takes on the values 1, 2, and 3 instead of 1, 2.

8.4.1   Model Fitting

Since there is interest in the interaction term, the preliminary model includes main effects and their interaction (saturated model). There is one parameter for the intercept (α), which is the reference parameter corresponding to the log odds of being cured if you have an uncomplicated diagnosis and are getting treatment C. The parameter β_1 is the increment for complicated diagnosis. The effect for treatment consists of two parameters: β_2 is the incremental effect for treatment A, and β_3 is the incremental effect for treatment B.

There is no particular reason to choose a parameterization that includes incremental effects for treatments A and B; you could choose to parameterize the model by including incremental effects for treatments A and C. Often, data analysts choose the reference parameter to be the control group, with incremental effects representing various exposure effects. However, it's important to note that an effect with L levels must be represented by $(L - 1)$ parameters.

The interaction effect is comprised of two additional parameters, β_4 and β_5, which represent the interaction terms for complicated diagnosis and treatment A, and complicated diagnosis and treatment B, respectively. When you are creating interaction terms from two effects, you create a number of terms equal to the product of the number of terms for both effects.

You can write this saturated model in matrix formulation as

$$
\begin{bmatrix}
\text{logit}(\theta_{11}) \\
\text{logit}(\theta_{12}) \\
\text{logit}(\theta_{13}) \\
\text{logit}(\theta_{21}) \\
\text{logit}(\theta_{22}) \\
\text{logit}(\theta_{23})
\end{bmatrix}
=
\begin{bmatrix}
\alpha + \beta_1 + \beta_2 \quad\;\; + \beta_4 \\
\alpha + \beta_1 \quad\;\; + \beta_3 \quad\;\;\; + \beta_5 \\
\alpha + \beta_1 \\
\alpha \quad\;\; + \beta_2 \\
\alpha \quad\quad\;\; + \beta_3 \\
\alpha
\end{bmatrix}
=
\begin{bmatrix}
1 & 1 & 1 & 0 & 1 & 0 \\
1 & 1 & 0 & 1 & 0 & 1 \\
1 & 1 & 0 & 0 & 0 & 0 \\
1 & 0 & 1 & 0 & 0 & 0 \\
1 & 0 & 0 & 1 & 0 & 0 \\
1 & 0 & 0 & 0 & 0 & 0
\end{bmatrix}
\begin{bmatrix}
\alpha \\
\beta_1 \\
\beta_2 \\
\beta_3 \\
\beta_4 \\
\beta_5
\end{bmatrix}
$$

Note that if you had parameterized the model so that there were three columns for treatment effects, each consisting of 1s corresponding to those logits representing the respective treatments, the columns would add up to a column of 1s. This would be redundant with the column of 1s for the intercept, and so PROC LOGISTIC would set the parameter corresponding to the third column of the effect equal to zero, since it is a linear combination of other columns. You could of course fit this model by creating indicator variables both for the incremental effects and for their interactions. You would need two indicator variables for the incremental effects for treatment A and treatment B, one indicator variable for complicated diagnosis, and two indicator variables for the interaction of diagnosis and treatment. However, you can perform this analysis much more easily by using a CLASS statement.

8.4.2 PROC LOGISTIC for Nominal Effects

The following DATA step creates SAS data set UTI.

```
data uti;
   input diagnosis : $13. treatment $ response $ count @@;
   datalines;
complicated     A  cured 78  complicated   A not 28
complicated     B  cured 101 complicated   B not 11
complicated     C  cured 68  complicated   C not 46
uncomplicated   A  cured 40  uncomplicated A not 5
uncomplicated   B  cured 54  uncomplicated B not 5
uncomplicated   C  cured 34  uncomplicated C not 6
;
run;
```

Since this model is saturated, the goodness-of-fit statistics don't apply; there are no available degrees of freedom because the number of groups and the number of parameters are the same (6). PROC LOGISTIC prints out near-zero values and zero df for saturated models. However, fitting this model does allow you to determine whether there is an interaction effect. Fitting the reduced model without the interaction terms and taking the difference in the deviances allows you to determine whether the interaction is meaningful. The following PROC LOGISTIC statements fit the full and reduced models.

```
ods select FitStatistics;
proc logistic;
   freq count;
   class diagnosis treatment /param=ref;
   model response = diagnosis|treatment;
run;
```

```
ods select FitStatistics GoodnessOfFit
          TypeIII OddsRatios;
proc logistic;
   freq count;
   class diagnosis treatment;
   model response = diagnosis treatment /
       scale=none aggregate;
run;
```

Output 8.19 contains $-2 \log L$ for the full model, and Output 8.20 contains the $-2 \log L$ for the reduced model.

Output 8.19   Log Likelihood for the Full Model

	Model Fit Statistics		
Criterion	Intercept Only	Intercept and Covariates	
AIC	494.029	459.556	
SC	498.194	484.549	
-2 Log L	492.029	447.556	

Output 8.20   Log Likelihood for the Reduced Model

	Model Fit Statistics		
Criterion	Intercept Only	Intercept and Covariates	
AIC	494.029	458.071	
SC	498.194	474.733	
-2 Log L	492.029	450.071	

The difference between 447.556 (full) and 450.071 (reduced) is 2.515; since the difference in the number of parameters in these models is 2, this value should be compared to a chi-square distribution with 2 df (you can use the PROBCHI function to compute the probability with the DATA step). Thus, the likelihood ratio test for the hypothesis that the additional terms in the expanded model are zero cannot be rejected. The interaction between treatment and diagnosis is not significant. This test also serves as the goodness-of-fit test for the reduced model, which is the main effects model; it supports the model's adequacy.

Output 8.21 contains the goodness-of-fit statistics Q_P and Q_L. Note that Q_L has the same value as the likelihood ratio statistic; thus, you could have simply fit the main effects model and used Q_L as the test for interaction, knowing that the two omitted terms were the two interaction terms.

Output 8.21   Goodness-of-Fit Statistics

```
        Deviance and Pearson Goodness-of-Fit Statistics

      Criterion      DF      Value     Value/DF    Pr > ChiSq

      Deviance        2      2.5147     1.2573       0.2844
      Pearson         2      2.7574     1.3787       0.2519

              Number of unique profiles: 6
```

The following "TYPE III Analysis of Effects" table is from the main effects model.

Output 8.22   Main Effects Model

```
                Type III Analysis of Effects

                                  Wald
        Effect        DF      Chi-Square    Pr > ChiSq

        diagnosis      1       10.2885        0.0013
        treatment      2       24.6219        <.0001
```

Note that in the previous examples, the Wald test for the interaction term could also be used as a goodness-of-fit test for the main effects model. However, since in this case the interaction consists of two terms, you can't get a test for the total interaction effect from this table. If both interaction terms are significant, you can often assume that the overall interaction is also significant (Section 8.4.3 shows how to construct a test for the total interaction using the CONTRAST statement).

Output 8.23 contains the odds ratio estimates and their confidence limits, which are the 95% Wald confidence limits. None of these limits contain the value 1, indicating that there are sigificant treatment and diagnosis effects.

Output 8.23   Odds Ratio Estimates

```
                    Odds Ratio Estimates

                                          Point        95% Wald
      Effect                            Estimate    Confidence Limits

      diagnosis complicated vs uncomplicated  0.382    0.212      0.688
      treatment A vs C                        1.795    1.069      3.011
      treatment B vs C                        4.762    2.564      8.847
```

You have 4.8 times higher odds of being cured if you get treatment B compared with treatment C, and 1.8 times higher odds of being cured if you get treatment A compared to treatment C. You have 0.38 times lower odds of being cured if you have a complicated diagnosis as compared to an uncomplicated diagnosis; you have $(1/0.382) = 2.6$ times higher odds of being cured if you have uncomplicated diagnosis compared with complicated diagnosis. Note that all these odds ratios have been adjusted for the other explanatory variable.

To confirm what these odds ratios represent, consider the the model-predicted probabilities and odds listed in Table 8.8. Taking the ratio of odds for complicated diagnosis and treatment A versus complicated diagnosis and treatment C yields e^{β_2}. A similar exercise for treatment B yields e^{β_3}. To determine the odds ratio for complicated diagnosis to uncomplicated diagnosis, take the ratio of the odds for complicated to uncomplicated diagnosis at any level of treatment. You should get e^{β_1}.

Table 8.8.　Model-Predicted Probabilities and Odds

Diagnosis	Treatment	Pr{Cured}	Odds of Cured
Complicated	A	$e^{\alpha+\beta_1+\beta_2}/(1 + e^{\alpha+\beta_1+\beta_2})$	$e^{\alpha+\beta_1+\beta_2}$
Complicated	B	$e^{\alpha+\beta_1+\beta_3}/(1 + e^{\alpha+\beta_1+\beta_3})$	$e^{\alpha+\beta_1+\beta_3}$
Complicated	C	$e^{\alpha+\beta_1}/(1 + e^{\alpha+\beta_1})$	$e^{\alpha+\beta_1}$
Uncomplicated	A	$e^{\alpha+\beta_2}/(1 + e^{\alpha+\beta_2})$	$e^{\alpha+\beta_2}$
Uncomplicated	B	$e^{\alpha+\beta_3}/(1 + e^{\alpha+\beta_3})$	$e^{\alpha+\beta_3}$
Uncomplicated	C	$e^{\alpha}/(1 + e^{\alpha})$	e^{α}

PROC LOGISTIC can also produce confidence limits for the odds ratios that are likelihood-ratio based. These are also known as profile likelihood confidence intervals. They are particularly desirable when the sample sizes are only moderately large rather than very large. The following PROC LOGISTIC invocation requests profile likelihood confidence intervals for the odds ratios with the CLODDS=PL option. It also requests profile likelihood confidence intervals for the regression parameters with the CLPARM=PL option (the CLPARM=WALD option specifies confidence intervals for the parameters based on asymptotic normality of the parameter estimates).

```
ods select ClparmPL CloddsPL;
proc logistic;
   freq count;
   class diagnosis treatment;
   model response = diagnosis treatment /
      scale=none aggregate clodds=pl clparm=pl;
run;
```

Output 8.24 displays the output produced by the CLODDS and CLPARM options.

Output 8.24   Confidence Limits for Odds Ratios

```
                  Profile Likelihood Confidence Interval for Parameters

            Parameter                      Estimate    95% Confidence Limits

            Intercept                        1.6528      1.3621      1.9751
            diagnosis     complicated       -0.4808     -0.7897     -0.1987
            treatment     A                 -0.1304     -0.4618      0.2052
            treatment     B                  0.8456      0.4756      1.2523


          Profile Likelihood Confidence Interval for Adjusted Odds Ratios

      Effect                              Unit    Estimate    95% Confidence Limits

      diagnosis complicated vs uncomplicated    1.0000      0.382     0.206     0.672
      treatment A vs C                          1.0000      1.795     1.074     3.031
      treatment B vs C                          1.0000      4.762     2.615     9.085
```

If you compare the confidence intervals in Output 8.24 and Output 8.23, you will find that they are similar.

8.4.3 Testing Hypotheses about the Parameters

In the previous analysis, the overall effect for treatment was significant and so were the individual incremental effects parameters for treatment A and treatment B. However, you may also be interested in determining whether the effect for treatment A is different from the effect for treatment B. In addition, you may also want the odds ratio for the comparison of treatment A and treatment B, because, as pointed out above, the odds ratios produced by default are for treatments A and B relative to treatment C. You can request both the comparison test and the odds ratio with the CONTRAST statement in PROC LOGISTIC.

If you recall the likelihood ratio test strategies, it should be clear that you can generate this test by computing the likelihood ratio test for the main effects model compared to the model containing the diagnosis effect only. In fact, if you do this, you will obtain a likelihood ratio test of $478.185 - 450.071 = 28.114$, clearly significant with 2 df.

In order to assess whether any of the treatments are similar, linear combinations of the parameters are tested to see if they are significantly different from zero.

$$H_0: \mathbf{L}\beta = 0$$

By choosing the appropriate elements of $\mathbf{L}$, you can construct linear combinations of the parameters that will produce the test of interest. The Wald statistic for a given linear combination $\mathbf{L}$ is computed as

$$Q_W = (\mathbf{L}\hat{\beta})'(\mathbf{L}\ \mathbf{V}(\hat{\beta})\mathbf{L}')^{-1}(\mathbf{L}\hat{\beta})$$

where $\hat{\beta}$ is the vector of parameter estimates. Q_W follows the chi-square distribution with degrees of freedom equal to the number of linearly independent rows of $\mathbf{L}$.

The test for whether treatment A is equivalent to treatment B is expressed as

$$H_0: \beta_2 - \beta_3 = 0$$

which correspond to $\mathbf{L} = [1 \; -1]$ for $\beta = [\beta_2, \beta_3]$. The test for whether treatment A is equivalent to treatment C is expressed as

$$H_0: \beta_2 = 0$$

since, according to the model parameterization, β_2 is an incremental effect for treatment A in reference to Treatment C. If β_2 equals zero, then treatment A is the same as treatment C, and the intercept represents the logit for uncomplicated diagnosis for either treatment A or treatment C. You follow the same logic to see whether treatment B is equivalent to treatment C. The $\mathbf{L}$ for this third contrast is $\mathbf{L} = [1 \; 0, 0 \; 1]$.

To compute the Wald test for the joint effect of treatment A and treatment B relative to treatment C (or the equality of treatments A, B, and C to one another), you test the hypothesis

$$H_0: \beta_2 = \beta_3 = 0$$

This is the hypothesis tested in the "Type III Analysis of Effects" table. The $\mathbf{L}$ for this third contrast is

$$\begin{bmatrix} 1 & 0 \\ 0 & 1 \end{bmatrix}$$

You specify these hypotheses in the CONTRAST statement. You list each hypothesis on a different statement, providing a name for the test within quotes. This can be up to 256 characters long. You then list the effect variable name and provide the coefficients for the $\mathbf{L}$ matrix. The following CONTRAST statements request the test comparing A and B, the individual test for A, and the joint test for A, B, and C.

```
ods select ContrastTest ContrastEstimate;
proc logistic;
   freq count;
   class diagnosis treatment /param=ref;
   model response = diagnosis treatment;
   contrast 'B versus A' treatment -1 1
         / estimate=exp;
   contrast 'A' treatment 1 0;
   contrast 'joint test' treatment 1 0,
                         treatment 0 1;
run;
```

The ESTIMATE= option in the CONTRAST statement requests the estimate of the linear combination $\mathbf{L}\beta$. The ESTIMATE=EXP option requests that the estimate be produced and exponentiated. Recall that the odds ratio for being cured for treatment B compared to treatment A is $e^{\beta_3 - \beta_2}$. Thus, the ESTIMATE=EXP option should produce the correct quantity.

Output 8.25 contains the results. With a Wald chi-square of 8.6919 and a p-value of 0.0032, clearly treatments A and B are significantly different. The joint test statistic has the value 24.6219, which is the same as displayed in the "Type III Analysis of Effects" for the treatment effect.

Output 8.25 Contrast Test Results

```
                     Contrast Test Results

                                 Wald
        Contrast       DF    Chi-Square    Pr > ChiSq

        B versus A      1        8.6919        0.0032
        A               1        4.9020        0.0268
        joint test      2       24.6219       <.0001
```

Output 8.26 contains the results of the contrast estimation.

Output 8.26 Contrast Estimation Results

```
          Contrast Rows Estimation and Testing Results

                                 Standard          Lower    Upper
  Contrast    Type    Row  Estimate    Error  Alpha  Limit    Limit

  B versus A  EXP      1    2.6539    0.8786   0.05  1.3870   5.0778

          Contrast Rows Estimation and Testing Results

                                      Wald
        Contrast    Type     Row  Chi-Square    Pr > ChiSq

        B versus A  EXP       1      8.6919        0.0032
```

The point estimate for the odds ratio is 2.6539, with a lower limit of 1.3870 and an upper limit of 5.0778. This means that those on treatment B have 2.65 times higher odds of being cured than those on treatment A.

8.5 Continuous and Ordinal Explanatory Variables

8.5.1 Goodness of Fit

Frequently, some or all of the explanatory variables in a logistic regression analysis are continuous. Analysis strategies are the same as those described in previous sections, except in the evaluation of goodness of fit.

The following data are from the same study on coronary artery disease as previously analyzed; in addition, the continuous variable AGE is an explanatory variable. The variable ECG is now treated as an ordinal variable, with values 0, 1, and 2. ECG is coded 0 if the ST segment depression is less than 0.1, 1 if it equals 0.1 or higher but less than 0.2, and 2 if the ST segment depression is greater than or equal to 0.2. The variable AGE is age in years.

```
data coronary;
   input sex ecg age ca @@  ;
   datalines;
0 0 28 0    1 0 42 1    0 1 46 0    1 1 45 0
0 0 34 0    1 0 44 1    0 1 48 1    1 1 45 1
0 0 38 0    1 0 45 0    0 1 49 0    1 1 45 1
0 0 41 1    1 0 46 0    0 1 49 0    1 1 46 1
0 0 44 0    1 0 48 0    0 1 52 0    1 1 48 1
0 0 45 1    1 0 50 0    0 1 53 1    1 1 57 1
0 0 46 0    1 0 52 1    0 1 54 1    1 1 57 1
0 0 47 0    1 0 52 1    0 1 55 0    1 1 59 1
0 0 50 0    1 0 54 0    0 1 57 1    1 1 60 1
0 0 51 0    1 0 55 0    0 2 46 1    1 1 63 1
0 0 51 0    1 0 59 1    0 2 48 0    1 2 35 0
0 0 53 0    1 0 59 1    0 2 57 1    1 2 37 1
0 0 55 1    1 1 32 0    0 2 60 1    1 2 43 1
0 0 59 0    1 1 37 0    1 0 30 0    1 2 47 1
0 0 60 1    1 1 38 1    1 0 34 0    1 2 48 1
0 1 32 1    1 1 38 1    1 0 36 1    1 2 49 0
0 1 33 0    1 1 42 1    1 0 38 1    1 2 58 1
0 1 35 0    1 1 43 0    1 0 39 0    1 2 59 1
0 1 39 0    1 1 43 1    1 0 42 0    1 2 60 1
0 1 40 0    1 1 44 1
;
run;
```

Look at the values listed for AGE. While some observations share the same AGE value, most of these values are unique. Thus, there will be only one observation in most of the cells created by the cross-classification of the explanatory variable values. In fact, the SEX by ECG by AGE cross-classification produces 68 groups from these 78 observations. This means that the sample size requirement for the use of the Pearson chi-square goodness-of-fit test and the likelihood ratio goodness-of-fit test—that each predicted cell count tends to be at least 5—is not met. This is almost always the case when you have continuous explanatory variables.

There are several alternative strategies. First, you can fit the desired model, fit an appropriate expanded model with additional explanatory variables, and look at the differences in the log-likelihood ratio statistics. This difference is distributed as chi-square with degrees of freedom equal to the difference in degrees in freedom of the two models (given sufficiently large samples to support approximate normal estimates from the expanded model).

The second strategy is to examine the residual score statistic, Q_{RS} (Breslow and Day 1980). This criterion is directed at the extent to which the residuals from the model are linearly associated with other potential explanatory variables. If there is an association, this is an indication that these variables should also be included in the model. Thus, to compute the residual score statistic, you need to have access to the variables that comprise the potential expansion. Q_{RS} is distributed as chi-square, with degrees of freedom equal to the difference in the number of parameters for the two models.

However, unlike computing the log-likelihood ratio statistic where you have to execute PROC LOGISTIC twice and form the difference of the log-likelihood ratio statistics, you can generate this score goodness-of-fit statistic with one invocation of PROC LOGISTIC. You do this by taking advantage of the LOGISTIC procedure's model-building capabilities. The SELECTION=FORWARD method adds variables to your model in the manner in which you specify, computing model assessment statistics for each of the models it fits. In addition, it prints a score statistic that assesses the joint contribution of the remaining model effects that have not yet been incorporated into the model. With the right choice of model effects in the MODEL statement, this is the score goodness-of-fit statistic. You can also generate the constituent one degree of freedom score tests by including the DETAILS option in the MODEL statement.

A third strategy is to compute an alternative goodness-of-fit statistic proposed by Hosmer and Lemeshow (1989). This test places subjects into deciles based on the model-predicted probabilities, then computes a Pearson chi-square test based on the observed and expected number of subjects in the deciles. The statistic is compared to a chi-square distribution with t degrees of freedom, where t is the number of decile groups minus 2. Depending on the number of observations, there may be less than ten groups. PROC LOGISTIC prints this statistic when you specify the LACKFIT option in the MODEL statement. You should note that this method may have low power for detecting departures from goodness of fit, and so some caution may be needed in its interpretation.

8.5.2 Fitting a Main Effects Model

A model of interest for these data is a main effects model with terms for sex, ECG, and age. To generate a score statistic, you need to choose the effects that constitute the expanded model. Your choice depends partially on the sample size. There should be at least 5 observations for the rarer outcome per parameter being considered in the expanded model. Some analysts would prefer at least 10. In this data set, there are 37 observations with no coronary artery disease and 41 observations with coronary artery disease. Thus, no coronary artery disease is the rarer event, and the quotient 37/5 suggests that 7–8 parameters can be supported.

For these data, an appropriate expanded model consists of all second-order terms, which are the squared terms for age and ECG plus all pairwise interactions. This creates eight parameters beyond the intercept. One might also include the third-order terms, but their inclusion would result in too few observations per parameter for the necessary sample size requirements for these statistics. If there did happen to be substantial third-order variation, this approach would not be appropriate.

The following PROC LOGISTIC statements fit the main effects model and compute the score test. The first- and second-order terms are listed on the right-hand side of the MODEL statement, with CA as the response variable. SELECTION=FORWARD is specified as a MODEL statement option after a '/'. The option INCLUDE=3 requests that the first three terms listed in the MODEL statement are to be included in each fitted model. PROC LOGISTIC first fits this model, which is the main effects model, and then produces the score goodness-of-fit statistic.

```
proc logistic descending;
model ca=sex ecg age ecg*ecg age*age
        sex*ecg sex*age ecg*age /
        selection=forward include=3 details lackfit;
run;
```

Note that 1 is the first ordered value, since the DESCENDING option was specified in the PROC statement, so the model is based on Pr{coronary artery disease}.

Output 8.27   Response Profile

```
                        Response Profile

                Ordered                      Total
                Value            ca        Frequency

                   1             1             41
                   2             0             37



                  Forward Selection Procedure


The following effects will be included in each model:

Intercept   sex   ecg   age

Step   0. The INCLUDE effects were entered.


NOTE: No (additional) effects met the 0.05 significance level for entry into
      the model.
```

After the "Response Profile" table, PROC LOGISTIC prints a list of the variables included in each model. Note that the score statistic printed in the table "Testing Global Null Hypothesis: BETA=0" is not the score goodness-of-fit statistic. This score statistic is strictly testing the hypothesis that the specified model effects are jointly equal to zero.

Output 8.28   Assessing Fit

```
              Testing Global Null Hypothesis: BETA=0

        Test                Chi-Square     DF     Pr > ChiSq

        Likelihood Ratio     21.1145        3      <.0001
        Score                18.5624        3      0.0003
        Wald                 14.4410        3      0.0024
```

The "Residual Chi-Square" is printed after the "Association of Predicted Probabilities and Observed Responses" table. This is the score goodness-of-fit statistic.

Output 8.29   Residual Chi-Square

```
                 Residual Chi-Square Test

         Chi-Square       DF      Pr > ChiSq

           2.3277          5        0.8022
```

Since the difference between the number of parameters for the expanded model and the main effects model is $9 - 4 = 5$, it has 5 degrees of freedom. Since $Q_{RS} = 2.3277$ and $p = 0.8022$, the main effects model fits adequately. The DETAILS option causes the "Analysis of Effects Not in the Model" table to be printed. These tests are the score tests for the addition of the single effects to the model. Each of these tests has one degree of freedom. As one might expect, all of these tests indicate that the single effects add little to the main effects model. Since the sample size requirements for the global test are very roughly met, the confirmation of goodness of fit with the single tests is reasonable, since sample size requirements for these individual expanded models are easily met.

Output 8.30   Analysis of Effects Not in the Model

```
              Analysis of Effects Not in the Model

                                    Score
          Effect      DF       Chi-Square      Pr > ChiSq

          ecg*ecg      1         0.3766          0.5394
          age*age      1         0.7712          0.3798
          sex*ecg      1         0.0352          0.8513
          sex*age      1         0.0290          0.8647
          ecg*age      1         0.8825          0.3475
```

Note that this testing process is conservative with respect to confirming model fit. Inadequate sample size may produce spuriously large chi-squares and correspondingly small p-values. However, this would mean that you decide that the fit is not adequate, and you search for another model. Small sample sizes will not misleadingly cause these methods to suggest that poor fit is adequate, although they would have the limitation of low power to detect real departures from a model.

You may have a concern with the evaluation of multiple tests to assess model goodness of fit. However, by requiring the global test and most single tests to be nonsignificant, the assessment of goodness of fit is more stringent. Also, the multiplicity can be evaluated relative to what might be expected by chance in an assessment of goodness of fit.

Output 8.31 displays the results produced by the LACKFIT option.

Output 8.31   Results from the LACKFIT Option

```
            Partition for the Hosmer and Lemeshow Test

                                  ca = 1              ca = 0
       Group      Total    Observed   Expected   Observed   Expected

          1          8          2       1.02          6       6.98
          2          8          1       1.80          7       6.20
          3          8          3       2.59          5       5.41
          4          8          3       3.42          5       4.58
          5          8          4       4.07          4       3.93
          6          9          6       5.38          3       3.62
          7          9          4       5.97          5       3.03
          8          8          7       5.99          1       2.01
          9          8          7       6.98          1       1.02
         10          4          4       3.77          0       0.23


            Hosmer and Lemeshow Goodness-of-Fit Test

            Chi-Square         DF       Pr > ChiSq

              4.7766            8          0.7812
```

The Hosmer and Lemeshow statistic has a value of 4.7766 with 8 df; $p = 0.7812$. Thus, this measure also supports the model's adequacy for these data. The output also includes the observed and expected counts for each predicted probability decile for each value of the response variable. This criterion can also be used as a measure of goodness of fit for the strictly qualitative explanatory variable situation.

The satisfactory goodness-of-fit statistics make it reasonable to examine the main effects parameter estimates.

Output 8.32   Main Effects Parameter Estimates

```
            Analysis of Maximum Likelihood Estimates

                                    Standard        Wald
       Parameter   DF    Estimate      Error    Chi-Square    Pr > ChiSq

       Intercept    1     -5.6418     1.8061      9.7572        0.0018
       sex          1      1.3564     0.5464      6.1616        0.0131
       ecg          1      0.8732     0.3843      5.1619        0.0231
       age          1      0.0929     0.0351      7.0003        0.0081


                      Odds Ratio Estimates

                        Point           95% Wald
            Effect     Estimate     Confidence Limits

            sex          3.882       1.330      11.330
            ecg          2.395       1.127       5.086
            age          1.097       1.024       1.175
```

The parameter estimates are all significant at the 0.05 level, as judged by the accompanying Wald statistics. Thus, the estimated equation for the log odds is

$$\text{logit}(\theta_{hi}) = -5.6418 + 1.3564 \text{ SEX} + 0.8732 \text{ ECG} + 0.0929 \text{ AGE}$$

Presence of coronary artery disease is positively associated with age and ST segment depression, and it is more likely for males in this population. The odds ratio listed for SEX, 3.882, is the odds of coronary disease presence for males relative to females adjusted for age and ST segment depression. The value listed for ECG, 2.395, is the extent to which the odds of coronary artery disease presence is higher per level increase in ST segment depression. The value 1.097 for AGE is the extent to which the odds is higher each year. A more desirable statistic may be the extent to which the odds of coronary artery disease increase per ten years of age; instead of exponentiating the parameter estimate 0.0929, you compute $e^{10 \times 0.0929}$ to obtain 2.53. Thus, the odds of coronary artery disease increase by a factor of 2.53 every ten years. However, note that this model is useful for prediction only for persons in the walk-in population who fall into the age range of those in this study—ages 28 to 60.

PROC LOGISTIC includes a UNITS statement that enables you to specify the units of change for which you want the odds ratios computed. To obtain the odds ratio for AGE for 10 year units of change, you specify

```
proc logistic descending;
   model ca=sex ecg age;
   units age=10;
run;
```

The following results agree with those calculated by hand.

Output 8.33   Odds Ratios for Units of 10

Adjusted Odds Ratios		
Effect	Unit	Estimate
age	10.0000	2.531

8.6   A Note on Diagnostics

While goodness-of-fit statistics can tell you how well a particular model fits the data, they tell you little about the lack of fit, or where a particular model fails to fit the data. Measures called regression diagnostics have long been useful tools to assess lack of fit for linear regression models, and in the 1980s researchers proposed similar measures for the analysis of binary data. In particular, work by Pregibon (1981) provided the theoretical basis of extending diagnostics used in linear regression to logistic regression. Both Hosmer and Lemeshow (1989) and Collett (1991) include lengthy discussions on model-checking for logistic regression models; Collett includes many references for recent work in this area. Standard texts on regression analysis like Draper and Smith (1981) discuss model-checking strategies for linear regression; Cook and Weisberg (1982) discuss residual analysis and diagnostics extensively.

This section presents a basic description of a few diagnostic tools and an example of their application with the urinary tract data set. The Pearson and deviance chi-square tests are two measures that assess overall model fit. It makes some sense that by looking at the individual components of these statistics, which are functions of the observed group counts and their model-predicted values, you will gain insight into a model's lack of fit.

Suppose that you have s groups, $i = 1, \ldots, s$, and n_i total subjects for the ith group. If y_i is the number of events (success, yes) for the ith group, and $\hat{\theta}_i$ denotes the predicted probability of success for the ith group, then define the ith residual as

$$e_i = \frac{y_i - n_i\hat{\theta}_i}{\sqrt{n_i\hat{\theta}_i(1 - \hat{\theta}_i)}}$$

These residuals are known as Pearson residuals, since the sum of their squares is Q_P. They compare the differences between observed counts and their predicted values, scaled by the observed count's standard deviation. By examining the e_i, you can determine how well the model fits the individual groups. Often, the residual values are considered to be indicative of lack of fit if they exceed 2 in size.

Similarly, the deviance residual is a component of the deviance statistic. The deviance residual is written

$$d_i = \text{sgn}(y_i - \hat{y}_i)\left[2y_i \log\left(\frac{y_i}{\hat{y}_i}\right) + 2(n_i - y_i)\log\left(\frac{n_i - y_i}{n_i - \hat{y}_i}\right)\right]^{\frac{1}{2}}$$

where $\hat{y}_i = n\hat{\theta}_i$. The sum of squares of the d_i values is the deviance statistic.

These residuals are often presented in tabular form; however, graphical display usually aids their inspection. One simple plot is called an *index plot*, in which the residuals are plotted against the corresponding observation number, the index. By examining these plots, you can determine if there are unusually large residuals, possibly indicative of outliers, or systematic patterns of variation, possibly indicative of a poor model choice.

These residuals are examined for the urinary tract data. As you will recall, the main effects model was considered to have an adequate fit. The INFLUENCE option requests that PROC LOGISTIC provide regression diagnostics.

Notice that the data are input differently than they were in Section 8.4. The variable RESPONSE is now the number of cures in a group, and the variable TRIALS is the total number of patients in that group, the sum of those who were cured and those who were not. The *events/trials* MODEL statement syntax allows you to specify the response as a ratio of two variables, the *events* variable and the *trials* variable. When the response is specified this way, developed to support the binomial trials framework, the residuals are calculated using an n_i that is based on the group size, which is desired. (If you specify a single response, called *actual model* syntax, when you compute residuals, the residuals are calculated using a group size of 1.)

```
data uti2;
   input diagnosis : $13. treatment $ response trials;
datalines;
complicated      A   78   106
complicated      B  101   112
complicated      C   68   114
uncomplicated    A   40    45
uncomplicated    B   54    59
uncomplicated    C   34    40
;
proc logistic data=uti2;
   class diagnosis treatment / param=ref;
   model response/trials = diagnosis treatment/
         influence;
run;
```

Output 8.34 displays the table of covariate profiles that is first printed in the diagnostics output. There should be one case for each group in your data.

Output 8.34 Covariates

```
                     Regression Diagnostics

                          Covariates

      Case                         treatment    treatment
      Number   diagnosiscomplicated      A            B

        1            1.0000         1.0000          0
        2            1.0000            0          1.0000
        3            1.0000            0            0
        4              0            1.0000          0
        5              0              0          1.0000
        6              0              0            0
```

Output 8.35 contains the Pearson and Deviance residuals for this model. The INFLUENCE option produces other diagnostics as well; these are not reproduced here.

Output 8.35 Residuals

```
                       Regression Diagnostics

          Pearson Residual                Deviance Residual

   Case            (1 unit = 0.16)                 (1 unit = 0.15)
   Number   Value   -8  -4  0 2 4 6 8     Value   -8  -4  0 2 4 6 8

     1     -0.0773   |        *       |   -0.0772   |       *|        |
     2      0.6300   |        |  *    |    0.6460   |        |    *   |
     3     -0.3453   |     *  |       |   -0.3445   |     *  |        |
     4      0.1609   |        |*      |    0.1624   |        |*       |
     5     -1.3020   |*       |       |   -1.1823   |*       |        |
     6      0.7171   |        |   *   |    0.7406   |        |     *  |
```

Note that the largest Pearson residual for the main effects model is -1.3020 for the fifth group (uncomplicated diagnosis, treatment B) and the largest deviance residual is -1.1823, also for the fifth group. The other residuals are all less than 1 (in absolute value). All these residuals are acceptable.

To see what happens in a model that doesn't fit, the model with the single main effect DIAGNOSIS is requested. The IPLOTS option is specified to produce index plots.

```
proc logistic;
   class diagnosis treatment / param=ref;
   model response/trials = diagnosis/
     scale=none aggregate=(treatment diagnosis) influence iplots;
run;
```

The goodness-of-fit tests are displayed in Output 8.36.

Output 8.36   Goodness-of-Fit Statistics

```
            Deviance and Pearson Goodness-of-Fit Statistics

        Criterion      DF        Value     Value/DF    Pr > ChiSq

        Deviance        4       30.6284      7.6571       <.0001
        Pearson         4       28.7265      7.1816       <.0001

               Number of unique profiles: 6
```

With values of 30.6284 and 28.7265, respectively, Q_L and Q_P clearly do not support the model.

The residuals for this model are displayed in Output 8.37.

Output 8.37   Residuals

```
                        Regression Diagnostics

                                       Pearson Residual
                  Covariates
       Case                                (1 unit = 0.48)
       Number   diagnosiscomplicated   Value    -8  -4  0 2 4 6 8

          1            1.0000         -0.1917   |          *        |
          2            1.0000          3.8267   |          |      *|
          3            1.0000         -3.6081   |*         |        |
          4               0          0.000076   |          *        |
          5               0           0.6445    |          |*       |
          6               0          -0.7825    |        * |        |

                        Regression Diagnostics

                         Deviance Residual

           Case                        (1 unit = 0.53)
           Number      Value       -8  -4  0 2 4 6 8

              1       -0.1911       |          *        |
              2        4.2166       |          |     *|
              3       -3.4358       | *        |        |
              4       0.000076      |          *        |
              5        0.6694       |          |*       |
              6       -0.7477       |        *|        |
```

This model appears to fit very poorly for groups 2 and 3; the Pearson residuals take the values 3.8267 and -3.6081, respectively, and the deviance residuals take the values 4.2166 and -3.4358 for the same groups. Output 8.38 displays the index plot for the Pearson residuals.

This display obviously makes it easy to spot those residuals that are outside a desirable range and then identify the corresponding group. The points for the second and third observations stand out clearly.

Output 8.38   Index Plot for Pearson Residuals

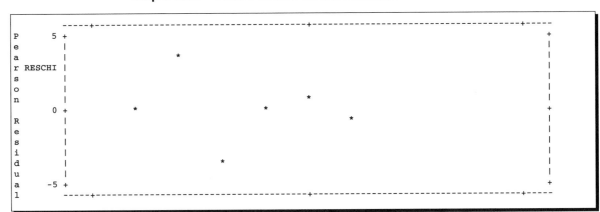

The Pearson and deviance residuals need to be used cautiously when the data contain continuous explanatory variables so that most of the group sizes are 1. This is for the same reason that Q_P and the deviance are inappropriate—the sample size requirements for approximate chi-square distributions are not met. However, these residuals are often considered useful as a rough indicator of model fit in this situation, and they are often examined.

Other types of diagnostics include changes in the Q_P and deviance when the ith observation is excluded; the ith leverage; and distances between estimated parameters and the estimated parameters when the ith observation is excluded. In addition, there are a variety of plots that have been devised to assist in evaluating model adequacy. Refer to the *SAS/STAT User's Guide, Version 8* for information on what diagnostics are provided by the LOGISTIC procedure. Diagnostics development is an active research area, particularly for assessing model fit for generalized linear models, and additional tools will be available in the future.

8.7 Maximum Likelihood Estimation Problems and Alternatives

If you perform enough logistic regressions, you will encounter data for which maximum likelihood estimation does not produce a unique solution for the parameters; you do not obtain convergence. In addition, for data with small cell counts, large sample theory may not be applicable and thus tests based on the asymptotic normality of the maximum likelihood estimates may be unreliable. This section discusses some of the situations in which maximum likelihood methods may not produce solutions and Section 8.8 discusses the alternative strategies based on exact methods.

8.7.1 Examples of Non-Convergence

To gain insight into the possible data configurations that result in non-convergence, consider the following table:

Table 8.9. Infinite Odds Ratio Example

Factor	Response=Yes	Response=No
Factor 1	15	0
Factor 2	0	34

Computing the odds ratio for these data results in the quantity

$$\frac{a \times d}{b \times c} = \frac{15 \times 34}{0 \times 0}$$

which is infinite. Since the odds ratio is e^{β}, where β is the parameter for the factor, this means that β is infinite.

The LOGISTIC procedure performs some checking to determine whether the input data have a configuration that leads to infinite parameter estimates. If convergence is not attained within eight iterations, PROC LOGISTIC computes the probability of allocating each observation to the correct response group. If this probability is equal to 1 for all observations, there is said to be *complete separation* of data points (this occurs if all the observations having unique covariate profiles have the same response outcome—for example, all the Factor=1 subjects responded yes, and all the Factor=2 subjects responded no). If complete separation is found, the iterative process is halted and a warning message is printed.

If nearly all the observations have a probability of 1 of being allocated to the correct response group, then the data configuration may be one of *quasicomplete separation*. (For quasicomplete separation to occur, the dispersion matrix also becomes unbounded.) Iteration also stops when this condition is detected, and a warning message is printed, since the parameter estimates are also infinite.

If neither of these conditions exists for the data, then they are said to be *overlapping*. The data points overlap so that observations with the same covariate profile have all possible responses. Maximum likelihood estimates exist and are unique for overlapping configurations. The problems of complete separation and quasi-complete separation generally occur for small data sets. Usually quasi-complete separation does not occur if you have a continuous explanatory variable; complete separation can always occur. Refer to Albert and Anderson (1984) for more information about infinite parameters and the data configurations that produce them; refer to Silvapulle (1981) for a discussion of the necessary and sufficient conditions for the existence of maximum likelihood estimators in binomial response models.

Earlier releases of PROC LOGISTIC did not check for these conditions, and if the data configuration produced no unique solutions, then the procedure printed a message saying that convergence was not attained. You then had the option of fine-tuning the estimation process by increasing the number of iterations or changing the convergence criterion, or assuming that the parameter estimates were infinite. Occasionally, unique parameter estimates exist that require more iterations or a different convergence criterion for their estimation than the default.

The following statements input a data set with several zero values for the response outcome counts.

```
data quasi;
   input treatA treatB response count @@;
   datalines;
0 0 0 0  0 0 1  0
0 1 0 2  0 1 1  0
1 0 0 0  1 0 1  8
1 1 0 6  1 1 1 21
;
proc logistic;
   freq count;
   model response = TreatA TreatB;
run;
```

Output 8.39 contains the results from PROC LOGISTIC. Since there is quasi-complete separation, the maximum likelihood solution may not exist.

Output 8.39   Quasi-Complete Separation Note

```
                             Response Profile

                       Ordered                     Total
                        Value      response      Frequency

                          1           0              8
                          2           1             29

NOTE: 4 observations having zero frequencies or weights were excluded since
      they do not contribute to the analysis.


                          Model Convergence Status

             Quasi-complete separation of data points detected.

WARNING: The maximum likelihood estimate may not exist.
WARNING: The LOGISTIC procedure continues in spite of the above warning.
         Results shown are based on the last maximum likelihood iteration.
         Validity of the model fit is questionable.
```

The next statements input a data set that also includes two dichotomous explanatory variables and the same number of zero counts; however, the placement of the zero counts results in complete separation of the data points.

```
data complete;
    input gender region count response @@;
    datalines;
0 0 0  1  0 0 5   0
0 1 1  1  0 1 0   0
1 0 0  1  1 0 175 0
1 1 53 1  1 1 0   0
;
proc logistic;
    freq count;
    model response = gender region;
run;
```

Output 8.40 contains the results. Since there is complete separation, the maximum likelihood solution does not exist.

Output 8.40 Complete Separation Note

```
                         Response Profile

                  Ordered                    Total
                  Value      response      Frequency

                     1           0            180
                     2           1             54

NOTE: 4 observations having zero frequencies or weights were excluded since
      they do not contribute to the analysis.


                      Model Convergence Status

            Complete separation of data points detected.

WARNING: The maximum likelihood estimate does not exist.
WARNING: The LOGISTIC procedure continues in spite of the above warning.
         Results shown are based on the last maximum likelihood iteration.
         Validity of the model fit is questionable.
```

Most of the time, the data generating non-unique infinite solutions will not be this simple.

8.8   Exact Methods in Logistic Regression

Until recently, there was no convenient alternative in situations where maximum likelihood estimation failed or small cell counts made the resulting maximum likelihood estimates inappropriate. However, it is now possible to compute parameter estimates, confidence intervals, and p-values for statistical tests using methodology based on exact permutation distributions. The key is conditioning on the appropriate sufficient statistic. The idea is not a new one, having been suggested by Cox (1970), but recent algorithmic advances in computing the exact distributions have made the methodology computationally feasible. Refer to Tritchler (1984) and Hirji, Mehta, and Patel (1987) for more details regarding these algorithms. See Appendix B in Chapter 10 for a brief overview of the methodology involved in deriving exact conditional distributions and computing tests and point estimates.

With Release 8.1, the SAS System provides exact logistic regression for binary outcomes in the LOGISTIC procedure. It provides an exact probability test and an exact score test for the hypotheses that parameters for the specified hypothesis are equal to zero; these tests produce an exact p-value which is the probability of obtaining a more extreme statistic than the one observed and a mid p-value, which adjusts for the discreteness of the distribution. Simultaneous tests can be specified. You can also request the point estimates of the parameters and the exponential (usually represents an odds ratio); these come with one or two-sided confidence limits and one or two-sided p-values for testing that the parameter estimate is zero.

Consider the following data in Table 8.10 from a study on liver function outcomes for high risk overdose patients in which antidote and historical control groups are compared. The data are stratified by time to hospital admission (Koch, Gillings, and Stokes 1980).

Table 8.10.   Liver Function Outcomes

Time	Antidote		Control	
to Hospital	Severe	Not Severe	Severe	Not Severe
Early	6	12	6	2
Delayed	3	4	3	0
Late	5	1	6	0

The small counts in many cells—seven of the twelve cells have values less than 5—make the applicability of large sample theory somewhat questionable.

The following DATA step inputs the data.

```
data liver;
   input time $ group $ status $ count @@;
   datalines;
early    antidote severe 6 early    antidote not 12
early    control  severe 6 early    control  not  2
delayed antidote severe 3 delayed antidote not 4
delayed control  severe 3 delayed control  not 0
late     antidote severe 5 late     antidote not 1
late     control  severe 6 late     control  not 0
;
run;
```

The following PROC LOGISTIC statements request an unconditional logistic regression analysis of the severity of the outcome with explanatory variables based on time to admission and treatment group. The early level for TIME is the reference level and the control level is the reference level for GROUP. The PARAM=REF option requests incremental effects parameterization.

The EXACT statement requests the exact analysis. (Note that if you include the EXACTONLY option in the PROC statement, only the exact analysis is performed.) You can include more than one EXACT statement, so you can provide an individual label for the output from each statement. Exact tests are performed for the variables listed in the statement; in this case, this includes the intercept, TIME, and GROUP. The tests are conditioned on any other variables included in either the MODEL statement or the EXACT statement. The option ESTIMATE=BOTH in the first EXACT statement specifies that point estimates for both the parameter and the exponentiated parameter be computed. Note that exponentiated parameters are computed for CLASS variables only if PARAM=REF is specified in the CLASS statement. The JOINT option in the second EXACT statement requests a joint test for variables TIME and GROUP.

```
proc logistic descending;
   freq count;
   class time(ref='early') group(ref='control') /param=ref;
   model status = time group / scale=none aggregate clparm=wald;
   exact 'Model 1' intercept time group /
       estimate=both;
   exact 'Joint Test' time group / joint;
run;
```

Output 8.41 contains the goodness-of-fit statistics, and Output 8.42 contains the resulting maximum likelihood parameter estimates.

Output 8.41 Goodness-of-Fit Statistics

```
          Deviance and Pearson Goodness-of-Fit Statistics

   Criterion        DF          Value       Value/DF     Pr > ChiSq

   Deviance          2         1.1728        0.5864         0.5563
   Pearson           2         0.7501        0.3750         0.6873

                 Number of unique profiles: 6
```

Output 8.42 MLE Estimates

```
            Analysis of Maximum Likelihood Estimates

                                      Standard        Wald
   Parameter         DF    Estimate      Error    Chi-Square    Pr > ChiSq

   Intercept          1      1.4132     0.7970       3.1439        0.0762
   time    delayed    1      0.7024     0.8344       0.7087        0.3999
   time    late       1      2.5533     1.1667       4.7893        0.0286
   group   antidote   1     -2.2170     0.8799       6.3480        0.0118
```

The 95% confidence limits for the parameter estimates are displayed in Output 8.43.

Output 8.43 Confidence Limits for Estimates

```
            Wald Confidence Interval for Parameters

   Parameter                Estimate    95% Confidence Limits

   Intercept                  1.4132     -0.1489      2.9754
   time    delayed            0.7024     -0.9330      2.3378
   time    late               2.5533      0.2666      4.8400
   group   antidote          -2.2170     -3.9417     -0.4924
```

Odds ratios and their 95% confidence limits are displayed in Output 8.44.

Output 8.44 Odds Ratio Estimates

```
                    Odds Ratio Estimates

                           Point        95% Wald
   Effect                Estimate    Confidence Limits

   time   delayed vs early    2.019      0.393      10.359
   time   late vs early      12.849      1.305     126.471
   group antidote vs control  0.109      0.019       0.611
```

Output 8.45 contains the exact test results. First, the exact tests corresponding to the first EXACT statement are printed, followed by the results for the second EXACT statement.

Output 8.45   Exact Tests

```
                    Exact Conditional Analysis

                Conditional Exact Tests for 'Model 1'

                                        --- p-Value ---
        Effect      Test        Statistic    Exact      Mid

        Intercept   Score          3.4724    0.1150   0.0922
                    Probability    0.0457    0.1150   0.0922
        time        Score          6.0734    0.0442   0.0418
                    Probability    0.00471   0.0442   0.0418
        group       Score          7.1656    0.0085   0.0050
                    Probability    0.00698   0.0085   0.0050


                Conditional Exact Tests for 'Joint Test'

                                        --- p-Value ---
        Effect    Test        Statistic    Exact      Mid

        Joint     Score         13.1459    0.0027   0.0027
                  Probability    0.000015  0.0015   0.0015
        time      Score          6.0734    0.0442   0.0418
                  Probability    0.00471   0.0442   0.0418
        group     Score          7.1656    0.0085   0.0050
                  Probability    0.00698   0.0085   0.0050
```

For the 'Model 1' results, both the exact score conditional test and the probability test are reported; in this instance, they both have p-values that are the same. This will not always be the case. For the time effect, the exact p-value is 0.0442, and for the group effect, the exact p-value is 0.0085. Again, these tests are conditioned on the other effects in the model. Note that, if an effect consists of two or more parameters, then this test is evaluating the hypthesis that all the relevant parameters are equal to zero simultaneously. For the 'Joint' results, the score test produces an exact p-value of 0.0027, and the probability test produces an exact p-value of 0.0015. Note that when you specify the JOINT option, you also generate the tests for the individual components.

Output 8.46 displays the parameter estimates and their 95% confidence limits. Note that the parameter estimates are fairly similar to those based on the large sample approximate methods. The exact p-values for the group effect parameter has a different value that those reported for the exact conditional tests. This is because the exact p-values for the single parameters are the results of likelihood ratio tests based on the conditional pdf used to estimate them. For the most part, you would rely on the exact p-values reported in the "Exact Conditional Analysis" table.

Output 8.46   Exact Parameter Estimates

```
                       Exact Conditional Analysis

               Exact Parameter Estimates for 'Model 1'

                                        95% Confidence
        Parameter              Estimate      Limits           p-Value

        Intercept               1.3695   -0.2361    3.6386    0.1140
        time      delayed       0.6675   -1.2071    2.6444    0.6667
                  late          2.4387    0.1364    6.4078    0.0331
        group     antidote     -2.0992   -4.5225   -0.3121    0.0154
```

Output 8.47 displays the odds ratio estimates and their 95% confidence limits.

Output 8.47   Exact Odds Ratio Estimates

```
                       Exact Conditional Analysis

                  Exact Odds Ratios for 'Model 1'

                                        95% Confidence
        Parameter              Estimate      Limits           p-Value

        Intercept               3.933     0.790     38.037    0.1140
        time      delayed       1.949     0.299     14.075    0.6667
                  late         11.458     1.146    606.546    0.0331
        group     antidote      0.123     0.011      0.732    0.0154
```

Table 8.11 provides a comparison of the unconditional maximum likelihood estimates and the exact conditional estimates:

Table 8.11.   Exact and Asymptotic Estimates

Variable	Inference Type	Estimate	Lower 95% CI bound	Upper 95% CI bound	p-value
Intercept	Asymptotic	1.4132	−0.1489	2.9754	0.0762
	Exact	1.3695	−0.2361	3.6386	0.1140
Delayed	Asymptotic	0.7024	−0.9330	2.3378	0.3999
	Exact	0.6675	−1.2071	2.6444	0.6667
Late	Asymptotic	2.5535	0.2666	4.8404	0.0286
	Exact	2.4387	0.1364	6.4078	0.0331
Antidote	Asymptotic	−2.2171	−3.9418	−0.4924	0.0118
	Exact	−2.0992	−4.5225	−0.3121	0.0154

For the exact computations performed with PROC LOGISTIC, the p-value listed is twice the one-sided p-value. Note that the exact methods do not produce a standard error for the estimate. For these data, you can see that exact logistic regression produces estimates that are different, although not substantially, from the maximum likelihood estimates. For each parameter, the p-values listed for the exact estimates are larger than those for the asymptotic estimates. Usually, the exact methods lead to more conservative results than the approximate methods. As a general rule, when the sample sizes are small and the approximate p-values are less than 0.10, it is a good idea to look at the exact results. If the approximate p-values are larger than 0.15, then the approximate methods are probably satisfactory in the sense that the exact results are likely to agree with them.

Besides being appropriate for data sets with small cell counts, exact methods often can produce estimates and corresponding confidence bounds for data sets for which maximum likelihood methods fail to converge.

Exact Confidence Limits for Common Odds Ratios for 2×2 Tables

Section 3.3 in Chapter 3, "Sets of 2×2 Tables," points out that while currently, the FREQ procedure doesn't provide exact confidence limits for the average odds ratio in a set of 2×2 tables, you can obtain them with exact logistic regression. You formulate the analysis as a regression where the column variable is the response variable and the row and stratification variables are the explanatory variables. Then, you condition on the stratification variable and estimate the odds ratio for the row variable. This odds ratio will be an average odds ratio.

Consider the data in Table 8.12. A small company initiated exercise programs at both of its locations, downtown and a satellite office in a nearby suburb. The office program consisted of directed aerobic activities such as running, walking, and bicycling, conducted under the guidance of an exercise counselor. The home program consisted of a range of activities that were self-monitored. Each employee signed an agreement to participate in a program and to check in monthly to ensure continual effort. After a year, participants and non-participants underwent a cardiovascular stress test to assess their fitness, and their result was recorded as good or not good depending on age-adjusted criteria. The exercise counselor was interested in whether type of program was associated with good test results.

Table 8.12. Cardiovascular Test Outcomes

Location	Program	Good	Not Good	Total
Downtown	Office	12	6	18
Downtown	Home	3	5	8
	Total	15	11	26
Satellite	Office	6	1	7
Satellite	Home	1	3	4
	Total	7	4	11

Interest lies in computing an odds ratio comparing good results for the office program compared to the home program. The sample sizes in these tables are too small to be able to justify the asymptotic confidence limits for the odds ratio produced by the FREQ procedure. However, you can obtain these odds ratios and exact confidence limits by performing an exact logistic regression.

The following DATA step inputs these data into SAS data set EXERCISE.

```
data exercise;
    input location $ program $ outcome $ count @@;
    datalines;
```

```
downtown office   good 12 downtown office not 5
downtown home     good 3 downtown  home   not 5
satellite office  good 6 satellite office not 1
satellite home    good 1 satellite home   not 3
;
run;
```

To perform the exact logistic regression, you put both LOCATION and PROGRAM in the CLASS statement and, in order to compare office to home, use the REF=FIRST option with the PROGRAM variable. PROC LOGISTIC only estimates odds ratio for CLASS variables for reference parameterization (incremental effects) so the PARAM=REF option is included in the CLASS statement as well. The response variable is OUTCOME in the MODEL statement; since 'good' outcome is the first alphanumerically ordered value (the other is 'not'), the model is based on the probability of good outcome.

You then specify the EXACT statement, requesting exact tests for the variable PROGRAM and also specifying the ESTIMATE=BOTH option to obtain both the parameter estimate and the odds ratio estimate.

```
proc logistic;
   freq count;
   class location program(ref=first) /param=ref;
   model outcome = location program;
   exact program / estimate=both;
run;
```

Output 8.48 displays the results of the exact tests for exercise program. Both the score and probability tests have an exact p-value of 0.0307, indicating significance at the $\alpha = 0.05$ level of significance.

Output 8.48   Exact Test Results

```
                  Exact Conditional Analysis

                  Conditional Exact Tests

                                        --- p-Value ---
      Effect   Test          Statistic    Exact     Mid

      program  Score           5.5739     0.0307    0.0215
               Probability     0.0183     0.0307    0.0215
```

Output 8.49 displays the exact parameter estimate and the exact odds ratio estimate comparing office program to home program; the odds ratio estimate takes the value 5.413 with 95% confidence limits of (1.049, 33.312). This means that those persons participating in the office exercise program had roughly 5 times the odds of a good test outcome as the odds of those participating in a home exercise program. However, note that the confidence limits are very wide, and that the lower bound is just beyond the value 1.

Output 8.49   Exact Estimates

```
                    Exact Conditional Analysis

                    Exact Parameter Estimates

                                     95% Confidence
        Parameter           Estimate       Limits          p-Value

        program   office    1.6889    0.0474    3.5059     0.0424


                         Exact Odds Ratios

                                     95% Confidence
        Parameter           Estimate       Limits          p-Value

        program   office    5.413     1.049    33.312      0.0424
```

Compare the exact results to those produced by the asymptotic analysis, which are displayed in Output 8.50.

Output 8.50   Odds Ratio for Asymptotic Analysis

```
                    Odds Ratio Estimates

                                Point        95% Wald
        Effect                Estimate   Confidence Limits

        location downtown vs satellit   0.758    0.151    3.803
        program   office vs home        6.111    1.331   28.062
```

The point estimate here is 6.111 and the 95% Wald confidence limits are (1.331, 28.062). Note that the Mantel-Haenszel estimator of the odds ratio (produced by PROC FREQ but not shown here) takes the value 5.8421 with confidence limits (1.3012, 26.2296).

Thus, using the exact method provides a somewhat more conservative picture than the inappropriate asymptotic method.

8.9   Using the CATMOD and GENMOD Procedures for Logistic Regression

The CATMOD and GENMOD procedures provide alternative ways to perform logistic regression in the SAS System. Prior to the inclusion of the CLASS statement in the LOGISTIC procedure, these procedures provided a way to perform logistic regression with classification variables without having to create a batch of indicator variables. Now that PROC LOGISTIC handles classification variables, there is less need to point out the availability of these procedures for binary logistic regression. However, the CATMOD procedure currently provides the only way in which to perform generalized logits regression in the SAS System (discussed in Chapter 9), and the GENMOD procedure performs logistic regression for correlated responses via the generalized estimating equations method (discussed in Chapter 15). Thus, this section provides an introduction to these procedures for the relatively simple case of binary logistic regression.

8.9.1   Performing Logistic Regression with the CATMOD Procedure

The CATMOD procedure is a general modeling procedure that fits linear models to functions of proportions with weighted least squares. In addition, it performs maximum likelihood estimation when the response functions are logits or generalized logits. To understand how the CATMOD procedure works, consider the following example.

Assuming that data set SENTENCE from Section 8.3 has been created, the following PROC CATMOD statements fit a main effects model.

```
proc catmod data=sentence order=data;
   weight count;
   model sentence = type prior;
run;
```

The following output is produced.

Output 8.51   PROC CATMOD

```
                            Data Summary

        Response              sentence    Response Levels     2
        Weight Variable       count       Populations         4
        Data Set              SENTENCE    Total Frequency   863
        Frequency Missing     0           Observations        8


                        Population Profiles

          Sample      type       prior     Sample Size
          ---------------------------------------------
               1      nrb        some              151
               2      nrb        none               92
               3      other      some              208
               4      other      none              412


                        Response Profiles

              Response      sentence
              --------------------------
                   1         y
                   2         n
```

The CATMOD procedure determines the model matrix structure from the explanatory variables listed in the MODEL statement. It forms a separate group, or 'sample,' for each combination of explanatory variable values. These groups are displayed in the "Population Profiles" table. The four different rows, or profiles, correspond to rows in Table 8.5. The sample sizes correspond to the row totals listed in that table. The "Response Profiles" table in the PROC CATMOD output lists the response variable and its values. The ORDER=DATA option was used to put the response value 'y' first, so that the model fit would be based on Pr{prison}. It is always important to check these tables in the PROC CATMOD output to make sure that you understand the internal order of the response variable values and the classification of explanatory effects.

After the iteration history is printed (not shown here), PROC CATMOD prints the ANOVA table and the table of maximum likelihood estimates, as displayed in Output 8.52. The ANOVA table contains Wald statistics for the model effects; the entry labeled "Likelihood Ratio" is the likelihood ratio goodness-of-fit test. Note that the likelihood ratio statistic and the Wald statistics are identical to those computed for the main effects model fit with PROC LOGISTIC, as displayed in Section 8.3.3.

Output 8.52   PROC CATMOD

```
                 Maximum Likelihood Analysis of Variance

            Source              DF      Chi-Square      Pr > ChiSq
            ----------------------------------------------------------
            Intercept            1         243.25         <.0001
            type                 1           9.05         0.0026
            prior                1           3.31         0.0687

            Likelihood Ratio     1           0.51         0.4762


                 Analysis of Maximum Likelihood Estimates

                                            Standard      Chi-
        Effect        Parameter   Estimate    Error      Square    Pr > ChiSq
        ----------------------------------------------------------------------
        Intercept         1       -1.4828     0.0951     243.25       <.0001
        type              2        0.2960     0.0984       9.05       0.0026
        prior             3        0.1735     0.0953       3.31       0.0687
```

If you compare the parameter estimates in this PROC CATMOD output to those for PROC LOGISTIC in Output 8.17, you will find that they are the same. This is because the default parameterization for PROC CATMOD is the same deviation from the mean parameterization that is the default for PROC LOGISTIC, as demonstrated in Section 8.3.3. See the discussion in that section concerning the parameterization and how to compute the odds ratio in this situation. You will have to do this yourself with the CATMOD procedure, as it does not produce the odds ratios for you.

However, the CATMOD procedure can be a convenient way to perform logistic regression when you have qualitative explanatory variables. It can also handle continuous variables that have a relatively limited number of unique values. You just specify the variable in the DIRECT statement, and PROC CATMOD inserts its values directly into the model matrix. This feature is discussed in Chapter 13, "Weighted Least Squares," which discusses the CATMOD procedure comprehensively for weighted least squares applications.

However, PROC CATMOD is not an appropriate procedure for logistic regression when you have continuous explanatory variables with many distinct values. The internal CATMOD machinery always creates a separate group for each distinct combination of explanatory variable values. With a very large number of continuous explanatory variables, the underlying table can become too sparse and you may get messages about dependent response functions and infinite parameter estimates.

8.9.2   Performing Logistic Regression with the GENMOD Procedure

Generalized Linear Models

The GENMOD procedure fits generalized linear models. Such models are a generalization of the general linear model that is fit by the GLM procedure. Generalized linear models include not only classical linear models but logistic and probit models for binary data, and Poisson regression models for Poisson data. You can also fit loglinear models for multinomial data indirectly through computational equivalences with fitting Poisson regression models. You can generate many other statistical models by the appropriate selection of a *link function* and the probability distribution of the response.

A generalized linear model has three components:

- a random sample of independent response variable $\{y_i\}$ with some probability distribution, $i = 1, 2, \ldots, n$

- a set of explanatory variables $\mathbf{x}_i$ and parameter vector $\boldsymbol{\beta}$

- a monotonic link function g that describes how the expected value of y_i, μ_i, is related to $\mathbf{x}_i'\boldsymbol{\beta}$:

$$g(\mu_i) = \mathbf{x}_i'\boldsymbol{\beta}$$

You construct a generalized linear model by choosing the appropriate link function and response probability distribution. In the classical linear model, the probability distribution is the normal and the usual link function is the identity: $g(\mu) = \mu$. For logistic regression, the distribution is the binomial and the usual link function is the logit:

$$g(\mu) = \log\left(\frac{\mu}{1-\mu}\right)$$

For Poisson regression, the distribution is Poisson and the link function is $g(\mu) = \log(\mu)$.

In the SAS System, the GENMOD procedure fits the generalized linear model and thus provides another mechanism for performing logistic regression analysis (SAS/INSIGHT software also fits generalized linear models). The following section describes how to perform logistic regression using PROC GENMOD. See Chapter 12, "Poisson Regression," for a discussion of Poisson regression and illustrations using the GENMOD procedure. For a comprehensive discussion of the generalized linear model, refer to McCullagh and Nelder (1989). For an introduction to the topic, refer to Dobson (1990) or Agresti (1995).

Fitting Logistic Regression Models with PROC GENMOD

Fitting logistic regression models with the GENMOD procedure is a relatively straightforward matter. PROC GENMOD includes a CLASS statement, so you simply list your classification variables in it, just as you now do in PROC LOGISTIC. The parameterization it produces is equivalent to the incremental effects parameterization implemented for most of the analyses performed in this chapter. In PROC GENMOD, the reference cell is the combination of the last sorted levels of the effects listed in the CLASS statement (and 0s for any continuous explanatory variables). Incremental effects parameters are estimated for the remaining levels.

Consider the urinary tract infection data analyzed in Section 8.4. If you sorted the values of TREATMENT and DIAGNOSIS, those observations that had an uncomplicated diagnosis and treatment C would become the reference cell.

The following statements produce an analysis using PROC GENMOD. The CLASS statement is a more basic version that the one available with PROC LOGISTIC; you can't specify choices of parameterization or reference levels. You need to specify LINK=LOGIT and DIST=BINOMIAL to request logistic regression with PROC GENMOD. The TYPE3 option requests tests of effects for the model.

```
proc genmod data=uti;
    freq count;
    class diagnosis treatment;
    model response = diagnosis treatment /
        link=logit dist=binomial type3
        aggregate=(diagnosis treatment);
run;
```

The output is displayed in Output 8.53.

Output 8.53   Goodness of Fit

```
                    Class Level Information

          Class           Levels     Values

          diagnosis          2       complicated uncomplicated
          treatment          3       A B C


                 Criteria For Assessing Goodness Of Fit

      Criterion                 DF          Value        Value/DF

      Deviance                   2          2.5147        1.2573
      Scaled Deviance            2          2.5147        1.2573
      Pearson Chi-Square         2          2.7574        1.3787
      Scaled Pearson X2          2          2.7574        1.3787
      Log Likelihood                     -225.0355
```

The table labeled "Criteria for Assessing Goodness of Fit" includes the Log Likelihood statistic, which has the value -225.0355. Note that if you multiply this value by two and reverse the sign, you get the same value as -2LOG L displayed in the output for the same model in the PROC LOGISTIC output. Other criteria displayed are approximate chi-square statistics. The Deviance is the log-likelihood statistic for the difference between this main effects model and the saturated model, the same as computed by PROC LOGISTIC.

The estimates displayed in Output 8.54 are identical to those produced with PROC LOGISTIC for the same model. However, those levels that become the reference levels under incremental effects coding, uncomplicated diagnosis and treatment C, are assigned 0s for the parameter estimate and related statistics.

The table labeled "LR Statistics For Type 3 Analysis" can be viewed as serving a similar role to that of an ANOVA table. It includes likelihood ratio tests for each of the effects. The effect for treatment, which has three levels, has 2 df. The effect for diagnosis, with two levels, has 1 df. Both tests are clearly significant, with values of 28.11 and 11.72, respectively.

Output 8.54 Parameter Estimates

```
                      Analysis Of Parameter Estimates

                                      Standard   Wald 95% Confidence    Chi-
 Parameter                 DF   Estimate   Error       Limits        Square   Pr > ChiSq

 Intercept                  1    1.4184   0.2987    0.8330    2.0038   22.55     <.0001
 diagnosis   complicated    1   -0.9616   0.2998   -1.5492   -0.3740   10.29     0.0013
 diagnosis   uncomplicated  0    0.0000   0.0000    0.0000    0.0000     .          .
 treatment   A              1    0.5847   0.2641    0.0671    1.1024    4.90     0.0268
 treatment   B              1    1.5608   0.3160    0.9415    2.1800   24.40     <.0001
 treatment   C              0    0.0000   0.0000    0.0000    0.0000     .          .
 Scale                      0    1.0000   0.0000    1.0000    1.0000

 NOTE: The scale parameter was held fixed.


                      LR Statistics For Type 3 Analysis

                                      Chi-
          Source             DF     Square    Pr > ChiSq

          diagnosis           1     11.72      0.0006
          treatment           2     28.11      <.0001
```

To assess whether any of the treatments are similar, linear combinations of the parameters are tested to see if they are significantly different from zero.

$$H_0\colon \mathbf{L}\boldsymbol{\beta} = \mathbf{0}$$

By choosing the right elements of $\mathbf{L}$, you can construct linear combinations of the parameters that will produce the appropriate test. By default, PROC GENMOD computes a likelihood ratio test; on request, it can produce the corresponding Wald test. The likelihood ratio test for a contrast is twice the difference between the log likelihood of the current fitted model and the log likelihood of the model fitted under the constraint that the linear function of the parameters defined by the contrast is equal to zero.

The test for whether treatment A is equivalent to treatment B is expressed as

$$H_0\colon \beta_A = \beta_B$$

and the test for whether treatment A is equivalent to treatment C is expressed as

$$H_0\colon \beta_A = \beta_C$$

You request these tests with the CONTRAST statement in PROC GENMOD. The following CONTRAST statement is required to produce the first test. You place an identifying name for the test in quotes, name the effect variable, and then list the appropriate coefficients for **L**. These coefficients are listed according to the order in which the levels of the variable are known to PROC GENMOD. When you use a CONTRAST statement, or specify the ITPRINT, COVB, CORRB, WALDCI, or LRCI options in the MODEL statement, the GENMOD output includes information on what levels of effects the parameters represent.

The CONTRAST statement is very similar to the CONTRAST statement in PROC GLM.

```
contrast 'A-B' treat 1 -1  0;
```

The following SAS code produces the tests of interest.

```
proc genmod data=uti;
   freq count;
   class diagnosis treatment;
   model response = diagnosis treatment /
      link=logit dist=binomial;
   contrast 'treatment' treatment 1 0 -1 ,
                        treatment 0 1 -1;
   contrast 'A-B' treatment 1 -1  0;
   contrast 'A-C' treatment 1  0 -1;
run;
```

Output 8.55 contains the information about what the parameters represent.

Output 8.55 Parameter Information

```
                    Parameter Information

        Parameter        Effect      diagnosis       treatment

        Prm1             Intercept
        Prm2             diagnosis   complicated
        Prm3             diagnosis   uncomplicated
        Prm4             treatment                   A
        Prm5             treatment                   B
        Prm6             treatment                   C
```

Output 8.56 contains the results of the hypothesis tests.

Output 8.56 Contrasts

```
                    Contrast Results

                         Chi-
        Contrast   DF    Square    Pr > ChiSq    Type

        treatment   2    28.11       <.0001       LR
        A-B         1     9.22       0.0024       LR
        A-C         1     4.99       0.0255       LR
```

$Q_L = 9.22$ for the test of whether treatment A and treatment B are the same; $Q_L = 4.99$ for the test of whether treatment A and treatment C are the same; both of these are clearly significant at the $\alpha = 0.05$ level of significance. Note that these tests are similar to those displayed in the analysis performed in Section 8.4. If you execute these same statements using the WALD option, you will obtain identical results to the Wald tests obtained from PROC LOGISTIC in Section 8.4.

Appendix A: Statistical Methodology for Dichotomous Logistic Regression

Consider the relationship of a dichotomous outcome variable to a set of explanatory variables. Such situations can arise from clinical trials where the explanatory variables are treatment, stratification variables, and background covariables; another common source of such analyses are observational studies where the explanatory variables represent factors for evaluation and background variables.

The model for θ, the probability of the event, can be specified as follows:

$$\theta = \frac{\exp(\alpha + \sum_{k=1}^{t} \beta_k x_k)}{1 + \exp(\alpha + \sum_{k=1}^{t} \beta_k x_k)}$$

It follows that the odds are written

$$\frac{\theta}{1 - \theta} = \exp(\alpha + \sum_{k=1}^{t} \beta_k x_k)$$

so the model for the logit is linear:

$$\log\left\{\frac{\theta}{1 - \theta}\right\} = \alpha + \sum_{k=1}^{t} \beta_k x_k$$

The $\exp(\beta_k)$ are the odds ratios for unit changes in x_k, that is, the amount by which $\theta/(1 - \theta)$ is multiplied per unit change in x_k.

You can apply the product binomial distribution when the data for the dichotomous outcome are from a sampling process equivalent to stratified simple random sampling from subpopulations according to the explanatory variables. Relative to this structure, the maximum likelihood estimates are obtained by iteratively solving the equations:

$$\sum_{i=1}^{s} n_{i+} \hat{\theta}_i (1, x_{i1}, \ldots, x_{it}) = \sum_{i=1}^{s} n_{i1} (1, x_{i1}, \ldots, x_{it})$$

where n_{i1} is the number of subjects who have the event corresponding to θ among n_i subjects with $(x_{i1}, \ldots, x_{it})$ status.

The quantity

$$\hat{\theta}_i = \frac{\exp\{\hat{\alpha} + \sum_{k=1}^{t} \hat{\beta}_k x_{ik}\}}{1 + \exp\{\hat{\alpha} + \sum_{k=1}^{t} \hat{\beta}_k x_{ik}\}}$$

is the model-predicted value for θ_i.

For sufficient sample size, the quantities $\hat{\alpha}$ and $\hat{\beta}_k$ have approximate multivariate normal distributions for which a consistent estimate of the covariance structure is available.

You can assess goodness of fit of the model with Pearson chi-square statistics when sample sizes are sufficiently large (80% of the $\{n_{i1}\}$ and the $\{n_i - n_{i1}\}$ are ≥ 5 and all others are ≥ 2).

$$Q_P = \sum_{i=1}^{s} \frac{(n_{i1} - n_{i+}\hat{\theta}_i)^2}{n_{i+}\hat{\theta}_i(1 - \hat{\theta}_i)}$$

is approximately chi-square with $(s - 1 - t)$ degrees of freedom.

You can also use log-likelihood ratio statistics to evaluate goodness of fit by evaluating the need for a model to include additional explanatory variables.

In the setting where you have continuous explanatory variables, you cannot use Q_P to assess goodness of fit because you no longer have sufficient sample sizes n_{i+}. However, you can still apply the strategy of fitting an expanded model and then verifying that the effects not in the original model are nonsignificant. If the model matrix for the original model $\mathbf{X}$ has rank t, then the expanded model $[\mathbf{X}, \mathbf{W}]$ has rank $t + w$, where w is the rank of $\mathbf{W}$. You can evaluate the significance of $\mathbf{W}$ with the difference of the log-likelihood statistics for the models $\mathbf{X}$ and $[\mathbf{X}, \mathbf{W}]$.

$$Q_{LR} = \sum_{i=1}^{2} \sum_{j=1}^{2} 2n_{ij} \log\left(\frac{m_{ij,w}}{m_{ij}}\right)$$

where s is the total number of groups with at least one subject, m_{ij} is the predicted value of n_{ij} for model $\mathbf{X}$ ($m_{i1} = n_i\hat{\theta}_i$ and $m_{i2} = n_i(1 - \hat{\theta}_i)$), and $m_{ij,w}$ is the predicted value of n_{ij} for model $[\mathbf{X}, \mathbf{W}]$. Q_{LR} has an approximate chi-square distribution with w degrees of freedom.

Another approach that doesn't involve fitting an expanded model is the score statistic for assessing the association of the residuals $(\mathbf{n}_{*1} - \mathbf{m}_{*1})$ with $\mathbf{W}$ via the linear functions $\mathbf{g} = \mathbf{W}'(\mathbf{n}_{*1} - \mathbf{m}_{*1})$. The score statistic is written

$$Q_S = \mathbf{g}'\{\mathbf{W}'[\mathbf{D}_\mathbf{v}^{-1} - \mathbf{D}_\mathbf{v}^{-1}\mathbf{X}_\mathbf{A}(\mathbf{X}_\mathbf{A}'\mathbf{D}_\mathbf{v}^{-1}\mathbf{X}_\mathbf{A})^{-1}\mathbf{X}_\mathbf{A}'\mathbf{D}_\mathbf{v}^{-1}]\mathbf{W}\}^{-1}\mathbf{g}$$

where $\mathbf{n}_{*1} = (n_{11}, n_{21}, \ldots, n_{s1})'$, $\mathbf{X}_\mathbf{A} = [\mathbf{1}, \mathbf{X}]$, $\mathbf{m}_{*1} = (m_{11}, m_{21}, \ldots, m_{s1})'$, and $\mathbf{D}_\mathbf{v}$ is a diagonal matrix with diagonal elements $v_i = [n_{i+}\hat{\theta}_i(1 - \hat{\theta}_i)]^{-1}$. Q_S approximately has a chi-square distribution with w degrees of freedom when the total sample size is large enough to support an approximately multivariate normal distribution for the linear functions $[\mathbf{X}_\mathbf{A}', \mathbf{W}']\mathbf{n}_{*1}$.

Chapter 9

Logistic Regression II: Polytomous Response

Chapter Table of Contents

Chapter 9
Logistic Regression II: Polytomous Response

9.1 Introduction

While the typical logistic regression analysis models a dichotomous response as discussed in Chapter 8, "Logistic Regression I: Dichotomous Response," logistic regression is also applicable to multilevel responses. The response may be ordinal (no pain, slight pain, substantial pain) or nominal (Democrats, Republicans, Independents). For ordinal response outcomes, you can model functions called *cumulative logits* by performing ordered logistic regression using the proportional odds model (McCullagh 1980). For nominal response outcomes, you form *generalized logits* and perform a logistic analysis similar to those described in the previous chapter, except that you model multiple logits per subpopulation. The analysis of generalized logits is a form of the loglinear model, discussed in Chapter 16, "Loglinear Models." The LOGISTIC procedure is used to model cumulative logits, and currently the CATMOD procedure is used to model generalized logits.

9.2 Ordinal Response: Proportional Odds Model

9.2.1 Methodology

Consider the arthritis pain data in Table 9.1. Male and female subjects received an active or placebo treatment for their arthritis pain, and the subsequent extent of improvement was recorded as marked, some, or none (Koch and Edwards 1988).

Table 9.1. Arthritis Data

Sex	Treatment	Improvement			Total
		Marked	Some	None	
Female	Active	16	5	6	27
Female	Placebo	6	7	19	32
Male	Active	5	2	7	14
Male	Placebo	1	0	10	11

One possible analysis strategy is to create a dichotomous response variable by combining two of the response categories, basing a model on either Pr{marked improvement} versus Pr{some or no improvement} or Pr{marked or some improvement} versus

Pr{no improvement}. However, since there is a natural ordering to these response levels, it makes sense to consider a strategy that takes advantage of this ordering.

Consider the quantities

$$\theta_{hi1} = \pi_{hi1}, \quad \theta_{hi2} = \pi_{hi1} + \pi_{hi2}$$

where π_{hi1} denotes the probability of marked improvement, π_{hi2} denotes the probability of some improvement, and π_{hi3} denotes the probability of no improvement. The $\{\theta_{hij}\}$ represent cumulative probabilities: θ_{hi1} is the probability of marked improvement, and θ_{hi2} is the probability of marked or some improvement ($h = 1$ for females, $h = 2$ for males; $i = 1$ for active treatment, $i = 2$ for placebo).

For a dichotomous response, you compute a logit function for each subpopulation. For a multi-level response, you create more than one logit function for each subpopulation. With ordinal data, you can compute *cumulative logits*, which are based on the cumulative probabilities. For three response levels, you compute two cumulative logits:

$$\text{logit}(\theta_{hi1}) = \log\left[\frac{\pi_{hi1}}{\pi_{hi2} + \pi_{hi3}}\right], \quad \text{logit}(\theta_{hi2}) = \log\left[\frac{\pi_{hi1} + \pi_{hi2}}{\pi_{hi3}}\right]$$

These cumulative logits are the log odds of marked improvement to none or some improvement and the log odds of marked or some improvement to no improvement, respectively. Both log odds focus on more favorable to less favorable response. The proportional odds model takes both of these odds into account.

Assuming that the data arise from a stratified simple random sample or are at least conceptually representative of a stratified population, they have the following likelihood:

$$\Pr\{n_{hij}\} = \prod_{h=1}^{2}\prod_{i=1}^{2} n_{hi+}! \prod_{j=1}^{3} \frac{\pi_{hij}^{n_{hij}}}{n_{hij}!}$$

where

$$\sum_{j=1}^{3}\pi_{hij} = 1$$

You could write a model that applies to both logits simultaneously for each combination of gender and treatment:

$$\text{logit}(\theta_{hik}) = \alpha_k + \mathbf{x}_{hi}'\boldsymbol{\beta}_k$$

where k indexes the two logits. This says that there are separate intercept parameters (α_k) and different sets of regression parameters ($\boldsymbol{\beta}_k$) for each logit.

If you take the difference in logits between two subpopulations for this model, you get

$$\text{logit}(\theta_{hik}) - \text{logit}(\theta_{hi'k}) = (\mathbf{x}_{hi} - \mathbf{x}_{hi'})'\boldsymbol{\beta}_k \text{ for } k = 1, 2$$

Thus, you would need to look at two differences in logits simultaneously to compare the response between two subpopulations. This is the same number of comparisons you would need to compare two subpopulations for a three-level nominal response, for example, in a test for association in a contingency table (that is, $r - 1$ where r is the number of response outcomes). Therefore, this model doesn't take the ordinality of the data into account.

The proportional odds assumption is that $\beta_k = \beta$ for all k, simplifying the model to

$$\text{logit}(\theta_{hik}) = \alpha_k + \mathbf{x}'_{hi}\beta$$

If you take the difference in logits for this model, you obtain the equations

$$\text{logit}(\theta_{hi1}) - \text{logit}(\theta_{hi'1}) = \log\left[\frac{\pi_{hi1}/(\pi_{hi2} + \pi_{hi3})}{\pi_{hi'1}/(\pi_{hi'2} + \pi_{hi'3})}\right] = (\mathbf{x}_{hi} - \mathbf{x}_{hi'})'\beta$$

$$\text{logit}(\theta_{hi2}) - \text{logit}(\theta_{hi'2}) = \log\left[\frac{(\pi_{hi1} + \pi_{hi2})/\pi_{hi3}}{(\pi_{hi'1} + \pi_{hi'2})/\pi_{hi'3}}\right] = (\mathbf{x}_{hi} - \mathbf{x}_{hi'})'\beta$$

This says that the log cumulative odds are proportional to the distance between the explanatory variable values and that the influence of the explanatory variables is independent of the cutpoint for the cumulative logit. In this case, there is a "cut" at marked improvement to form $\text{logit}(\theta_{hi1})$ and a cut at some improvement to form $\text{logit}(\theta_{hi2})$. This proportionality is what gives the proportional odds model its name. For a single continuous explanatory variable, the regression lines would be parallel to each other, their relative position determined by the values of the intercept parameter.

This model can also be stated as

$$\theta_{hik} = \frac{\exp(\alpha_k + \mathbf{x}'_{hi}\beta)}{1 + \exp(\alpha_k + \mathbf{x}'_{hi}\beta)}$$

and is written in summation notation as

$$\theta_{hik} = \frac{\exp\{\alpha_k + \sum_{g=1}^{t} \beta_g x_{hig}\}}{1 + \exp\{\alpha_k + \sum_{g=1}^{t} \beta_g x_{hig}\}}$$

where $g = (1, 2, \ldots, t)$ references the explanatory variables. This model is similar to the previous logistic regression models and is also fit with maximum likelihood methods. You can determine the values for π_{hij} from this model by performing the appropriate subtractions of the θ_{hik}.

$$\pi_{hi1} = \theta_{hi1}$$
$$\pi_{hi2} = \theta_{hi2} - \theta_{hi1}$$
$$\pi_{hi3} = 1 - \theta_{hi2}$$

The main effects model is an appropriate starting point for the analysis of the arthritis data. You can write this model in matrix notation as

$$
\begin{bmatrix}
\text{logit}(\theta_{111}) \\
\text{logit}(\theta_{112}) \\
\text{logit}(\theta_{121}) \\
\text{logit}(\theta_{122}) \\
\text{logit}(\theta_{211}) \\
\text{logit}(\theta_{212}) \\
\text{logit}(\theta_{221}) \\
\text{logit}(\theta_{222})
\end{bmatrix}
=
\begin{bmatrix}
\alpha_1 & + \beta_1 + \beta_2 \\
& \alpha_2 + \beta_1 + \beta_2 \\
\alpha_1 & + \beta_1 \\
& \alpha_2 + \beta_1 \\
\alpha_1 & + \beta_2 \\
& \alpha_2 + \beta_2 \\
\alpha_1 & \\
& \alpha_2
\end{bmatrix}
=
\begin{bmatrix}
1 & 0 & 1 & 1 \\
0 & 1 & 1 & 1 \\
1 & 0 & 1 & 0 \\
0 & 1 & 1 & 0 \\
1 & 0 & 0 & 1 \\
0 & 1 & 0 & 1 \\
1 & 0 & 0 & 0 \\
0 & 1 & 0 & 0
\end{bmatrix}
\begin{bmatrix}
\alpha_1 \\
\alpha_2 \\
\beta_1 \\
\beta_2
\end{bmatrix}
$$

This is very similar to the models described in Chapter 8 except that there are two intercept parameters corresponding to the two cumulative logit functions being modeled for each group. The parameter α_1 is the intercept for the first cumulative logit, α_2 is the intercept for the second cumulative logit, β_1 is an incremental effect for females, and β_2 is an incremental effect for active. Males on placebo comprise the reference cell.

Table 9.2 contains the cell probabilities for marked improvement and no improvement based on this model. Table 9.3 contains the odds. The cell probabilities for marked improvement are based on the model for the first logit function, and the probabilities for no improvement are based on the model for the second logit function (these probabilities are computed from $1 - \theta_{hi2}$). Since the probabilities for all three levels sum to 1, you can determine the cell probabilities for some improvement through subtraction.

The odds ratio for females versus males is e^{β_1}, and the odds ratio for active treatment versus placebo is e^{β_2}. The odds ratios are computed in the same manner as for the logistic regression analysis for a dichotomous response—you form the ratio of the appropriate odds.

Table 9.2.   Formulas for Cell Probabilities

Sex	Treatment	Improvement	
		Marked	None
Female	Active	$e^{\alpha_1+\beta_1+\beta_2}/(1+e^{\alpha_1+\beta_1+\beta_2})$	$1/(1+e^{\alpha_2+\beta_1+\beta_2})$
Female	Placebo	$e^{\alpha_1+\beta_1}/(1+e^{\alpha_1+\beta_1})$	$1/(1+e^{\alpha_2+\beta_1})$
Male	Active	$e^{\alpha_1+\beta_2}/(1+e^{\alpha_1+\beta_2})$	$1/(1+e^{\alpha_2+\beta_2})$
Male	Placebo	$e^{\alpha_1}/(1+e^{\alpha_1})$	$1/(1+e^{\alpha_2})$

Table 9.3. Formulas for Model Odds

Sex	Treatment	Improvement	
		Marked Versus Some or None	Marked or Some Versus None
Female	Active	$e^{\alpha_1+\beta_1+\beta_2}$	$e^{\alpha_2+\beta_1+\beta_2}$
Female	Placebo	$e^{\alpha_1+\beta_1}$	$e^{\alpha_2+\beta_1}$
Male	Active	$e^{\alpha_1+\beta_2}$	$e^{\alpha_2+\beta_2}$
Male	Placebo	e^{α_1}	e^{α_2}

For example, the odds of marked improvement versus some or no improvement for females compared to males is

$$\frac{e^{\alpha_1+\beta_1+\beta_2}}{e^{\alpha_1+\beta_2}} = e^{\beta_1}$$

As constrained by the proportional odds model, this is also the odds ratio for marked or some improvement versus no improvement.

9.2.2 Fitting the Proportional Odds Model with PROC LOGISTIC

PROC LOGISTIC fits the proportional odds model whenever the response variable has more than two levels. Thus, you need to ensure that you do, indeed, have an ordinal response variable because PROC LOGISTIC will assume that you do. Note that the GENMOD and PROBIT procedures also fit the proportional odds model with maximum likelihood estimation.

The following SAS statements create the data set ARTHRITIS. Note that these data are in the form of counts, so a variable named COUNT is created to contain the frequencies for each table cell. The variable IMPROVE is a character variable that takes the values marked, some, or none to indicate the subject's extent of improvement of arthritic pain. The variable SEX takes the values male and female, and the variable TREATMENT takes the values active and placebo.

```
data arthritis;
   length treatment $7. sex $6.;
   input sex $ treatment $ improve $ count @@;
   datalines;
female active  marked 16 female active  some 5 female active  none  6
female placebo marked  6 female placebo some 7 female placebo none 19
male    active  marked  5 male    active  some 2 male    active  none  7
male    placebo marked  1 male    placebo some 0 male    placebo none 10
;
run;
```

The use of PROC LOGISTIC is identical to previous invocations for dichotomous response logistic regression. The response variable is listed on the left-hand side of the equal sign and the explanatory variables are listed on the right-hand side. Since ORDER=DATA is specified in the PROC statement, the values for IMPROVE are ordered in the sequence in which PROC LOGISTIC encounters them in the data, which is marked, some, and none. (Another legitimate ordering would be none, some, and marked.) It is very important to ensure that the ordering is correct when you are using ordinal data strategies. The procedure still performs an analysis if the response values are ordered incorrectly, but the results will be erroneous. The burden is on the user to specifiy the correct order and then to check the results.

The following code requests that PROC LOGISTIC fit a proportional odds model.

```
proc logistic order=data;
   freq count;
   class treatment sex / param=reference;
   model improve = sex treatment / scale=none aggregate;
run;
```

The "Response Profile" table in Output 9.1 shows that the response variable values are ordered correctly in terms of decreasing improvement. Thus, the cumulative logits modeled are based on more to less improvement. The procedure also prints out a note that a zero count observation has been encountered. For these data, this is not a problem since the total row counts are acceptably large. Computationally, zero counts are discarded. The model still produces predicted values for the cell that corresponds to the zero cell, males on placebo who showed some improvement.

Output 9.1   Response Profiles

```
                        Response Profile

            Ordered                      Total
            Value      improve         Frequency

               1       marked             28
               2       some               14
               3       none               42

NOTE: 1 observation having zero frequency or weight was excluded since it does
      not contribute to the analysis.
```

The procedure next prints the "Class Level Information" table, which shows that the parameterization takes the form of incremental effects for active treatment and females.

Output 9.2 Class Level Information

```
                    Class Level Information

                                          Design
                                         Variables
           Class          Value              1

           treatment      active             1
                          placebo            0

           sex            female             1
                          male               0
```

Next, PROC LOGISTIC prints out a test for the appropriateness of the proportional odds assumption. The test performed is a score test that determines whether, if you fit a different set of explanatory variable parameters β_k for each logit function, those sets of parameters are equivalent. Thus, the model considered is

$$\text{logit}(\theta_{hik}) = \alpha_k + \mathbf{x}'_{hi}\boldsymbol{\beta}_k$$

The hypothesis tested is that there is a common parameter vector β instead of distinct β_k. The hypothesis can be stated as $\beta_k = \beta$ for all k. Thus, if you reject the null hypothesis, you reject the assumption of proportional odds and you need to consider a different approach, such as modeling generalized logits, discussed in Section 9.3. If the null hypothesis is not rejected, then the test supports the assumption of proportional odds. Since the test is comparing t parameters for the t explanatory variables across $(r-1)$ logits, where r is the number of response levels, it has $t * (r-2)$ degrees of freedom.

The sample size requirements for this test are moderately demanding; you need approximately five observations at each outcome at each level of each main effect, or roughly the same sample size as if you were fitting a generalized logit model. Small samples may artificially make the statistic large, meaning that any resulting significance needs to be interpreted cautiously. However, nonsignificant results are always informative.

The partial proportional odds model is an alternative model that can be fit when the proportionality assumption does not hold for all explanatory variables, but there is proportionality for some. Refer to Koch, Amara, and Singer (1985) for a discussion of this model and refer to Section 15.12 for an example of fitting this type of model using the generalized estimating equations approach. When there appears to be no proportionality, the best approach may be to treat the data as nominal and fit a model to the generalized logits, as discussed in Section 9.3.

Output 9.3 displays this score test.

Output 9.3 Proportional Odds Test

```
        Score Test for the Proportional Odds Assumption

           Chi-Square         DF        Pr > ChiSq

             1.8833            2           0.3900
```

Q_{RS} takes the value 1.883 with 2 df. This is clearly nonsignificant, and so the assumption of proportional odds is a reasonable one for these data.

The evaluation of goodness of fit for the proportional odds model is similar to the evaluation of goodness of fit for the dichotomous response logistic regression model. If you have sufficient sample size, with 80% of the observed cell counts at least 5, then you can use counterparts of Q_P and Q_L. Q_P is distributed as chi-square with df = $\{(r-1)(s-1) - t\}$, where t is the number of explanatory variables, r is the number of response levels, and s is the number of subpopulations. Output 9.4 contains these statistics. With values of 2.7121 and 1.9099, respectively, and 4 df, Q_L and Q_P support the adequacy of the model. The 4 df come from $(3-1)(4-1) - 2 = 4$.

Output 9.4 Goodness of Fit Statistics

```
             Deviance and Pearson Goodness-of-Fit Statistics

        Criterion         DF        Value      Value/DF     Pr > ChiSq

        Deviance           4        2.7121      0.6780        0.6071
        Pearson            4        1.9099      0.4775        0.7523

                   Number of unique profiles: 4
```

The tests for assessing model fit through explanatory capability are also supportive of the model; the likelihood ratio test has a value of 19.8865 with 2 df and the score test has a value of 17.8677 with 2 df, as displayed in Output 9.5.

Output 9.5 Global Tests

```
                 Testing Global Null Hypothesis: BETA=0

        Test                   Chi-Square        DF      Pr > ChiSq

        Likelihood Ratio         19.8865          2        <.0001
        Score                    17.8677          2        0.0001
        Wald                     16.7745          2        0.0002
```

You can also investigate goodness of fit by performing the score test for a set of additional terms not in the model. In this case, this effect would simply be the treatment × sex interaction. The following code requests that PROC LOGISTIC fit a main effects model and then perform a score test for the other effect listed in the MODEL statement, which is the interaction.

```
proc logistic order=data;
   freq count;
   class sex treatment / param=reference;
   model improve = sex treatment sex*treatment /
              selection=forward start=2;
run;
```

The score test of interest is labeled "Residual Chi-Square" and is printed after the "Testing Global Null Hypothesis: BETA=0" table; it is displayed in Output 9.6. The value is 0.2801 (1 df since you are testing the addition of one term to the model) with $p = 0.5967$. This indicates that the main effects model is adequate.

Output 9.6   Score Statistic to Evaluate Goodness of Fit

Residual Chi-Square Test		
Chi-Square	DF	Pr > ChiSq
0.2801	1	0.5967

An alternative goodness-of-fit test is the difference in the likelihood ratios for the main effects model and the saturated model. Although the output is not displayed here, the difference in these statistics is $(150.029 - 149.721) = 0.308$. This is also clearly nonsignificant, compared to a chi-square distribution with 1 df. (Again, note that whenever you form a test statistic based on the difference in likelihoods, then the corresponding degrees of freedom are equal to the difference in the number of parameters for the two models.)

Output 9.7 contains the "Type III Analysis of Effects" table. Both sex and treatment are influential effects. Since these effects have 1 df each, the tests are the same as printed for the parameter estimates listed in Output 9.8.

Output 9.7   Type III Analysis of Effects

Type III Analysis of Effects			
Effect	DF	Wald Chi-Square	Pr > ChiSq
sex	1	6.2096	0.0127
treatment	1	14.4493	0.0001

Output 9.8   Parameter Estimates

Analysis of Maximum Likelihood Estimates						
Parameter		DF	Estimate	Standard Error	Wald Chi-Square	Pr > ChiSq
Intercept		1	-2.6671	0.5997	19.7800	<.0001
Intercept2		1	-1.8127	0.5566	10.6064	0.0011
sex	female	1	1.3187	0.5292	6.2096	0.0127
treatment	active	1	1.7973	0.4728	14.4493	0.0001

Table 9.4 displays the parameter interpretations.

Table 9.4. Parameter Estimates

Parameter	Estimate(SE)	Interpretation
α_1	$-2.667(0.600)$	log odds of marked improvement versus some or no improvement for males receiving placebo
α_2	$-1.813(0.557)$	log odds of marked or some improvement versus no improvement for males receiving placebo
β_1	$1.319(0.529)$	increment for both types of log odds due to female sex
β_2	$1.797(0.473)$	increment for both types of log odds due to active drug

Females have $e^{1.319}$=3.7 times higher odds of showing improvement as males, both for marked improvement versus some or no improvement and for marked or some improvement versus no improvement. Those subjects receiving the active drug have $e^{1.8} = 6$ times higher odds of showing improvement as those on placebo, both for marked improvement versus some or no improvement and for some or marked improvement versus no improvement. These odds ratio estimates are displayed in Output 9.9.

Output 9.9 Odds Ratio Estimates

```
                        Odds Ratio Estimates

                                  Point          95% Wald
        Effect                  Estimate     Confidence Limits

        sex        female vs male    3.739        1.325      10.547
        treatment  active vs placebo 6.033        2.388      15.241
```

9.2.3   Multiple Qualitative Explanatory Variables

The inclusion of multiple explanatory variables in a proportional odds model produces no additional problems. The data in Table 9.5 are from an epidemiological study of chronic respiratory disease analyzed in Semenya and Koch (1980). Researchers collected information on subjects' exposure to general air pollution, exposure to pollution in their jobs, and whether they smoked. The response measured was chronic respiratory disease status. Subjects were assigned to one of four possible categories.

- Level I: no symptoms
- Level II: cough or phlegm less than three months a year
- Level III: cough or phlegm more than three months a year
- Level IV: cough and phlegm plus shortness of breath more than three months a year

Table 9.5. Chronic Respiratory Disease Data

Air Pollution	Job Exposure	Smoking Status	Response Level I	II	III	IV	Total
Low	No	Non	158	9	5	0	172
Low	No	Ex	167	19	5	3	194
Low	No	Current	307	102	83	68	560
Low	Yes	Non	26	5	5	1	37
Low	Yes	Ex	38	12	4	4	58
Low	Yes	Current	94	48	46	60	248
High	No	Non	94	7	5	1	107
High	No	Ex	67	8	4	3	82
High	No	Current	184	65	33	36	318
High	Yes	Non	32	3	6	1	42
High	Yes	Ex	39	11	4	2	56
High	Yes	Current	77	48	39	51	215

The outcome is clearly ordinal, although there is no obvious distance between adjacent levels. You could combine response categories and fit the set of models that compared Level I versus Level II, III, and IV; Levels I and II versus Levels III and IV; and Levels I, II, and III versus Level IV. Note that if you did this, you would be computing models for the individual cumulative logits. The proportional odds model addresses these cumulative logits simultaneously by assuming that the slope parameters for the explanatory variables are the same regardless of the cumulative logit cutpoints.

From these data, you form three cumulative logits:

$$\text{logit}(\theta_{i1}) = \log\left[\frac{\pi_{i1}}{\pi_{i2} + \pi_{i3} + \pi_{i4}}\right]$$

$$\text{logit}(\theta_{i2}) = \log\left[\frac{\pi_{i1} + \pi_{i2}}{\pi_{i3} + \pi_{i4}}\right]$$

$$\text{logit}(\theta_{i3}) = \log\left[\frac{\pi_{i1} + \pi_{i2} + \pi_{i3}}{\pi_{i4}}\right]$$

where $i = 1, 2, \ldots, 12$ references the 12 populations determined by the levels of air pollution, job exposure, and smoking status, as ordered in Table 9.5. These cumulative logits are the log odds of a Level I response to a Level II, III, or IV response; the log odds of a Level I or II response to a Level III or IV response; and the log odds of a Level I, II, or III response to a Level IV response, respectively.

However, if you are more interested in the odds of more severe responses to less severe responses, you may want to order the cumulative logits in the opposite direction.

$$\text{logit}(\theta_{i1}) = \log\left[\frac{\pi_{i4}}{\pi_{i3} + \pi_{i2} + \pi_{i1}}\right]$$

$$\text{logit}(\theta_{i2}) = \log\left[\frac{\pi_{i4} + \pi_{i3}}{\pi_{i2} + \pi_{i1}}\right]$$

$$\text{logit}(\theta_{i3}) = \log\left[\frac{\pi_{i4} + \pi_{i3} + \pi_{i2}}{\pi_{i1}}\right]$$

You can generate this ordering in PROC LOGISTIC by using the DESCENDING option in the PROC statement, as shown in the following analysis.

The primary model of interest for these data is a main effects model. Besides three intercept terms α_1, α_2, and α_3 for the three cumulative logits, the main effects model includes the parameters β_1, β_2, β_3, and β_4 for incremental effects for air pollution exposure, job pollution exposure, ex-smoker status, and current smoking status, respectively.

The following SAS statements create the data set RESPIRE.

```
data respire;
   input air $ exposure $ smoking $ level count @@;
   datalines;
low no non   1 158 low  no   non 2   9
low no ex    1 167 low  no   ex  2  19
low no cur   1 307 low  no   cur 2 102
low yes non  1  26 low  yes  non 2   5
low yes ex   1  38 low  yes  ex  2  12
low yes cur  1  94 low  yes  cur 2  48
high no non  1  94 high no   non 2   7
high no ex   1  67 high no   ex  2   8
high no cur  1 184 high no   cur 2  65
high yes non 1  32 high yes  non 2   3
high yes ex  1  39 high yes  ex  2  11
high yes cur 1  77 high yes  cur 2  48
low  no   non 3  5  low  no non  4  0
low  no   ex  3  5  low  no ex   4  3
low  no   cur 3 83  low  no cur  4 68
low  yes non 3  5  low  yes non 4  1
low  yes ex  3  4  low  yes ex   4  4
low  yes cur 3 46  low  yes cur 4 60
high no   non 3  5  high no non  4  1
high no   ex  3  4  high no ex   4  3
```

```
high no   cur 3 33   high no   cur 4 36
high yes  non 3  6   high yes  non 4  1
high yes  ex  3  4   high yes  ex  4  2
high yes  cur 3 39   high yes  cur 4 51
;
run;
```

The following PROC LOGISTIC code requests a main effects proportional odds model. The statements are set up to specify a score statistic for the goodness of fit of the expanded model containing all pairwise interaction terms. The SCALE=NONE and AGGREGATE=(AIR EXPOSURE SMOKING) options request the goodness-of-fit tests based on the 12 subpopulations. The REF='no' option specified for the EXPOSURE variable in the CLASS statement causes no exposure to be the reference level.

```
proc logistic descending;
    freq count;
    class air exposure(ref='no') smoking / param=reference;
    model level = air exposure smoking
            air*exposure air*smoking exposure*smoking /
            selection=forward include=3 scale=none
            aggregate=(air exposure smoking);
run;
```

Output 9.10 shows the internal ordered values that PROC LOGISTIC uses. Since the response variable LEVEL has numeric values, the DESCENDING option causes PROC LOGISTIC to sort the values numerically, then reverses them to form the ordered values.

Output 9.10   Response Profile

Ordered Value	level	Total Frequency
1	4	230
2	3	239
3	2	337
4	1	1283

The score test for the proportional odds assumption takes the value $Q_{R\,S} = 12.0745$ ($p = 0.1479$) with 8 df $(4(4-2))$. Thus, the proportional odds assumption is not contradicted.

Output 9.11   Test for Proportionality

Score Test for the Proportional Odds Assumption

Chi-Square	DF	Pr > ChiSq
12.0745	8	0.1479

The three intercepts and four indicator variables representing the main effects are first entered into the model. The residual chi-square has a value of 2.7220 with 5 df and $p = 0.7428$, so this measure of goodness of fit suggests that model-predicted cell proportions are acceptably close to the observed proportions.

Output 9.12   Assessment of Fit

```
                    Residual Chi-Square Test

          Chi-Square        DF        Pr > ChiSq

             2.7220          5           0.7428
```

Output 9.13 displays the goodness-of-fit statistics. $Q_L = 29.9969$ and $Q_P = 28.0796$, both with $(r-1)(s-1) - t = 29$ df ($r = 4$, $s = 12$, $t = 4$). Model adequacy is again supported.

Output 9.13   Goodness-of-Fit Statistics

```
          Deviance and Pearson Goodness-of-Fit Statistics

      Criterion        DF        Value      Value/DF      Pr > ChiSq

      Deviance         29       29.9969      1.0344         0.4142
      Pearson          29       28.0796      0.9683         0.5137

              Number of unique profiles: 12
```

The "Type III Analysis of Effects" table displayed in Output 9.14 suggests a strong effect for job pollution exposure but no significant effect for outside air pollution ($p = 0.675$). The smoking effect is also highly significant.

Output 9.14   Type III Analysis of Effects

```
                  Type III Analysis of Effects

                                   Wald
      Effect            DF      Chi-Square      Pr > ChiSq

      air                1         0.1758         0.6750
      exposure           1        82.0603         <.0001
      smoking            2       209.8507         <.0001
```

The parameter estimates are displayed in Output 9.15.

Output 9.15 Parameter Estimates

```
                Analysis of Maximum Likelihood Estimates

                                       Standard      Wald
Parameter                  DF  Estimate   Error   Chi-Square   Pr > ChiSq

Intercept                   1   -3.8938  0.1779   479.2836      <.0001
Intercept2                  1   -2.9696  0.1693   307.7931      <.0001
Intercept3                  1   -2.0884  0.1633   163.5861      <.0001
air          high           1   -0.0393  0.0937     0.1758      0.6750
exposure     yes            1    0.8648  0.0955    82.0603      <.0001
smoking      cur            1    1.8527  0.1650   126.0383      <.0001
smoking      ex             1    0.4000  0.2019     3.9267      0.0475
```

The predicted odds ratios illustrate this model's conclusions. Those with job exposure have $e^{0.8648} = 2.374$ times higher odds of having serious problems to less serious problems compared to those not exposed on the job. Current smokers have $e^{1.8527} = 6.377$ times higher odds of having serious problems to less serious problems compared to nonsmokers. Both of these odds ratios have been adjusted for the other variables in the model.

Output 9.16 Odds Ratios

```
                    Odds Ratio Estimates

                           Point        95% Wald
Effect                    Estimate   Confidence Limits

air       high vs low      0.961     0.800     1.155
exposure  yes vs no        2.374     1.969     2.863
smoking   cur vs non       6.377     4.615     8.812
smoking   ex vs non        1.492     1.004     2.216
```

Note that if you fit the same model without reversing the order of the cumulative logits with the DESCENDING option, you fit an equivalent model. The intercepts will be in the opposite order and have opposite signs; that is, INTERCEP3 will have the value of this model's INTERCEP1 with the opposite sign. The parameters for the effects will have opposite signs, and the odds ratios will be inverted since they would represent the odds of less serious response to more serious response.

9.3   Nominal Response: Generalized Logits Model

9.3.1   Methodology

When you have nominal response variables, you can also use logistic regression to model your data. Instead of fitting a model to cumulative logits, you fit a model to generalized logits. Table 9.6 redisplays the data analyzed in Section 6.3.3. Recall that schoolchildren in experimental learning settings were surveyed to determine which program they preferred. Investigators were interested in whether their response was associated with their school and their school day, which could be a standard school day or include afterschool care.

Table 9.6. School Program Data

School	Program	Learning Style Preference		
		Self	Team	Class
1	Regular	10	17	26
1	After	5	12	50
2	Regular	21	17	26
2	After	16	12	36
3	Regular	15	15	16
3	After	12	12	20

Since the levels of the response variable (self, team, and class) have no inherent ordering, the proportional odds model is not an appropriate mechanism for their analysis. You could form logits comparing self to (team or class) or (self or team) to class, but that collapses the original structure of the response levels, which you may want to keep in your analysis. You can model a nominal response variable with more than two levels by performing a logistic analysis on the generalized logits.

The generalized logit is defined as

$$\text{logit}_{hij} = \log\left[\frac{\pi_{hij}}{\pi_{hir}}\right]$$

for $j = 1, 2, \ldots, (r - 1)$. A logit is formed for the probability of each succeeding category over the last response category.

Thus, the generalized logits for a three-level response like that displayed in Table 9.6 is

$$\text{logit}_{hi1} = \log\left[\frac{\pi_{hi1}}{\pi_{hi3}}\right], \quad \text{logit}_{hi2} = \log\left[\frac{\pi_{hi2}}{\pi_{hi3}}\right]$$

for $h = 1, 2, 3$ for the schools, $i = 1$ for regular program, and $i = 2$ for afterschool program.

The model you fit for generalized logits is the model discussed in Section 9.2.1.

$$\text{logit}_{hik} = \alpha_k + \mathbf{x}'_{hi}\beta_k$$

where k indexes the two logits. This says that there are separate intercept parameters (α_k) and different sets of regression parameters (β_k) for each logit. The matrix $\mathbf{x}_{hi}$ is the set of explanatory variable values for the hith group. Instead of estimating one set of parameters for one logit function, as in logistic regression for a dichotomous response variable, you are estimating sets of parameters for multiple logit functions. Whereas for the proportional odds model you estimated multiple intercept parameters for the cumulative logit functions but only one set of parameters corresponding to the explanatory variables, for the generalized logits model you are estimating multiple sets of parameters for both the intercept terms and the explanatory variables.

This poses no particular problems. Since there are multiple response functions being modeled per subpopulation, there are more degrees of freedom associated with each effect. Since the model matrix needs to account for multiple response functions, it takes a more complicated form. However, the modeling proceeds as usual; you fit your specified model, examine goodness-of-fit statistics, and possibly perform model reduction. Note that since you are predicting more than one response function per subpopulation, the sample size needs to be large enough to support the number of functions you are modeling. Sometimes, in those situations where there isn't enough data to justify the analysis of generalized logits, you will also encounter problems with parameter estimation and the software will print out notes about infinite parameters. In those situations, you can often simplify the response structure to a reasonable dichotomy and proceed with a binary logistic regression.

9.3.2 Fitting Models to Generalized Logits with PROC CATMOD

The CATMOD procedure is used to perform an analysis of generalized logits; the LOGISTIC procedure currently does not perform this analysis, although it will have the capability for fitting the generalized logits model for nominal data in a future release. You should read Section 8.9.1 "Performing Logistic Regression with the CATMOD Procedure," before continuing with this section.

The following SAS statements request the desired analysis. First, the data set SCHOOL is created, then the CATMOD procedure is invoked. Since the generalized logit is the default response for PROC CATMOD, and maximum likelihood is the default estimation method for generalized logit functions, all you have to specify are the WEIGHT and MODEL statements. PROC CATMOD constructs two generalized logits per group from the levels of the variable STYLE; it composes six groups based on the unique values of the explanatory variables, SCHOOL and PROGRAM.

```
data school;
    input school program $ style $ count @@;
    datalines;
1 regular  self 10  1 regular  team 17 1 regular class   26
1 after     self  5  1 after     team 12 1 after    class   50
2 regular  self 21  2 regular  team 17 2 regular class   26
2 after     self 16  2 after     team 12 2 after    class   36
3 regular  self 15  3 regular  team 15 3 regular class   16
3 after     self 12  3 after     team 12 3 after    class   20
;
run;
proc catmod order=data;
    weight count;
    model style=school program school*program;
run;
```

Output 9.17 contains the population profiles for the data. There are six groups formed based on the values of SCHOOL and PROGRAM. This table also contains the sample sizes for each group. They are all moderately sized.

Output 9.17   Population Profiles

```
                    Population Profiles

      Sample    school    program    Sample Size
      ---------------------------------------------
         1        1       regular          53
         2        1       after            67
         3        2       regular          64
         4        2       after            64
         5        3       regular          46
         6        3       after            44
```

Output 9.18 contains the response profiles. While the analysis does not require the response values to be ordered in any particular way, unlike the proportional odds model analyses, it is often useful to order the levels in a manner that facilitates interpretation. Since the ORDER=DATA option was specified, the response variable levels are in the order self, team, and class. This means that generalized logits are formed for the probability of self with respect to class, and for the probability of team with respect to class.

Output 9.18   Response Profiles

```
                Response Profiles

           Response        style
           -----------------------
              1            self
              2            team
              3            class
```

The iteration history follows in the output but is not displayed here. Convergence took four iterations, which is not unusual for these types of applications.

The analysis of variance table is displayed in Output 9.19.

Output 9.19   ANOVA Table

```
         Maximum Likelihood Analysis of Variance

   Source              DF    Chi-Square    Pr > ChiSq
   -------------------------------------------------------
   Intercept            2       40.05        <.0001
   school               4       14.55        0.0057
   program              2       10.48        0.0053
   school*program       4        1.74        0.7827

   Likelihood Ratio     0          .            .
```

Since the model is saturated, with as many response functions being modeled as there are groups or subpopulations, the likelihood ratio test does not apply and PROC CATMOD prints missing values and 0 df. The school $\times$ program interaction is nonsignificant, with a Wald chi-square of 1.74 with 4 df. Note that the degrees of freedom for modeling two generalized logits are twice what you would expect for modeling one logit: instead of 1 df for the intercept you have 2 df; instead of 2 df for SCHOOL, which has three levels, you have 4 df. This is because you are simultaneously modeling two response functions instead of one; you are doubling the number of parameters being estimated since you have to estimate parameters for both logits. To determine the correct number of degrees of freedom for effects in models using generalized logits, multiply the number you would expect for modeling one logit (the usual logistic regression for a dichotomous outcome) by $r - 1$, where r is the number of response levels.

Since the interaction is nonsignificant, the main effects model is fit.

$$
\begin{bmatrix}
\text{logit}_{111} \\
\text{logit}_{112} \\
\text{logit}_{121} \\
\text{logit}_{122} \\
\text{logit}_{211} \\
\text{logit}_{212} \\
\text{logit}_{221} \\
\text{logit}_{222} \\
\text{logit}_{311} \\
\text{logit}_{312} \\
\text{logit}_{321} \\
\text{logit}_{322}
\end{bmatrix}
=
\begin{bmatrix}
1 & 0 & 1 & 0 & 0 & 0 & 1 & 0 \\
0 & 1 & 0 & 1 & 0 & 0 & 0 & 1 \\
1 & 0 & 1 & 0 & 0 & 0 & -1 & 0 \\
0 & 1 & 0 & 1 & 0 & 0 & 0 & -1 \\
1 & 0 & 0 & 0 & 1 & 0 & 1 & 0 \\
0 & 1 & 0 & 0 & 0 & 1 & 0 & 1 \\
1 & 0 & 0 & 0 & 1 & 0 & -1 & 0 \\
0 & 1 & 0 & 0 & 0 & 1 & 0 & -1 \\
1 & 0 & -1 & 0 & -1 & 0 & 1 & 0 \\
0 & 1 & 0 & -1 & 0 & -1 & 0 & 1 \\
1 & 0 & -1 & 0 & -1 & 0 & -1 & 0 \\
0 & 1 & 0 & -1 & 0 & -1 & 0 & -1
\end{bmatrix}
\begin{bmatrix}
\alpha_1 \\
\alpha_2 \\
\beta_1 \\
\beta_2 \\
\beta_3 \\
\beta_4 \\
\beta_5 \\
\beta_6
\end{bmatrix}
$$

Essentially, this model matrix has the same structure as one for modeling a single response function, except that it models two response functions. Thus, the odd rows are for the first logit, and the even rows are for the second logit. Similarly, the odd columns correspond to parameters for the first logit, and the even columns correspond to parameters for the second logit. Note that PROC CATMOD uses differential effects by default, as opposed to the incremental effects fit with PROC LOGISTIC for the proportional odds model discussed in the previous sections (see Section 8.9 for further discussion). See Table 9.7 for the interpretation of these parameters.

The following statements produce the main effects model.

```
proc catmod order=data;
    weight count;
    model style=school program;
run;
```

Output 9.20 contains the ANOVA table for the main effects model.

Output 9.20   ANOVA Table

```
          Maximum Likelihood Analysis of Variance

          Source           DF    Chi-Square   Pr > ChiSq
          -------------------------------------------------
          Intercept         2       39.88       <.0001
          school            4       14.84       0.0050
          program           2       10.92       0.0043

          Likelihood Ratio  4        1.78        0.7766
```

The likelihood ratio statistic has a value of 1.78 with 4 df, which is indicative of a good fit. The tests for the school and program effects are also significant; SCHOOL has a Wald chi-square value of 14.84 with 4 df, and PROGRAM has a Wald chi-square value of 10.92 with 2 df.

The parameter estimates and tests for individual parameters are displayed in Output 9.21. The order of these parameters corresponds to the order in which the response variable and explanatory variable levels are listed in the response profiles table and the population profiles table.

Output 9.21   Parameter Estimates

```
           Analysis of Maximum Likelihood Estimates

                                         Standard    Chi-
       Effect       Parameter  Estimate    Error    Square   Pr > ChiSq
       ----------------------------------------------------------------
       Intercept        1      -0.7979    0.1465    29.65      <.0001
                        2      -0.6589    0.1367    23.23      <.0001
       school           3      -0.7992    0.2198    13.22      0.0003
                        4      -0.2786    0.1867     2.23      0.1356
                        5       0.2836    0.1899     2.23      0.1352
                        6      -0.0985    0.1892     0.27      0.6028
       program          7       0.3737    0.1410     7.03      0.0080
                        8       0.3713    0.1353     7.53      0.0061
```

For example, since the order of the response values is self, team, and class, Parameter 1 is the intercept for logit_{hi1} and Parameter 2 is the intercept for logit_{hi2}. For the SCHOOL effect, Parameter 3 is the parameter for School 1 for the first logit, and Parameter 4 is the parameter for School 1 for the second logit. Parameters 5 and 6 are the corresponding parameters for School 2, and parameters 7 and 8 are the parameters for logit_{hi1} and logit_{hi2}, respectively, for the regular school program. If you go down the list of parameters in order, the response function varies most quickly, and the groups vary the same way that they vary in the population profiles table. Table 9.7 summarizes the correspondence between the PROC CATMOD numbered parameters in the output and the model parameters.

Table 9.7. Parameter Interpretations

CATMOD Parameter	Model Parameter	Interpretation
1	α_1	Intercept for $logit_{hi1}$
2	α_2	Intercept for $logit_{hi2}$
3	β_1	Differential Effect for School 1 for $logit_{hi1}$
4	β_2	Differential Effect for School 1 for $logit_{hi2}$
5	β_3	Differential Effect for School 2 for $logit_{hi1}$
6	β_4	Differential Effect for School 2 for $logit_{hi2}$
7	β_5	Differential Effect for Regular School for $logit_{hi1}$
8	β_6	Differential Effect for Regular School for $logit_{hi2}$

Table 9.8 contains the parameter estimates arranged according to the logits they reference. This is often a useful way to display the results from an analysis of generalized logits.

Table 9.8. Coefficients from Final Model

	logit(self/class)		logit(team/class)	
Variable	Coefficient	Standard Error	Coefficient	Standard Error
Intercept	$-0.798\ (\hat{\alpha}_1)$	0.146	$-0.659\ (\hat{\alpha}_2)$	0.137
School 1	$-0.799\ (\hat{\beta}_1)$	0.220	$-0.279\ (\hat{\beta}_2)$	0.187
School 2	$0.284\ (\hat{\beta}_3)$	0.190	$-0.099\ (\hat{\beta}_4)$	0.189
Program	$0.374\ (\hat{\beta}_5)$	0.141	$0.371\ (\hat{\beta}_6)$	0.135

School 1 has the largest effect of the schools, particularly for the logit comparing self to class. Program has a nearly similar effect on both logits.

Odds ratios can also be used in models for generalized logits to facilitate model interpretation. Table 9.9 contains the odds corresponding to each logit function for each subpopulation in the data. However, unlike the proportional odds model where the form of the odds ratio was the same regardless of the cumulative logit being considered, the formulas for the odds ratio for the generalized logits model depend on which generalized logit is being considered.

Table 9.9. Model-Predicted Odds

School	Program	Odds Self/Class	Odds Team/Class
1	Regular	$e^{\alpha_1+\beta_1+\beta_5}$	$e^{\alpha_2+\beta_2+\beta_6}$
1	After	$e^{\alpha_1+\beta_1-\beta_5}$	$e^{\alpha_2+\beta_2-\beta_6}$
2	Regular	$e^{\alpha_1+\beta_3+\beta_5}$	$e^{\alpha_2+\beta_4+\beta_6}$
2	After	$e^{\alpha_1+\beta_3-\beta_5}$	$e^{\alpha_2+\beta_4-\beta_6}$
3	Regular	$e^{\alpha_1-\beta_1-\beta_3+\beta_5}$	$e^{\alpha_2-\beta_2-\beta_4+\beta_6}$
3	After	$e^{\alpha_1-\beta_1-\beta_3-\beta_5}$	$e^{\alpha_2-\beta_2-\beta_4-\beta_6}$

To determine the odds ratio of self to class for school program, you compute (for School 1)

$$\frac{e^{\alpha_1+\beta_1+\beta_5}}{e^{\alpha_1+\beta_1-\beta_5}} = e^{2\beta_5}$$

Thus, the odds were $e^{(2)(0.374)} = 2.11$ times higher of choosing the self-learning style over the class learning style if students attended the regular school program versus the afterschool program. Note that you obtain the same result if you do the comparison for School 2 or School 3. If you work through the exercise for the odds ratio of team to class, you find that the odds were $e^{(2)(0.371)} = 2.10$ times higher of choosing the team learning style as the class learning style if students attended the regular school program versus the afterschool program.

Comparing the odds ratio for School 1 to School 2 produces a slightly more complicated form for the odds ratio. You form the ratio of the odds for School 1, regular school program to School 2, regular school program (afterschool program would also work) to obtain (for self/class logit)

$$\frac{e^{\alpha_1+\beta_1+\beta_5}}{e^{\alpha_1+\beta_3+\beta_5}} = e^{\beta_1-\beta_3}$$

Thus, the subjects from School 1 were $e^{-0.799-0.284} = 0.33$ times as likely to choose the self-learning style over the class learning style as those students from School 2.

9.3.3   Activity Limitation Data Example

The data in Table 9.10 are derived from an analysis of activity limitation data that were obtained in the 1989 National Health Interview Study (Lafata, Koch, and Weissert 1994). Researchers were interested in generating estimates of activity limitation for the civilian population of the United States for small areas such as individual states. The original data included children, adults, and the elderly; a possible consequence of providing such estimates is better health care resources to assist such persons. The data in Table 9.10 are those corresponding to the older children's age groups.

Table 9.10. Activity Limitation Study

Age	Sex	Race	Poverty	Activity Limitation		
				Major	Other	None
10–14	males	whites	low	5.361	1.329	102.228
10–14	males	whites	medium	20.565	13.952	336.160
10–14	males	whites	high	21.299	5.884	284.931
10–14	males	other	low	53.314	16.402	827.900
10–14	males	other	medium	102.076	36.551	1518.796
10–14	males	other	high	52.338	21.105	666.909
10–14	females	whites	low	1.172	1.199	87.292
10–14	females	whites	medium	11.169	2.945	304.234
10–14	females	whites	high	15.286	3.665	302.511
10–14	females	other	low	21.882	16.979	846.270
10–14	females	other	medium	52.354	33.106	1452.895
10–14	females	other	high	28.203	11.455	687.109
15–19	males	whites	low	0.915	1.711	91.071
15–19	males	whites	medium	12.591	8.026	326.930
15–19	males	whites	high	21.059	6.993	313.633
15–19	males	other	low	36.384	27.558	888.833
15–19	males	other	medium	85.974	42.755	1509.87
15–19	males	other	high	40.112	23.493	725.004
15–19	females	whites	low	5.876	2.550	115.968
15–19	females	whites	medium	8.772	6.922	344.076
15–19	females	whites	high	17.385	2.354	286.68
15–19	females	other	low	42.741	31.025	817.478
15–19	females	other	medium	72.688	35.979	1499.816
15–19	females	other	high	26.296	29.321	716.860

The counts are weighted because the data were collected as part of a complex survey design and various sampling-related adjustments were performed. Normally, you would use methods that account for the sample design in your analysis of such data, but for illustrative purposes, stratified simple random sampling is assumed here.

Since there is some kind of order to the levels of the response variable (major limitation is more than other limitation is more than no limitation), it may seem that activity limitation could be treated as an ordinal response. However, the necessary proportional odds assumption is not met by these data, so ordered regression is not a potential analysis strategy. Thus, analyzing the generalized logits is the strategy of choice.

The logits of interest are

$$\text{logit}_{hijk1} = \log\left[\frac{\pi_{hijk1}}{\pi_{hijk3}}\right], \quad \text{logit}_{hijk2} = \log\left[\frac{\pi_{hijk2}}{\pi_{hijk3}}\right]$$

where π_{hijk1} is the $\Pr\{\text{major limitation}\}$, π_{hijk2} is the $\Pr\{\text{other limitation}\}$, and π_{hijk3} is the $\Pr\{\text{no limitation}\}$ for the $hijk$th group ($h = 1$ for ages 10–14, $h = 2$ for ages 15–19; $i = 1$ for males and $i = 2$ for females; $j = 1$ for whites and $j = 2$ for other; $k = 1$ for low poverty, $k = 2$ for medium poverty, and $k = 3$ for high poverty). The logits are comparing major activity limitation to no activity limitation and other activity limitation to no activity limitation.

The first model fit is the model with all pairwise interactions. The following SAS statements input the data.

```
data survey;
   input age sex race poverty function $ count @@;
   datalines;
1 0 0 0 major 5.361     1 0 0 0 other    1.329   1 0 0 0 not 102.228
1 0 0 1 major 20.565    1 0 0 1 other   13.952   1 0 0 1 not  336.160
1 0 0 2 major 21.299    1 0 0 2 other    5.884   1 0 0 2 not  284.931
1 0 1 0 major 53.314    1 0 1 0 other   16.402   1 0 1 0 not 827.900
1 0 1 1 major 102.076   1 0 1 1 other   36.551   1 0 1 1 not 1518.796
1 0 1 2 major 52.338    1 0 1 2 other   21.105   1 0 1 2 not 666.909
1 1 0 0 major 1.172     1 1 0 0 other    1.199   1 1 0 0 not  87.292
1 1 0 1 major 11.169    1 1 0 1 other    2.945   1 1 0 1 not 304.234
1 1 0 2 major 15.286    1 1 0 2 other    3.665   1 1 0 2 not 302.511
1 1 1 0 major 21.882    1 1 1 0 other   16.979   1 1 1 0 not 846.270
1 1 1 1 major 52.354    1 1 1 1 other   33.106   1 1 1 1 not 1452.895
1 1 1 2 major 28.203    1 1 1 2 other   11.455   1 1 1 2 not 687.109
2 0 0 0 major .915      2 0 0 0 other    1.711   2 0 0 0 not  91.071
2 0 0 1 major 12.591    2 0 0 1 other    8.026   2 0 0 1 not 326.930
2 0 0 2 major 21.059    2 0 0 2 other    6.993   2 0 0 2 not 313.633
2 0 1 0 major 36.384    2 0 1 0 other   27.558   2 0 1 0 not 888.833
2 0 1 1 major 85.974    2 0 1 1 other   42.755   2 0 1 1 not 1509.87
2 0 1 2 major 40.112    2 0 1 2 other   23.493   2 0 1 2 not 725.004
2 1 0 0 major 5.876     2 1 0 0 other    2.550   2 1 0 0 not 115.968
2 1 0 1 major  8.772    2 1 0 1 other    6.922   2 1 0 1 not 344.076
2 1 0 2 major 17.385    2 1 0 2 other    2.354   2 1 0 2 not 286.68
2 1 1 0 major 42.741    2 1 1 0 other   31.025   2 1 1 0 not 817.478
2 1 1 1 major 72.688    2 1 1 1 other   35.979   2 1 1 1 not 1499.816
2 1 1 2 major 26.296    2 1 1 2 other   29.321   2 1 1 2 not 716.860
;
run;
```

The following PROC CATMOD invocation requests the model with all pairwise interactions. Note the use of the @2 notation to request all pairwise interactions. The fact that the counts are not integers is not a problem for the CATMOD procedure; it accepts all counts that are non-negative. It does not truncate the values.

```
proc catmod order=data;
   direct poverty;
   weight count;
   model function=age|sex|race|poverty@2;
run;
```

Output 9.22 contains the population profiles for these data. With two levels each for AGE, SEX, and RACE, and three categories for POVERTY, 24 groups are formed. The counts have been preserved in their noninteger form.

Output 9.22   Population Profiles

```
                         Population Profiles

                                               Sample
            Sample   age   sex   race   poverty   Size
            -----------------------------------------------
               1      1     0     0       0      108.918
               2      1     0     0       1      370.677
               3      1     0     0       2      312.114
               4      1     0     1       0      897.616
               5      1     0     1       1     1657.423
               6      1     0     1       2      740.352
               7      1     1     0       0       89.663
               8      1     1     0       1      318.348
               9      1     1     0       2      321.462
              10      1     1     1       0      885.131
              11      1     1     1       1     1538.355
              12      1     1     1       2      726.767
              13      2     0     0       0       93.697
              14      2     0     0       1      347.547
              15      2     0     0       2      341.685
              16      2     0     1       0      952.775
              17      2     0     1       1     1638.599
              18      2     0     1       2      788.609
              19      2     1     0       0      124.394
              20      2     1     0       1      359.77
              21      2     1     0       2      306.419
              22      2     1     1       0      891.244
              23      2     1     1       1     1608.483
              24      2     1     1       2      772.477
```

Output 9.23 contains the response profiles. Since the ORDER=DATA option is specified, the values of FUNCTION are ordered internally as major, other, and none. This means that the desired generalized logit functions are formed, since PROC CATMOD forms the logit for the first level compared to the third level and the second level compared to the third level.

Output 9.23   Response Profiles

```
                    Response Profiles

                 Response      function
                 --------------------
                    1          major
                    2          other
                    3          not
```

The ANOVA table and likelihood ratio goodness-of-fit statistic are displayed in Output 9.24. $Q_L = 25.66$ with 26 df and $p = 0.4818$, which is supportive of model fit. Note the degrees of freedom associated with each of the effects in the table. Since AGE, SEX, and RACE would be associated with 1 df in the dichotomous outcome case, they are associated with 2 df since two logits are being modeled. Similarly, POVERTY is associated with 2 df, since it is entered on the DIRECT statement and its values are used directly in a column of the model matrix. If you have interactions, you determine their df by multiplying the df of each constituent effect that would be associated with a dichotomous outcome, then multiplying that by the number of functions being modeled.

A look at the *p*-values associated with these effects shows that the model could be simplified by eliminating terms.

Output 9.24　ANOVA Table

```
              Maximum Likelihood Analysis of Variance

       Source              DF    Chi-Square    Pr > ChiSq
       ----------------------------------------------------
       Intercept            2      1518.29        <.0001
       age                  2         1.21        0.5467
       sex                  2         2.89        0.2357
       age*sex              2        14.72        0.0006
       race                 2         4.53        0.1037
       age*race             2         1.27        0.5296
       sex*race             2         4.00        0.1355
       poverty              2         8.95        0.0114
       poverty*age          2         0.72        0.6979
       poverty*sex          2         1.53        0.4650
       poverty*race         2         5.52        0.0632

       Likelihood Ratio    26        25.66        0.4818
```

A reduced model is then fit, with only the AGE*SEX, SEX*RACE, and RACE*POVERTY interactions being retained.

```
proc catmod order=data;
   direct poverty;
   weight count;
   model function=age sex race poverty
      age*sex sex*race race*poverty;
run;
```

The ANOVA table pertaining to this model is displayed in Output 9.25.

Output 9.25　ANOVA Table for Reduced Model

```
              Maximum Likelihood Analysis of Variance

       Source              DF    Chi-Square    Pr > ChiSq
       ----------------------------------------------------
       Intercept            2      1527.92        <.0001
       age                  2         6.61        0.0367
       sex                  2        18.47        <.0001
       race                 2         4.62        0.0991
       poverty              2         9.64        0.0081
       age*sex              2        15.10        0.0005
       sex*race             2         4.44        0.1088
       poverty*race         2         5.28        0.0713

       Likelihood Ratio    32        29.67        0.5848
```

Since the model has been reduced by 6 parameters, the df for the likelihood ratio test is increased by 6 df. However, since $Q_L = 29.67$ with 32 df, the reduced model has an adequate fit. The only further model reduction that seems warranted at this point is the removal of the SEX*RACE term. It would appear that RACE*POVERTY is marginally influential, and that may change once the SEX*RACE effect is removed.

```
proc catmod order=data;
   direct poverty;
   weight count;
   model function=age sex race poverty
      age*sex race*poverty /pred=freq;
run;
```

Output 9.26 displays the results when the SEX*RACE term is removed.

Output 9.26   ANOVA Table for Final Reduced Model

```
              Maximum Likelihood Analysis of Variance

      Source            DF    Chi-Square    Pr > ChiSq
      -------------------------------------------------
      Intercept          2      1536.94       <.0001
      age                2         6.51       0.0385
      sex                2        26.44       <.0001
      race               2         4.71       0.0950
      poverty            2         9.59       0.0083
      age*sex            2        15.09       0.0005
      poverty*race       2         5.22       0.0736

      Likelihood Ratio  34        34.27       0.4548
```

The goodness of fit is adequate for this model ($Q_L = 34.27$ with 34 df and $p = 0.4548$). Since RACE*POVERTY remains marginally influential, it is kept in the model. Thus, this model includes main effects for age, sex, race, and poverty, along with interactions for age and sex and poverty and race. You could compute odds ratios to aid in model interpretation.

The PRED=FREQ option in PROC CATMOD requests the computation of model-predicted frequencies for this model (you can also generate model-predicted probabilities with the PRED=PROB option). Output 9.27 contains a partial listing of these predicted frequencies from the PROC CATMOD output. Table 9.11 contains the complete set of predicted values. You can compare these to the original frequencies to see how well the model works.

Output 9.27   Partial Listing of Predicted Frequencies

```
                Maximum Likelihood Predicted Values for Frequencies

                               -----Observed-----  -----Predicted----
                                          Standard            Standard
        age sex race poverty function Frequency    Error Frequency    Error Residual
        ------------------------------------------------------------------------------
        1   0   0    0       major       5.361 2.257682   4.07892 0.849289  1.28208
                             other       1.329 1.145768   2.376429 0.652845 -1.04743
                             not       102.228 2.50581  102.4627 1.055218 -0.23465

        1   0   0    1       major      20.565 4.407274  19.95014 2.191072  0.614855
                             other      13.952 3.664268   7.036412 1.140648  6.915588
                             not       336.16  5.594892 343.6904 2.433505 -7.53044

        1   0   0    2       major      21.299 4.454833  23.95679 2.814102 -2.65779
                             other       5.884 2.402722   5.11516 1.108229  0.76884
                             not       284.931 4.98152  283.0421 2.980315  1.888946

        1   0   1    0       major      53.314 7.081484  52.88666 4.391243  0.427342
                             other      16.402 4.012766  20.97862 2.684635 -4.57662
                             not       827.9   8.018809 823.7507 5.054315  4.149279

        1   0   1    1       major     102.076 9.787207 104.3194 6.566059 -2.24344
                             other      36.551 5.978708  40.65949 4.206884 -4.10849
                             not      1518.796 11.27086 1512.444 7.647375  6.351926

        1   0   1    2       major      52.338 6.974098  49.76105 4.154098  2.576947
                             other      21.105 4.528064  19.05689 2.474509  2.048109
                             not       666.909 8.133723 671.5341 4.731773 -4.62506
```

Table 9.11.   Activity Limitation Study

| Age | Sex | Race | Poverty | Predicted % Limitation | | |
				Major	Other	None
10-14	males	whites	low	4.0789	2.376	102.463
10-14	males	whites	medium	19.950	7.0364	343.690
10-14	males	whites	high	23.957	5.115	283.042
10-14	males	other	low	52.887	20.979	823.751
10-14	males	other	medium	104.319	40.659	1512.444
10-14	males	other	high	49.761	19.057	671.534
10-14	females	whites	low	1.774	1.481	86.408
10-14	females	whites	medium	9.117	4.610	304.621
10-14	females	whites	high	13.272	4.063	304.127
10-14	females	other	low	27.854	15.839	841.438
10-14	females	other	medium	51.835	28.963	1457.557
10-14	females	other	high	26.215	14.393	686.159
15-19	males	whites	low	2.635	2.315	88.747
15-19	males	whites	medium	14.109	7.505	325.933
15-19	males	whites	high	19.907	6.410	315.368
15-19	males	other	low	42.376	25.351	885.048
15-19	males	other	medium	77.923	45.803	1514.873
15-19	males	other	high	40.085	23.152	725.372
15-19	females	whites	low	3.170	3.0737	118.15
15-19	females	whites	medium	13.254	7.779	338.738
15-19	females	whites	high	16.227	5.766	284.426
15-19	females	other	low	35.984	23.753	831.508
15-19	females	other	medium	69.457	45.050	1493.976
15-19	females	other	high	35.666	22.730	714.081

Chapter 10
Conditional Logistic Regression

Chapter Table of Contents

Chapter 10
Conditional Logistic Regression

10.1 Introduction

Sometimes, the usual maximum likelihood approach to estimation in logistic regression is not appropriate. As discussed in Chapter 8, "Logistic Regression I: Dichotomous Response," there may be insufficient sample size for logistic regression, particularly if the data are highly stratified and there are a small number of subjects in each stratum. In these situations, you have a small sample size relative to the number of parameters being estimated since you will be estimating parameters for the stratification effects. For the maximum likelihood estimates to be valid, you need a large sample size relative to the number of parameters.

Often, highly stratified data come from a design with cluster sampling, that is, designs with two or more observations for each primary sampling unit or cluster. Common examples of such data are paired observations, such as fraternal twins (or litter mates), right and left sides of the body in a dermatology study, or two occasions for an expression of an opinion. Ordinary logistic regression may be inappropriate for such data, since you have insufficient sample size to estimate the pair effect (family, litter, patient, respondent) without bias. However, by using conditioning arguments, you can eliminate the pair effect and estimate the other effects in which you are interested.

The appropriate form of logistic regression for these types of data is called *conditional logistic regression*. It takes the stratification into account by basing the maximum likelihood estimation of the model parameters on a conditional likelihood. You can fit these models in the SAS System with the PHREG procedure. In the following sections, the conditional likelihood for paired observations from small clusters is derived, and the methodology is illustrated with data from a randomized clinical trial and a two-period crossover design study. Then, the more general stratified situation is discussed and illustrated with data from a three-period crossover study and a repeated measurements study.

Matched case control studies in epidemiology also produce highly stratified data. In these studies, you match cases (those persons with a disease or condition) to controls (those persons without the disease or condition) on the basis of variables thought to be potential confounders such as age, race, and sex. The use of conditional logistic regression for matched studies in epidemiological work is discussed and illustrated with two examples.

Finally, exact logistic regression is discussed for the stratified setting.

10.2   Paired Observations from a Highly Stratified Cohort Study

Consider a randomized clinical trial where $i = 1, 2, \ldots, q$ centers are randomly selected, and, at each center, one randomly selected patient is placed on treatment, and another randomly selected patient is placed on placebo. Interest lies in whether the patients improve; thus, improvement is the event of interest. Since there are only two observations per center, it is not possible to estimate a center effect (pair effect) without bias. As a general rule, you need each possible outcome to have five observations per explanatory variable in the model for valid estimation to proceed.

Suppose $y_{ij} = 1$ if improvement occurs and $y_{ij} = 0$ if it does not ($j = 1$ for treatment and $j = 2$ for the placebo; $i = 1, 2, \ldots, q$). Suppose $x_{ij} = 1$ for treatment and $x_{ij} = 0$ for placebo, and $\mathbf{z}_{ij} = (z_{ij1}, z_{ij2}, \ldots, z_{ijt})'$ represents the t explanatory variables.

The usual logistic likelihood for $\{y_{ij}\}$ is written

$$\Pr\{y_{ij}\} = \pi_{ij} = \frac{\exp\{\alpha_i + \beta x_{ij} + \boldsymbol{\gamma}' \mathbf{z}_{ij}\}}{1 + \exp\{\alpha_i + \beta x_{ij} + \boldsymbol{\gamma}' \mathbf{z}_{ij}\}}$$

where α_i is the effect of the ith center, β is the treatment parameter, and $\boldsymbol{\gamma}' = (\gamma_1, \gamma_2, \ldots, \gamma_t)$ is the parameter vector for the covariates $\mathbf{z}$. Since there are only two observations per center, you can't estimate these parameters without bias. However, you can fit a model based on conditional probabilities that condition away the center effects, which results in a model that contains substantially fewer parameters. In this context, the α_i are known as *nuisance parameters*. It is useful to describe these data with a model that considers the probability of a pair's treatment patient improving and the pair's placebo patient not improving, compared to the probability that one of them improved.

You can write a conditional probability for $\{y_{ij}\}$ as the ratio of the joint probability of a pair's treatment patient improving and the pair's placebo patient not improving to the joint probability that either the treatment patient or the placebo patient improved.

$$\Pr\left\{y_{i1}{=}1, y_{i2}{=}0 \middle| y_{i1}{=}1, y_{i2}{=}0 \text{ or } y_{i1}{=}0, y_{i2}{=}1\right\} =$$

$$\frac{\Pr\{y_{i1}{=}1\} \Pr\{y_{i2}{=}0\}}{\Pr\{y_{i1}{=}1\} \Pr\{y_{i2}{=}0\} + \Pr\{y_{i1}{=}0\} \Pr\{y_{i2}{=}1\}}$$

If you write the probabilities in terms of the logistic model,

$$\Pr\{y_{i1}{=}1\} \Pr\{y_{i2}{=}0\} = \frac{\exp\{\alpha_i + \beta + \boldsymbol{\gamma}' \mathbf{z}_{i1}\}}{1 + \exp\{\alpha_i + \beta + \boldsymbol{\gamma}' \mathbf{z}_{i1}\}} \cdot \frac{1}{1 + \exp\{\alpha_i + \boldsymbol{\gamma}' \mathbf{z}_{i2}\}}$$

and

$$\Pr\{y_{i1}{=}1\} \Pr\{y_{i2}{=}0\} + \Pr\{y_{i1}{=}0\} \Pr\{y_{i2}{=}1\} =$$

$$\frac{\exp\{\alpha_i + \beta + \boldsymbol{\gamma}'\mathbf{z}_{i1}\}}{1 + \exp\{\alpha_i + \beta + \boldsymbol{\gamma}'\mathbf{z}_{i1}\}} \cdot \frac{1}{1 + \exp\{\alpha_i + \boldsymbol{\gamma}'\mathbf{z}_{i2}\}}$$

$$+ \frac{1}{1 + \exp\{\alpha_i + \beta + \boldsymbol{\gamma}'\mathbf{z}_{i1}\}} \cdot \frac{\exp\{\alpha_i + \boldsymbol{\gamma}'\mathbf{z}_{i2}\}}{1 + \exp\{\alpha_i + \boldsymbol{\gamma}'\mathbf{z}_{i2}\}}$$

If you form their ratio, you obtain

$$\frac{\exp\{\alpha_i + \beta + \boldsymbol{\gamma}'\mathbf{z}_{i1}\}}{\exp\{\alpha_i + \beta + \boldsymbol{\gamma}'\mathbf{z}_{i1}\} + \exp\{\alpha_i + \boldsymbol{\gamma}'\mathbf{z}_{i2}\}}$$

since the denominators cancel out.

This expression reduces to

$$\frac{\exp\{\beta + \boldsymbol{\gamma}'(\mathbf{z}_{i1} - \mathbf{z}_{i2})\}}{1 + \exp\{\beta + \boldsymbol{\gamma}'(\mathbf{z}_{i1} - \mathbf{z}_{i2})\}}$$

which no longer contains the $\{\alpha_i\}$. Thus, by focusing on modeling a meaningful conditional probability, you develop a model with a reduced number of parameters that can be estimated without bias.

The conditional likelihood for the entire data is written

$$\prod_{i=1}^{q} \left\{ \frac{\exp\{\beta + \boldsymbol{\gamma}'(\mathbf{z}_{i1} - \mathbf{z}_{i2})\}}{1 + \exp\{\beta + \boldsymbol{\gamma}'(\mathbf{z}_{i1} - \mathbf{z}_{i2})\}} \right\}^{y_{i1}(1-y_{i2})} \left\{ \frac{1}{1 + \exp\{\beta + \boldsymbol{\gamma}'(\mathbf{z}_{i1} - \mathbf{z}_{i2})\}} \right\}^{(1-y_{i1})y_{i2}}$$

This is the unconditional likelihood for the usual logistic model, except that the intercept is now β, the effect for treatment, and each observation represents a pair of observations from a center where the response is 1 if the pair represents the combination $\{y_{i1}=1 \text{ and } y_{i2}=0\}$ and 0 if the pair has the combination $\{y_{i1}=0 \text{ and } y_{i2}=1\}$. The explanatory variables are the differences in values of the explanatory variables for the treatment patient and the placebo patient. Since the likelihood conditioned on the discordant pairs, the concordant pairs (the observations where $\{y_{i1}=1 \text{ and } y_{i2}=1\}$ and $\{y_{i1}=0 \text{ and } y_{i2}=0\}$) are noninformative and thus can be ignored.

Note that the ratio above can also be rewritten as

$$\frac{\exp\{\beta + \boldsymbol{\gamma}'\mathbf{z}_{i1}\}}{\exp\{\beta + \boldsymbol{\gamma}'\mathbf{z}_{i1}\} + \exp\{\boldsymbol{\gamma}'\mathbf{z}_{i2}\}}$$

and the corresponding likelihood for the entire data as

$$\prod_{i=1}^{q} \left\{ \frac{\exp\{\beta + \boldsymbol{\gamma}'\mathbf{z}_{i1}\}}{\exp\{\beta + \boldsymbol{\gamma}'\mathbf{z}_{i1}\} + \exp\{\boldsymbol{\gamma}'\mathbf{z}_{i2}\}} \right\}^{y_{i1}(1-y_{i2})} \left\{ \frac{\exp\{\beta + \boldsymbol{\gamma}'\mathbf{z}_{i2}\}}{\exp\{\beta + \boldsymbol{\gamma}'\mathbf{z}_{i1}\} + \exp\{\boldsymbol{\gamma}'\mathbf{z}_{i2}\}} \right\}^{(1-y_{i1})y_{i2}}$$

This is the same likelihood that applies to paired data in a simple case of the Cox regression model, or proportional hazards model, which is used in the analysis of survival times. This means that the computing machinery that fits the proportional hazards model can also be used for conditional logistic regression.

Note that if there are no covariates for such a study, so that the data represent a 2×2 table where the responses for treatment are cross-classified with the responses for placebo, then testing $\beta = 0$ is equivalent to McNemar's test. Also, it can be shown that e^β is estimated by n_{12}/n_{21}, where n_{12} and n_{21} are the off-diagonal counts from this table.

10.3 Clinical Trials Study Analysis

Researchers studying the effect of a new treatment on a skin condition collected information from 79 clinics. In each clinic, one patient received the treatment, and another patient received a placebo. Variables collected included age, sex, and an initial grade for the skin condition, which ranged from 1 to 4 for mild to severe. The response variable was whether the skin condition improved. Using conditional logistic regression is suitable for the analysis of such data, and, in this example, both the LOGISTIC and PHREG procedures are implemented. The PROC LOGISTIC analysis demonstrates that the case of paired observations can be handled with the computations of the unconditional logistic regression on the differences of the explanatory variables; this is useful to understand. The PHREG procedure is used for subsequent analyses because it is more straightforward to apply.

10.3.1 Analysis Using the LOGISTIC Procedure

Each data line in the following input data includes two observations from each clinic: one for the patient receiving the treatment and one for the patient receiving the placebo. The DATA step creates indicator variables for sex and treatment, which is required in order to create their differences. Then, it creates various interaction terms. These must be created before the observations are combined, so you can't use the interaction term construction in the MODEL statement. Then, the DATA step combines the paired observations and creates variables whose values are the differences in their respective values. Since observations where the response variable IMPROVE had the same value for both treatment and placebo do not affect the analysis, only those observations with discordant responses are output to the data set TRIAL.

```
data trial;
   drop center1 i_sex1 age1 initial1 improve1 trtsex1 trtinit1
      trtage1 isexage1 isexint1 iageint1;
   retain center1 i_sex1 age1 initial1 improve1 trtsex1 trtinit1
      trtage1 isexage1 isexint1 iageint1 0;
   input center treat $ sex $ age improve initial @@;
   /* compute model terms for each observation */
   i_sex=(sex='m');        i_trt=(treat='t');
   trtsex=i_sex*i_trt;     trtinit=i_trt*initial;
   trtage=i_trt*age;       isexage=i_sex*age;
   isexinit=i_sex*initial;iageinit=age*initial;
   /* compute differences for paired observation */
   if (center=center1) then do;
```

```
        pair=10*improve + improve1;
        i_sex=i_sex1-i_sex;
        age=age1-age;
        initial=initial1-initial;
        trtsex=trtsex1-trtsex;
        trtinit=trtinit1-trtinit;
        trtage=trtage1-trtage;
        isexage=isexage1-isexage;
        isexinit=isexint1-isexinit;
        iageinit=iageint1-iageinit;
        if (pair=10 or pair=1) then do;
            /* output discordant pair observations */
            improve=(pair=1); output trial; end;
        end;
    else do;
        center1=center; age1=age;
        initial1=initial; i_sex1=i_sex; improve1=improve;
        trtsex1=trtsex; trtinit1=trtinit; trtage1=trtage;
        isexage1=isexage; isexint1=isexinit; iageint1=iageinit;
        end;
    datalines;
1   t f 27 0 1    1 p f 32 0 2 41 t f 13 1 2 41 p m 22 0 3
2   t f 41 1 3    2 p f 47 0 1 42 t m 31 1 1 42 p f 21 1 3
3   t m 19 1 4    3 p m 31 0 4 43 t f 19 1 3 43 p m 35 1 3
4   t m 55 1 1    4 p m 24 1 3 44 t m 31 1 3 44 p f 37 0 2
5   t f 51 1 4    5 p f 44 0 2 45 t f 44 0 1 45 p f 41 1 1
6   t m 23 0 1    6 p f 44 1 3 46 t m 41 1 2 46 p m 41 0 1
7   t m 31 1 2    7 p f 39 0 2 47 t m 41 1 2 47 p f 21 0 4
8   t m 22 0 1    8 p m 54 1 4 48 t f 51 1 2 48 p m 22 1 1
9   t m 37 1 3    9 p m 63 0 2 49 t f 62 1 3 49 p f 32 0 3
10  t m 33 0 3   10 p f 43 0 3 50 t m 21 0 1 50 p m 34 0 1
11  t f 32 1 1   11 p m 33 0 3 51 t m 55 1 3 51 p f 35 1 2
12  t m 47 1 4   12 p m 24 0 4 52 t f 61 0 1 52 p m 19 0 1
13  t m 55 1 3   13 p f 38 1 1 53 t m 43 1 2 53 p m 31 0 2
14  t f 33 0 1   14 p f 28 1 2 54 t f 44 1 1 54 p f 41 1 1
15  t f 48 1 1   15 p f 42 0 1 55 t m 67 1 2 55 p m 41 0 1
16  t m 55 1 3   16 p m 52 0 1 56 t m 41 0 2 56 p m 21 1 4
17  t m 30 0 4   17 p m 48 1 4 57 t f 51 1 3 57 p m 51 0 2
18  t f 31 1 2   18 p m 27 1 3 58 t m 62 1 3 58 p m 54 1 3
19  t m 66 1 3   19 p f 54 0 1 59 t m 22 0 1 59 p f 22 0 1
20  t f 45 0 2   20 p f 66 1 2 60 t m 42 1 2 60 p f 29 1 2
21  t m 19 1 4   21 p f 20 1 4 61 t f 51 1 1 61 p f 31 0 1
22  t m 34 1 4   22 p f 31 0 1 62 t m 27 0 2 62 p m 32 1 2
23  t f 46 0 1   23 p m 30 1 2 63 t m 31 1 1 63 p f 21 0 1
24  t m 48 1 3   24 p f 62 0 4 64 t m 35 0 3 64 p m 33 1 3
25  t m 50 1 4   25 p m 45 1 4 65 t m 67 1 2 65 p m 19 0 1
26  t m 57 1 3   26 p f 43 0 3 66 t m 41 0 2 66 p m 62 1 4
27  t f 13 0 2   27 p m 22 1 3 67 t f 31 1 2 67 p m 45 1 3
28  t m 31 1 1   28 p f 21 0 1 68 t m 34 1 1 68 p f 54 0 1
29  t m 35 1 3   29 p m 35 1 3 69 t f 21 0 1 69 p m 34 1 4
30  t f 36 1 3   30 p f 37 0 3 70 t m 64 1 3 70 p m 51 0 1
31  t f 45 0 1   31 p f 41 1 1 71 t f 61 1 3 71 p m 34 1 3
32  t m 13 1 2   32 p m 42 0 1 72 t m 33 0 1 72 p f 43 0 1
33  t m 14 0 4   33 p f 22 1 2 73 t f 36 0 2 73 p m 37 0 3
```

```
34 t f 15 1 2 34 p m 24 0 1 74 t m 21 1 1 74 p m 55 0 1
35 t f 19 1 3 35 p f 31 0 1 75 t f 47 0 2 75 p f 42 1 3
36 t m 20 0 2 36 p m 32 1 3 76 t f 51 1 4 76 p m 44 0 2
37 t m 23 1 3 37 p f 35 0 1 77 t f 23 1 1 77 p m 41 1 3
38 t f 23 0 1 38 p m 21 1 1 78 t m 31 0 2 78 p f 23 1 4
39 t m 24 1 4 39 p m 30 1 3 79 t m 22 0 1 79 p m 19 1 4
40 t m 57 1 3 40 p f 43 1 3
;
```

The following PROC LOGISTIC invocation requests forward selection, forcing all the main effects in the model with the INCLUDE=3 option and making all interaction terms available for consideration.

```
proc logistic data=trial descending;
   model improve = initial age i_sex
      isexage isexinit iageinit
      trtsex trtinit trtage /
      selection=forward include=3 details;
run;
```

The response profiles shown in Output 10.1 indicate that 34 pairs of observations have the (1,0) profile (treatment improved, placebo did not) and 20 pairs have the (0,1) profile (treatment didn't improve, placebo did).

Output 10.1   Response Profiles

	Response Profile	
Ordered Value	improve	Total Frequency
1	1	34
2	0	20

Output 10.2 shows the residual score statistic ($Q_{RS} = 4.7214$ with 6 df and $p = 0.5800$) as well as the score statistics for the addition of the individual terms into the model. Since there are 20 observations with the less prevalent response, this model can support about $20/5 = 4$ terms. Thus, there are possibly too many terms to rely entirely on the residual score statistic to assess goodness of fit. However, considering the residual test as well as the individual tests provides reasonable confidence that the model fits adequately. All of the individual tests have p-values greater than 0.08, and most of them have p-values greater than 0.5. This model doesn't require the addition of any interaction terms. (Note that you could have assessed goodness of fit by taking the difference of $-2 \text{ LOG } L$ for this model and for one that included all the interaction terms.)

Output 10.2 Score Statistics

```
                    Residual Chi-Square Test

          Chi-Square        DF      Pr > ChiSq

            4.7213           6         0.5800


           Analysis of Effects Not in the Model

                                 Score
         Effect      DF      Chi-Square    Pr > ChiSq

         isexage      1         0.6593       0.4168
         isexinit     1         0.1775       0.6736
         iageinit     1         2.9194       0.0875
         trtsex       1         0.2681       0.6046
         trtinit      1         0.0121       0.9125
         trtage       1         0.4336       0.5102
```

Output 10.3 contains the maximum likelihood estimates of the parameters. Recall that, in this model, the treatment effect is represented by the intercept. It takes the value 0.7024, which is nearly significant with $p = 0.0511$. Neither age nor sex appear to be very influential but are left in the model as covariates. The effect for initial score is highly significant ($p = 0.0011$).

Output 10.3 Maximum Likelihood Estimates

```
          Analysis of Maximum Likelihood Estimates

                              Standard
     Parameter   DF   Estimate    Error   Chi-Square   Pr > ChiSq

     Intercept   1    0.7024     0.3601     3.8053       0.0511
     initial     1    1.0915     0.3351    10.6105       0.0011
     age         1    0.0248     0.0224     1.2252       0.2683
     i_sex       1    0.5312     0.5545     0.9176       0.3381
```

Output 10.4 contains the odds ratios and their 95% confidence limits.

Output 10.4 Odds Ratios

```
                  Odds Ratio Estimates

                    Point          95% Wald
       Effect     Estimate     Confidence Limits

       initial      2.979      1.545      5.745
       age          1.025      0.981      1.071
       i_sex        1.701      0.574      5.043
```

Note that the LOGISTIC procedure does not print the odds ratio for the INTERCEPT, as usually it is of no interest. However, since the intercept represents the treatment effect in this model, it is appropriate to determine the odds ratio. The odds of improving for those patients receiving the treatment is $e^{0.7024} = 2.019$ times higher than for those patients receiving the placebo. The odds of improvement also increase by a factor of 2.979 for each unit increase in the initial grade. Note that the confidence limits for this odds ratio are (1.545, 5.745). Thus, even adjusting for the effect of initial grade, treatment has a nearly significant effect. And, performing this stratified analysis has taken into account the effect of center.

Consider the model where the intercept is the only term.

```
proc logistic data=trial descending;
   model improve = ;
run;
```

Output 10.5 contains the parameter estimates.

Output 10.5 Treatment Effect Only Model

Analysis of Maximum Likelihood Estimates					
Parameter	DF	Estimate	Standard Error	Chi-Square	Pr > ChiSq
Intercept	1	0.5306	0.2818	3.5457	0.0597

Note that $e^{\beta} = e^{0.5306} = 1.70$. In addition, the Wald test for the intercept takes the value 3.5457 with $p = 0.0597$, which is nearly significant.

Table 10.1 displays the crosstabulation of pairs by treatment and response.

Table 10.1. Pairs Breakdown

Placebo Response	Treatment Response	
	No	Yes
No	7	34
Yes	20	18

Thus, McNemar's test statistic is computed as

$$\frac{(34 - 20)^2}{(34 + 20)} = 3.63$$

which is also nearly significant. As the sample size grows, the Wald statistic for the intercept and McNemar's test statistic become asymptotically equivalent. In addition, note that $n_{12}/n_{21} = 1.7$, which is the same as $e^{\hat{\beta}} = e^{0.5306}$.

10.3.2 Analysis Using the PHREG Procedure

The advantage of using the PHREG procedure for conditional logistic regression is that you can operate directly on the actual observations; you don't have to create difference observations. The following DATA step creates SAS data set TRIAL2. In order to fit into the proportional hazards computing framework of PROC PHREG, the outcome of interest must be ordered first. Thus, the value of response variable IMPROVE is subtracted from 2. Indicator variables and interactions are created the same as in the PROC LOGISTIC analysis.

```
data trial2;
    input center treat $ sex $ age improve initial @@;
    /* create indicator variables for sex and interaction terms */
    improve=2-improve;
    isex=(sex='m');
    itreat=(treat='t');
    sex_age=isex*age;
    treat_age=itreat*age;
    sex_treat=isex*itreat;
    sex_initial=isex*initial;
    treat_initial=itreat*initial;
    age_initial=age*initial;
    datalines;
1   t f 27 0 1   1 p f 32 0 2  41 t f 13 1 2  41 p m 22 0 3
2   t f 41 1 3   2 p f 47 0 1  42 t m 31 1 1  42 p f 21 1 3
3   t m 19 1 4   3 p m 31 0 4  43 t f 19 1 3  43 p m 35 1 3
4   t m 55 1 1   4 p m 24 1 3  44 t m 31 1 3  44 p f 37 0 2
5   t f 51 1 4   5 p f 44 0 2  45 t f 44 0 1  45 p f 41 1 1
6   t m 23 0 1   6 p f 44 1 3  46 t m 41 1 2  46 p m 41 0 1
7   t m 31 1 2   7 p f 39 0 2  47 t m 41 1 2  47 p f 21 0 4
8   t m 22 0 1   8 p m 54 1 4  48 t f 51 1 2  48 p m 22 1 1
9   t m 37 1 3   9 p m 63 0 2  49 t f 62 1 3  49 p f 32 0 3
10  t m 33 0 3  10 p f 43 0 3  50 t m 21 0 1  50 p m 34 0 1
11  t f 32 1 1  11 p m 33 0 3  51 t m 55 1 3  51 p f 35 1 2
12  t m 47 1 4  12 p m 24 0 4  52 t f 61 0 1  52 p m 19 0 1
13  t m 55 1 3  13 p f 38 1 1  53 t m 43 1 2  53 p m 31 0 2
14  t f 33 0 1  14 p f 28 1 2  54 t f 44 1 1  54 p f 41 1 1
15  t f 48 1 1  15 p f 42 0 1  55 t m 67 1 2  55 p m 41 0 1
16  t m 55 1 3  16 p m 52 0 1  56 t m 41 0 2  56 p m 21 1 4
17  t m 30 0 4  17 p m 48 1 4  57 t f 51 1 3  57 p m 51 0 2
18  t f 31 1 2  18 p m 27 1 3  58 t m 62 1 3  58 p m 54 1 3
19  t m 66 1 3  19 p f 54 0 1  59 t m 22 0 1  59 p f 22 0 1
20  t f 45 0 2  20 p f 66 1 2  60 t m 42 1 2  60 p f 29 1 2
21  t m 19 1 4  21 p f 20 1 4  61 t f 51 1 1  61 p f 31 0 1
22  t m 34 1 4  22 p f 31 0 1  62 t m 27 0 2  62 p m 32 1 2
23  t f 46 0 1  23 p m 30 1 2  63 t m 31 1 1  63 p f 21 0 1
24  t m 48 1 3  24 p f 62 0 4  64 t m 35 0 3  64 p m 33 1 3
25  t m 50 1 4  25 p m 45 1 4  65 t m 67 1 2  65 p m 19 0 1
26  t m 57 1 3  26 p f 43 0 3  66 t m 41 0 2  66 p m 62 1 4
27  t f 13 0 2  27 p m 22 1 3  67 t f 31 1 2  67 p m 45 1 3
28  t m 31 1 1  28 p f 21 0 1  68 t m 34 1 1  68 p f 54 0 1
29  t m 35 1 3  29 p m 35 1 3  69 t f 21 0 1  69 p m 34 1 4
30  t f 36 1 3  30 p f 37 0 3  70 t m 64 1 3  70 p m 51 0 1
```

```
31 t f 45 0 1 31 p f 41 1 1 71 t f 61 1 3 71 p m 34 1 3
32 t m 13 1 2 32 p m 42 0 1 72 t m 33 0 1 72 p f 43 0 1
33 t m 14 0 4 33 p f 22 1 2 73 t f 36 0 2 73 p m 37 0 3
34 t f 15 1 2 34 p m 24 0 1 74 t m 21 1 1 74 p m 55 0 1
35 t f 19 1 3 35 p f 31 0 1 75 t f 47 0 2 75 p f 42 1 3
36 t m 20 0 2 36 p m 32 1 3 76 t f 51 1 4 76 p m 44 0 2
37 t m 23 1 3 37 p f 35 0 1 77 t f 23 1 1 77 p m 41 1 3
38 t f 23 0 1 38 p m 21 1 1 78 t m 31 0 2 78 p f 23 1 4
39 t m 24 1 4 39 p m 30 1 3 79 t m 22 0 1 79 p m 19 1 4
40 t m 57 1 3 40 p f 43 1 3
;
```

The same model is specified with the PHREG procedure. Since you are conditioning on center, the variable CENTER is listed in the STRATA statement. The conditional logistic model does not include an intercept, so the variable ITREAT is now included in the MODEL statement. The PHREG procedure also supports model building, so the SELECTION=FORWARD option is specified with the main effects model selected as the first model fit.

```
proc phreg data=trial2 nosummary;
   strata center;
   model improve = initial age isex itreat
   sex_age sex_initial age_initial
   sex_treat treat_initial treat_age / ties=discrete
   selection=forward include=4 details;
run;
```

The PHREG procedure first produces a table of model fit statistics, displayed in Output 10.6. It then produces a table of global fit statistics that test whether the parameters are jointly equal to zero.

Output 10.6   Model Fit Statistics

	Model Fit Statistics	
Criterion	Without Covariates	With Covariates
-2 LOG L	74.860	50.562
AIC	74.860	58.562
SBC	74.860	70.813

Output 10.7   Global Fit Statistics

Testing Global Null Hypothesis: BETA=0			
Test	Chi-Square	DF	Pr > ChiSq
Likelihood Ratio	24.2976	4	<.0001
Score	19.8658	4	0.0005
Wald	13.0099	4	0.0112

Output 10.8 contains the parameter estimates. Note that this analysis produces no intercept term.

Output 10.8 Parameter Estimates from PROC PHREG

```
                 Analysis of Maximum Likelihood Estimates

                     Parameter    Standard                          Hazard
    Variable    DF    Estimate       Error   Chi-Square  Pr > ChiSq   Ratio

    initial      1     1.09148     0.33508     10.6104      0.0011    2.979
    age          1     0.02483     0.02243      1.2252      0.2683    1.025
    isex         1     0.53115     0.55451      0.9175      0.3381    1.701
    itreat       1     0.70244     0.36009      3.8052      0.0511    2.019
```

The parameter estimates and standard errors are identical to those produced with the LOGISTIC procedure on data consisting of the explanatory variable differences for each case.

10.4 Crossover Design Studies

Conditional logistic regression is a useful technique in the analysis of the *crossover design study*, also called the *changeover study*. In these designs, often used in clinical trials, the study is divided into periods and patients receive a different treatment during each period. Thus, the patients act as their own controls. Interest lies in comparing the efficacy of the treatments, adjusting for period effects and carryover effects. The basic crossover design is a two-period design, but designs with three or more periods are also implemented. This section describes the use of conditional logistic regression for both two- and three- period designs.

10.4.1 Two-Period Crossover Design

A two-period crossover study can be considered another example of paired data. Table 10.2 contains data from a two-period crossover design clinical trial (Koch et al. 1977). Patients were stratified according to two age groups and then assigned to one of three treatment sequences. Responses were measured as favorable (F) or unfavorable (U); thus, FF indicates a favorable response in both Period 1 and Period 2.

Table 10.2. Two-Period Crossover Study

Age	Sequence	Response Profiles FF	FU	UF	UU	Total
older	A:B	12	12	6	20	50
older	B:P	8	5	6	31	50
older	P:A	5	3	22	20	50
younger	B:A	19	3	25	3	50
younger	A:P	25	6	6	13	50
younger	P:B	13	5	21	11	50

Sequence A:B means that Drug A was administered during the first period and Drug B was administered during the second period. The value P indicates Placebo. There are six possible sequences over the two age groups; each sequence occurs for one set of 50 patients.

These data can be considered paired data in the sense that there is a response for both Period 1 and Period 2. One strategy for analyzing these data is to model the probability of improvement for each patient in the first period (and not the second) versus the probability of improvement in either the first or second period but not both. This can be expressed as the conditional probability

$$\frac{\Pr\{\text{Period1}=F\}\,\Pr\{\text{Period2}=U\}}{\Pr\{\text{Period1}=F\}\,\Pr\{\text{Period2}=U\} + \Pr\{\text{Period1}=U\}\,\Pr\{\text{Period2}=F\}}$$

Thus, the analysis strategy can proceed in the same manner as for the highly stratified paired data. In that example, the analysis adjusted out center-to-center variability (intercenter variability) and concentrated on intracenter variability. In this example, you are conditioning away, or adjusting out, patient-to-patient variability (interpatient variability) and concentrating on intrapatient information. This allows you to perform analyses that may not be possible with population-averaging methods (such as ordinary logistic regression) because of small sample size, although the resulting strategy may not be as efficient. These conditioning methods also lead to results with different interpretation; for example, the resulting odds ratios apply to each patient individually in the study rather than to patients on average.

The effects of interest are the period effect, effects for drugs A and B, and a carryover effect for drugs A and B from Period 1 to Period 2. Table 10.3 and Table 10.4 display the effects for Period 1 and Period 2, using incremental effects parameterization.

Table 10.3. Period 1 Data

Age	Treatment	Period1	Period × Age	Drug A	Drug B	CarryA	CarryB
older	A	1	1	1	0	0	0
older	B	1	1	0	1	0	0
older	P	1	1	0	0	0	0
younger	B	1	0	0	1	0	0
younger	A	1	0	1	0	0	0
younger	P	1	0	0	0	0	0

Table 10.4. Period 2 Data

Age	Treatment	Period1	Period × Age	Drug A	Drug B	CarryA	CarryB
older	B	0	0	0	1	1	0
older	P	0	0	0	0	0	1
older	A	0	0	1	0	0	0
younger	A	0	0	1	0	0	1
younger	P	0	0	0	0	1	0
younger	B	0	0	0	1	0	0

To fit the paired observations paradigm, you subtract the values for the variables for Period 1 and Period 2 and proceed as in the first example. Or, you could use the general conditional likelihood approach and proceed with the PHREG procedure, which is more straightforward.

Note that there are six response functions, logits based on FU versus UF, and thus six degrees of freedom with which to work. If you include the two effects for drugs A and B, the age × period effect, and the period effect, then there are two degrees of freedom left over. These can be used to explore the carryover effects or the age × drug effects. The two degree-of-freedom tests for both sets of effects are identical.

The model employed includes the carryover effects. You can write this model as

$$\Pr\{FU|FU \text{ or } UF\} = \frac{\exp\{\beta + \boldsymbol{\tau}'\mathbf{z}\}}{1 + \exp\{\beta + \boldsymbol{\tau}'\mathbf{z}\}}$$

where $\mathbf{z}$ consists of the difference between the two periods for period × age, Drug A, Drug B, CarryA, and CarryB. The parameter β is the effect for period, τ_0 is the effect for period × age, τ_1 and τ_2 are the effects for Drug A and Drug B, respectively, and τ_3 and τ_4 are the effects for CarryA and CarryB, respectively.

The following DATA step inputs the cell counts of the table one response profile at a time.

```
data cross1 (drop=count);
    input age $ sequence $ time1 $ time2 $ count;
    do i=1 to count;
        output;
    end;
datalines;
older AB F F 12
older AB F U 12
older AB U F 6
older AB U U 20
older BP F F 8
older BP F U 5
older BP U F 6
older BP U U 31
older PA F F 5
older PA F U 3
older PA U F 22
older PA U U 20
younger BA F F 19
younger BA F U 3
younger BA U F 25
younger BA U U 3
younger AP F F 25
younger AP F U 6
younger AP U F 6
younger AP U U 13
younger PB F F 13
younger PB F U 5
younger PB U F 21
younger PB U U 11
;
```

The DATA step that creates SAS data set CROSS2 creates a series of indicator variables for the observations coming from either Period 1 or Period 2. It creates indicator variables DRUGA and DRUGB for whether the patient received those respective drugs and also creates indicator variables CARRYA and CARRYB to indicate whether the previous period included treatment with either of those drugs. Obviously, these variables have the value 0 if they have come from Period 1. The response variable RESPONSE is re-defined to take the value 1 for a favorable response. The PHREG procedure does not handle character-valued response variables.

```
data cross2; set cross1;
   subject=_n_;
   period1=1;
      druga = (substr(sequence, 1, 1)='A');
      drugb = (substr(sequence, 1, 1)='B');
      carrya=0;
      carryb=0;
      response =(time1='F');
      output;
   period1=0;
      druga =  (substr(sequence, 2, 1)='A');
      drugb =  (substr(sequence, 2, 1)='B');
      carrya = (substr(sequence, 1, 1)='A');
      carryb = (substr(sequence, 1, 1)='B');
      response =(time2='F');
      output;
run;
```

Finally, the DATA step that creates SAS data set CROSS3 creates indicator variables for the desired interaction terms and also creates an indicator variable for age; the PHREG procedure doesn't yet handle CLASS variables or the full model specification of the GLM procedure. The variable RESPONSE is redefined to be subtracted from 2 so that the procedure will model the probability of favorable response (alphanumeric ordering).

```
data cross3;
   set cross2;
   response=2-response;
   older=(age='older');
   druga_older=druga*older;
   drugb_older=drugb*older;
   period1_older=period1*older;
run;
```

The following PROC PHREG statements request the desired analysis. The strata variable is specified in the STRATA statement; these are the effects that will be conditioned out. The NOSUMMARY option in the PROC PHREG statement eliminates the usual events and trials listing (part of the survival machinery) from the output. The TIES=DISCRETE option is necessary for proper estimation (since there are only two subjects in a stratum, the default TIES=BRESLOW produces the same results, but getting into the habit of specifying TIES=DISCRETE is a good idea).

```
proc phreg data=cross3 nosummary;
   strata subject;
   model response = period1 druga drugb period1_older
   carrya carryb / ties=discrete;
run;
```

Output 10.9 displays the fit statistics.

Output 10.9   Model Fit Statistics

	Model Fit Statistics	
Criterion	Without Covariates	With Covariates
-2 LOG L	166.355	117.579
AIC	166.355	129.579
SBC	166.355	155.961

The table of maximum likelihood estimates in Output 10.10 indicates that neither carryover effect is influential. There appears to be a significant period effect and a significant Drug A effect.

Output 10.10   Maximum Likelihood Estimates

			Analysis of Maximum Likelihood Estimates			
Variable	DF	Parameter Estimate	Standard Error	Chi-Square	Pr > ChiSq	Hazard Ratio
period1	1	-1.43698	0.70258	4.1832	0.0408	0.238
druga	1	1.24669	0.68066	3.3547	0.0670	3.479
drugb	1	-0.00190	0.64116	0.0000	0.9976	0.998
period1_older	1	0.69125	0.46544	2.2056	0.1375	1.996
carrya	1	-0.19029	1.11248	0.0293	0.8642	0.827
carryb	1	-0.56532	1.15562	0.2393	0.6247	0.568

The reduced model that excludes the carryover effects is fit next. Since the period × age effect is modestly suggestive, it is kept in the model. The following PROC PHREG invocation fits this model. It also includes a test for whether Drug A and Drug B have similar effects.

```
proc phreg data=cross3 nosummary;
   strata subject;
   model response = period1 druga drugb
   period1_older / ties=discrete;
   A_B: test druga=drugb;
run;
```

Output 10.11 displays the model fit statistics for the reduced model. If you take the difference in -2 LOG L for the full and reduced models, $117.826 - 117.579$, you get the log likelihood ratio test for the carryover effects. Since $Q_L = 0.247$ with 2 df, this test is nonsignificant. (If you fit the model with age and drug interactions and perform a similar model reduction, this test would have the same value.)

Output 10.11   Model Assessment Statistics

	Model Fit Statistics	
Criterion	Without Covariates	With Covariates
-2 LOG L	166.355	117.826
AIC	166.355	125.826
SBC	166.355	143.413

The maximum likelihood estimates are displayed in Output 10.12. The period effect remains clearly significant ($Q_W = 12.9534$, $p = 0.0003$). Drug A appears to be strongly significant relative to placebo, while Drug B appears to be nonsignificant.

Output 10.12   Maximum Likelihood Estimates

		Analysis of Maximum Likelihood Estimates				
Variable	DF	Parameter Estimate	Standard Error	Chi-Square	Pr > ChiSq	Hazard Ratio
period1	1	-1.19052	0.33078	12.9534	0.0003	0.304
druga	1	1.34622	0.32894	16.7497	<.0001	3.843
drugb	1	0.26618	0.32334	0.6777	0.4104	1.305
period1_older	1	0.71017	0.45757	2.4088	0.1207	2.034

The period $\times$ age effect is still suggestive. Whether you remove this effect from the model depends on your approach to the analysis. If you think of the study as two separate studies of older and younger people, then you probably will want to keep this effect in the model. If your general structural purpose did not include the distinction of older and younger groups, then you will probably want to remove this effect.

Output 10.13 contains the results of the test comparing the Drug B effect and the Drug A effect. The test is clearly significant; the drugs have different effects.

Output 10.13   Drug A versus Drug B

	Linear Hypotheses Testing Results		
Label	Wald Chi-Square	DF	Pr > ChiSq
A_B	10.9220	1	0.0010

10.4.2 Three-Period Crossover Study

The three-period crossover study provides additional challenges. Consider the data from an exercise study in which participants with chronic respiratory conditions were exposed to low, medium, and high levels of air pollution while exercising on a stationary bike. The outcome was the level of respiratory distress as measured on a scale from 0 for none to 3 for severe. A dichotomous baseline reading of 0 for no distress and 1 for some distress was also recorded before each subject began bicycling. There was a two-week washout period between each of the sessions. As in the two-period crossover study, there is interest in examining carryover effects as well as period effects. The subjects were randomized to one of six sequences: HLM, HML, LHM, LMH, MHL, and MLH, where L, M, and H correspond to low, medium, and high amounts of air pollution. These data are loosely based on an example discussed in Tudor, Koch, and Catellier (2000), which is a useful discussion of biostatistical data from crossover studies.

The conditional analysis of these data provides a way to detect within-subject effects, namely the pollution effect, and also investigates the period and carryover effects. The response of interest is dichotomous—whether the subject had severe distress versus no distress, the 3s versus all other outcomes. Table 10.5 contains the data organized by randomization sequence scheme.

Table 10.5. Randomization Frequencies

Sequence	Frequencies	Percent
HLM	72	16.00
HML	78	17.33
LHM	72	16.00
LMH	72	16.00
MHL	60	13.33
MLH	96	21.33

Consider the possible outcome profiles similar to those discussed above for the two-period crossover study. On page 284, the likelihood conditioned on the discordant pairs, or those response profiles where y_{i1} and y_{i2} were different. Those cases where $y_{i1} = y_{i2}$ were considered noninformative. The sum $\sum_j^2 y_{ij}$ implies equal levels of y_{i1} and y_{i2} when it is is 0 (0,0) and 2 (1,1) but not when the sum is 1. In conditional logistic regression, you are conditioning on the $\{\sum_j^r y_{ij}\}$, which are the sufficient statistics for the $\{\alpha_i\}$.

For the three-period case, $r = 3$ and eight possible profiles exist, two of which are noninformative, when $\{\sum_j^3 y_{ij}\} = 0$ or 3. When $\{\sum_j^3 y_{ij}\} = 1$ or 2, there are three possible patterns for (y_{i1}, y_{i2}, y_{i3}).

The contributions to the conditional likelihood are:

$$\frac{\Pr\{y_{ij} = 1, y_{ij'} = 0 \text{ for all } j' \neq j\}}{\Pr\{y_{i1} + y_{i2} + y_{i3} = 1\}} = \frac{\exp(\mathbf{x}'_{ij}\boldsymbol{\beta})}{\sum_{j'}^3 \exp(\mathbf{x}'_{ij'}\boldsymbol{\beta})} \text{ for } j = 1, 2, 3$$

and

$$\frac{\Pr\{y_{ij}=0, y_{ij'}=1 \text{ for all } j' \neq j\}}{\Pr\{y_{i1}+y_{i2}+y_{i3}=2\}} = \frac{\exp(\sum_{j'=1}^{3} \mathbf{x}'_{ij'}\boldsymbol{\beta} - \mathbf{x}'_{ij}\boldsymbol{\beta})}{\sum_{j'=1}^{3} \exp(\sum_{j'=1}^{3} \mathbf{x}'_{ij'}\boldsymbol{\beta} - \mathbf{x}'_{ij}\boldsymbol{\beta})}$$

This likelihood structure turns out to be the same as the trichotomous loglinear extension of logistic regression, and you can use unconditional maximum likelihood to estimate $\boldsymbol{\beta}$. The functions you would analyze are generalized logits, which can be fit with the CATMOD procedure. However, the PHREG procedure is much more straightforward in fitting the conditional logistic model, and is used in the analysis of these data.

For the exercise data, there is interest in evaluating whether pollution has an effect on respiratory distress and whether there are period effects and carryover effects. One could consider carryover effects from one level to another from Period 1 to Period 2, and then from Period 2 to Period 3. However, there are not enough degrees of freedom for these data to pursue such a strategy. Instead, the analysis focuses on whether there is a carryover effect from a medium pollution period to another period and a carryover effect from a high pollution period to another period. This is a reasonable strategy.

The following DATA step inputs the exercise data. There is one observation per subject per period. The variable SEQUENCE contains the sequence information, for example, observations with the value 'HML' received the sequence high in the first period, medium in the second period, and low in the third period. The indicator variables HIGH and MEDIUM take the value 1 if the exposure is high or medium, respectively, for that period. The variable ID is the subject ID within sequence group, PERIOD1 and PERIOD2 are indicator variables for whether the observation is from Period 1 or Period 2, and CARRYHIGH and CARRYMEDIUM are indicator variables for whether the previous period was high exposure or medium exposure. The variable BASELINE takes the value 1 for respiratory distress at the beginning of the study.

The variable STRATA, a unique identifier for each subject based on a combination of SEQUENCE and ID, is defined in this DATA step as well.

```
data exercise;
   input Sequence $ ID $ Period1 Period2 High Medium Baseline
      Response CarryHigh CarryMedium @@;
   strata=sequence||id;
   DichotResponse = 2-(Response >0);
   datalines;
HML 1    1 0 1 0 0 3 0 0  HML 1    0 1 0 1 0 1 1 0  HML 1    0 0 0 0 0 0 0 1
HML 2    1 0 1 0 0 3 0 0  HML 2    0 1 0 1 0 2 1 0  HML 2    0 0 0 0 0 0 0 1
HML 3    1 0 1 0 1 3 0 0  HML 3    0 1 0 1 0 2 1 0  HML 3    0 0 0 0 0 0 0 1
HML 4    1 0 1 0 0 2 0 0  HML 4    0 1 0 1 0 0 1 0  HML 4    0 0 0 0 0 2 0 1
HML 5    1 0 1 0 0 3 0 0  HML 5    0 1 0 1 0 0 1 0  HML 5    0 0 0 0 0 1 0 1
HML 6    1 0 1 0 1 2 0 0  HML 6    0 1 0 1 0 1 1 0  HML 6    0 0 0 0 0 2 0 1
HML 7    1 0 1 0 0 3 0 0  HML 7    0 1 0 1 0 1 1 0  HML 7    0 0 0 0 0 2 0 1
HML 8    1 0 1 0 0 3 0 0  HML 8    0 1 0 1 0 2 1 0  HML 8    0 0 0 0 0 1 0 1
HML 9    1 0 1 0 1 2 0 0  HML 9    0 1 0 1 0 1 1 0  HML 9    0 0 0 0 0 1 0 1
HML 10   1 0 1 0 0 1 0 0  HML 10   0 1 0 1 0 1 1 0  HML 10   0 0 0 0 0 0 0 1
HML 11   1 0 1 0 0 2 0 0  HML 11   0 1 0 1 0 0 1 0  HML 11   0 0 0 0 0 0 0 1
```

```
HML 12   1 0 1 0 0 3 0 0   HML 12   0 1 0 1 0 0 1 0   HML 12   0 0 0 0 0 0 0 1
HML 13   1 0 1 0 0 1 0 0   HML 13   0 1 0 1 0 3 1 0   HML 13   0 0 0 0 0 1 0 1
HML 14   1 0 1 0 0 2 0 0   HML 14   0 1 0 1 0 2 1 0   HML 14   0 0 0 0 0 0 0 1
HML 15   1 0 1 0 1 2 0 0   HML 15   0 1 0 1 0 2 1 0   HML 15   0 0 0 0 0 0 0 1
HML 16   1 0 1 0 1 2 0 0   HML 16   0 1 0 1 0 2 1 0   HML 16   0 0 0 0 0 0 0 1
HML 17   1 0 1 0 1 2 0 0   HML 17   0 1 0 1 0 2 1 0   HML 17   0 0 0 0 0 3 0 1
HML 18   1 0 1 0 0 2 0 0   HML 18   0 1 0 1 0 2 1 0   HML 18   0 0 0 0 0 2 0 1
HML 19   1 0 1 0 0 2 0 0   HML 19   0 1 0 1 0 0 1 0   HML 19   0 0 0 0 0 2 0 1
HML 20   1 0 1 0 0 3 0 0   HML 20   0 1 0 1 0 0 1 0   HML 20   0 0 0 0 0 1 0 1
HML 21   1 0 1 0 0 1 0 0   HML 21   0 1 0 1 0 1 1 0   HML 21   0 0 0 0 0 0 0 1
HML 22   1 0 1 0 0 2 0 0   HML 22   0 1 0 1 0 0 1 0   HML 22   0 0 0 0 0 2 0 1
HML 23   1 0 1 0 0 2 0 0   HML 23   0 1 0 1 0 1 1 0   HML 23   0 0 0 0 0 1 0 1
HML 24   1 0 1 0 1 3 0 0   HML 24   0 1 0 1 0 1 1 0   HML 24   0 0 0 0 0 0 0 1
HML 25   1 0 1 0 0 3 0 0   HML 25   0 1 0 1 0 2 1 0   HML 25   0 0 0 0 0 1 0 1
HML 26   1 0 1 0 0 3 0 0   HML 26   0 1 0 1 0 2 1 0   HML 26   0 0 0 0 0 0 0 1
HLM 1    1 0 1 0 0 3 0 0   HLM 1    0 1 0 0 0 0 1 0   HLM 1    0 0 0 1 0 2 0 0
HLM 2    1 0 1 0 0 2 0 0   HLM 2    0 1 0 0 0 1 1 0   HLM 2    0 0 0 1 0 1 0 0
HLM 3    1 0 1 0 1 2 0 0   HLM 3    0 1 0 0 0 2 1 0   HLM 3    0 0 0 1 0 2 0 0
HLM 4    1 0 1 0 0 2 0 0   HLM 4    0 1 0 0 0 2 1 0   HLM 4    0 0 0 1 0 1 0 0
HLM 5    1 0 1 0 0 3 0 0   HLM 5    0 1 0 0 0 1 1 0   HLM 5    0 0 0 1 0 1 0 0
HLM 6    1 0 1 0 1 1 0 0   HLM 6    0 1 0 0 0 0 1 0   HLM 6    0 0 0 1 0 1 0 0
HLM 7    1 0 1 0 0 2 0 0   HLM 7    0 1 0 0 0 1 1 0   HLM 7    0 0 0 1 0 1 0 0
HLM 8    1 0 1 0 0 2 0 0   HLM 8    0 1 0 0 0 1 1 0   HLM 8    0 0 0 1 0 1 0 0
HLM 9    1 0 1 0 1 2 0 0   HLM 9    0 1 0 0 0 1 1 0   HLM 9    0 0 0 1 0 0 0 0
HLM 10   1 0 1 0 0 2 0 0   HLM 10   0 1 0 0 0 0 1 0   HLM 10   0 0 0 1 0 2 0 0
HLM 11   1 0 1 0 0 3 0 0   HLM 11   0 1 0 0 0 0 1 0   HLM 11   0 0 0 1 0 1 0 0
HLM 12   1 0 1 0 0 1 0 0   HLM 12   0 1 0 0 0 1 1 0   HLM 12   0 0 0 1 0 0 0 0
HLM 13   1 0 1 0 0 0 0 0   HLM 13   0 1 0 0 0 1 1 0   HLM 13   0 0 0 1 0 0 0 0
HLM 14   1 0 1 0 0 3 0 0   HLM 14   0 1 0 0 0 0 1 0   HLM 14   0 0 0 1 0 2 0 0
HLM 15   1 0 1 0 1 0 0 0   HLM 15   0 1 0 0 0 2 1 0   HLM 15   0 0 0 1 0 0 0 0
HLM 16   1 0 1 0 1 3 0 0   HLM 16   0 1 0 0 0 0 1 0   HLM 16   0 0 0 1 0 1 0 0
HLM 17   1 0 1 0 1 2 0 0   HLM 17   0 1 0 0 0 0 1 0   HLM 17   0 0 0 1 0 1 0 0
HLM 18   1 0 1 0 0 3 0 0   HLM 18   0 1 0 0 0 1 1 0   HLM 18   0 0 0 1 0 1 0 0
HLM 19   1 0 1 0 0 3 0 0   HLM 19   0 1 0 0 0 0 1 0   HLM 19   0 0 0 1 0 1 0 0
HLM 20   1 0 1 0 0 2 0 0   HLM 20   0 1 0 0 0 1 1 0   HLM 20   0 0 0 1 0 2 0 0
HLM 21   1 0 1 0 0 2 0 0   HLM 21   0 1 0 0 0 0 1 0   HLM 21   0 0 0 1 0 2 0 0
HLM 22   1 0 1 0 0 1 0 0   HLM 22   0 1 0 0 0 1 1 0   HLM 22   0 0 0 1 0 3 0 0
HLM 23   1 0 1 0 0 3 0 0   HLM 23   0 1 0 0 0 1 1 0   HLM 23   0 0 0 1 0 2 0 0
HLM 24   1 0 1 0 1 2 0 0   HLM 24   0 1 0 0 0 1 1 0   HLM 24   0 0 0 1 0 2 0 0
MHL 1    1 0 0 1 0 1 0 0   MHL 1    0 1 1 0 0 2 0 1   MHL 1    0 0 0 0 0 0 1 0
MHL 2    1 0 0 1 0 0 0 0   MHL 2    0 1 1 0 0 3 0 1   MHL 2    0 0 0 0 0 1 1 0
MHL 3    1 0 0 1 1 2 0 0   MHL 3    0 1 1 0 0 2 0 1   MHL 3    0 0 0 0 0 0 1 0
MHL 4    1 0 0 1 0 1 0 0   MHL 4    0 1 1 0 0 3 0 1   MHL 4    0 0 0 0 0 1 1 0
MHL 5    1 0 0 1 0 1 0 0   MHL 5    0 1 1 0 0 2 0 1   MHL 5    0 0 0 0 0 2 1 0
MHL 6    1 0 0 1 1 0 0 0   MHL 6    0 1 1 0 0 3 0 1   MHL 6    0 0 0 0 0 0 1 0
MHL 7    1 0 0 1 0 2 0 0   MHL 7    0 1 1 0 0 1 0 1   MHL 7    0 0 0 0 0 1 1 0
MHL 8    1 0 0 1 0 1 0 0   MHL 8    0 1 1 0 0 1 0 1   MHL 8    0 0 0 0 0 2 1 0
MHL 9    1 0 0 1 1 0 0 0   MHL 9    0 1 1 0 0 2 0 1   MHL 9    0 0 0 0 0 0 1 0
MHL 10   1 0 0 1 0 0 0 0   MHL 10   0 1 1 0 0 2 0 1   MHL 10   0 0 0 0 0 0 1 0
MHL 11   1 0 0 1 0 2 0 0   MHL 11   0 1 1 0 0 1 0 1   MHL 11   0 0 0 0 0 0 1 0
MHL 12   1 0 0 1 0 0 0 0   MHL 12   0 1 1 0 0 1 0 1   MHL 12   0 0 0 0 0 0 1 0
MHL 13   1 0 0 1 0 1 0 0   MHL 13   0 1 1 0 0 2 0 1   MHL 13   0 0 0 0 0 1 1 0
MHL 14   1 0 0 1 0 1 0 0   MHL 14   0 1 1 0 0 3 0 1   MHL 14   0 0 0 0 0 1 1 0
MHL 15   1 0 0 1 1 0 0 0   MHL 15   0 1 1 0 0 3 0 1   MHL 15   0 0 0 0 0 1 1 0
```

```
MHL 16   1 0 0 1 1 1 0 0   MHL 16   0 1 1 0 0 3 0 1   MHL 16   0 0 0 0 0 0 1 0
MHL 17   1 0 0 1 1 0 0 0   MHL 17   0 1 1 0 0 2 0 1   MHL 17   0 0 0 0 0 1 1 0
MHL 18   1 0 0 1 0 1 0 0   MHL 18   0 1 1 0 0 2 0 1   MHL 18   0 0 0 0 0 2 1 0
MHL 19   1 0 0 1 0 2 0 0   MHL 19   0 1 1 0 0 3 0 1   MHL 19   0 0 0 0 0 1 1 0
MHL 20   1 0 0 1 0 1 0 0   MHL 20   0 1 1 0 0 3 0 1   MHL 20   0 0 0 0 0 0 1 0
MLH 1    1 0 0 1 0 1 0 0   MLH 1    0 1 0 0 0 0 0 1   MLH 1    0 0 1 0 0 2 0 0
MLH 2    1 0 0 1 0 1 0 0   MLH 2    0 1 0 0 0 1 0 1   MLH 2    0 0 1 0 0 3 0 0
MLH 3    1 0 0 1 1 0 0 0   MLH 3    0 1 0 0 0 1 0 1   MLH 3    0 0 1 0 0 2 0 0
MLH 4    1 0 0 1 0 0 0 0   MLH 4    0 1 0 0 0 0 0 1   MLH 4    0 0 1 0 0 3 0 0
MLH 5    1 0 0 1 0 1 0 0   MLH 5    0 1 0 0 0 0 0 1   MLH 5    0 0 1 0 0 3 0 0
MLH 6    1 0 0 1 1 2 0 0   MLH 6    0 1 0 0 0 1 0 1   MLH 6    0 0 1 0 0 3 0 0
MLH 7    1 0 0 1 0 0 0 0   MLH 7    0 1 0 0 0 0 0 1   MLH 7    0 0 1 0 0 2 0 0
MLH 8    1 0 0 1 0 2 0 0   MLH 8    0 1 0 0 0 1 0 1   MLH 8    0 0 1 0 0 2 0 0
MLH 9    1 0 0 1 1 0 0 0   MLH 9    0 1 0 0 0 0 0 1   MLH 9    0 0 1 0 0 3 0 0
MLH 10   1 0 0 1 0 2 0 0   MLH 10   0 1 0 0 0 2 0 1   MLH 10   0 0 1 0 0 1 0 0
MLH 11   1 0 0 1 0 1 0 0   MLH 11   0 1 0 0 0 2 0 1   MLH 11   0 0 1 0 0 1 0 0
MLH 12   1 0 0 1 0 1 0 0   MLH 12   0 1 0 0 0 0 0 1   MLH 12   0 0 1 0 0 3 0 0
MLH 13   1 0 0 1 0 1 0 0   MLH 13   0 1 0 0 0 0 0 1   MLH 13   0 0 1 0 0 0 0 0
MLH 14   1 0 0 1 0 1 0 0   MLH 14   0 1 0 0 0 2 0 1   MLH 14   0 0 1 0 0 3 0 0
MLH 15   1 0 0 1 1 0 0 0   MLH 15   0 1 0 0 0 1 0 1   MLH 15   0 0 1 0 0 2 0 0
MLH 16   1 0 0 1 1 0 0 0   MLH 16   0 1 0 0 0 0 0 1   MLH 16   0 0 1 0 0 1 0 0
MLH 17   1 0 0 1 1 2 0 0   MLH 17   0 1 0 0 0 0 0 1   MLH 17   0 0 1 0 0 3 0 0
MLH 18   1 0 0 1 0 1 0 0   MLH 18   0 1 0 0 0 1 0 1   MLH 18   0 0 1 0 0 2 0 0
MLH 19   1 0 0 1 0 1 0 0   MLH 19   0 1 0 0 0 1 0 1   MLH 19   0 0 1 0 0 3 0 0
MLH 20   1 0 0 1 0 1 0 0   MLH 20   0 1 0 0 0 0 0 1   MLH 20   0 0 1 0 0 2 0 0
MLH 21   1 0 0 1 0 0 0 0   MLH 21   0 1 0 0 0 0 0 1   MLH 21   0 0 1 0 0 2 0 0
MLH 22   1 0 0 1 0 0 0 0   MLH 22   0 1 0 0 0 1 0 1   MLH 22   0 0 1 0 0 3 0 0
MLH 23   1 0 0 1 0 1 0 0   MLH 23   0 1 0 0 0 2 0 1   MLH 23   0 0 1 0 0 2 0 0
MLH 24   1 0 0 1 1 1 0 0   MLH 24   0 1 0 0 0 2 0 1   MLH 24   0 0 1 0 0 2 0 0
MLH 25   1 0 0 1 0 1 0 0   MLH 25   0 1 0 0 0 1 0 1   MLH 25   0 0 1 0 0 2 0 0
MLH 26   1 0 0 1 0 1 0 0   MLH 26   0 1 0 0 0 2 0 1   MLH 26   0 0 1 0 0 2 0 0
MLH 27   1 0 0 1 0 0 0 0   MLH 27   0 1 0 0 0 1 0 1   MLH 27   0 0 1 0 0 2 0 0
MLH 28   1 0 0 1 0 0 0 0   MLH 28   0 1 0 0 0 1 0 1   MLH 28   0 0 1 0 0 3 0 0
MLH 29   1 0 0 1 0 2 0 0   MLH 29   0 1 0 0 0 1 0 1   MLH 29   0 0 1 0 0 2 0 0
MLH 30   1 0 0 1 1 1 0 0   MLH 30   0 1 0 0 0 1 0 1   MLH 30   0 0 1 0 0 3 0 0
MLH 31   1 0 0 1 0 1 0 0   MLH 31   0 1 0 0 0 1 0 1   MLH 31   0 0 1 0 0 1 0 0
MLH 32   1 0 0 1 0 1 0 0   MLH 32   0 1 0 0 0 1 0 1   MLH 32   0 0 1 0 0 2 0 0
LHM 1    1 0 0 0 0 2 0 0   LHM 1    0 1 1 0 0 2 0 0   LHM 1    0 0 0 1 0 2 1 0
LHM 2    1 0 0 0 0 0 0 0   LHM 2    0 1 1 0 0 3 0 0   LHM 2    0 0 0 1 0 1 1 0
LHM 3    1 0 0 0 1 0 0 0   LHM 3    0 1 1 0 0 2 0 0   LHM 3    0 0 0 1 0 1 1 0
LHM 4    1 0 0 0 0 2 0 0   LHM 4    0 1 1 0 0 3 0 0   LHM 4    0 0 0 1 0 0 1 0
LHM 5    1 0 0 0 0 2 0 0   LHM 5    0 1 1 0 0 1 0 0   LHM 5    0 0 0 1 0 1 1 0
LHM 6    1 0 0 0 1 1 0 0   LHM 6    0 1 1 0 0 0 0 0   LHM 6    0 0 0 1 0 1 1 0
LHM 7    1 0 0 0 0 1 0 0   LHM 7    0 1 1 0 0 0 0 0   LHM 7    0 0 0 1 0 1 1 0
LHM 8    1 0 0 0 0 2 0 0   LHM 8    0 1 1 0 0 3 0 0   LHM 8    0 0 0 1 0 1 1 0
LHM 9    1 0 0 0 1 0 0 0   LHM 9    0 1 1 0 0 3 0 0   LHM 9    0 0 0 1 0 1 1 0
LHM 10   1 0 0 0 0 2 0 0   LHM 10   0 1 1 0 0 3 0 0   LHM 10   0 0 0 1 0 2 1 0
LHM 11   1 0 0 0 0 0 0 0   LHM 11   0 1 1 0 0 1 0 0   LHM 11   0 0 0 1 0 2 1 0
LHM 12   1 0 0 0 0 2 0 0   LHM 12   0 1 1 0 0 1 0 0   LHM 12   0 0 0 1 0 1 1 0
LHM 13   1 0 0 0 0 0 0 0   LHM 13   0 1 1 0 0 2 0 0   LHM 13   0 0 0 1 0 2 1 0
LHM 14   1 0 0 0 0 1 0 0   LHM 14   0 1 1 0 0 3 0 0   LHM 14   0 0 0 1 0 1 1 0
LHM 15   1 0 0 0 1 2 0 0   LHM 15   0 1 1 0 0 2 0 0   LHM 15   0 0 0 1 0 2 1 0
LHM 16   1 0 0 0 1 0 0 0   LHM 16   0 1 1 0 0 3 0 0   LHM 16   0 0 0 1 0 1 1 0
LHM 17   1 0 0 0 1 0 0 0   LHM 17   0 1 1 0 0 2 0 0   LHM 17   0 0 0 1 0 1 1 0
```

```
LHM 18  1 0 0 0 0 1 0 0  LHM 18  0 1 1 0 0 3 0 0  LHM 18  0 0 0 1 0 0 1 0
LHM 19  1 0 0 0 0 1 0 0  LHM 19  0 1 1 0 0 0 0 0  LHM 19  0 0 0 1 0 0 1 0
LHM 20  1 0 0 0 0 1 0 0  LHM 20  0 1 1 0 0 2 0 0  LHM 20  0 0 0 1 0 1 1 0
LHM 21  1 0 0 0 0 1 0 0  LHM 21  0 1 1 0 0 1 0 0  LHM 21  0 0 0 1 0 1 1 0
LHM 22  1 0 0 0 0 2 0 0  LHM 22  0 1 1 0 0 1 0 0  LHM 22  0 0 0 1 0 0 1 0
LHM 23  1 0 0 0 0 2 0 0  LHM 23  0 1 1 0 0 2 0 0  LHM 23  0 0 0 1 0 1 1 0
LHM 24  1 0 0 0 1 2 0 0  LHM 24  0 1 1 0 0 3 0 0  LHM 24  0 0 0 1 0 1 1 0
LMH 1   1 0 0 0 0 0 0 0  LMH 1   0 1 0 1 0 2 0 0  LMH 1   0 0 1 0 0 3 0 1
LMH 2   1 0 0 0 0 1 0 0  LMH 2   0 1 0 1 0 2 0 0  LMH 2   0 0 1 0 0 3 0 1
LMH 3   1 0 0 0 1 1 0 0  LMH 3   0 1 0 1 0 1 0 0  LMH 3   0 0 1 0 0 2 0 1
LMH 4   1 0 0 0 0 1 0 0  LMH 4   0 1 0 1 0 2 0 0  LMH 4   0 0 1 0 0 3 0 1
LMH 5   1 0 0 0 0 1 0 0  LMH 5   0 1 0 1 0 1 0 0  LMH 5   0 0 1 0 0 2 0 1
LMH 6   1 0 0 0 1 2 0 0  LMH 6   0 1 0 1 0 0 0 0  LMH 6   0 0 1 0 0 3 0 1
LMH 7   1 0 0 0 0 3 0 0  LMH 7   0 1 0 1 0 0 0 0  LMH 7   0 0 1 0 0 2 0 1
LMH 8   1 0 0 0 0 2 0 0  LMH 8   0 1 0 1 0 2 0 0  LMH 8   0 0 1 0 0 3 0 1
LMH 9   1 0 0 0 1 1 0 0  LMH 9   0 1 0 1 0 1 0 0  LMH 9   0 0 1 0 0 2 0 1
LMH 10  1 0 0 0 0 3 0 0  LMH 10  0 1 0 1 0 2 0 0  LMH 10  0 0 1 0 0 3 0 1
LMH 11  1 0 0 0 0 1 0 0  LMH 11  0 1 0 1 0 1 0 0  LMH 11  0 0 1 0 0 3 0 1
LMH 12  1 0 0 0 0 1 0 0  LMH 12  0 1 0 1 0 2 0 0  LMH 12  0 0 1 0 0 3 0 1
LMH 13  1 0 0 0 0 0 0 0  LMH 13  0 1 0 1 0 1 0 0  LMH 13  0 0 1 0 0 1 0 1
LMH 14  1 0 0 0 0 0 0 0  LMH 14  0 1 0 1 0 2 0 0  LMH 14  0 0 1 0 0 2 0 1
LMH 15  1 0 0 0 1 0 0 0  LMH 15  0 1 0 1 0 1 0 0  LMH 15  0 0 1 0 0 1 0 1
LMH 16  1 0 0 0 1 1 0 0  LMH 16  0 1 0 1 0 2 0 0  LMH 16  0 0 1 0 0 0 0 1
LMH 17  1 0 0 0 1 2 0 0  LMH 17  0 1 0 1 0 1 0 0  LMH 17  0 0 1 0 0 0 0 1
LMH 18  1 0 0 0 0 1 0 0  LMH 18  0 1 0 1 0 0 0 0  LMH 18  0 0 1 0 0 1 0 1
LMH 19  1 0 0 0 0 1 0 0  LMH 19  0 1 0 1 0 2 0 0  LMH 19  0 0 1 0 0 2 0 1
LMH 20  1 0 0 0 0 0 0 0  LMH 20  0 1 0 1 0 2 0 0  LMH 20  0 0 1 0 0 3 0 1
LMH 21  1 0 0 0 0 0 0 0  LMH 21  0 1 0 1 0 1 0 0  LMH 21  0 0 1 0 0 1 0 1
LMH 22  1 0 0 0 0 0 0 0  LMH 22  0 1 0 1 0 2 0 0  LMH 22  0 0 1 0 0 2 0 1
LMH 23  1 0 0 0 0 0 0 0  LMH 23  0 1 0 1 0 1 0 0  LMH 23  0 0 1 0 0 2 0 1
LMH 24  1 0 0 0 1 0 0 0  LMH 24  0 1 0 1 0 2 0 0  LMH 24  0 0 1 0 0 2 0 1
;
```

The STRATA statement defines the strata; note that the specification TIES=DISCRETE is required in order to produce the correct estimates. You use the TEST statement to specify tests concerning the parameter estimates: here, the joint test for both the carryover and period effects are requested.

```
proc phreg data=exercise nosummary;
   strata strata;
   model DichotResponse = period1 period2 high medium baseline
      CarryHigh CarryMedium / ties=discrete;
   Reduce: test CarryHigh=CarryMedium=period1=period2=0;
run;
```

Output 10.14 contains the parameter estimates from this analysis. If you look at the p-values for the explanatory variables, you see that the carryover and period effects appear to be non-influential, and so is the baseline measurement.

Output 10.14   Parameter Estimates

		Parameter	Standard			Hazard
Variable	DF	Estimate	Error	Chi-Square	Pr > ChiSq	Ratio
Period1	1	-0.03277	0.50326	0.0042	0.9481	0.968
Period2	1	0.07135	0.32635	0.0478	0.8269	1.074
High	1	2.14726	0.42840	25.1227	<.0001	8.561
Medium	1	0.57396	0.32985	3.0278	0.0818	1.775
Baseline	1	-0.64124	0.56340	1.2954	0.2550	0.527
CarryHigh	1	-0.32336	0.48447	0.4455	0.5045	0.724
CarryMedium	1	-0.29047	0.60431	0.2310	0.6308	0.748

Analysis of Maximum Likelihood Estimates

Output 10.15 displays the results of the joint test for whether the period effects and carryover effects are equal to zero. This test has 4 df since they each have two parameters associated with them. The joint test has a value of 0.6386 with $p = 0.9587$. This test confirms the impression you get from the parameter estimates table and justifies a model reduced by those four terms.

Output 10.15   Joint Tests

Linear Hypotheses Testing Results

	Wald		
Label	Chi-Square	DF	Pr > ChiSq
Reduce	0.6386	4	0.9587

The model with treatment effects and baseline is fit next; the forward selection method is used to include the treatment effects and then evaluate whether the baseline variable also enters the model. The RL option requests that confidence limits be produced for the odds ratios, which are labeled "Hazard Ratio" in the PROC PHREG output. The TEST statement tests the hypothesis that the effects for high pollution and medium pollution are the same. The ODS SELECT statement is used to restrict the produced results to the parameter estimates table, the score test, and the test statement results.

```
ods select ResidualChiSq ParameterEstimates TestStmts;
proc phreg data=exercise nosummary;
   strata strata;
   model DichotResponse = high medium baseline
      / selection=forward rl include=2 details ties=discrete;
   Treat: test high = medium = 0;
run;
```

Output 10.16 contains the score test, labeled "Residual Chi-Square Test," which has the value 1.2087 and indicates adequate goodness of fit with 1 df and $p = 0.2716$. Thus, the model containing only treatment effects is reasonable.

Output 10.16 Score Test

Residual Chi-Square Test		
Chi-Square	DF	Pr > ChiSq
1.2087	1	0.2716

Output 10.17 displays the parameter estimates for the final model.

Output 10.17 Final Model Parameter Estimates

		Analysis of Maximum Likelihood Estimates			
Variable	DF	Parameter Estimate	Standard Error	Chi-Square	Pr > ChiSq
High	1	2.26349	0.39863	32.2410	<.0001
Medium	1	0.66054	0.25264	6.8360	0.0089

	Analysis of Maximum Likelihood Estimates		
Variable	Hazard Ratio	95% Hazard Ratio Confidence Limits	
High	9.617	4.403	21.006
Medium	1.936	1.180	3.176

With the other effects out of the model, both the parameter for high and the parameter for medium pollution levels are significant at the $\alpha = 0.05$ level of significance.

Output 10.18 displays the test results for whether the high pollution and medium pollution effects are equivalent.

Output 10.18 Test Results for Treatment

Linear Hypotheses Testing Results			
Label	Wald Chi-Square	DF	Pr > ChiSq
Treat	33.2115	2	<.0001

With a Wald chi-square statistic of 33.2115, 2 df, and $p < 0.0001$, this hypothesis is strongly rejected. High pollution has a much stronger effect on response than medium pollution.

The odds ratios are listed in the output as "Hazard Ratios" because this procedure is really set up for survival analysis. The odds ratio listed for high level of pollution means that those with a high level of pollution have roughly ten times higher odds of experiencing severe respiratory distress than those at the low level of pollution. The 95% confidence limits for this odds ratio are (4.403, 21.006). Similarly, those subjects exposed to medium pollution levels have roughly two times higher odds of experiencing severe respiratory distress than those subjects exposed to low levels, with 95% confidence limits of (1.180, 3.176).

10.5   General Conditional Logistic Regression

When you have paired data or responses that comprise profiles considered to come from a trinomial distribution, you can easily write down the possible response profiles and identify the informative and noninformative observations. In the case of the paired response, you can construct a likelihood that is similar to the unconditional likelihood for a dichotomous response; in the case of a trinomial outcome, you can create a model and likelihood that is based on generalized logits and is an extension of the loglinear model, which is discussed in Chapter 16, "Loglinear Models."

However, for more complicated data situations, equivalent unconditional strategies are not conveniently available. Consider the case of the diagnosis data that are analyzed below. Researchers studied subjects at two times under two conditions. You can treat these data as repeated measurements, which is done in Chapter 14, "Modeling Repeated Measurements Data with WLS," and Chapter 15, "Generalized Estimating Equations," or you can perform a conditional logistic regression, considering each patient to be a separate stratum. This is a reasonable strategy if you are only interested in within-subject effects since you are conditioning out subject to subject variability. In addition, the resulting odds ratios apply to subjects individually instead of on average. Such models are called subject-specific models versus population-averaged models, which are discussed in Chapter 15. If one of the purposes of your analysis is to come with a prediction model, such as one you might use in a clinical setting to determine on treatments for patients, the subject-specific model may be appealing. Random effect models, such as the mixed models you fit with the MIXED procedure, are also examples of subject-specific models.

If you consider the diagnosis data, you see that there are two possible outcomes at four different combinations of condition and time. Only two profiles are noninformative, the case where all of the responses are 'no' and the case where all of the responses are 'yes'. There are, however, fourteen other profiles (2^4 total profiles): four in which only one 'yes' is recorded, six in which two 'yes's are recorded, and four profiles in which three 'yes's are recorded.

Consider the general model for stratified logistic regression:

$$\log\left\{\frac{\theta}{1-\theta}\right\} = \alpha_i + \mathbf{x}\beta$$

The α_i are stratum-specific parameters for each stratum, $i = 1, \dots, s$. In conditional inference, you treat the α_i as nuisance parameters and eliminate them from the likelihood function by conditioning on their sufficient statistic.

The sufficient statistic T for $\boldsymbol{\alpha}$ is

$$t = \sum_{i=1}^{n} y_i$$

Recall the attention to the sum of the y_{ij}s in the previous discussions of the various profiles possible in the paired case and the three-period crossover. They were the sufficient statistics in those cases. The analysis, and the elimination of the αs in the likelihood, involved conditioning on those sufficient statistics.

Now, consider the model

$$\text{logit}(\pi) = \mathbf{X}\beta$$

Partition the $t \times 1$ vector β into two components:

β_0, the $s \times 1$ vector of stratum-specific intercepts

β_1, the $t \times 1$ vector of stratum-specific parameters

Also, consider the partition of $\mathbf{X}$ accordingly into $\mathbf{X}_0$ and $\mathbf{X}_1$.

You can write the sufficient statistic T_j for β_j as

$$T_j = \sum_{i=1}^{n} x_{ij} y_i, j = 1, \ldots, t + s$$

The conditional probability density function of $\mathbf{T}_1$ given $\mathbf{T}_0 = t_0$ is

$$f_{\beta_1}(t_1 | t_0) = \frac{C(t_0, t_1) \exp(t_1' \beta_1)}{\sum_{u_1} C(t_0, \mathbf{u}_1) \exp(\mathbf{u}_1' \beta_1)}$$

where $C(t_0, \mathbf{u}_1)$ are the number of ys such that $\{\mathbf{X}_0' \mathbf{y} = t_0, \mathbf{X}_1' \mathbf{y} = t_1\}$.

You then create a conditional likelihood function and apply an algorithm such as Newton-Raphson to obtain maximum likelihood estimates. Currently, the PHREG procedure provides these estimates.

See Mehta and Patel (1995) for more detail.

10.5.1 Analyzing Diagnostic Data

MacMillan et al. (1981) analyze data from a one population ($s = 1$) observational study involving 793 subjects. For each subject, two diagnostic procedures (standard and test) were carried out at each of two times. The results of the four evaluations were classified as positive or negative. Since a dichotomous response ($c = 2$) was measured at $t = 4$ occasions, there are $r = 2^4 = 16$ response profiles. Table 10.6 displays the resulting data.

You can consider each of the subjects in this study to be a separate stratum, with four measurements in each stratum. By performing a conditional logistic regression, you are eliminating subject-to-subject variability. The effects of interest, time and treatment, are within-subject effects, which can be handled by conditional logistic regression. Note that other strategies would be required if between-subject effects were of interest, such as age, clinic, and sex.

Table 10.6.   Diagnostic Test Results for 793 Subjects

Time 1		Time 2		No. of
Standard	Test	Standard	Test	Subjects
Negative	Negative	Negative	Negative	509
Negative	Negative	Negative	Positive	4
Negative	Negative	Positive	Negative	17
Negative	Negative	Positive	Positive	3
Negative	Positive	Negative	Negative	13
Negative	Positive	Negative	Positive	8
Negative	Positive	Positive	Negative	0
Negative	Positive	Positive	Positive	8
Positive	Negative	Negative	Negative	14
Positive	Negative	Negative	Positive	1
Positive	Negative	Positive	Negative	17
Positive	Negative	Positive	Positive	9
Positive	Positive	Negative	Negative	7
Positive	Positive	Negative	Positive	4
Positive	Positive	Positive	Negative	9
Positive	Positive	Positive	Positive	170

The following DATA step creates SAS data set DIAGNOSIS:

```
data diagnosis;
   input std1 $ test1 $ std2 $ test2 $ count;
   do i=1 to count;
      output;
   end;
   datalines;
Neg Neg Neg Neg 509
Neg Neg Neg Pos   4
Neg Neg Pos Neg  17
Neg Neg Pos Pos   3
Neg Pos Neg Neg  13
Neg Pos Neg Pos   8
Neg Pos Pos Neg   0
Neg Pos Pos Pos   8
Pos Neg Neg Neg  14
Pos Neg Neg Pos   1
Pos Neg Pos Neg  17
Pos Neg Pos Pos   9
Pos Pos Neg Neg   7
Pos Pos Neg Pos   4
Pos Pos Pos Neg   9
Pos Pos Pos Pos 170
;
run;
```

The next two DATA steps create one record per measurement per subject, indicator variables for time and test procedure, a numerical response variable with the value 1 for a negative response and 2 for a positive response, and an indicator variable for the time and procedure interaction. They also create a unique SUBJECT value for each subject in the study.

```
data diagnosis2;
   set diagnosis;
   drop std1 test1 std2 test2;
   subject=_n_;
   time=0; procedure=0;
   response=std1; output;
   time=0; procedure=1;
   response=test1; output;
   time=1; procedure=0;
   response=std2; output;
   time=1; procedure=1;
   response=test2; output;
run;

data diagnosis3;
   set diagnosis2;
   outcome = 2 - (response='Neg');
   time_procedure=time*procedure;
```

The following PROC PHREG invocation requests the model including the variables TIME, TREATMENT, and their interaction. The variable SUBJECT is placed in the STRATA statement so that the estimation process conditions on subject.

```
proc phreg data=diagnosis3 nosummary;
   strata subject;
   model outcome=time procedure
      time_procedure /ties=discrete;
run;
```

A look at the table of parameter estimates indicates that the interaction is not important.

Output 10.19 Parameter Estimates for Full Model

Analysis of Maximum Likelihood Estimates						
Variable	DF	Parameter Estimate	Standard Error	Chi-Square	Pr > ChiSq	Hazard Ratio
time	1	-0.06249	0.25001	0.0625	0.8026	0.939
procedure	1	0.38478	0.25439	2.2879	0.1304	1.469
time_procedure	1	0.47248	0.36297	1.6944	0.1930	1.604

The model is refit with just the variables TIME and PROCEDURE. The options SELECTION=FORWARD, INCLUDE=2, and DETAILS are specified to obtain a score test to serve as a goodness-of-fit test for the model. The RL option is specified to produce confidence limits for the odds ratios.

```
proc phreg data=diagnosis3 nosummary;
   strata subject;
      model outcome=time procedure time_procedure
         /ties=discrete selection=forward include=2 details rl;
run;
```

Output 10.20 displays the score statistic based on the remaining influence of the time $\times$ procedure interaction. It takes the value 1.7002 with $p = 0.1923$, indicating an adequate model fit.

Output 10.20   Score Statistic

Residual Chi-Square Test		
Chi-Square	DF	Pr > ChiSq
1.7002	1	0.1923

Output 10.21 contains the parameter estimates and odds ratio. The test procedure is highly significant, and the odds of a positive response are almost twice as much for the test procedure as for the standard procedure. The confidence limits for this odds ratio are (1.292, 2.653).

Output 10.21   Parameter Estimates for Main Effects Model

Analysis of Maximum Likelihood Estimates					
Variable	DF	Parameter Estimate	Standard Error	Chi-Square	Pr > ChiSq
time	1	0.16273	0.18066	0.8114	0.3677
procedure	1	0.61592	0.18359	11.2551	0.0008

Analysis of Maximum Likelihood Estimates		
Variable	Hazard Ratio	95% Hazard Ratio Confidence Limits
time	1.177	0.826 1.677
procedure	1.851	1.292 2.653

10.6   Paired Observations in a Retrospective Matched Study

Epidemiological investigations often involve the use of retrospective, or case-control studies, where a person known to have the event of interest (case) is paired, or matched, with a person who doesn't have the event (control). The idea is to determine whether the exposure factor is associated with the event; this is presumably made less complicated by using matching to control for possible covariates.

- In a 1:1 matched study, the matched set consists of one case and one control from each stratum. This is the most common situation.

- In a 1:m matched study, the matched set consists of one case and m controls. Usually, m ranges between 2 and 5.

- In the m:n matched study, the matched set consists of n cases with m controls, where usually both m and n are between 1 and 5.

Then, data are collected to determine whether the case and control were exposed to certain risk factors, as measured by the explanatory variables. Through the use of a conditional likelihood, you can define a model that allows you to predict the odds for the event given the explanatory variables. This involves setting up the probabilities for having the exposure given the event and then using Bayes' Theorem to determine a relevant conditional probability concerning the event. You derive the conditional likelihood by first focusing on the conditional probability of observing the explanatory variables given the outcome (event or not). The derivation of the likelihood in the matched pairs setting is discussed in Appendix A in this chapter. This likelihood is similar to that seen in the preceding sections for highly stratified data.

Note that the conditional likelihood for the matched pairs data is the unconditional likelihood for a logistic regression model where the response is always equal to 1, the covariate values are equal to the differences between the values for the case and the control, and there is no intercept. This means that you can use standard logistic regression computer programs by configuring your data appropriately and eliminating the intercept term. You need to do the following:

- Make the sampling unit the matched pair by creating one record per matched set and making the explanatory variables the differences between the case values and the control values.

- Set the response variable equal to 1 (or any constant value).

- Set the model intercept equal to zero.

Through a similar process, you can show that the conditional likelihood for the 1:m matched setting is

$$\prod_{h=1}^{q}\left[1 + \sum_{i=1}^{m}\exp\left\{\beta'(\mathbf{x}_{hi} - \mathbf{x}_{h0})\right\}\right]^{-1}$$

where $i = 1, 2, \ldots, m$ indexes the controls and $i = 0$ corresponds to the case. However, this is not equivalent to any unconditional form, so you have to use special computer programs to fit models for the cases of 1:m as well as m:n matched data.

Similar to the previous examples, the LOGISTIC procedure can be used to fit conditional logistic models for 1:1 matching. In addition, the PHREG procedure fits conditional logistic models and must be used for the case of 1:m and m:n matching. The following sections illustrate the use of PROC LOGISTIC and PROC PHREG in applications of conditional logistic regression. Refer to Breslow and Day (1980) and Collett (1991) for more detail on conditional logistic regression.

10.6.1   1:1 Conditional Logistic Regression

Researchers studied women in a retirement community in the 1970s to determine if there was an association between the use of estrogen and the incidence of endometrial cancer (Mack et al. 1976).* Cases were matched to controls who were within a year of the same age, had the same marital status, and were living in the same community at the time of the diagnosis of the case. Information was also collected on obesity, hypertension, gallbladder disease history, and non-estrogen drug use. The data used here is a subset of the actual data. There are 63 matched pairs, with the variable CASE=1 indicating a case and CASE=0 indicating a control. The goal of the analysis is to determine whether the presence of endometrial disease is associated with any of the explanatory variables.

Each matched pair is transformed into a single observation, where the explanatory variable value is the difference between the corresponding values for the case and the control. The outcome variable CASE has the value 0 for all paired observations; the value does not matter as long as it is constant.

```
data match1;
    drop   id1 gall1 hyper1 age1 est1 nonest1 gallest1;
    retain id1 gall1 hyper1 age1 est1 nonest1 gallest1 0;
    input id case age est gall hyper nonest @@;
    gallest=est*gall;
    if (id = id1) then do;
        gall=gall1-gall; hyper=hyper1-hyper; age=age1-age;
        est=est1-est;   nonest=nonest1-nonest;
        gallest=gallest1-gallest;
        output;
    end;
    else do;
        id1=id; gall1=gall; hyper1=hyper; age1=age;
        est1=est; nonest1=nonest; gallest1=gallest;
    end;
    datalines;
 1 1 74 1 0 0   1  1 0 75 0 0 0   0
 2 1 67 1 0 0   1  2 0 67 0 0 1   1
 3 1 76 1 0 1   1  3 0 76 1 0 1   1
 4 1 71 1 0 0   0  4 0 70 1 1 0   1
 5 1 69 1 1 0   1  5 0 69 1 0 1   1
 6 1 70 1 0 1   1  6 0 71 0 0 0   0
 7 1 65 1 1 0   1  7 0 65 0 0 0   0
 8 1 68 1 1 1   1  8 0 68 0 0 1   1
 9 1 61 0 0 0   1  9 0 61 0 0 0   1
10 1 64 1 0 0   1 10 0 65 0 0 0   0
11 1 68 1 1 0   1 11 0 69 1 1 0   0
12 1 74 1 0 0   1 12 0 74 1 0 0   0
13 1 67 1 1 0   1 13 0 68 1 0 1   1
14 1 62 1 1 0   1 14 0 62 0 1 0   0
15 1 71 1 1 0   1 15 0 71 1 0 1   1
16 1 83 1 0 1   1 16 0 82 0 0 0   0
17 1 70 0 0 0   1 17 0 70 0 0 1   1
```

---

*Data provided by Norman Breslow.

```
18 1 74 1 0 0   1 18 0 75 0 0 0   0
19 1 70 1 0 0   1 19 0 70 0 0 0   0
20 1 66 1 0 1   1 20 0 66 1 0 0   1
21 1 77 1 0 0   1 21 0 77 1 1 1   1
22 1 66 1 0 1   1 22 0 67 0 0 1   1
23 1 71 1 0 1   0 23 0 72 0 0 0   0
24 1 80 1 0 0   1 24 0 79 0 0 0   0
25 1 64 1 0 0   1 25 0 64 1 0 0   1
26 1 63 1 0 0   1 26 0 63 1 0 1   1
27 1 72 0 1 0   1 27 0 72 0 0 1   0
28 1 57 1 0 0   0 28 0 57 1 0 1   1
29 1 74 0 1 0   1 29 0 74 0 0 0   1
30 1 62 1 0 1   1 30 0 62 1 0 0   1
31 1 73 1 0 1   1 31 0 72 1 0 0   1
32 1 71 1 0 1   1 32 0 71 1 0 1   1
33 1 64 0 0 1   1 33 0 65 1 0 0   1
34 1 63 1 0 0   1 34 0 64 0 0 0   1
35 1 79 1 1 1   1 35 0 78 1 1 1   1
36 1 80 1 0 0   1 36 0 81 0 0 1   1
37 1 82 1 0 1   1 37 0 82 0 0 0   1
38 1 71 1 0 1   1 38 0 71 0 0 1   1
39 1 83 1 0 1   1 39 0 83 0 0 0   1
40 1 61 1 0 1   1 40 0 60 0 0 0   1
41 1 71 1 0 0   1 41 0 71 0 0 0   0
42 1 69 1 0 1   1 42 0 69 0 1 0   1
43 1 77 1 0 0   1 43 0 76 1 0 1   1
44 1 64 1 0 0   0 44 0 64 1 0 0   0
45 1 79 0 1 0   0 45 0 82 1 0 0   1
46 1 72 1 0 0   1 46 0 72 1 0 0   1
47 1 82 1 1 1   1 47 0 81 0 0 0   0
48 1 73 1 0 1   1 48 0 74 1 0 0   1
49 1 69 1 0 0   1 49 0 68 0 0 0   1
50 1 79 1 0 1   1 50 0 79 0 0 0   1
51 1 72 1 0 0   0 51 0 71 1 0 1   1
52 1 72 1 0 1   1 52 0 72 1 0 1   1
53 1 65 1 0 1   1 53 0 67 0 0 0   0
54 1 67 1 0 1   1 54 0 66 1 0 0   1
55 1 64 1 1 0   1 55 0 63 0 0 0   1
56 1 62 1 0 0   0 56 0 63 0 0 0   0
57 1 83 0 1 1   1 57 0 83 0 1 0   0
58 1 81 1 0 0   1 58 0 79 0 0 0   0
59 1 67 1 0 0   1 59 0 66 1 0 1   1
60 1 73 1 1 1   1 60 0 72 1 0 0   1
61 1 67 1 1 0   1 61 0 67 1 1 0   1
62 1 74 1 0 1   1 62 0 75 0 0 0   1
63 1 68 1 1 0   1 63 0 69 1 0 0   1
;
```

The following PROC LOGISTIC invocation requests forward model selection; in addition, the NOINT option is specified so that no intercept term is included.

```
proc logistic;
   model case = gall est hyper age nonest /
      noint selection=forward details;
run;
```

Output 10.22 contains the response profiles. Note that since all responses have been assigned the value 0, there is effectively one profile.

Output 10.22   Response Profile

Response Profile		
Ordered Value	case	Total Frequency
1	0	63

In the model selection process, only EST and GALL are entered into the model. Output 10.23 displays the residual score statistic, which has a value of 0.2077 with 3 df, indicating an adequate fit. Output 10.24 displays the score statistic for each variable's entry into the model; since all of these are strongly nonsignificant, the model goodness of fit is supported.

Output 10.23   Residual Chi-Square

Residual Chi-Square Test		
Chi-Square	DF	Pr > ChiSq
0.2077	3	0.9763

Output 10.24   Model Selection Results

Analysis of Effects Not in the Model			
Effect	DF	Score Chi-Square	Pr > ChiSq
hyper	1	0.0186	0.8915
age	1	0.1432	0.7051
nonest	1	0.0370	0.8474

Output 10.25 displays the statistics that assess the model's explanatory capacity.

Output 10.25 Explanatory Capacity

```
            Testing Global Null Hypothesis: BETA=0

    Test                 Chi-Square      DF      Pr > ChiSq

    Likelihood Ratio      33.6457        2         <.0001
    Score                 27.0586        2         <.0001
    Wald                  15.3291        2         0.0005
```

Output 10.26 contains the parameter estimates for the model containing main effects EST and GALL. Output 10.27 contains the odds ratios.

Output 10.26 Parameter Estimates

```
            Analysis of Maximum Likelihood Estimates

                                    Standard        Wald
    Parameter    DF    Estimate      Error      Chi-Square    Pr > ChiSq

    gall          1     1.6551      0.7980        4.3017        0.0381
    est           1     2.7786      0.7605       13.3492        0.0003
```

Parameter estimates for both GALL and EST are significant. The odds ratio for GALL indicates that those persons with gallbladder disease history have 5.234 times higher odds of contracting endometrial cancer as those persons without it, adjusting for estrogen use. The odds ratio for EST indicates that those women who used estrogen have 16.096 times higher odds for contracting endometrial cancer as those women who don't use estrogen, adjusting for gallbladder disease history.

Output 10.27 Odds Ratios

```
                    Odds Ratio Estimates

                      Point          95% Wald
         Effect     Estimate    Confidence Limits

         gall         5.234      1.095      25.006
         est         16.096      3.626      71.457
```

10.6.2   Conditional Logistic Regression Using PROC PHREG

The PHREG procedure can analyze data from 1:1 matched designs without your first having to create a data set with difference variables; it also allows you to analyze data from 1:m and m:n matched studies. You must specify the TIES=DISCRETE option in the MODEL statement (for the 1:1 case, all TIES= options are equivalent, so using the default is adequate).

In this section, the same analysis of the endometrial cancer data is performed using PROC PHREG. The following statements create the data set MATCH2. The variable CASE is redefined so it has the value 1 if the observation is a case and the value 2 if the observation is a control. This is required so that the probability of being a case is modeled.

```
data match2;
    input id case age est gall hyper nonest @@;
    case=2-case;
    datalines;
 1 1 74 1 0 0   1  1 0 75 0 0 0   0
 2 1 67 1 0 0   1  2 0 67 0 0 1   1
 3 1 76 1 0 1   1  3 0 76 1 0 1   1
 4 1 71 1 0 0   0  4 0 70 1 1 0   1
 5 1 69 1 1 0   1  5 0 69 1 0 1   1
 6 1 70 1 0 1   1  6 0 71 0 0 0   0
 7 1 65 1 1 0   1  7 0 65 0 0 0   0
 8 1 68 1 1 1   1  8 0 68 0 0 1   1
 9 1 61 0 0 0   1  9 0 61 0 0 0   1
10 1 64 1 0 0   1 10 0 65 0 0 0   0
11 1 68 1 1 0   1 11 0 69 1 1 0   0
12 1 74 1 0 0   1 12 0 74 1 0 0   0
13 1 67 1 1 0   1 13 0 68 1 0 1   1
14 1 62 1 1 0   1 14 0 62 0 1 0   0
15 1 71 1 1 0   1 15 0 71 1 0 1   1
16 1 83 1 0 1   1 16 0 82 0 0 0   0
17 1 70 0 0 0   1 17 0 70 0 0 1   1
18 1 74 1 0 0   1 18 0 75 0 0 0   0
19 1 70 1 0 0   1 19 0 70 0 0 0   0
20 1 66 1 0 1   1 20 0 66 1 0 0   1
21 1 77 1 0 0   1 21 0 77 1 1 1   1
22 1 66 1 0 1   1 22 0 67 0 0 1   1
23 1 71 1 0 1   0 23 0 72 0 0 0   0
24 1 80 1 0 0   1 24 0 79 0 0 0   0
25 1 64 1 0 0   1 25 0 64 1 0 0   1
26 1 63 1 0 0   1 26 0 63 1 0 1   1
27 1 72 0 1 0   1 27 0 72 0 0 1   0
28 1 57 1 0 0   0 28 0 57 1 0 1   1
29 1 74 0 1 0   1 29 0 74 0 0 0   1
30 1 62 1 0 1   1 30 0 62 1 0 0   1
31 1 73 1 0 1   1 31 0 72 1 0 0   1
32 1 71 1 0 1   1 32 0 71 1 0 1   1
33 1 64 0 0 1   1 33 0 65 1 0 0   1
34 1 63 1 0 0   1 34 0 64 0 0 0   1
35 1 79 1 1 1   1 35 0 78 1 1 1   1
36 1 80 1 0 0   1 36 0 81 0 0 1   1
37 1 82 1 0 1   1 37 0 82 0 0 0   1
38 1 71 1 0 1   1 38 0 71 0 0 1   1
39 1 83 1 0 1   1 39 0 83 0 0 0   1
40 1 61 1 0 1   1 40 0 60 0 0 0   1
41 1 71 1 0 0   1 41 0 71 0 0 0   0
42 1 69 1 0 1   1 42 0 69 0 1 0   1
43 1 77 1 0 0   1 43 0 76 1 0 1   1
44 1 64 1 0 0   0 44 0 64 1 0 0   0
45 1 79 0 1 0   0 45 0 82 1 0 0   1
46 1 72 1 0 0   1 46 0 72 1 0 0   1
47 1 82 1 1 1   1 47 0 81 0 0 0   0
48 1 73 1 0 1   1 48 0 74 1 0 0   1
49 1 69 1 0 0   1 49 0 68 0 0 0   1
50 1 79 1 0 1   1 50 0 79 0 0 0   1
```

```
51 1 72 1 0 0   0 51 0 71 1 0 1   1
52 1 72 1 0 1   1 52 0 72 1 0 1   1
53 1 65 1 0 1   1 53 0 67 0 0 0   0
54 1 67 1 0 1   1 54 0 66 1 0 0   1
55 1 64 1 1 0   1 55 0 63 0 0 0   1
56 1 62 1 0 0   0 56 0 63 0 0 0   0
57 1 83 0 1 1   1 57 0 83 0 1 0   0
58 1 81 1 0 0   1 58 0 79 0 0 0   0
59 1 67 1 0 0   1 59 0 66 1 0 1   1
60 1 73 1 1 1   1 60 0 72 1 0 0   1
61 1 67 1 1 0   1 61 0 67 1 1 0   1
62 1 74 1 0 1   1 62 0 75 0 0 0   1
63 1 68 1 1 0   1 63 0 69 1 0 0   1
;
```

The following statements request the PHREG procedure to perform the conditional logistic analysis. You use the name of the variable identifying the matched set as the STRATA variable.

```
proc phreg;
   strata id;
   model case = gall est hyper age nonest /
      selection=forward details rl;
run;
```

Output 10.28 contains a partial listing of the data. For 1:1 matching, each stratum contains two observations, the case and the control. You can check this table to see that the strata have been set up correctly and contain the correct number of subjects.

Output 10.28   Summary Table

					Percent
Stratum	id	Total	Event	Censored	Censored
1	1	2	2	0	0.00
2	2	2	2	0	0.00
3	3	2	2	0	0.00
4	4	2	2	0	0.00
5	5	2	2	0	0.00
6	6	2	2	0	0.00
7	7	2	2	0	0.00
8	8	2	2	0	0.00
9	9	2	2	0	0.00
10	10	2	2	0	0.00
11	11	2	2	0	0.00
12	12	2	2	0	0.00
13	13	2	2	0	0.00
14	14	2	2	0	0.00
15	15	2	2	0	0.00
16	16	2	2	0	0.00
17	17	2	2	0	0.00
18	18	2	2	0	0.00
19	19	2	2	0	0.00
20	20	2	2	0	0.00

Summary of the Number of Event and Censored Values

Output 10.29 and Output 10.30 contain the model-fitting results. Variables EST and GALL were entered into the model, and HYPER, AGE, and NONEST were not. The residual score statistic has the value 0.2077, with 3 df, and $p = 0.9763$. These are the same results as produced with the PROC LOGISTIC analysis.

Output 10.29   Residual Score Statistic

Residual Chi-Square Test		
Chi-Square	DF	Pr > ChiSq
0.2077	3	0.9763

Output 10.30   Analysis of Variables Not Entered

Analysis of Variables Not in the Model		
Variable	Score Chi-Square	Pr > ChiSq
hyper	0.0186	0.8915
age	0.1432	0.7051
nonest	0.0370	0.8474

Output 10.31 contains the table of statistics that assess the explanatory capability of the model; note that the score test, with a value of 27.0586 and 2 df, is the same as the score test printed in the PROC LOGISTIC output.

Output 10.31   Explanatory Capacity

Testing Global Null Hypothesis: BETA=0			
Test	Chi-Square	DF	Pr > ChiSq
Likelihood Ratio	33.6457	2	<.0001
Score	27.0586	2	<.0001
Wald	15.3291	2	0.0005

The parameter estimates table is identical to the one printed by the PROC LOGISTIC analysis as well. The "Hazard Ratio" column contains the odds ratio estimates.

Output 10.32   Parameter Estimates

```
           Analysis of Maximum Likelihood Estimates

                         Parameter      Standard
        Variable    DF    Estimate        Error     Chi-Square    Pr > ChiSq

        gall         1     1.65509       0.79799       4.3017        0.0381
        est          1     2.77856       0.76049      13.3492        0.0003

           Analysis of Maximum Likelihood Estimates

                              Hazard      95% Hazard Ratio
             Variable          Ratio     Confidence Limits

             gall              5.234       1.095      25.006
             est              16.096       3.626      71.457
```

The PHREG procedure also performs analyses for 1:*m* and *m*:*n* matching; you follow the same steps as previously described. For both of these cases, you must request the option TIES=DISCRETE in the MODEL statement to obtain the correct results.

10.7   1:*m* Conditional Logistic Regression

Researchers in a midwestern county tracked flu cases requiring hospitalization in those residents aged 65 and older during a two-month period in one winter. They matched each case with two controls according to sex and age and also determined whether the cases and controls had a flu vaccine shot and whether they had lung disease. Vaccines were then verified by county health and individual medical practice records. Researchers were interested in whether vaccination had a protective influence on the odds of getting a severe case of flu.

This study is an example of a 1:2 matched study since two controls were chosen for each case. Thus, in order to analyze these data with the SAS System, you need to use the PHREG procedure. The following DATA step reads the data and computes the frequency of vaccine and lung disease for both cases and non-cases. The variable LUNG_VAC is the interaction of lung and vaccination. The variable OUTCOME is redefined so that the probability of being a case is modeled.

```
data matched;
    input id outcome lung vaccine @@;
    outcome=2-outcome;
    lung_vac=lung*vaccine;
    datalines;
 1 1 0 0   1 0 1 0   1 0 0 0   2 1 0 0   2 0 0 0   2 0 1 0
 3 1 0 1   3 0 0 1   3 0 0 0   4 1 1 0   4 0 0 0   4 0 1 0
 5 1 1 0   5 0 0 1   5 0 0 1   6 1 0 0   6 0 0 0   6 0 0 1
 7 1 0 0   7 0 0 0   7 0 0 1   8 1 1 1   8 0 0 0   8 0 0 1
 9 1 0 0   9 0 0 1   9 0 0 0  10 1 0 0  10 0 1 0  10 0 0 0
11 1 1 0  11 0 0 1  11 0 0 0  12 1 1 1  12 0 0 1  12 0 0 0
13 1 0 0  13 0 0 1  13 0 1 0  14 1 0 0  14 0 0 0  14 0 0 1
15 1 1 0  15 0 0 0  15 0 0 1  16 1 0 1  16 0 0 1  16 0 0 1
17 1 0 0  17 0 1 0  17 0 0 0  18 1 1 0  18 0 0 1  18 0 0 1
```

```
19 1 1 0  19 0 0 1  19 0 0 1   20 1 0 0   20 0 0 0   20 0 0 0
21 1 0 0  21 0 0 1  21 0 0 1   22 1 0 1   22 0 0 0   22 0 1 0
23 1 1 1  23 0 0 0  23 0 0 0   24 1 0 0   24 0 0 1   24 0 0 1
25 1 1 0  25 0 1 0  25 0 0 0   26 1 1 1   26 0 0 0   26 0 0 0
27 1 1 0  27 0 0 1  27 0 0 0   28 1 0 1   28 0 1 0   28 0 0 0
29 1 0 0  29 0 0 0  29 0 1 1   30 1 0 0   30 0 0 0   30 0 0 0
31 1 0 0  31 0 0 0  31 0 0 1   32 1 1 0   32 0 0 0   32 0 0 0
33 1 0 1  33 0 0 0  33 0 0 0   34 1 0 0   34 0 1 0   34 0 0 0
35 1 1 0  35 0 1 1  35 0 0 0   36 1 0 1   36 0 0 0   36 0 0 1
37 1 0 1  37 0 0 0  37 0 0 1   38 1 1 1   38 0 0 1   38 0 0 0
39 1 0 0  39 0 0 1  39 0 0 1   40 1 0 0   40 0 0 0   40 0 1 1
41 1 1 0  41 0 0 0  41 0 0 1   42 1 1 0   42 0 0 0   42 0 0 0
43 1 0 0  43 0 0 1  43 0 0 0   44 1 1 0   44 0 0 0   44 0 0 0
45 1 1 0  45 0 0 0  45 0 0 0   46 1 1 0   46 0 1 1   46 0 0 0
47 1 0 1  47 0 0 0  47 0 0 1   48 1 0 0   48 0 0 0   48 0 0 0
49 1 1 0  49 0 1 0  49 0 1 1   50 1 1 1   50 0 0 0   50 0 0 1
51 1 1 0  51 0 0 1  51 0 0 1   52 1 0 1   52 0 0 0   52 0 0 0
53 1 0 1  53 0 0 1  53 0 0 1   54 1 1 0   54 0 0 0   54 0 0 0
55 1 0 0  55 0 0 1  55 0 0 0   56 1 0 0   56 0 0 0   56 0 1 0
57 1 1 1  57 0 1 0  57 0 0 0   58 1 1 0   58 0 0 1   58 0 0 1
59 1 0 0  59 0 0 0  59 0 1 1   60 1 0 0   60 0 0 0   60 0 0 1
61 1 0 1  61 0 0 0  61 0 0 1   62 1 0 0   62 0 0 0   62 0 0 1
63 1 0 0  63 0 0 1  63 0 0 0   64 1 0 0   64 0 1 0   64 0 0 0
65 1 1 1  65 0 0 0  65 0 1 0   66 1 1 1   66 0 0 1   66 0 1 0
67 1 0 0  67 0 0 0  67 0 0 1   68 1 0 0   68 0 0 1   68 0 0 1
69 1 1 1  69 0 0 1  69 0 0 1   70 1 0 0   70 0 0 1   70 0 1 1
71 1 0 0  71 0 0 0  71 0 0 1   72 1 1 0   72 0 0 0   72 0 0 0
73 1 1 0  73 0 0 1  73 0 0 0   74 1 0 0   74 0 0 0   74 0 0 1
75 1 0 0  75 0 0 1  75 0 0 0   76 1 0 0   76 0 0 0   76 0 0 0
77 1 0 1  77 0 0 0  77 0 0 1   78 1 0 0   78 0 0 1   78 0 0 0
79 1 1 0  79 0 0 1  79 0 0 1   80 1 0 1   80 0 0 0   80 0 0 0
81 1 0 0  81 0 1 1  81 0 0 1   82 1 1 1   82 0 1 0   82 0 0 0
83 1 0 1  83 0 0 0  83 0 0 1   84 1 0 0   84 0 0 0   84 0 0 1
85 1 1 0  85 0 0 0  85 0 0 0   86 1 0 0   86 0 1 1   86 0 1 0
87 1 1 1  87 0 0 0  87 0 0 0   88 1 0 0   88 0 0 0   88 0 0 0
89 1 0 0  89 0 0 1  89 0 1 1   90 1 0 0   90 0 0 0   90 0 0 0
91 1 0 1  91 0 0 0  91 0 0 1    92 1 0 0    92 0 1 1    92 0 0 0
93 1 0 1  93 0 0 0  93 0 1 0    94 1 1 0    94 0 0 0    94 0 0 0
95 1 1 1  95 0 0 1  95 0 0 0    96 1 1 0    96 0 0 1    96 0 0 1
97 1 1 1  97 0 0 0  97 0 0 1    98 1 0 0    98 0 0 0    98 0 1 1
99 1 0 1  99 0 1 1  99 0 0 1   100 1 1 0   100 0 0 0   100 0 0 0
101 1 0 0  101 0 0 0  101 0 0 0  102 1 0 1  102 0 0 0  102 0 0 0
103 1 0 1  103 0 0 0  103 0 0 0  104 1 1 0  104 0 0 1  104 0 1 0
105 1 1 0  105 0 1 0  105 0 0 0  106 1 0 0  106 0 0 0  106 0 0 1
107 1 0 0  107 0 0 1  107 0 0 1  108 1 1 1  108 0 0 0  108 0 0 1
109 1 0 1  109 0 0 0  109 0 0 0  110 1 0 0  110 0 0 0  110 0 0 0
111 1 1 0  111 0 0 1  111 0 0 1  112 1 0 0  112 0 0 1  112 0 0 0
113 1 0 1  113 0 0 0  113 0 1 0  114 1 1 1  114 0 0 1  114 0 0 1
115 1 1 1  115 0 0 1  115 0 0 1  116 1 0 0  116 0 0 1  116 0 1 0
117 1 0 1  117 0 0 0  117 0 0 0  118 1 1 0  118 0 1 0  118 0 0 0
119 1 1 0  119 0 0 0  119 0 0 0  120 1 1 0  120 0 0 0  120 0 0 1
121 1 0 0  121 0 0 1  121 0 0 0  122 1 0 1  122 0 0 0  122 0 0 0
123 1 1 0  123 0 0 0  123 0 1 1  124 1 0 0  124 0 0 1  124 0 0 0
125 1 1 0  125 0 1 0  125 0 0 0  126 1 1 1  126 0 0 0  126 0 0 0
```

```
127 1 1 0 127 0 0 1 127 0 0 0 128 1 0 1 128 0 1 0 128 0 0 0
129 1 0 0 129 0 0 0 129 0 1 1 130 1 0 0 130 0 0 0 130 0 0 0
131 1 0 0 131 0 0 0 131 0 0 1 132 1 1 0 132 0 0 0 132 0 0 1
133 1 0 1 133 0 0 0 133 0 0 0 134 1 0 0 134 0 1 0 134 0 0 1
135 1 1 0 135 0 1 1 135 0 0 0 136 1 0 0 136 0 0 0 136 0 0 0
137 1 0 0 137 0 0 0 137 0 0 1 138 1 1 0 138 0 0 0 138 0 0 0
139 1 0 0 139 0 0 0 139 0 0 0 140 1 0 0 140 0 0 1 140 0 1 1
141 1 1 1 141 0 0 0 141 0 0 1 142 1 1 0 142 0 0 0 142 0 0 0
143 1 0 0 143 0 0 1 143 0 1 1 144 1 1 1 144 0 0 1 144 0 0 1
145 1 1 0 145 0 0 1 145 0 0 0 146 1 1 0 146 0 1 0 146 0 0 0
147 1 0 1 147 0 0 0 147 0 0 1 148 1 0 0 148 0 0 1 148 0 0 0
149 1 1 0 149 0 1 0 149 0 1 0 150 1 1 1 150 0 0 0 150 0 0 1
;
```

The following PROC FREQ statements request crosstabulations of vaccine by outcome status and lung disease by outcome status.

```
proc freq;
   tables outcome*lung outcome*vaccine /nocol nopct;
run;
```

Output 10.33 contains the frequencies of vaccine and lung disease for both cases and controls. In these data, 16% of the controls had lung disease, and 42% of the cases had lung disease. Also, 39% of the controls and 31% of the cases had been vaccinated.

Output 10.33   Frequencies of Vaccine and Smoking by Cases and Controls

```
                      Table of outcome by lung

          outcome       lung

          Frequency|
          Row Pct  |       0|       1|  Total
          ---------+--------+--------+
                1  |     87 |     63 |    150
                   |  58.00 |  42.00 |
          ---------+--------+--------+
                2  |    252 |     48 |    300
                   |  84.00 |  16.00 |
          ---------+--------+--------+
          Total         339      111      450


                     Table of outcome by vaccine

          outcome       vaccine

          Frequency|
          Row Pct  |       0|       1|  Total
          ---------+--------+--------+
                1  |    103 |     47 |    150
                   |  68.67 |  31.33 |
          ---------+--------+--------+
                2  |    183 |    117 |    300
                   |  61.00 |  39.00 |
          ---------+--------+--------+
          Total         286      164      450
```

The following statements request the conditional logistic regression analysis. The
SELECTION=FORWARD option is specified to request forward selection model building.

```
proc phreg;
   strata id;
   model outcome = lung vaccine lung_vac /
      selection=forward details ties=discrete;
run;
```

Output 10.34   Model Building Results

```
Step  1. Variable lung is entered.  The model contains the following
         explanatory variables:

         lung


                     Testing Global Null Hypothesis: BETA=0

            Test                   Chi-Square      DF     Pr > ChiSq

            Likelihood Ratio        32.9707         1       <.0001
            Score                   34.1798         1       <.0001
            Wald                    30.3502         1       <.0001


                   Analysis of Maximum Likelihood Estimates

                      Parameter    Standard                               Hazard
    Variable    DF     Estimate      Error    Chi-Square   Pr > ChiSq     Ratio

    lung         1      1.28198     0.23270     30.3502      <.0001        3.604


                   Analysis of Variables Not in the Model

                                     Score
                     Variable    Chi-Square    Pr > ChiSq

                     vaccine        3.2528        0.0713
                     lung_vac       0.5916        0.4418


                      Residual Chi-Square Test

                   Chi-Square        DF     Pr > ChiSq

                     3.2981           2       0.1922


NOTE: No (additional) variables met the 0.05 level for entry into the model.
```

The variable LUNG is entered into the model, but variables VACCINE and LUNG_VAC
are not. However, the p-value of 0.0713 for VACCINE is suggestive, so the model
including LUNG and VACCINE is fit next. The interaction LUNG_VAC is included to
obtain the residual score test as a measure of goodness of fit.

```
proc phreg;
   strata id;
   model outcome = lung vaccine lung_vac /
      selection=forward details include=2 ties=discrete rl;
run;
```

Output 10.35 displays the residual score statistic. With a value of 0.0573 and $p = 0.8107$, this statistic supports goodness of fit. Since there is only one variable, LUNG_VAC, being considered by this test, it has the same value as the individual test for LUNG_VAC that is displayed in the table "Analysis of Variables Not in the Model."

Output 10.35   Residual Score Statistic

```
            Analysis of Variables Not in the Model

                           Score
            Variable    Chi-Square    Pr > ChiSq

            lung_vac      0.0573        0.8107


                 Residual Chi-Square Test

         Chi-Square        DF      Pr > ChiSq

           0.0573           1        0.8107
```

Output 10.36 includes the parameter estimates.

Output 10.36   Parameter Estimates

```
            Analysis of Maximum Likelihood Estimates

                     Parameter    Standard
     Variable    DF   Estimate      Error    Chi-Square   Pr > ChiSq

     lung         1    1.30515     0.23483     30.8899      <.0001
     vaccine      1   -0.40078     0.22328      3.2220      0.0727

            Analysis of Maximum Likelihood Estimates

                      Hazard     95% Hazard Ratio
          Variable    Ratio     Confidence Limits

          lung        3.688       2.328     5.844
          vaccine     0.670       0.432     1.038
```

The odds ratio for getting a case of flu resulting in hospitalization is $e^{-0.40078} = 0.67$ for those with vaccine versus those without vaccine. Thus, study participants with vaccination reduced their odds of getting hospitalizable flu by 33% compared to their nonvaccinated matched counterparts. This means that vaccination had a protective effect, controlling for lung disease status (and age and sex, via matching). The confidence limits for this odds ratio are (0.432,1.038).

Note: The usage of the PHREG procedure described in this chapter is correct but computationally inefficient, although the inefficiency is unlikely to produce noticeable differences in time requirements. The usage described is the most straightforward for most users and that is why it is illustrated here. If you are familiar with the PHREG procedure for survival analysis, you may want to proceed more efficiently. If an outcome variable named OUTCOME takes the value 1 for case and 0 for control, create a variable TIME that is $2-$ OUTCOME. Then use the syntax

```
TIME*OUTCOME(0)
```

as your response in the MODEL statement, remembering to use the TIES=DISCRETE option for $m{:}n$ matching. The computations are more efficient, and the results are identical.

10.8   Exact Conditional Logistic Regression in the Stratified Setting

While conditional logistic regression often serves to counterbalance the small counts in a strata by conditioning away the strata effect, sometimes the data are so sparse that these methods also become inappropriate. The conditional exact inference described in Chapter 8 also applies to the stratified setting. Note that in the asymptotic logistic regression setting, the methodology for the unstratified and stratified analysis is different (the former is based on an unconditional likelihood and the latter is based on a conditional likelihood). In the exact setting, you use the same methodology (involving conditioning). The only difference is that, in the unstratified case, you don't have stratification variables and you are conditioning away only explanatory variables; in the stratified case, you are conditioning away both stratification variables and explanatory variables.

The following example is from Luta et al. (1998), which describes methods for analyzing clustered binary data with exact methods. The data are from a cardiovascular study of eight animals who received various drug treatments. Researchers then arrested coronary flow, which led to the development of regional ischemia, and they recorded whether an adverse cardiovascular event occurred during an eight-minute interval. The heart was reperfused for 50 minutes to allow the heart to return to normal, and then another treatment was tested. Thus, there are up to five repeated measurements on eight clusters, or animals. For various reasons, no animal received all of the five possible treatments. Because of the sequences of treatments used by the investigators, the investigation was not assumed to be a crossover study. Because of the reperfusion, the period and carryover effects were considered to be ignorable.

The data include relatively small counts so a reasonable strategy is exact stratified logistic regression, conditioning on the animals. The following DATA step inputs the data for this analysis. Only the observations corresponding to drug treatments are included; those observations corresponding to the shunt treatment are eliminated (the shunt is simply the placement of the intracoronary artery catheter). The treatments are control (C) which is no drug, test drug and counteracting agent (DA), low-dose test drug (D1), and high-dose test drug (D2). For this analysis, the drug effect is assumed to be ordinal with equally spaced intervals; the variable ORDTREAT is coded as 1 for control to 4 for high-dose drug. The variable ANIMAL takes the values from 1 to 8, and the variable RESPONSE is 1 if an event was observed and 0 otherwise.

```
data animal;
   input animal treatment $ response $ @@;
   if treatment='S' then delete;
   else if treatment='C'  then ordtreat=1;
   else if treatment='DA' then ordtreat=2;
   else if treatment='D1' then ordtreat=3;
   else if treatment='D2' then ordtreat=4;
   datalines;
1 S no  1 C  no  1 C  no  1 D2 yes 1 D1 yes
2 S no  2 D2 yes 2 C  no  2 D1 yes
3 S no  3 C  yes 3 D1 yes 3 DA no  3 C  no
4 S no  4 C  no  4 D1 yes 4 DA no  4 C  no
5 S yes 5 C  no  5 DA no  5 D1 no  5 C  no
6 S no  6 C  no  6 D1 yes 6 DA no  6 C  no
7 S no  7 C  no  7 D1 yes 7 DA no  7 C  no
8 S yes 8 C  yes 8 D1 yes
;
```

In order to request the exact analysis, you specify a MODEL statement that includes both the stratification variable, ANIMAL, and the explanatory variable of interest, ORDTREAT, as well as the intercept. Then, you specify the variable ORDTREAT in the EXACT statement; the analysis will condition on the remaining variable in the MODEL statement, ANIMAL. The ESTIMATE=PARM option requests that the point estimate of the parameter for the treatment effect be produced. The EXACTONLY option in the PROC statement restricts the analysis to the exact analysis.

```
proc logistic data=animal descending exactonly;
   class animal /param=ref;
   model response = animal ordtreat;
   exact 'parm' ordtreat / estimate=parm;
run;
```

When the EXACTONLY option is specified, PROC LOGISTIC prints the "Model Information" and "Response Profile" tables (not shown here) and then prints the results of the exact conditional analysis. Output 10.37 displays the exact tests for the treatment effect.

Output 10.37 Exact Tests

```
                  Exact Conditional Analysis

              Conditional Exact Tests for 'parm'

                                       --- p-Value ---
      Effect     Test         Statistic    Exact      Mid

      ordtreat   Score        10.4411    0.0009    0.0005
                 Probability  0.000723   0.0009    0.0005
```

Both the score and probability test have exact p-values of 0.0009, which is highly significant.

Output 10.38 displays the point estimate of the drug effect, 1.9420 and a 95% confidence interval (0.4824, 5.2932).

Output 10.38   Exact Parameter Estimate

```
                    Exact Conditional Analysis

                Exact Parameter Estimates for 'parm'

                                       95% Confidence
          Parameter    Estimate           Limits              p-Value

          ordtreat      1.9420        0.4824      5.2932        0.0017
```

Compare the score test for the exact stratified analysis to the score test for the asymptotic stratified analysis. In order to perform this analysis, you need to change the order of the response values, convert them to numeric values, and use PROC PHREG with the TIES=DISCRETE option. The following statements create SAS data set ANIMAL2.

```
data animal2;
   set animal;
   if response = 'yes' then event = 1;
   else event = 2;
run;
```

The following PROC PHREG statements request the conditional analysis. Note that the options SELECTION=FORWARD and SLENTRY=.05 are used to produce the score statistic for the drug effect. The clustering variable, ANIMAL, is put in the STRATA statement. The ODS SELECT statement restricts the output to the score chi-square and the parameter estimates table.

```
ods select ResidualChiSq ParameterEstimates;
proc phreg data=animal2;
   strata animal;
   model event = ordtreat /selection=forward
      details ties=discrete slentry=.05;
run;
```

Output 10.39 displays the residual score statistic for treatment, which has the value 10.4411, and, with 1 df, a p-value of 0.0012.

Output 10.39   Residual Score Test

```
                  Residual Chi-Square Test

          Chi-Square        DF      Pr > ChiSq

            10.4411          1          0.0012
```

Output 10.40 contains the parameter estimate, which is 1.94213 with a p-value of 0.0297 for the Wald chi-square. The estimate is very close to the estimate in the exact analysis. Note that the asymptotic p-value is larger than the exact one, which is a bit unusual since most often you find that the exact p-value is larger than the asymptotic p-value.

Output 10.40   Parameter Estimate

Analysis of Maximum Likelihood Estimates						
Variable	DF	Parameter Estimate	Standard Error	Chi-Square	Pr > ChiSq	Hazard Ratio
ordtreat	1	1.94213	0.89323	4.7275	0.0297	6.974

Note that, if you run an unstratified asymptotic analysis on these data, that is, use PROC LOGISTIC and regress RESPONSE on ANIMAL and ORDTREAT, you would get the following messages from PROC LOGISTIC:

```
Quasicomplete separation of data points detected.

WARNING: The maximum likelihood estimate may not exist.
WARNING: The LOGISTIC procedure continues in spite of the above
warning. Results shown are based on the last maximum likelihood
iteration. Validity of the model fit is questionable.
```

These messages are a red flag that your data are probably not suitable for asymptotic analysis and that you should consider exact methods. In addition, in the table for global fit in the asymptotic analysis (not shown here), the score statistic has the chi-square value 19.1924 ($p = 0.0139$) and the Wald statistic has the value 6.4478 ($p = 0.972$). Whenever these tests indicate very different results, it's another sign that your data may be most suitable for exact methods.

Printing More Digits

Occasionally, you may want to generate more digits for the p-value than are printed according to the default format in the LOGISTIC procedure. (PROC LOGISTIC actually computes the number of digits that machine accuracy allows.) You can do this in a fairly straightforward manner with ODS. You alter the default template for the LOGISTIC procedure to incorporate the new format and use that template to print the new table with a DATA _NULL_ step. Appendix C in this chapter contains the programming statements required to accomplish this task; Output 10.41 contains the results for the exact stratified analysis printed with the default format and the same results printed with additional decimal places.

Output 10.41   Exact Test Results

```
                        Exact Conditional Analysis

                   Conditional Exact Tests for 'parm'

                                              --- p-Value ---
            Effect       Test          Statistic    Exact      Mid

            ordtreat     Score          10.4411    0.0009    0.0005
                         Probability    0.000723   0.0009    0.0005




            Listing of ExactTests Using a Customized Template
                                                  ---- p-Value ----
                                                     Exact
    Label    Effect       Test         Statistic    p-value      Mid

    parm     ordtreat     Score         10.4411    0.000868    0.0005
    parm                  Probability   0.000723   0.000868    0.0005
```

These results make it easier to compare the results reported in the Luta et al. paper; refer to that paper for additional analyses performed on these data.

Appendix A: Theory for the Case-Control Retrospective Setting

Suppose that you have q matched pairs, $h = 1, 2, \ldots, q$, and θ_{hi} is the probability of the ith subject in the hth matched pair having the event ($i = 1, 2$). Suppose that $\mathbf{x}_{hi}$ represents the set of explanatory variables for the ith subject in the hth matched pair.

The likelihood for the vector of explanatory variables being $\mathbf{x}_{h1}$ given that subject $h1$ is the case (e) and being $\mathbf{x}_{h2}$ given that subject $h2$ is the control ($\bar{e}$) is

$$\Pr\{\mathbf{x}_{h1}|e\} \Pr\{\mathbf{x}_{h2}|\bar{e}\}$$

The sum of this likelihood and that for its reverse counterpart, the likelihood for the vector of explanatory variables being $\mathbf{x}_{h1}$ given the control and being $\mathbf{x}_{h2}$ given the case, is

$$\Pr\{\mathbf{x}_{h1}|e\} \Pr\{\mathbf{x}_{h2}|\bar{e}\} + \Pr\{\mathbf{x}_{h1}|\bar{e}\} \Pr\{\mathbf{x}_{h2}|e\}$$

and thus the conditional likelihood for a particular matched pair having the observed pairing of explanatory variables $\mathbf{x}_{h1}$ with the case e and the explanatory variables $\mathbf{x}_{h2}$ with the control $\bar{e}$ is

$$\frac{\Pr\{\mathbf{x}_{h1}|e\} \Pr\{\mathbf{x}_{h2}|\bar{e}\}}{\Pr\{\mathbf{x}_{h1}|e\} \Pr\{\mathbf{x}_{h2}|\bar{e}\} + \Pr\{\mathbf{x}_{h1}|\bar{e}\} \Pr\{\mathbf{x}_{h2}|e\}}$$

Applying Bayes' Theorem, $(P(A|B) = P(B|A)P(A)/P(B))$, to each of the six terms in the above expression, you can rewrite the preceding as

$$\frac{\Pr\{e|\mathbf{x}_{h1}\} \Pr\{\bar{e}|\mathbf{x}_{h2}\}}{\Pr\{e|\mathbf{x}_{h1}\} \Pr\{\bar{e}|\mathbf{x}_{h2}\} + \Pr\{\bar{e}|\mathbf{x}_{h1}\} \Pr\{e|\mathbf{x}_{h2}\}}$$

Thus, the conditional probabilities have been reversed so that they are the probabilities of the event given the explanatory variables.

If you assume a logistic model for θ_{hi}, the probability of the ith subject in the hth matched pair having the event, then you can make the appropriate substitutions into the conditional likelihood. The following is the logistic model for θ_{hi}.

$$\theta_{hi} = \frac{\exp\{\alpha_h + \boldsymbol{\beta}'\mathbf{x}_{hi}\}}{1 + \exp\{\alpha_h + \boldsymbol{\beta}'\mathbf{x}_{hi}\}}$$

where α_h is an effect for the hth stratum, or pair, the x_{hik} are the $k = 1, 2, \ldots, t$ explanatory variables for the ith subject in the hth matched pair, and the β_k are the corresponding parameters.

Substituting θ_{hi} for $\Pr\{e|\mathbf{x}_{hi}\}$ and $(1 - \theta_{hi})$ for $\Pr\{\bar{e}|\mathbf{x}_{hi}\}$ produces

$$\frac{\exp\{\alpha_h + \boldsymbol{\beta}'\mathbf{x}_{h1}\}}{\exp\{\alpha_h + \boldsymbol{\beta}'\mathbf{x}_{h1}\} + \exp\{\alpha_h + \boldsymbol{\beta}'\mathbf{x}_{h2}\}}$$

which is equivalent to

$$\frac{\exp\{\boldsymbol{\beta}'(\mathbf{x}_{h1} - \mathbf{x}_{h2})\}}{1 + \exp\{\boldsymbol{\beta}'(\mathbf{x}_{h1} - \mathbf{x}_{h2})\}}$$

Note that the α_h have dropped out and thus you have eliminated the stratum-specific parameters.

The conditional likelihood for the entire data is the product of the likelihoods for the individual strata.

$$\prod_{h=1}^{q} \frac{\exp\{\boldsymbol{\beta}'(\mathbf{x}_{h1} - \mathbf{x}_{h2})\}}{1 + \exp\{\boldsymbol{\beta}'(\mathbf{x}_{h1} - \mathbf{x}_{h2})\}}$$

For this conditional likelihood, matched pairs with $x_{h1k} = x_{h2k}$ for all k are noninformative (that is, their contribution to the likelihood is the constant 0.5), and so these matched pairs can be excluded from the analysis.

Through a similar process, you can show that the conditional likelihood for the $1:m$ matched setting is

$$\prod_{h=1}^{q} \left[1 + \sum_{i=1}^{m} \exp\left\{ \boldsymbol{\beta}'(\mathbf{x}_{hi} - \mathbf{x}_{h0}) \right\} \right]^{-1}$$

where $i = 1, 2, \ldots, m$ indexes the controls and $i = 0$ corresponds to the case. However, this is not equivalent to any unconditional form, so you have to use special computer programs to fit models for the cases of $1:m$ as well as $m:n$ matched data. The PHREG procedure currently performs these analyses in the SAS System.

Appendix B: Theory for Exact Conditional Inference

Section 10.5 provides a brief overview of the methodological ideas behind conditional asymptotic inference. For the model

$$\log\left\{\frac{\theta}{1-\theta}\right\} = \mathbf{X}\beta$$

you partition the $s + t$ vector β into components β_0, an $s \times 1$ vector of stratum-specific intercepts, and β_1, a $t \times 1$ vector of parameters of interest. For this discussion, consider β_0 to include the stratum-specific intercepts and/or any other nuisance parameters, that is, parameters that correspond to explanatory variables beyond the ones of interest. Consider β_1 to be a vector of parameters of interest. Partition $\mathbf{X}$ into a corresponding $\mathbf{X}_0$ and $\mathbf{X}_1$.

The sufficient statistics T_j for the β_j are

$$T_j = \sum_{i=1}^{n} y_i x_{ij} \quad j = 1, \ldots, t + s$$

If $\mathbf{T}_0$ and $\mathbf{T}_1$ are the sufficient statistics corresponding to β_0 and β_1, then you can define the conditional probability density function of $\mathbf{T}_1$ conditional on $\mathbf{T}_0$ as

$$f_{\beta_1}(\boldsymbol{t}_1|\boldsymbol{t}_0) = \frac{C(\boldsymbol{t})\exp(\boldsymbol{t}_1'\beta_1)}{\sum_u C(\mathbf{u},\boldsymbol{t}_0)\exp(\mathbf{u}'\beta_1)}$$

$C(\mathbf{u}, \boldsymbol{t}_0)$ are the number of vectors $\mathbf{y}$ such that $\mathbf{y}'\mathbf{X}_1 = \mathbf{u}$ and $\mathbf{y}'\mathbf{X}_0 = \boldsymbol{t}_0$. The function $f_{\beta_1}(\boldsymbol{t}_1|\boldsymbol{t}_0)$ is also the conditional likelihood function for β given $\mathbf{T}_0 = \boldsymbol{t}_0$. You can maximize this likelihood to obtain MLEs and conditional tests in a similar fashion to the way you would proceed with the unconditional likelihood.

Conditional exact inference involves generating the conditional permutational distribution $f_{\beta_1}(\boldsymbol{t}_1|\boldsymbol{t}_0)$ for the sufficient statistics of the parameter or parameters of interest. You could proceed by completely enumerating the joint distribution of $(\boldsymbol{t}_1, \boldsymbol{t}_0)$ but that becomes computationally infeasible after a handful of observations. Hirji, Mehta, and Patel (1987) devised the multivariate shift algorithm, a network algorithm, which makes the creation of the exact joint distribution computationally possible. Refer to Derr (2000) for an overview of how the algorithm works for a simple data set.

You can test hypotheses $H_0: \beta_1 = 0$ conditional on $\mathbf{T}_0 = \boldsymbol{t}_0$ with the exact probability test or the exact conditional score test. Under H_0, the statistic for the exact probability test is

$$f_{\beta_1=0}(\boldsymbol{t}_1|\boldsymbol{t}_0)$$

and the p-value is the probability of getting a more extreme statistic:

$$p(\boldsymbol{t}_1\boldsymbol{t}_0) = \sum_{u \in \Re_p} f_0(\mathbf{u}|\boldsymbol{t}_0)$$

where $u \in \Re_p$ are the u such that $\mathbf{y}$ exist with $\mathbf{y}'\mathbf{X}_1 = \mathbf{u}$, $\mathbf{y}'\mathbf{X}_0 = \boldsymbol{t}_0$, and

$$f_0(\mathbf{u}|\boldsymbol{t}_0) \leq f_0(\boldsymbol{t}_1|\boldsymbol{t}_0)$$

For the exact conditional score test, you define the conditional mean $\boldsymbol{\mu}_1$ and variance matrix $\mathbf{V}_1$ of $\mathbf{T}_1$ (conditional on $\mathbf{T}_0 = \boldsymbol{t}_0$) and compute the score statistic

$$s = (\boldsymbol{t}_1 - \boldsymbol{\mu}_1)'\mathbf{V}_1^{-1}(\boldsymbol{t}_1 - \boldsymbol{\mu}_1)$$

and compare it to the score for each member of the distribution

$$S = (\mathbf{T}_1 - \boldsymbol{\mu}_1)'\mathbf{V}_1^{-1}(\mathbf{T}_1 - \boldsymbol{\mu}_1)$$

The p-value is

$$p(\boldsymbol{t}_1|\boldsymbol{t}_0) = \Pr(S \geq s) = \sum_{u \in \Re_s} f_0(\mathbf{u}|\boldsymbol{t}_0)$$

where $\mathbf{u} \in \Re_s$ are the $\mathbf{u}$ such that $\mathbf{y}$ exist with $\mathbf{y}'\mathbf{X}_1 = \mathbf{u}$, $\mathbf{y}'\mathbf{X}_0 = \boldsymbol{t}_0$, and $S(u) \geq s$.

You obtain exact parameter estimates β_j by considering all the other parameters as nuisance parameters, forming the conditional pdf, and using Newton-Raphson to find the maximum exact conditional likelihood estimates. Likelihood ratio tests based on the conditional pdf are used to test $H_0: \beta_j = 0$.

Refer to Derr (2000) for more detail on the methods employed by the LOGISTIC procedure, including a basic illustration of how the network algorithm works. Refer to Mehta and Patel (1995) for a complete discussion of exact logistic regression methodology and numerous applications.

Appendix C: ODS Macro

The following code updates the default template for the exact tests output in the PROC LOGISTIC procedure to produce six decimal places.

```
ods output ExactTests=try1 ExactParmEst=try2;
proc logistic data=animal descending;
   class animal /param=ref;
   model response = animal ordtreat;
   exact 'parm' ordtreat / estimate=both;
run;
proc template;
   define table ExactTests2;
   parent=Stat.Logistic.Exacttests;
   column Label Effect Test Statistic ExactPValue MidPValue;
   define ExactPValue;
   parent =Stat.Logistic.ExactPValue;
```

```
      format=D8.6;
      end;
      end;
data _null_;
title2 'Listing of ExactTests Using a Customized Template';
      set try1;
      file print ods=(template='ExactTests2');
      put _ods_;
run;
```

Chapter 11
Quantal Bioassay Analysis

Chapter Table of Contents

Chapter 11
Quantal Bioassay Analysis

11.1 Introduction

Bioassay is the process of determining the potency or strength of a reagent or stimuli based on the response it elicits in biological organisms. Often, the reagent is a new drug and the subjects are experimental animals. Other possible stimuli include radiation and environmental exposures, and other possible subjects include humans and bacteria. Researchers are interested in the tolerance of the subjects to the stimulus or drug, where tolerance is defined as the amount of the stimulus required to produce a response. They are also interested in the relative potency of a new drug to a standard drug. In a direct assay, you steadily increase the doses until you generate the desired reaction. In an indirect assay, you observe the reaction of groups of subjects to specified sets of doses.

The measured response to the drug in an indirect assay can be either quantitative or quantal. An example of a quantitative response is red blood cells per milliliter of blood, and an example of a quantal response is death or survival. This chapter is concerned with quantal responses, which are analyzed with categorical data analysis strategies. Refer to Tsutakawa (1982) for an overview of general bioassay methods, and refer to Finney (1978) and Govindarajulu (1988) for textbook discussion of these areas.

11.2 Estimating Tolerance Distributions

Table 11.1 displays data from an experiment in which animals were exposed to bacterial challenges after having one-quarter of their spleen removed (splenectomy). After 96 hours, their survival status was assessed. The stimulus is the bacterial challenge, and interest lies in assessing the tolerances of the animals' immune systems to the bacterial challenge after they have had partial splenectomies (Koch and Edwards 1985).

Table 11.1. Status 96 Hours After Bacterial Challenge

Bacterial Dose	Status Dead	Status Alive
1.2×10^3	0	5
1.2×10^4	0	5
1.2×10^5	2	3
1.2×10^6	4	2
1.2×10^7	5	1
1.2×10^8	5	0

In bioassay analysis, you make the assumption that responses of subjects are determined through a tolerance distribution. This means that at certain levels of the dose (bacterial challenge in this case) the animals will die; that is, death will occur if dose exceeds the tolerance, and survival will occur when dose is below tolerance. Historically, the tolerances have been assumed to follow a normal distribution. This allows you to write the probability of death at a level x_i of the bacterial challenge as

$$p_i = \Phi\left(\frac{x_i - \mu}{\sigma}\right)$$

where Φ is the cumulative distribution function for the standard normal distribution with mean 0 and variance 1; the parameter μ is the mean (or median) of the tolerance distribution, and σ is the standard deviation.

If $\alpha = -\mu/\sigma$ and $\beta = 1/\sigma$, then

$$p_i = \Phi(\alpha + \beta x_i)$$

and

$$\Phi^{-1}(p_i) = \alpha + \beta x_i$$

The function $\Phi^{-1}(p_i)$ is called the *probit* (or *normit*), and its analysis is called probit analysis. Sometimes the value 5 is added to $\Phi^{-1}(p_i)$ in order to have positive values for all p_i.

Berkson (1951) pointed out that the logistic distribution also works well as a tolerance distribution, generating essentially the same results as the normal distribution. This is particularly true for values of p_i in the middle of the (0, 1) range and when the median μ of the tolerance distribution is the primary parameter. While sometimes a probit analysis of a data set is of more interest to researchers in some disciplines (for example, growth and development) because of the correspondence of its parameters to the mean and standard deviation of the underlying tolerance distribution, the focus in this chapter is on logistic analysis. Note that the measures discussed are also relevant to a model based on the probit.

If you assume the logistic distribution for the tolerances,

$$p_i = \frac{\exp\{\alpha + \beta x_i\}}{1 + \exp\{\alpha + \beta x_i\}}$$

and

$$\log\left\{\frac{p_i}{1 - p_i}\right\} = \alpha + \beta x_i$$

The parameters α and β are estimated with maximum likelihood estimation. Usually, the log of the tolerances is most likely to have a logistic distribution, so frequently you work with the log of the drug or concentration under investigation as the x_i.

One parameter of interest for estimation is the median of the tolerance distribution, or the dose at which 50% of the subjects produce a response. When the response is death, this estimate is called the LD50, for lethal dose. Otherwise, this measure is called the ED50, for effective dose. If you are working with log dose levels, you compute the log LD50 and then exponentiate it if you are also interested in the actual LD50.

Suppose x_{50} represents the log LD50 and p_{50} represents the probability of response at the median of the tolerance distribution.

$$\log\left\{\frac{p_{50}}{1-p_{50}}\right\} = \log\left\{\frac{.5}{.5}\right\}$$
$$= 0$$

Thus, the logistic parameters $\hat{\alpha}$ and $\hat{\beta}x_{50}$ can be set to zero to obtain

$$\hat{x}_{50} = \frac{-\hat{\alpha}}{\hat{\beta}}$$

An approximate form of the variance of $\hat{x}_{50}$ for situations where β is clearly different from 0 is written

$$\mathrm{var}\{\hat{x}_{50}\} = \{\hat{x}_{50}\}^2 \left\{\frac{V(\hat{\alpha})}{\hat{\alpha}^2} - \frac{2V(\hat{\alpha}, \hat{\beta})}{\hat{\alpha}\hat{\beta}} + \frac{V(\hat{\beta})}{\hat{\beta}^2}\right\}$$

where $V(\hat{\alpha})$, $V(\hat{\alpha}, \hat{\beta})$, and $V(\hat{\beta})$ represent the variance of $\hat{\alpha}$, the covariance of $\hat{\alpha}$ and $\hat{\beta}$, and the variance of $\hat{\beta}$, respectively. (Refer to page 333 for references on using Fieller's theorem to compute confidence intervals for these measures.)

This allows you to express the confidence interval for log LD50 as

$$\hat{x}_{50} \pm z_{1-\alpha/2}\sqrt{\mathrm{var}\{\hat{x}_{50}\}}$$

In order to compute the LD50, the actual dosage at which 50% of the subjects die, you exponentiate $\hat{x}_{50}$ (and its confidence limits). Sometimes analysts work on the log log scale for LD50 to produce more stable computations. In that case, you would use

$$\frac{\mathrm{var}(\hat{x}_{50})}{\hat{x}_{50}^2}$$

as the applicable variance for log $\hat{x}_{50}$, and you would double exponentiate the results to generate the estimate of the actual LD50 and its confidence interval.

The LOGISTIC procedure is used to fit these bioassay models. In the following section, a logistic model is fit to the data in Table 11.1, and the log LD50 is computed.

11.2.1   Analyzing the Bacterial Challenge Data

The following SAS statements input the data from Table 11.1 and compute two additional variables: LDOSE is the log dose (natural log), and SQ_LDOSE is the square of LDOSE. Using the log scale results in more evenly spaced dose levels. The variable SQ_LDOSE is used as a quadratic term in the model to help assess goodness of fit.

```
data bacteria;
   input dose status $ count @@;
   ldose=log(dose);
   sq_ldose=ldose*ldose;
   datalines;
1200        dead   0    1200       alive 5
12000       dead   0    12000      alive 5
120000      dead   2    120000     alive 3
1200000     dead   4    1200000    alive 2
12000000    dead   5    12000000   alive 1
120000000   dead   5    120000000  alive 0
;
proc print;
run;
```

In the PROC LOGISTIC specification, both LDOSE and SQ_LDOSE are listed in the MODEL statement. The SELECTION=FORWARD option is specified so that a score statistic for the quadratic term is computed. The COVB option requests that PROC LOGISTIC print the covariance matrix for the parameter estimates, quantities necessary to compute the confidence interval for the log LD50.

```
proc logistic data=bacteria descending;
   freq count;
   model status = ldose sq_ldose / scale=none aggregate
         selection=forward start=1 details covb;
run;
```

Output 11.1 displays the data, including values for the created variables LDOSE and SQ_LDOSE.

Output 11.1   Data Listing

Obs	dose	status	count	ldose	sq_ldose
1	1200	dead	0	7.0901	50.269
2	1200	alive	5	7.0901	50.269
3	12000	dead	0	9.3927	88.222
4	12000	alive	5	9.3927	88.222
5	120000	dead	2	11.6952	136.779
6	120000	alive	3	11.6952	136.779
7	1200000	dead	4	13.9978	195.939
8	1200000	alive	2	13.9978	195.939
9	12000000	dead	5	16.3004	265.704
10	12000000	alive	1	16.3004	265.704
11	120000000	dead	5	18.6030	346.072
12	120000000	alive	0	18.6030	346.072

Since the option START=1 is specified, the first model fit includes the intercept and the LDOSE term. The residual score statistic for the SQ_LDOSE term is not significant with $Q_S = 0.2580$ and $p = 0.6115$, so clearly this term makes no contribution to the model. This result supports the satisfactory fit of the intercept and slope model; the residual score test serves as a goodness-of-fit test for this model.

Output 11.2   Residual Score Statistic

Analysis of Effects Not in the Model

Effect	DF	Score Chi-Square	Pr > ChiSq
sq_ldose	1	0.2580	0.6115

The Pearson and deviance goodness-of-fit statistics also indicate that the model provides an adequate fit, as displayed in Output 11.3. However, note that the sampling requirements for these statistics are minimally met; certainly the expected values for all cell counts are not greater than 4 for several cells. In such cases, it is better to support assessment of fit with methods such as the residual score statistic for the addition of the quadratic term.

Output 11.3   Goodness of Fit Statistics

Deviance and Pearson Goodness-of-Fit Statistics

Criterion	DF	Value	Value/DF	Pr > ChiSq
Deviance	4	1.7508	0.4377	0.7815
Pearson	4	1.3379	0.3345	0.8549

Number of unique profiles: 6

Output 11.4 contains the maximum likelihood estimates for α and β. The estimate $\hat{\beta} = 0.7071$ has $p = 0.0027$ for the test of its significance. The level of bacterial challenge has a significant effect on survival. The intercept $\hat{\alpha} = -9.2680$.

Output 11.4   Maximum Likelihood Estimates

```
              Analysis of Maximum Likelihood Estimates

                                    Standard        Wald
  Parameter     DF     Estimate       Error      Chi-Square    Pr > ChiSq

  Intercept      1      -9.2680       3.1630       8.5857         0.0034
  ldose          1       0.7071       0.2354       9.0223         0.0027
```

Output 11.5 contains the estimated covariance matrix for the parameter estimates. The variance of $\hat{\alpha}$ is 10.0046, the variance of $\hat{\beta}$ is 0.05542, and the covariance of $\hat{\alpha}$ and $\hat{\beta}$ is -0.7334. Taking the square root of the variances produces the standard errors displayed in Output 11.4.

Output 11.5   Estimated Covariance Matrix

```
                  Estimated Covariance Matrix

        Variable        Intercept           ldose

        Intercept       10.00458          -0.73338
        ldose           -0.73338           0.055418
```

To compute the log LD50, use the estimated values of $\hat{\alpha}$ and $\hat{\beta}$.

$$\log \text{LD50} = \frac{-\hat{\alpha}}{\hat{\beta}} = \frac{9.2680}{0.7071} = 13.1070$$

Using the covariances from Output 11.5 in the formula for var$\{x_{50}\}$ yields the value 0.6005. Thus, a confidence interval for the log LD50 is written

$$13.1070 \pm 1.96\sqrt{0.6005}$$

so that the confidence interval is (11.588, 14.626). To determine the LD50 on the actual dose scale, you exponentiate the LD50 for the log scale.

$$\text{actual LD50} = e^{13.1070} = 4.9238 \times 10^5$$

To determine its confidence interval, exponentiate both bounds of the confidence interval to obtain $(1.0780 \times 10^5, 2.2490 \times 10^6)$. This confidence interval describes the location of the median bacterial challenge for the death of animals with one-fourth of the spleen removed.

11.3 Comparing Two Drugs

Bioassay often involves the comparison of two drugs, usually a new drug versus a standard drug. Consider the data in Table 11.2. Researchers studied the effects of the peptides neurotensin and somatostatin in potentiating nonlethal doses of the barbiturate pentobarbital. Groups of mice were administered various dose levels of either neurotensin or somatostatin (Nemeroff et al. 1977; analyzed in Imrey, Koch, and Stokes 1982).

Many times, one drug acts as a dilution of another drug. If this is the case, then the dose response relationship is parallel on the logit scale. Assays that are designed for the dilution assumption are called *parallel lines assays*. The quantity that describes the relationship of such drugs to one another through the ratio of doses of the two drugs that produce the same response is called the *relative potency*.

Table 11.2. N and S Comparison

Dose	Drug	Status Dead	Alive	Total
0.01	N	0	30	30
0.03	N	1	29	30
0.10	N	1	9	10
0.30	N	1	9	10
0.30	S	0	10	10
1.00	N	4	6	10
1.00	S	0	10	10
3.00	N	4	6	10
3.00	S	1	9	10
10.00	N	5	5	10
10.00	S	4	6	10
30.00	S	5	5	10
30.00	N	7	3	10
100.00	S	8	2	10

The dilution assumption for doses z_s of somatostatin and z_n of neurotensin can be stated as

$$z_s = \rho z_n$$

which means that the doses with comparable response for the two drugs are related by the constant ρ, the relative potency; that is, ρ units of neurotensin produce the same behavior as one unit of somatostatin. If x_n and x_s represent log doses, then the dilution assumption also implies that

$$x_s = \log \rho + x_n$$

Thus, assuming the logistic model structure for somatostatin is

$$p_s(x_{si}) = \{1 + \exp(-\alpha_s - \beta x_{si})\}^{-1}$$

where x_{si} denotes log dose levels of somatostatin, you can write the implied structure for log dose levels x_{ni} of neurotensin as

$$p_n(x_{ni}) = p_s(\log \rho + x_{ni}) = \{1 + \exp(-\alpha_s - \beta \log \rho - \beta x_{ni})\}^{-1}$$

$$= \{1 + \exp(-\alpha_n - \beta x_{ni})\}^{-1}$$

where $\alpha_n = \alpha_s + \beta \log \rho$.

By forming

$$\frac{p_n(x_{ni})}{1 - p_n(x_{ni})}$$

you obtain the result

$$\log\left\{\frac{p_n(x_{ni})}{1 - p_n(x_{ni})}\right\} = \{\alpha_s + \beta \log \rho\} + \beta x_{ni}$$
$$= \alpha_n + \beta x_{ni}$$

and

$$\log\left\{\frac{p_s(x_{si})}{1 - p_s(x_{si})}\right\} = \alpha_s + \beta x_{si}$$

Thus, the dilution assumption can be tested by fitting a model with separate intercepts and slopes and then testing for a common slope.

The constant ρ is the relative potency, and since

$$\alpha_n = \alpha_s + \beta \log \rho$$

then

$$\rho = \exp\left\{\frac{\alpha_n - \alpha_s}{\beta}\right\}$$

This means that ρ units of somatostatin produce the same reaction as one unit of neurotensin.

Fieller's theorem can be used to produce confidence intervals for the relative potency. This theorem is a general result that enables confidence intervals to be computed for the ratio of two normally distributed random variables. Fieller's theorem can also be used to produce confidence intervals for the LD50. Refer to Read (1983) for a description of Fieller's formula, and refer to Collett (1991) for a discussion of how to apply it to LD50s and relative potency. Zerbe (1978) describes a matrix implementation of Fieller's formula for use with the general linear model as illustrated in the following analysis.

11.3.1   Analysis of the Peptide Data

The following DATA step creates data set ASSAY for use with PROC LOGISTIC. Indicator variables INT_S and INT_N are created to form the intercepts for each drug, and indicator variables LDOSE_N and LDOSE_S are created to form separate dose columns (slopes) for each drug. LDOSE is dose on the log scale. Since the cell counts in Table 11.2 are small, the squared terms SQLDOSE_S and SQLDOSE_N are created so that a test of quadratic terms can be performed to help assess goodness of fit.

```
data assay;
    input drug $ dose status $ count;
    int_n=(drug='n');
    int_s=(drug='s');
    ldose=log(dose);
    ldose_n=int_n*ldose;
    ldose_s=int_s*ldose;
    sqldose_n=int_n*ldose*ldose;
    sqldose_s=int_s*ldose*ldose;
    datalines;
n 0.01    dead   0
n 0.01    alive 30
n  .03    dead   1
n  .03    alive 29
n  .10    dead   1
n  .10    alive  9
n  .30    dead   1
n  .30    alive  9
n 1.00    dead   4
n 1.00    alive  6
n 3.00    dead   4
n 3.00    alive  6
n 10.00   dead   5
n 10.00   alive  5
n 30.00   dead   7
n 30.00   alive  3
s  .30    dead   0
s  .30    alive 10
s 1.00    dead   0
s 1.00    alive 10
s 3.00    dead   1
s 3.00    alive  9
```

```
s  10.00   dead   4
s  10.00   alive  6
s  30.00   dead   5
s  30.00   alive  5
s 100.00   dead   8
s 100.00   alive  2
;
```

The following PROC LOGISTIC statements request the two intercepts and two slopes model. The NOINT option must be specified to suppress the default intercept. The TEST statement requests a test for equality of the two slope parameters β_n and β_s.

```
proc logistic data=assay descending;
   freq count;
   model status = int_n int_s ldose_n ldose_s
                  sqldose_n sqldose_s  / noint
                  scale=none aggregate
                  start=4 selection=forward details;
   eq_slope: test ldose_n=ldose_s;
run;
```

Output 11.6 contains a listing of the response profile. There are 14 groups based on the drug and dose level combinations, and the model is estimating the probability of death.

Output 11.6   Response Profiles

	Response Profile	
Ordered Value	status	Total Frequency
1	dead	41
2	alive	139

Output 11.7 displays the goodness-of-fit statistics. With values of 4.4144 and 3.6352 for Q_L and Q_P, respectively, these statistics support an adequate model fit.

Output 11.7   Goodness-of-Fit Statistics

Deviance and Pearson Goodness-of-Fit Statistics				
Criterion	DF	Value	Value/DF	Pr > ChiSq
Deviance	10	4.4144	0.4414	0.9267
Pearson	10	3.6352	0.3635	0.9623
Number of unique profiles: 14				

Output 11.8 contains the results for the residual score test for the two quadratic terms. It is nonsignificant, as are each of the individual tests. These results support the goodness of fit of the model.

Output 11.8   Tests for Quadratic Terms

```
                      Residual Chi-Square Test

          Chi-Square          DF      Pr > ChiSq

             1.4817            2         0.4767


            Analysis of Effects Not in the Model

                                    Score
          Effect         DF    Chi-Square    Pr > ChiSq

          sqldose_n       1        0.9311       0.3346
          sqldose_s       1        0.5506       0.4581
```

The parameter estimates are all significant, as seen in the "Analysis of Maximum Likelihood Estimates" table displayed in Output 11.9. However, if you examine the slope estimates (labeled LDOSE_N and LDOSE_S) and their standard errors, you see that it is possible that these two slopes can be represented by one slope. The Wald statistic for the hypothesis test $H_0: \beta_n = \beta_s$ bears this out with a nonsignificant $p = 0.1490$, displayed in Output 11.10.

Output 11.9   Maximum Likelihood Estimates

```
              Analysis of Maximum Likelihood Estimates

                                   Standard      Wald
         Parameter    DF   Estimate    Error   Chi-Square   Pr > ChiSq

         int_n         1    -1.1301   0.2948     14.6983       0.0001
         int_s         1    -3.3782   0.8797     14.7479       0.0001
         ldose_n       1     0.6199   0.1240     24.9907      <.0001
         ldose_s       1     1.0615   0.2798     14.3914       0.0001
```

Output 11.10   Equal Slopes Hypothesis Test Results

```
                Linear Hypotheses Testing Results

                             Wald
           Label         Chi-Square     DF    Pr > ChiSq

           eq_slope         2.0820       1       0.1490
```

Thus, it appears that a parallel lines model fits these data, and the following PROC LOGISTIC statements request this model. The COVB option in the MODEL statement requests that the covariances of the parameters be printed, and the OUTEST=ESTIMATE and COVOUT options request that they be placed into a SAS data set for further processing. Without the specification of the COVOUT option, only the parameter estimates are placed in the OUTEST data set. For convenience, the _LINK_ and _LNLIKE_ variables placed in the OUTEST data set by default are dropped.

```
proc logistic data=assay descending outest=estimate
                  (drop= intercept _link_ _lnlike_) covout;
    freq count;
    model status = int_n int_s ldose /
                  noint scale=none aggregate covb;
run;
```

Output 11.11 contains the goodness-of-fit statistics for this model, and they indicate that the model is adequate.

Output 11.11   Goodness-of-Fit Results

```
            Deviance and Pearson Goodness-of-Fit Statistics

         Criterion      DF        Value      Value/DF     Pr > ChiSq

         Deviance       11       6.8461       0.6224         0.8114
         Pearson        11       5.6480       0.5135         0.8958

                    Number of unique profiles: 14
```

Output 11.12 contains the parameter estimates; all of them are clearly significant.

Output 11.12   Maximum Likelihood Estimates

```
              Analysis of Maximum Likelihood Estimates

                                    Standard        Wald
         Parameter    DF    Estimate    Error    Chi-Square    Pr > ChiSq

         int_n         1     -1.1931    0.3158     14.2781        0.0002
         int_s         1     -2.4476    0.4532     29.1632       <.0001
         ldose         1      0.7234    0.1177     37.7681       <.0001
```

Output 11.13 contains the estimated covariance matrix.

Output 11.13   Covariance Matrix

```
                    Estimated Covariance Matrix

         Variable        int_n          int_s          ldose

         int_n         0.099702       0.025907       -0.00984
         int_s         0.025907       0.20542        -0.03648
         ldose        -0.00984       -0.03648         0.013856
```

The estimated log LD50s from this model are

$$\log \text{LD50}_n = \frac{-\hat{\alpha}_n}{\hat{\beta}} = \frac{1.1931}{0.7234} = 1.65$$

and

$$\log \mathrm{LD50}_s = \frac{-\hat{\alpha}_s}{\hat{\beta}} = \frac{2.4476}{0.7234} = 3.38$$

The log relative potency is estimated as

$$\log \hat{\rho} = \frac{\hat{\alpha}_n - \hat{\alpha}_s}{\hat{\beta}} = \frac{-1.1931 - (-2.4476)}{0.7234} = 1.73$$

You can compute approximate confidence intervals for these quantities using the linearized Taylor series, as in the previous section for the log LD50, or you can produce confidence intervals based on Fieller's theorem. The following SAS/IML code produces confidence intervals based on Fieller's theorem for ratios of estimates from a general linear model (Zerbe 1978).

```
proc iml;
  use estimate;
  start fieller;
  title 'Confidence Intervals';
  use estimate;
  read all into beta where (_type_='PARMS');
  beta=beta';
  read all into cov where (_type_='COV');
  ratio=(k'*beta) / (h'*beta);
  a=(h'*beta)**2-(3.84)*(h'*cov*h);
  b=2*(3.84*(k'*cov*h)-(k'*beta)*(h'*beta));
  c=(k'*beta)**2 -(3.84)*(k'*cov*k);
  disc=((b**2)-4*a*c);
  if (disc<=0 | a<=0) then do;
  print "confidence interval can't be computed", ratio;
  stop; end;
  sroot=sqrt(disc);
  l_b=((-b)-sroot)/(2*a);
  u_b=((-b)+sroot)/(2*a);
  interval=l_b||u_b;
  lname={"l_bound", "u_bound"};
  print "95 % ci for ratio based on fieller", ratio interval[colname=lname];
  finish fieller;
  k={ 1 -1 0 }';
  h={ 0  0 1 }';
  run fieller;
  k={-1 0 0 }';
  h={ 0 0 1 }';
  run fieller;
  k={ 0 -1 0 }';
  h={ 0 0 1 }';
  run fieller;
```

You specify coefficients for vectors **k** and **h** that premultiply the parameter vector to form the numerator and the denominator of the ratio of interest. For example, if

$\boldsymbol{\beta} = \{\alpha_n, \alpha_s, \beta\}$, $\mathbf{k} = \{1, -1, 0\}$, and $\mathbf{h} = \{0, 0, 1\}$, then

$$\frac{\mathbf{k}'\boldsymbol{\beta}}{\mathbf{h}'\boldsymbol{\beta}} = \frac{\alpha_n - \alpha_s}{\beta}$$

which is the relative potency. Other choices of coefficients produce $\log \mathrm{LD50}_n$ and $\log \mathrm{LD50}_s$. The program inputs the covariance matrix for the parameters and applies the appropriate manipulations to produce the corresponding 95% confidence intervals for the ratios that are specified. Executing the SAS/IML code produces the output in Output 11.14.

The ratio estimates for the log potency, $\log \mathrm{LD50}_n$, and $\log \mathrm{LD50}_s$ are displayed, and the lower and upper bounds of their confidence intervals appear under "l_bound" and "u_bound," respectively.

Output 11.14   Confidence Intervals Based on Fieller's Theorem

```
                    Confidence Intervals

               95 % ci for ratio based on fieller


                            INTERVAL
                 RATIO    l_bound    u_bound

            1.7341215  0.4262151  2.9994194


               95 % ci for ratio based on fieller


                            INTERVAL
                 RATIO    l_bound    u_bound

            1.6493371  0.8237277  2.6875216


               95 % ci for ratio based on fieller


                            INTERVAL
                 RATIO    l_bound    u_bound

            3.3834586  2.4863045  4.4505794
```

Table 11.3 contains these results. Thus, a dose of somatostatin must be 5.64 times higher than a dose of neurotensin to have the same effect, with a 95% confidence interval of (1.53, 20.07).

Table 11.3.   Estimated Measures from Parallel Assay

Estimate	Value	95% Confidence Interval	Exponentiated Value	Exponentiated Confidence Interval
log(Potency)	1.73	(0.4262, 2.9994)	5.64	(1.53, 20.07)
$\log \mathrm{LD50}_n$	1.65	(0.8237, 2.6875)	5.21	(2.28, 14.69)
$\log \mathrm{LD50}_s$	3.38	(2.4863, 4.4506)	29.37	(12.02, 85.68)

11.4   Analysis of Pain Study

Researchers investigated a new drug for pain relief by studying its effect on groups of subjects with two different diagnoses. The drug was administered at five dosages, and the outcome measured was whether the subjects reported adverse effects. Table 11.4 contains the data. Interest lies in investigating the association of adverse effects with dose and diagnosis; in addition, there is interest in describing the influence of dose and diagnosis on whether there are adverse effects with a statistical model.

Table 11.4.   Pain Study

Dose	Diagnosis I		Diagnosis II	
	Adverse	Not	Adverse	Not
1	3	26	6	26
5	7	26	20	12
10	10	22	26	6
12	14	18	28	4
15	18	14	31	1

Unlike the previous bioassay analysis, this study does not compare the tolerance distributions of two drugs and is not strictly concerned with estimating the tolerance distribution for either drug. But even though the study does not completely fall into the usual realm of bioassay, it has a bioassay flavor. Its analysis also serves to illustrate the blend of hypothesis testing and model fitting that is often desired in a statistical analysis of categorical data.

Mantel-Haenzsel statistics are computed to determine if there is an association between adverse effects and dose, adverse effects and diagnosis, and adverse effects and dose, controlling for diagnosis. A logistic model is then fit to describe the influence of dose and diagnosis on adverse effects, and ED50s are estimated for both diagnosis groups.

The following DATA step statements input the data and create indicator variables to be used later for the PROC LOGISTIC runs.

```
data adverse;
   input diagnos $ dose status $ count @@;
   i_diagII=(diagnos='II');
   i_diagI= (diagnos='I');
   doseI=i_diagI*dose;
   doseII=i_diagII*dose;
   diagdose=i_diagII*dose;
   if doseI > 0 then ldoseI=log(doseI); else ldoseI=0;
   if doseII > 0 then ldoseII=log(doseII); else ldoseII=0;
   datalines;
I    1    adverse  3 I    1 no 26
I    5    adverse  7 I    5 no 26
I    10   adverse 10 I   10 no 22
I    12   adverse 14 I   12 no 18
I    15   adverse 18 I   15 no 14
II   1    adverse  6 II   1 no 26
```

```
II   5    adverse 20 II   5 no 12
II   10   adverse 26 II  10 no  6
II   12   adverse 28 II  12 no  4
II   15   adverse 31 II  15 no  1
;

proc freq data=adverse;
   weight count;
   tables dose*status diagnos*status diagnos*dose*status /
          nopct nocol cmh;
run;
```

Output 11.15 contains the crosstabulation for DOSE × STATUS. There is a positive association between dose level and proportion of adverse effects.

Output 11.15   Table of DOSE × STATUS

```
              Table of dose by status

       dose        status

       Frequency|
       Row Pct  |adverse |no      |  Total
       ---------+--------+--------+
              1 |      9 |     52 |     61
                |  14.75 |  85.25 |
       ---------+--------+--------+
              5 |     27 |     38 |     65
                |  41.54 |  58.46 |
       ---------+--------+--------+
             10 |     36 |     28 |     64
                |  56.25 |  43.75 |
       ---------+--------+--------+
             12 |     42 |     22 |     64
                |  65.63 |  34.38 |
       ---------+--------+--------+
             15 |     49 |     15 |     64
                |  76.56 |  23.44 |
       ---------+--------+--------+
       Total        163      155      318
```

Output 11.16 contains the Mantel-Haenszel statistics. Since the dose levels are numeric, the 1 df correlation statistic is appropriate. $Q_{CS} = 55.7982$, which is strongly significant. As the dose increases, the proportion of subjects who experienced adverse effects also increases.

Output 11.16   Mantel-Haenszel Statistics

```
              Summary Statistics for dose by status

       Cochran-Mantel-Haenszel Statistics (Based on Table Scores)

       Statistic   Alternative Hypothesis    DF      Value      Prob
       ------------------------------------------------------------------
           1       Nonzero Correlation        1     55.7982    <.0001
           2       Row Mean Scores Differ     4     57.1403    <.0001
           3       General Association        4     57.1403    <.0001
```

Output 11.17 displays the crosstabulation for DIAGNOS $\times$ STATUS.

Output 11.17   DIAGNOS $\times$ STATUS Table

```
                Table of diagnos by status

         diagnos      status

         Frequency|
         Row Pct  |adverse |no      |  Total
         ---------+--------+--------+
         I        |     52 |    106 |    158
                  |  32.91 |  67.09 |
         ---------+--------+--------+
         II       |    111 |     49 |    160
                  |  69.38 |  30.63 |
         ---------+--------+--------+
         Total         163      155      318
```

Output 11.18 contains the Mantel-Haenszel statistics. $Q_{MH} = 42.1732$ with 1 df, which is also strongly significant. Subjects with diagnosis II were more likely to experience adverse effects.

Output 11.18   Mantel-Haenszel Statistics

```
             Summary Statistics for diagnos by status

      Cochran-Mantel-Haenszel Statistics (Based on Table Scores)

   Statistic    Alternative Hypothesis    DF      Value      Prob
   ----------------------------------------------------------------
       1        Nonzero Correlation        1     42.1732    <.0001
       2        Row Mean Scores Differ     1     42.1732    <.0001
       3        General Association        1     42.1732    <.0001
```

Output 11.19 contains the extended Mantel-Haenszel statistics for the association of dose and status after adjusting for diagnosis. The correlation statistic is appropriate, and $Q_{CS} = 65.5570$ with 1 df, which is clearly significant.

Output 11.19   DIAGNOS*DOSE*STATUS

```
              Summary Statistics for dose by status
                    Controlling for diagnos

       Cochran-Mantel-Haenszel Statistics (Based on Table Scores)

    Statistic    Alternative Hypothesis    DF      Value      Prob
    ----------------------------------------------------------------
        1        Nonzero Correlation        1     65.5570    <.0001
        2        Row Mean Scores Differ     4     67.4362    <.0001
        3        General Association        4     67.4362    <.0001
```

The following PROC LOGISTIC statements fit a model that contains separate intercepts and slopes for the two diagnoses. First, the actual dose is used.

```
proc logistic data=adverse outest=estimate
                  (drop= intercept _link_ _lnlike_) covout;
   freq count;
   model status = i_diagI i_diagII doseI doseII  /
             noint scale=none aggregate;
   eq_slope: test doseI=doseII;
run;
```

Output 11.20 contains the response profiles and goodness-of-fit statistics. The model fit appears to be quite good.

Output 11.20　　Response Profiles and Goodness of Fit

```
                        Response Profile

                Ordered                         Total
                Value         status          Frequency

                  1           adverse            163
                  2           no                 155


         Deviance and Pearson Goodness-of-Fit Statistics

        Criterion      DF         Value     Value/DF    Pr > ChiSq

        Deviance        6         2.7345      0.4557       0.8414
        Pearson         6         2.7046      0.4508       0.8449

              Number of unique profiles: 10
```

Output 11.21 contains the model parameters, and Output 11.22 contains the test for a common slope. The hypothesis of a common slope is rejected at the $\alpha = 0.05$ level of significance.

Output 11.21　　Parameter Estimates

```
           Analysis of Maximum Likelihood Estimates

                                 Standard      Wald
         Parameter    DF   Estimate    Error  Chi-Square   Pr > ChiSq

         i_diagI       1    -2.2735   0.4573    24.7197       <.0001
         i_diagII      1    -1.4341   0.3742    14.6887       0.0001
         doseI         1     0.1654   0.0414    15.9478       <.0001
         doseII        1     0.3064   0.0486    39.8186       <.0001
```

Output 11.22   Hypothesis Test

```
                 Linear Hypotheses Testing Results

                              Wald
            Label          Chi-Square      DF      Pr > ChiSq

            eq_slope         4.8787          1        0.0272
```

Next, the model based on log doses is fit.

```
proc logistic data=adverse;
   freq count;
   model status = i_diagI i_diagII ldoseI ldoseII  /
                  noint scale=none aggregate;
   eq_slope: test ldoseI=ldoseII;
run;
```

Output 11.23 contains the goodness-of-fit tests, which are not as supportive of this model as they are for the model based on actual dose; however, they are still entirely satisfactory.

Output 11.23   Goodness-of-Fit Tests

```
            Deviance and Pearson Goodness-of-Fit Statistics

        Criterion       DF       Value     Value/DF    Pr > ChiSq

        Deviance         6      4.8774      0.8129       0.5596
        Pearson          6      4.4884      0.7481       0.6109

                 Number of unique profiles: 10
```

Output 11.24 contains the results for the test that the slopes are equal.

Output 11.24   Hypothesis Test Results

```
                 Linear Hypotheses Testing Results

                              Wald
            Label          Chi-Square      DF      Pr > ChiSq

            eq_slope         2.4034          1        0.1211
```

With $p = 0.1211$, you would not usually reject the hypothesis that the slopes are equal.

Thus, both models do fit the data, and one model offers the possibility of a parallel lines model. Frequently, you do encounter different model choices in your analyses and need to make a decision about which model to present. Since this is not a true bioassay, in the sense of a study comparing two drugs, the fact that you can fit a model with a common slope has less motivation. Potency in this setting means only that the shape of the tolerance distribution of the analgesic is similar for the two diagnoses, which may not be as important as simply determining that the drug works differently for the two diagnoses.

The model with the actual dose is used, since it fits very well and since there is no *a priori* reason to need to use log doses. (One very good reason might be to compare results with other studies if they worked with dose on the log scale.) It is of interest to compute ED50s for both diagnoses, to help describe the median impact on adverse effects for the two diagnoses. The SAS/IML routine is again used to compute the ED50s and to produce a confidence interval based on Fieller's formula. (The entire module is not displayed again.)

The required coefficients are

```
k={ -1 0  0 0}';
h={  0 0  1 0}';
```

and

```
k={ 0 -1 0 0}';
h={ 0  0 0 1}';
run fieller;
```

Output 11.25 contains the results. You need 13.74 units of the analgesic to produce adverse effects in 50% of the subjects with Diagnosis I; you only need 4.68 units of the drug to produce adverse effects in 50% of the subjects with Diagnosis II. The respective confidence intervals are $(11.5095, 18.2537)$ and $(2.9651, 6.0377)$.

Output 11.25   ED50s

```
                    Confidence Intervals

           95 % ci for ratio based on fieller


                         INTERVAL
                 RATIO   l_bound    u_bound

            13.741832 11.509478 18.253683


           95 % ci for ratio based on fieller


                         INTERVAL
                 RATIO   l_bound    u_bound

            4.6799535 2.9651466 6.0377151
```

This example illustrates that bioassay methods can be used for the analysis of data that are not strictly bioassay data but are concerned with the investigation of drug responses. Bioassay methods can also be extended to other application areas as well, such as child development studies. For example, concepts like ED50 can be applied to describe the median ages at which certain physical developments occur. Understanding the strategies that are designed for certain specialty areas can lead to useful applications in nonrelated areas. Refer to Bock and Jones (1968) and Bock (1975) for some statistical methodology related to child development and behavioral areas. Refer to Landis and Koch (1979) for examples of categorical analysis of behavioral data.

Chapter 12
Poisson Regression

Chapter Table of Contents

Chapter 12
Poisson Regression

12.1 Introduction

Categorical data often appear as discrete counts that are considered to be distributed as Poisson. Examples include colony counts for bacteria or viruses, accidents, equipment failures, insurance claims, and incidence of disease. Interest often lies in estimating a rate or incidence (bacteria counts per unit volume or cancer deaths per person-months of exposure to a carcinogen) and determining its relationship to a set of explanatory variables. Poisson regression became popularized as an analysis method in the 1970s and 1980s (Frome, Kutner, and Beauchamp 1973; Charnes, Frome, and Yu 1976; and Frome 1981), although Cochran pointed out the possibilities in a 1940 paper (Cochran 1940), along with the suggestion of the appropriateness of the loglinear model. Currently, Poisson regression is a widely-used modeling technique; recent uses of Poisson regression include a homicide incidence study (Shahpar and Guohua 1999), a study of injuries incurred by electrical utility workers (Loomis et al. 1999), and an evaluation of the risk of endometrial cancer as related to occupational physical activity (Moradi et al. 1998).

This chapter describes the methodology of Poisson regression in Section 12.2 and illustrates the use of the strategy with applications in the next two sections. Section 12.5 describes the issue of overdispersion with Poisson data and offers one technique for adjusting for it. Poisson regression is discussed in other chapters in this book as well. Chapter 15, "Generalized Estimating Equations," describes the analysis of Poisson-distributed correlated data with the GEE method in Section 15.9 and describes a GEE-based approach for managing overdispersion in Poisson regression for a univariate outcome in Section 15.14. There is a proportionality relationship between the likelihood for Poisson regression and the likelihood for the loglinear model, so the loglinear model can be fit using Poisson regression methods. This is illustrated in Section 16.3 of Chapter 16, "Loglinear Models." In addition, the likelihoods for Poisson regression and the piecewise exponential model for analyzing time-to-event data are proportional, so the former can be used to fit the latter. This is illustrated in Chapter 17, "Analyzing Time-to-Event Data."

12.2 Methodology for Poisson Regression

Suppose that a response variable Y is distributed as Poisson and has expected value μ. Recall that the variance of a Poisson variable is also μ. If you have a single explanatory variable x, you can write a regression model for μ as

$$g(\mu) = \alpha + x\beta$$

where g is a link function, in terms of a GLM (generalized linear model). Usually, g is taken to be the log function. If so, you have a loglinear model

$$\log(\mu) = \alpha + x\beta$$

You can rewrite this model as

$$\mu = e^{\alpha} e^{x\beta}$$

If you increase the explanatory variable x by one unit, it has a multiplicative effect of e^{β} on μ. Since this model is specified as a GLM, with a log link and a Poisson distribution, you can fit it with the GENMOD procedure and use the usual deviance and likelihood ratio tests to assess model fit and use Wald or score statistics to assess the model effects.

Frequently, discrete counts represent information collected over time (days, years) or in space (volume for bacteria counts) and interest lies in modeling rates. If the exposure time or volume is denoted as N, you write the rate as Y/N and write the expected value as μ/N Modeling this rate with a loglinear model is written

$$\log \frac{\mu}{N} = \alpha + x\beta$$

which can be rearranged as

$$\log \mu = \alpha + x\beta + \log(N)$$

The term $\log(N)$ is called an *offset* and must be accounted for in the estimation process. Note that if you exponentiate both sides of this expression you obtain

$$\mu = \exp\{\alpha + x\boldsymbol{\beta} + \log(N)\} = Ne^{\alpha} e^{x\beta}$$

which means that the mean is proportional to N. Holding everything else constant, if you multiplied N by some number, you would be multiplying the expected mean by the same number.

More generally, when you have multiple explanatory variables, you can write the model in matrix terms

$$\mu(\mathbf{x}) = \{N(\mathbf{x})\}\{g(\boldsymbol{\beta}|\mathbf{x})\}$$

where $\mu(\mathbf{x})$ is the expected value of the number of events $n(\mathbf{x})$, $\mathbf{x}$ is the vector of explanatory variables, $\mathbf{x} = (x_1, x_2, \ldots, x_t)'$, and $N(\mathbf{x})$ is the known total exposure to risk in the units in which the events occur (subject-days, for example). The rate for incidence is written

$$\lambda(\mathbf{x}) = \mu(\mathbf{x}) / N(\mathbf{x})$$

The loglinear model is written as

$$\log\left\{\frac{\mu(\mathbf{x})}{N(\mathbf{x})}\right\} = \mathbf{x}'\beta$$

for counts $n(\mathbf{x})$ with independent Poisson distributions. An equivalent form is

$$\mu(\mathbf{x}) = \{N(\mathbf{x})\}\{\exp(\mathbf{x}'\beta)\}$$

If you have s independent groups referenced by $i = 1, 2, \ldots, s$, each with a vector $\mathbf{x}_i = (x_{i1}, x_{i2}, \ldots, x_{it})$ of t explanatory variables, you can write a likelihood function for the data as

$$\Phi(\mathbf{n}|\boldsymbol{\mu}) = \prod_{i=1}^{s} \mu_i^{n_i} \{\exp(-\mu_i)\} / n_i!$$

where $\mathbf{n} = (n_1, n_2, \ldots, n_s)'$ and $\boldsymbol{\mu} = (\mu_1, \mu_2, \ldots, \mu_s)'$.

The loglinear Poisson model is often written as

$$\log\{n_i\} = \log\{N_i\} + \mathbf{x}_i'\beta$$

in the generalized linear models framework, where the quantity $\log\{N_i\}$ is the offset. For more information on Poisson regression, refer to Koch, Atkinson, and Stokes (1986).

12.3   Simple Poisson Counts Example

The following data come from a cross-sectional study of 400 patients who had malignant melanoma (Roberts et al. 1981). The site of the tumor and the histological type were recorded. These data are also analyzed in Chapter 16, "Loglinear Models," from the perspective of the loglinear model framework, in which you consider all dimensions for classification to be response variables (not response and explanatory variables) and you determine if there is association within that set of variables.

It is reasonable to consider the tumor counts to be distributed as Poisson and to determine whether those counts are influenced by site and type of tumor. By taking the log of the counts, you can fit a loglinear model.

Table 12.1. Malignant Melanoma Data

Tumor Type	Tumor Site			Total
	Head and Neck	Trunk	Extremities	
Hutchinson's melanotic freckle	22	2	10	34
Superficial spreading melanoma	16	54	115	185
Nodular	19	33	73	125
Indeterminate	11	17	28	56
Total	68	106	226	400

The following DATA step inputs the melanoma count data into the SAS data set MELANOMA:

```
data melanoma;
   input type $ site $ count;
   datalines;
Hutchinson's   Head&Neck     22
Hutchinson's   Trunk          2
Hutchinson's   Extremities   10
Superficial    Head&Neck     16
Superficial    Trunk         54
Superficial    Extremities  115
Nodular        Head&Neck     19
Nodular        Trunk         33
Nodular        Extremities   73
Indeterminate Head&Neck     11
Indeterminate Trunk         17
Indeterminate Extremities   28
;
run;
```

Since Poisson regression is a form of the generalized linear model, you perform the analysis with the GENMOD procedure. As described in Section 8.9.2, the GENMOD procedure fits generalized linear models in the SAS System. See that section for a discussion of using PROC GENMOD to fit the logistic model. In order to perform Poisson regression, you specify the log link function with the LINK=LOG option and specify the Poisson distribution with the DIST=POISSON option. Note that you could leave off the LINK=LOG option because the default canonical link function is the log.

```
proc genmod;
   class type site;
   model count=type|site / dist=poisson link=log type3;
run;
```

Output 12.1 displays the model information for this analysis and Output 12.2 displays the CLASS variable levels for the explanatory variables.

Output 12.1   Model Information

```
                        Model Information

          Data Set                    WORK.MELANOMA
          Distribution                      Poisson
          Link Function                         Log
          Dependent Variable                  count
          Observations Used                      12
```

Output 12.2   Class Variable Information

```
                   Class Level Information

        Class      Levels    Values

        type          4      Hutchins Indeterm Nodular Superfic
        site          3      Extremit Head&Nec Trunk
```

Output 12.3 contains the table with the goodness-of-fit statistics. Since this model fits twelve parameters to twelve response functions, the log counts, it is saturated and no goodness-of-fit statistics are defined.

Output 12.3   Assessment of Fit

```
              Criteria For Assessing Goodness Of Fit

        Criterion              DF         Value      Value/DF

        Deviance                0        0.0000          .
        Scaled Deviance         0        0.0000          .
        Pearson Chi-Square       0        0.0000          .
        Scaled Pearson X2        0        0.0000          .
        Log Likelihood                 1150.2247
```

The Type 3 analysis is displayed in Output 12.4. The type × site interaction is highly significant, and in this model, both main effects appear to be significant as well.

Output 12.4   Type 3 Analysis

```
              LR Statistics For Type 3 Analysis

                                  Chi-
              Source      DF     Square    Pr > ChiSq

              type         3      85.07      <.0001
              site         2      33.34      <.0001
              type*site    6      51.80      <.0001
```

Further investigation might involve determining whether site is best included in the model differently for Hutchinson's type than the other types, since the cell counts seem to suggest a different distribution of site for that type. However, this is not done here.

12.4   Poisson Regression for Incidence Densities

Most of the time, Poisson regression is performed when you have counts plus some
measure of exposure. The next example also concerns data on melanoma cases, but
includes information on exposure. Thus, you are interested in fitting a model to the log
rate, or incidence densities, of melanoma exposure. This involves including an offset
variable in the model.

Consider Table 12.2. The counts n_{hi} are the number of new melanoma cases reported in
1969–1971 for white males in two areas (Gail 1978 and Koch, Imrey et al. 1985). The
totals N_{hi} are the sizes of the estimated populations at risk; they may represent counts of
people or counts of exposure units. Researchers were interested in whether the rates
n_{hi}/N_{hi}, which are incidence densities, varied across age groups or region ($h = 1$ for
Northern region, $h = 2$ for Southern region; $i = 1, 2, 3, 4, 5, 6$ for ascending age groups).

Table 12.2.   New Melanoma Cases Among White Males: 1969-1971

Region	Age Group	Cases	Total
Northern	< 35	61	2880262
Northern	35–44	76	564535
Northern	45–54	98	592983
Northern	55–64	104	450740
Northern	65–74	63	270908
Northern	> 75	80	161850
Southern	< 35	64	1074246
Southern	35–44	75	220407
Southern	45–54	68	198119
Southern	55–64	63	134084
Southern	65–74	45	70708
Southern	> 75	27	34233

For this application of Poisson regression, the model of interest includes incremental
effects for age levels and region. The following DATA step inputs the melanoma data.

```
data melanoma;
   input age $ region $ cases total;
   ltotal=log(total);
   datalines;
35-44 south 75   220407
45-54 south 68   198119
55-64 south 63   134084
65-74 south 45    70708
75+   south 27    34233
<35   south 64 1074246
35-44 north 76   564535
45-54 north 98   592983
55-64 north 104 450740
65-74 north 63   270908
75+   north 80   161850
<35   north 61 2880262

;
```

The next statements invoke the PROC GENMOD procedure.

```
proc genmod data=melanoma order=data;
   class age region;
   model cases = age region
      / dist=poisson link=log offset=ltotal;
run;
```

In PROC GENMOD, the last sorted value of the CLASS variable determines the reference cell. In order for the reference cell to be those subjects from the North who are less than 35 years old, the data are entered so that those less than 35 appear last for each region, and the data for the South appear before the data for the North. Then, the ORDER=DATA option is specified in the PROC GENMOD statement.

The MODEL statement specifies that a main effects model be fit; CASES is the response variable, and AGE and REGION are the effects. The option DIST=POISSON specifies the Poisson distribution, and the option LINK=LOG specifies that the link function is the log function. The variable LTOTAL is to be treated as the offset. If you look in the preceding DATA step, you see that LTOTAL is the log of TOTAL. Thus, you are fitting a loglinear model to the ratio of cancer incidence to exposure.

Output 12.5 contains model specification information, and Output 12.6 contains information about the sort levels of the CLASS variables. This confirms that the reference level for the parameterization are those persons from the northern region who are younger than 35.

Output 12.5 Model Information

```
                 Model Information

        Data Set              WORK.MELANOMA
        Distribution                Poisson
        Link Function                   Log
        Dependent Variable            cases
        Offset Variable              ltotal
        Observations Used                12
```

Output 12.6 Class Variable Information

```
                Class Level Information

        Class      Levels     Values

        age           6        35-44 45-54 55-64 65-74 75+ <35
        region        2        south north
```

Output 12.7 contains information on assessment of fit. Since $Q_P = 6.1151$ and the deviance has the value 6.2149, each with 5 df for their approximately chi-square distributions, the fit is satisfactory.

Output 12.7 Assessment of Fit

```
            Criteria For Assessing Goodness Of Fit

    Criterion              DF         Value       Value/DF

    Deviance                5         6.2149        1.2430
    Scaled Deviance         5         6.2149        1.2430
    Pearson Chi-Square      5         6.1151        1.2230
    Scaled Pearson X2       5         6.1151        1.2230
    Log Likelihood                 2694.9262
```

Output 12.8 contains the table of estimated model parameters. The log incidence density increases over each of the age intervals and also increases for the southern region.

Output 12.8 Estimated Model Parameters

```
                    Analysis Of Parameter Estimates

                              Standard   Wald 95% Confidence    Chi-
Parameter          DF  Estimate   Error       Limits          Square   Pr > ChiSq

Intercept           1  -10.6583   0.0952  -10.8449  -10.4718  12538.4    <.0001
age       35-44     1    1.7974   0.1209    1.5604    2.0344   220.92    <.0001
age       45-54     1    1.9131   0.1184    1.6810    2.1452   260.90    <.0001
age       55-64     1    2.2418   0.1183    2.0099    2.4737   358.89    <.0001
age       65-74     1    2.3657   0.1315    2.1080    2.6235   323.56    <.0001
age       75+       1    2.9447   0.1320    2.6859    3.2035   497.30    <.0001
age       <35       0    0.0000   0.0000    0.0000    0.0000      .         .
region    south     1    0.8195   0.0710    0.6803    0.9587   133.11    <.0001
region    north     0    0.0000   0.0000    0.0000    0.0000      .         .
Scale               0    1.0000   0.0000    1.0000    1.0000

NOTE: The scale parameter was held fixed.
```

You can exponentiate these parameters to express incidence density ratios in a similar manner to exponentiating parameters in logistic regression to obtain odds ratios. For example, exponentiating the parameter estimate for the increment for ages 45–54, $e^{1.9131} = 6.774$, gives you the ratio of the incidence of melanoma for those aged 45–54 relative to those less than 35. Similarly, $e^{0.8195} = 2.269$ is the ratio of the incidence of melanoma for those from the southern region relative to those in the northern region.

12.5 Overdispersion in Lower Respiratory Infection Example

Researchers studying the incidence of lower respiratory illness in infants took repeated observations of infants over one year. They studied 284 children and examined them every two weeks. Explanatory variables evaluated included passive smoking (one or more smokers in the household), socioeconomic status, and crowding. Refer to LaVange et al. (1994) for more information on the study and a discussion of the analysis of incidence densities. One outcome of interest was the total number of times, or counts, of lower respiratory infection recorded for the year. The strategy was to model these counts with Poisson regression. However, it is reasonable to expect that the children experiencing colds are more likely to have other infections; therefore, there may be some additional variance, or overdispersion, in these data.

Overdispersion

Section 8.2.7 mentions overdispersion in the case of logistic regression. Overdispersion occurs when the observed variance is larger than the nominal variance for a particular distribution. It occurs with some regularity in the analysis of proportions and discrete counts. This is not surprising for the assumed distributions (binomial and Poisson, respectively) because the respective variances are fixed by a single parameter, the mean. When present, overdispersion can have a major impact on inference so it needs to be taken into account. Note that underdispersion also occurs. Refer to McCullagh and Nelder (1989) and Dean (1998) for more detail on overdispersion.

One way to manage the overdispersion is to assume a more flexible distribution, such as the negative binomial in the case of overdispersed Poisson data. You can also adjust the covariance matrix of a Poisson-based analysis with a scaling factor. You expect that the goodness-of-fit chi-squares have values close to their degrees of freedom with this distribution; an indication of overdispersion is when their ratio is greater than 1. One way to manage this is to allow the variance function to have a multiplicative factor, that is, the variance is assumed to be $\phi\mu$ instead of μ. The chi-square statistic value divided by its degrees of freedom is used as the scaling factor ϕ. The covariance matrix is pre-multiplied by the scaling factor, and the scaled deviance and the log likelihoods are divided by ϕ, as is the profile likelihood function used in computing the confidence limits. Note that when there are indications of overdispersion, you also have to consider other causes besides overdispersion such as outliers and a misspecified model.

The following DATA step inputs the data into a SAS data set named LRI.[*]

```
data lri;
    input id count risk passive crowding ses agegroup race @@;
    logrisk =log(risk/52);
    datalines;
 1 0 42 1 0 2 2 0  96 1 41 1 0 1 2 0   191 0 44 1 0 0 2 0
 2 0 43 1 0 0 2 0  97 1 26 1 1 2 2 0   192 0 45 0 0 0 2 1
 3 0 41 1 0 1 2 0  98 0 36 0 0 0 2 0   193 0 42 0 0 0 2 0
 4 1 36 0 1 0 2 0  99 0 34 0 0 0 2 0   194 1 31 0 0 0 2 1
 5 1 31 0 0 0 2 0 100 1  3 1 1 2 3 1   195 0 35 0 0 0 2 0
 6 0 43 1 0 0 2 0 101 0 45 1 0 0 2 0   196 1 35 1 0 0 2 0
 7 0 45 0 0 0 2 0 102 0 38 0 0 1 2 0   197 1 27 1 0 1 2 0
 8 0 42 0 0 0 2 1 103 0 41 1 1 1 2 1   198 1 33 0 0 0 2 0
 9 0 45 0 0 0 2 1 104 1 37 0 1 0 2 0   199 0 39 1 0 1 2 0
10 0 35 1 1 0 2 0 105 0 40 0 0 0 2 0   200 3 40 0 1 2 2 0
11 0 43 0 0 0 2 0 106 1 35 1 0 0 2 0   201 4 26 1 0 1 2 0
12 2 38 0 0 0 2 0 107 0 28 0 1 2 2 0   202 0 14 1 1 1 1 1
13 0 41 0 0 0 2 0 108 3 33 0 1 2 2 0   203 0 39 0 1 1 2 0
14 0 12 1 1 0 1 0 109 0 38 0 0 0 2 0   204 0  4 1 1 1 3 0
15 0  6 0 0 0 3 0 110 0 42 1 1 2 2 1   205 1 27 1 1 1 2 1
16 0 43 0 0 0 2 0 111 0 40 1 1 2 2 0   206 0 36 1 0 0 2 1
17 2 39 1 0 1 2 0 112 0 38 0 0 0 2 0   207 0 30 1 0 2 2 1
18 0 43 0 1 0 2 0 113 2 37 0 1 1 2 0   208 0 34 0 1 0 2 0
19 2 37 0 0 0 2 1 114 1 42 0 1 0 2 0   209 1 40 1 1 1 2 0
20 0 31 1 1 1 2 0 115 5 37 1 1 1 2 1   210 0  6 1 0 1 1 1
```

---

[*]Data provided by Lisa LaVange.

```
21 0 45 0 1 0 2 0   116 0 38 0 0 0 2 0   211 1 40 1 1 1 2 0
22 1 29 1 1 1 2 1   117 0  4 0 0 0 3 0   212 2 43 0 1 0 2 0
23 1 35 1 1 1 2 0   118 2 37 1 1 1 2 0   213 0 36 1 1 1 2 0
24 3 20 1 1 2 2 0   119 0 39 1 0 1 2 0   214 0 35 1 1 1 2 1
25 1 23 1 1 1 2 0   120 0 42 1 1 0 2 0   215 1 35 1 1 2 2 0
26 1 37 1 0 0 2 0   121 0 40 1 0 0 2 0   216 0 43 1 0 1 2 0
27 0 49 0 0 0 2 0   122 0 36 1 0 0 2 0   217 0 33 1 1 2 2 0
28 0 35 0 0 0 2 0   123 1 42 0 1 1 2 0   218 0 36 0 1 1 2 1
29 3 44 1 1 1 2 0   124 1 39 0 0 0 2 0   219 1 41 0 0 0 2 0
30 0 37 1 0 0 2 0   125 2 29 0 0 0 2 0   220 0 41 1 1 0 2 1
31 2 39 0 1 1 2 0   126 3 37 1 1 2 2 1   221 1 42 0 0 0 2 1
32 0 41 0 0 0 2 0   127 0 40 1 0 0 2 0   222 0 33 0 1 2 2 1
33 1 46 1 1 2 2 0   128 0 40 0 0 0 2 0   223 0 40 1 1 2 2 0
34 0  5 1 1 2 3 1   129 0 39 0 0 0 2 0   224 0 40 1 1 1 2 1
35 1 29 0 0 0 2 0   130 0 40 1 0 1 2 0   225 0 40 0 0 2 2 0
36 0 31 0 1 0 2 0   131 1 32 0 0 0 2 0   226 0 28 1 0 1 2 0
37 0 22 1 1 2 2 0   132 0 46 1 0 1 2 0   227 0 47 0 0 0 2 1
38 1 22 1 1 2 2 1   133 4 39 1 1 0 2 0   228 0 18 1 1 2 2 1
39 0 47 0 0 0 2 0   134 0 37 0 0 0 2 0   229 0 45 1 0 0 2 0
40 1 46 1 1 1 2 1   135 0 51 0 0 1 2 0   230 0 35 0 0 0 2 0
41 0 37 0 0 0 2 0   136 1 39 1 1 0 2 0   231 1 17 1 0 1 1 1
42 1 39 0 0 0 2 0   137 1 34 1 1 0 2 0   232 0 40 0 0 0 2 0
43 0 33 0 1 1 2 1   138 1 14 0 1 0 1 0   233 0 29 1 1 2 2 0
44 0 34 1 0 1 2 0   139 2 15 1 0 0 2 0   234 1 35 1 1 1 2 0
45 3 32 1 1 1 2 0   140 1 34 1 1 0 2 1   235 0 40 0 0 2 2 0
46 3 22 0 0 0 2 0   141 0 43 0 1 0 2 0   236 1 22 1 1 1 2 0
47 1  6 1 0 2 3 0   142 1 33 0 0 0 2 0   237 0 42 0 0 0 2 0
48 0 38 0 0 0 2 0   143 3 34 1 0 0 2 1   238 0 34 1 1 1 2 1
49 1 43 0 1 0 2 0   144 0 48 0 0 0 2 0   239 6 38 1 0 1 2 0
50 2 36 0 1 0 2 0   145 4 26 1 1 0 2 0   240 0 25 0 0 1 2 1
51 0 43 0 0 0 2 0   146 0 30 0 1 2 2 1   241 0 39 0 1 0 2 0
52 0 24 1 0 0 2 0   147 0 41 1 1 1 2 0   242 1 35 0 1 2 2 1
53 0 25 1 0 1 2 1   148 0 34 0 1 1 2 0   243 1 36 1 1 1 2 1
54 0 41 0 0 0 2 0   149 0 43 0 1 0 2 0   244 0 23 1 0 0 2 0
55 0 43 0 0 0 2 0   150 1 31 1 0 1 2 0   245 4 30 1 1 1 2 0
56 2 31 0 1 1 2 0   151 0 26 1 0 1 2 0   246 1 41 1 1 1 2 1
57 3 28 1 1 1 2 0   152 0 37 0 0 0 2 0   247 0 37 0 1 1 2 0
58 1 22 0 0 1 2 1   153 0 44 0 0 0 2 0   248 0 46 1 1 0 2 0
59 1 11 1 1 1 1 0   154 0 40 1 0 0 2 0   249 0 45 1 1 0 2 1
60 3 41 0 1 1 2 0   155 0  8 1 1 1 3 1   250 1 38 1 1 1 2 0
61 0 31 0 0 1 2 0   156 0 40 1 1 1 2 1   251 0 10 1 1 1 1 0
62 0 11 0 0 1 1 1   157 1 45 0 0 0 2 0   252 0 30 1 1 2 2 0
63 0 44 0 1 0 2 0   158 0  4 0 0 2 3 0   253 0 32 0 1 2 2 0
64 0  9 1 0 0 3 1   159 1 36 0 1 0 2 0   254 0 46 1 0 0 2 0
65 0 36 1 1 1 2 0   160 3 37 1 1 1 2 0   255 5 35 1 1 2 2 1
66 0 29 1 0 0 2 0   161 0 15 1 0 0 1 0   256 0 44 0 0 0 2 0
67 0 27 0 1 0 2 1   162 1 27 1 0 1 2 1   257 0 41 0 1 1 2 0
68 0 36 0 1 0 2 0   163 2 31 0 1 0 2 0   258 2 36 1 0 1 2 0
69 1 33 1 0 0 2 0   164 0 42 0 0 0 2 0   259 0 34 1 1 1 2 1
70 2 13 1 1 2 1 1   165 0 42 1 0 0 2 0   260 1 30 0 1 0 2 1
71 0 38 0 0 0 2 0   166 1 38 0 0 0 2 0   261 1 27 1 0 0 2 0
72 0 41 0 0 0 2 1   167 0 44 1 0 0 2 0   262 0 48 1 0 0 2 0
73 0 41 1 0 2 2 0   168 0 45 0 0 0 2 0   263 1  6 0 1 2 3 1
74 0 35 0 0 1 2 0   169 0 34 0 1 0 2 0   264 0 38 1 1 0 2 1
```

```
75  0 45 0 0 0 2 0   170 2 41 0 0 0 2 0   265 0 29 1 1 1 2 1
76  4 38 1 0 2 2 1   171 2 30 1 1 1 2 0   266 1 43 0 1 2 2 1
77  1 42 1 0 0 2 1   172 0 44 0 0 0 2 0   267 0 43 0 1 0 2 0
78  1 42 1 1 2 2 1   173 0 40 1 0 0 2 0   268 0 37 1 0 2 2 0
79  6 36 1 1 0 2 0   174 2 31 0 0 0 2 0   269 1 23 1 1 0 2 1
80  2 23 1 1 1 2 1   175 0 41 1 0 0 2 0   270 0 44 0 0 1 2 0
81  1 32 0 0 1 2 0   176 0 41 0 0 0 2 0   271 0  5 0 1 1 3 1
82  0 41 0 1 0 2 0   177 0 39 1 0 0 2 0   272 0 25 1 0 2 2 0
83  0 50 0 0 0 2 0   178 0 40 1 0 0 2 0   273 0 25 1 0 1 2 0
84  0 42 1 1 1 2 1   179 2 35 1 0 2 2 0   274 1 28 1 1 1 2 1
85  1 30 0 0 0 2 0   180 1 43 1 0 0 2 0   275 0  7 0 1 0 3 1
86  2 47 0 1 0 2 0   181 2 39 0 0 0 2 0   276 0 32 0 0 0 2 0
87  1 35 1 1 2 2 0   182 0 35 1 1 0 2 0   277 0 41 0 0 0 2 0
88  1 38 1 0 1 2 1   183 0 37 0 0 0 2 0   278 1 33 1 1 2 2 1
89  1 38 1 1 1 2 1   184 3 37 0 0 0 2 0   279 2 36 1 1 2 2 0
90  1 38 1 1 1 2 1   185 0 43 0 0 0 2 0   280 0 31 0 0 0 2 0
91  0 32 1 1 1 2 0   186 0 42 0 0 0 2 0   281 0 18 0 0 0 2 0
92  1  3 1 0 1 3 1   187 0 42 0 0 0 2 0   282 1 32 1 0 2 2 0
93  0 26 1 0 0 2 1   188 0 38 0 0 0 2 0   283 0 22 1 1 2 2 1
94  0 35 1 0 0 2 0   189 0 36 1 0 0 2 0   284 0 35 0 0 0 2 1
95  3 37 1 0 0 2 0   190 0 39 0 1 0 2 0
;

proc genmod data=lri;
   class ses id race agegroup;
   model count = passive crowding ses race agegroup /
      dist=poisson offset=logrisk type3;
run;
```

Output 12.9 contains the general model information.

Output 12.9 Model Information

```
                     Model Information

           Data Set               WORK.LRI
           Distribution            Poisson
           Link Function               Log
           Dependent Variable        count
           Offset Variable          logrisk
           Observations Used           284
```

Output 12.10 contains the goodness-of-fit statistics, along with the ratios of their values to their degrees of freedom. With values of 1.4788 for the Deviance/df and 1.7951 for Pearson/df, there is evidence of overdispersion. The model-based estimates of standard errors may not be appropriate and therefore any inference is questionable. (When such ratios are close to 1, you conclude that little evidence of over- or underdispersion exists).

Output 12.10   Fit Statistics

```
                Criteria For Assessing Goodness Of Fit

        Criterion            DF          Value        Value/DF

        Deviance             276        408.1549        1.4788
        Scaled Deviance      276        408.1549        1.4788
        Pearson Chi-Square   276        495.4494        1.7951
        Scaled Pearson X2    276        495.4494        1.7951
        Log Likelihood                 -260.4117
```

The model is refit with a scaling factor specified to adjust for the overdispersion. This is requested with the SCALE=PEARSON option, which computes a scaling factor that is the Pearson Q statistic divided by its degrees of freedom.

```
proc genmod data=lri;
   class ses id race agegroup;
   model count = passive crowding ses race agegroup /
      dist=poisson offset=logrisk type3 scale=pearson;
run;
```

Output 12.11 displays the goodness-of-fit statistics. Note that the scaled deviance and the scaled Pearson chi-square have different values because they have been divided by the scaling factor. The scaled Pearson chi-square is now 1 because the scaling factor requested was the Pearson chi-square value divided by the df.

Output 12.11   Assessment of Fit

```
                Criteria For Assessing Goodness Of Fit

        Criterion            DF          Value        Value/DF

        Deviance             276        408.1549        1.4788
        Scaled Deviance      276        227.3708        0.8238
        Pearson Chi-Square   276        495.4494        1.7951
        Scaled Pearson X2    276        276.0000        1.0000
        Log Likelihood                 -145.0676
```

Output 12.12 contains the results of the Type 3 analysis. The chi-square statistics have different values than in the previous analysis because of the scaling adjustment. Note that this table also includes F statistics; the chi-square approximation to the likelihood ratio test may have a less clear basis in this situation. Refer to *SAS/STAT User's Guide, Version 8* for more detail about their computation.

Output 12.12 Type 3 Analysis

```
                  LR Statistics For Type 3 Analysis

                                                   Chi-
  Source       Num DF   Den DF   F Value   Pr > F   Square   Pr > ChiSq

  passive        1       276      3.89     0.0494    3.89      0.0484
  crowding       1       276      5.86     0.0162    5.86      0.0155
  ses            2       276      1.22     0.2966    2.44      0.2950
  race           1       276      0.38     0.5408    0.38      0.5403
  agegroup       2       276      1.07     0.3443    2.14      0.3429
```

Both passive smoking and crowding are strongly significant. Social economic status and race do not appear to be influential, and neither does age group.

Finally, Output 12.13 contains the parameter estimates. The standard errors are adjusted due to the scaling factor, and they are larger than the standard errors for the unadjusted model, which are displayed in Output 12.14.

Output 12.13 Estimated Model Parameters

```
                    Analysis Of Parameter Estimates

                             Standard   Wald 95% Confidence    Chi-
  Parameter       DF  Estimate  Error        Limits          Square  Pr > ChiSq

  Intercept        1   0.6047   0.7304   -0.8269    2.0362     0.69    0.4077
  passive          1   0.4310   0.2214   -0.0029    0.8649     3.79    0.0515
  crowding         1   0.5199   0.2166    0.0953    0.9444     5.76    0.0164
  ses        0     1  -0.3970   0.2886   -0.9627    0.1687     1.89    0.1690
  ses        1     1  -0.0681   0.2627   -0.5830    0.4469     0.07    0.7956
  ses        2     0   0.0000   0.0000    0.0000    0.0000      .       .
  race       0     1   0.1402   0.2309   -0.3123    0.5928     0.37    0.5436
  race       1     0   0.0000   0.0000    0.0000    0.0000      .       .
  agegroup   1     1  -0.4792   0.9043   -2.2516    1.2931     0.28    0.5962
  agegroup   2     1  -0.9919   0.6858   -2.3361    0.3522     2.09    0.1481
  agegroup   3     0   0.0000   0.0000    0.0000    0.0000      .       .
  Scale            0   1.3398   0.0000    1.3398    1.3398

NOTE: The scale parameter was estimated by the square root of Pearson's Chi-Square/DOF.
```

Output 12.14 Estimated Model Parameters for Unadjusted Model

```
                    Analysis Of Parameter Estimates

                             Standard   Wald 95% Confidence    Chi-
  Parameter       DF  Estimate  Error        Limits          Square  Pr > ChiSq

  Intercept        1   0.6047   0.5452   -0.4638    1.6732     1.23    0.2673
  passive          1   0.4310   0.1652    0.1072    0.7548     6.81    0.0091
  crowding         1   0.5199   0.1617    0.2030    0.8367    10.34    0.0013
  ses        0     1  -0.3970   0.2154   -0.8192    0.0252     3.40    0.0653
  ses        1     1  -0.0681   0.1961   -0.4524    0.3163     0.12    0.7285
  ses        2     0   0.0000   0.0000    0.0000    0.0000      .       .
  race       0     1   0.1402   0.1723   -0.1975    0.4780     0.66    0.4158
  race       1     0   0.0000   0.0000    0.0000    0.0000      .       .
  agegroup   1     1  -0.4792   0.6749   -1.8020    0.8436     0.50    0.4777
  agegroup   2     1  -0.9919   0.5119   -1.9951    0.0113     3.76    0.0526
  agegroup   3     0   0.0000   0.0000    0.0000    0.0000      .       .
  Scale            0   1.0000   0.0000    1.0000    1.0000

NOTE: The scale parameter was held fixed.
```

See Section 15.14 in Chapter 15 for another method to adjust for overdispersion in these data.

Chapter 13
Weighted Least Squares

Chapter Table of Contents

Chapter 13
Weighted Least Squares

13.1   Introduction

Previous chapters discussed statistical modeling of categorical data with logistic regression. Maximum likelihood estimation (ML) was used to estimate parameters for models based on logits and cumulative logits. Logistic regression is suitable for many situations, particularly for dichotomous response outcomes. However, there are situations where modeling techniques other than logistic regression are of interest. You may be interested in modeling functions besides logits, such as mean scores, proportions, or more complicated functions of the responses. In addition, the analysis framework may dictate a different modeling approach, such as in the case of repeated measurements studies.

Weighted least squares (WLS) estimation provides a methodology for modeling a wide range of categorical data outcomes. This chapter focuses on the application of weighted least squares for the modeling of mean scores and proportions in the stratified simple random sampling framework, as well as for the modeling of estimates produced by more complex sampling mechanisms, such as those required for complex sample surveys. The methodology is explained in the context of a basic example.

The CATMOD procedure is a general procedure for modeling categorical data. It performs logistic regression analysis using maximum likelihood estimation when the response functions are generalized logits, and it performs weighted least squares estimation for a variety of other response functions. This chapter discusses the use of PROC CATMOD for numerous applications of weighted least squares analyses. Chapter 14, "Modeling Repeated Measurements Data with WLS," discusses the use of weighted least squares for the advanced topic of repeated measurements analysis.

You should be familiar with the material in Chapter 8, "Logistic Regression I: Dichotomous Response," and Chapter 9, "Logistic Regression II: Polytomous Response," before proceeding with this chapter.

13.2   Weighted Least Squares Methodology

To motivate the discussion of weighted least squares methodology, consider the following example. Epidemiologists investigating air pollution effects conducted a study of childhood respiratory disease (Stokes 1986). Investigators visited groups of children two times and recorded whether they were exhibiting symptoms of colds. The children were recorded as having no periods with a cold, one period with a cold, or two periods with a cold. Investigators were interested in determining whether sex or residence affected the distribution of colds. These data are displayed in Table 13.1.

Table 13.1. Colds in Children

Sex	Residence	Periods with Colds			Total
		0	1	2	
Female	Rural	45	64	71	180
Female	Urban	80	104	116	300
Male	Rural	84	124	82	290
Male	Urban	106	117	87	310

As previously discussed, statistical modeling addresses the question of how a response outcome is distributed across the various levels of the explanatory variables. In the standard linear model, this is done by fitting a model to the response mean. In logistic regression, the function modeled is the logit or cumulative logit. For these data, a response measure of interest is the mean number of periods with colds. However, because there are a small, discrete number of response values, it is unlikely that the normality assumptions usually required for the standard linear model are met. However, weighted least squares methodology provides a useful strategy for analyzing these data.

13.2.1   Weighted Least Squares Framework

Underlying most types of weighted least squares methods for categorical data analysis is a contingency table. The general idea is to model the distribution of the response variable, represented in the columns of the table, across the levels of the explanatory variables, represented by the rows of the table. These rows are determined by the cross-classification of the levels, or values, of the explanatory variables. The contingency table for the colds data has four rows and three columns. There are four rows since there are four combinations of sex and residence; there are three columns because the response variable has three possible outcomes: 0, 1, and 2.

The general contingency table is displayed in Table 13.2, where s represents the number of rows, or groups, in the table and r represents the number of responses. The rows of the table are also referred to as subpopulations.

Table 13.2. Underlying Contingency Table

Group	Response				Total
	1	2	$\cdots$	r	
1	n_{11}	n_{12}	$\cdots$	n_{1r}	n_{1+}
2	n_{21}	n_{22}	$\cdots$	n_{2r}	n_{2+}
$\cdots$	$\cdots$	$\cdots$	$\cdots$	$\cdots$	$\cdots$
s	n_{s1}	n_{s2}	$\cdots$	n_{sr}	n_{s+}

The proportion of subjects in each group who have each response is written

$$p_{ij} = n_{ij}/n_{i+}$$

where n_{ij} is the number of subjects in the ith group who have the jth response. For example, $p_{11} = 45/180$ in Table 13.1. You can put the proportions for one group together in a proportion vector that describes the response distribution for that group. For the colds data, it looks like the following:

$$\mathbf{p}_i = (p_{i1}, p_{i2}, p_{i3})'$$

You can then form a proportion vector for each group in the contingency table. The proportions for each group add up to 1. All the functions that can be modeled with weighted least squares methodology are generated from these proportion vectors.

The rows of the contingency table are considered to be simple random samples from the multinomial distribution; since the rows are independent, the entire table is distributed as product multinomial. You can write the covariance matrix for the proportions in the ith row as

$$\mathbf{V}_i = \frac{1}{n_{i+}} \begin{bmatrix} p_{i1}(1 - p_{i1}) & -p_{i1}p_{i2} & \cdots & -p_{i1}p_{ir} \\ -p_{i2}p_{i1} & p_{i2}(1 - p_{i2}) & \cdots & -p_{i2}p_{ir} \\ \vdots & \vdots & \vdots & \vdots \\ -p_{ir}p_{i1} & -p_{ir}p_{i2} & \cdots & p_{ir}(1 - p_{ir}) \end{bmatrix}$$

and then write the covariance matrix for the entire table as

$$\mathbf{V_p} = \begin{bmatrix} \mathbf{V}_1 & \mathbf{0} & \cdots & \mathbf{0} \\ \mathbf{0} & \mathbf{V}_2 & \cdots & \mathbf{0} \\ \vdots & \vdots & \vdots & \vdots \\ \mathbf{0} & \mathbf{0} & \cdots & \mathbf{V}_s \end{bmatrix}$$

where $\mathbf{V}_i$ is the covariance matrix for the ith row.

13.2.2 Weighted Least Squares Estimation

Once the proportion vector and covariance matrix are computed, the modeling phase begins with the choice of a response function. You can model the proportions themselves; mean scores, which are simple linear functions of the proportions; logits, which are constructed by taking a linear function (difference) of the log proportions; and a number of more complicated functions that are created by combinations of various transformations of the proportions, such as the kappa statistic for observer agreement (refer to Landis and Koch 1977) or rank measures of association (refer to Koch and Edwards 1988).

For the colds data, the response function is the mean number of periods with a cold. You construct these means from the proportions of responses in each row of Table 13.1 and then apply a statistical model that determines the effect of sex and residence on their distribution. Table 13.3 displays the row proportions.

Table 13.3.   Colds in Children

Sex	Residence	Periods with Colds 0	1	2	Total
Female	Rural	0.25	0.36	0.39	1.00
Female	Urban	0.27	0.35	0.39	1.00
Male	Rural	0.29	0.43	0.28	1.00
Male	Urban	0.34	0.38	0.28	1.00

For example, to compute the mean number of periods of colds for females in a rural residence, you would perform the following computation.

$$\text{mean colds} = 0 * p_{11} + 1 * p_{12} + 2 * p_{13}$$
$$= 0 * (0.25) + 1 * (0.36) + 2 * (0.39)$$
$$= 1.14$$

In matrix terms, you have multiplied the proportion vector by a linear transformation matrix $\mathbf{A}$.

$$\mathbf{A}\mathbf{p}_1 = \begin{bmatrix} 0 & 1 & 2 \end{bmatrix} \begin{bmatrix} 0.25 \\ 0.36 \\ 0.39 \end{bmatrix} = 1.14$$

Means are generated for each sex $\times$ residence group to produce a total of four functions for the table. The ith function is denoted $F(\mathbf{p}_i)$. The following expression shows how you generate a function vector by applying a linear transformation matrix to the total proportion vector $\mathbf{p} = (\mathbf{p}_1', \mathbf{p}_2', \mathbf{p}_3', \mathbf{p}_4')'$ to produce the four means of interest.

$$F(\mathbf{p}) = \mathbf{A}\mathbf{p} = \begin{bmatrix} 0 & 1 & 2 & 0 & 0 & 0 & 0 & 0 & 0 & 0 & 0 & 0 \\ 0 & 0 & 0 & 0 & 1 & 2 & 0 & 0 & 0 & 0 & 0 & 0 \\ 0 & 0 & 0 & 0 & 0 & 0 & 0 & 1 & 2 & 0 & 0 & 0 \\ 0 & 0 & 0 & 0 & 0 & 0 & 0 & 0 & 0 & 0 & 1 & 2 \end{bmatrix} \mathbf{p} = \begin{bmatrix} 1.14 \\ 1.12 \\ 0.99 \\ 0.94 \end{bmatrix}$$

If the groups have sufficient sample size, usually $n_{i+} \geq 25$, then the variation among the response functions can be investigated by fitting linear regression models with weighted least squares.

$$E_A\{\mathbf{F}(\mathbf{p})\} = \mathbf{F}(\boldsymbol{\pi}) = \mathbf{X}\boldsymbol{\beta}$$

E_A denotes asymptotic expectation, and $\boldsymbol{\pi} = E\{\mathbf{p}\}$ denotes the vector of population probabilities for all the populations together. The vector $\boldsymbol{\beta}$ contains the parameters that describe the variation among the response functions, and $\mathbf{X}$ is the model specification matrix. The equations for WLS estimation are similar to those for least squares estimation.

$$\mathbf{b} = (\mathbf{X}'\mathbf{V}_{\mathbf{F}}^{-1}\mathbf{X})^{-1}\mathbf{X}'\mathbf{V}_{\mathbf{F}}^{-1}\mathbf{F}$$

$\mathbf{V}_{\mathbf{F}}$ is the covariance matrix for the vector of response functions and is usually nonsingular when the sample sizes n_{i+} are sufficiently large (for example, $n_{i+} \geq 25$ and at least two $n_{ij} \geq 1$ in each row). This is the weight matrix component of weighted least squares estimation. Its form depends on the nature of the response functions. In the case of the colds data, where the response functions are means computed as $\mathbf{Ap}$, the covariance matrix is computed as

$$\mathbf{V}_{\mathbf{F}} = \mathbf{A}\mathbf{V}_{\mathbf{p}}\mathbf{A}'$$

The covariance matrix for $\mathbf{b}$ is written

$$V(\mathbf{b}) = (\mathbf{X}'\mathbf{V}_{\mathbf{F}}^{-1}\mathbf{X})^{-1}$$

Model adequacy is assessed with Wald goodness-of-fit statistics. They are computed as

$$Q_W = (\mathbf{F} - \mathbf{X}\mathbf{b})'\mathbf{V}_{\mathbf{F}}^{-1}(\mathbf{F} - \mathbf{X}\mathbf{b})$$

Q_W is distributed as chi-square for moderately large sample sizes (for example, all $n_{i+} \geq 25$), and its degrees of freedom are equal to the difference between the number of rows of $F(\mathbf{p})$ and the number of parameters. If only one response function is created per row of the contingency table, then this is the number of table rows minus the number of estimated parameters.

You can address questions about the parameters with the use of hypothesis tests. Each hypothesis is written in the form

$$H_0: \mathbf{C}\beta = \mathbf{0}$$

and can investigate whether specified linear combinations of the parameters are equal to zero. The test statistic employed is a Wald statistic that is expressed as

$$Q_C = (\mathbf{C}\mathbf{b})'[\mathbf{C}(\mathbf{X}'\mathbf{V}_{\mathbf{F}}^{-1}\mathbf{X})^{-1}\mathbf{C}']^{-1}(\mathbf{C}\mathbf{b})$$

Q_C is distributed as chi-square with degrees of freedom equal to the number of linearly independent rows in $\mathbf{C}$.

You can also generate predicted values $\hat{\mathbf{F}} = \mathbf{X}\mathbf{b}$ of the response functions and their covariance matrix $\mathbf{V}_{\hat{\mathbf{F}}} = \mathbf{X}V(\mathbf{b})\mathbf{X}'$. See Appendix A in this chapter for more statistical theory concerning weighted least squares estimation.

13.2.3   Model Parameterization

The preliminary model of interest for a WLS analysis is often the *saturated* model, in which all the variation is explained by the parameters. In a saturated model, there are as many parameters in the model as there are response functions. For these data, the saturated model is written

$$
\begin{bmatrix} F(\mathbf{p}_1) \\ F(\mathbf{p}_2) \\ F(\mathbf{p}_3) \\ F(\mathbf{p}_4) \end{bmatrix} = \begin{bmatrix} \alpha + \beta_1 + \beta_2 + \beta_3 \\ \alpha + \beta_1 - \beta_2 - \beta_3 \\ \alpha - \beta_1 + \beta_2 - \beta_3 \\ \alpha - \beta_1 - \beta_2 + \beta_3 \end{bmatrix} = \begin{bmatrix} 1 & 1 & 1 & 1 \\ 1 & 1 & -1 & -1 \\ 1 & -1 & 1 & -1 \\ 1 & -1 & -1 & 1 \end{bmatrix} \begin{bmatrix} \alpha \\ \beta_1 \\ \beta_2 \\ \beta_3 \end{bmatrix}
$$

Here, α is a centered intercept, β_1 is the differential effect for sex, β_2 is the differential effect for residence, and β_3 represents their interaction. The intercept is the mean number of colds averaged over all the groups. The differential effects represent average deviations from the mean; β_1 is the amount you need to add to the average of the mean periods with colds to compute the mean number of colds for females (averaged over residence); it is also the amount you need to subtract from the average of the mean periods with colds to compute the mean number of colds for males (averaged over residence).

As discussed in Section 8.8, this type of parameterization is the default for the CATMOD procedure and is called deviation from the mean parameterization; it is a *full rank* parameterization. This imposes restrictions on the parameters, unlike the GLM procedure, which uses an overparameterized model that does not place restrictions on the parameters. In PROC CATMOD, if an effect such as sex or residence has s levels, then it is represented by $s - 1$ parameters. The same effect would be represented by s parameters in PROC GLM. To understand the restrictions imposed by PROC CATMOD, consider the sex effect for the colds data and consider a model that contains only the intercept and the sex effect. You could write such a model as

$$
E\{F(\mathbf{p}_i)\} = \alpha + \tau_i
$$

where α represents the overall mean, and τ_i represents the *i*th level of the main effect.

$$
E\{F(\mathbf{p}_1)\} = \alpha + \tau_1, \quad E\{F(\mathbf{p}_2)\} = \alpha + \tau_1
$$

and

$$
E\{F(\mathbf{p}_3)\} = \alpha + \tau_2, \quad E\{F(\mathbf{p}_4)\} = \alpha + \tau_2
$$

In matrix terms, this model would be written

$$
\begin{bmatrix} F(\mathbf{p}_1) \\ F(\mathbf{p}_2) \\ F(\mathbf{p}_3) \\ F(\mathbf{p}_4) \end{bmatrix} = \begin{bmatrix} \alpha + \tau_1 \\ \alpha + \tau_1 \\ \alpha + \tau_2 \\ \alpha + \tau_2 \end{bmatrix} = \begin{bmatrix} 1 & 1 & 0 \\ 1 & 1 & 0 \\ 1 & 0 & 1 \\ 1 & 0 & 1 \end{bmatrix} \begin{bmatrix} \alpha \\ \tau_1 \\ \tau_2 \end{bmatrix}
$$

If you add these equations, you obtain

$$E\left\{\sum_{i=1}^{4} F(\mathbf{p}_i)\right\} = 4\alpha + 2(\tau_1 + \tau_2)$$

or

$$E\{\bar{F}\} = \alpha + (\tau_1 + \tau_2)/2$$

Since

$$E\{\bar{F}\} = \alpha$$

and α is the overall mean, there is an implied restriction that

$$\tau_1 + \tau_2 = 0$$

or that $\tau_1 = -\tau_2$. Thus, τ_2 would be redundant in the model since it is a linear combination of other model parameters, and it can be eliminated. If you have an effect with two levels, it is represented in PROC CATMOD with one parameter. Similarly, if you have an effect that has s levels, then that effect is represented with $s - 1$ parameters. Understanding the parameterization is important in understanding what the model coefficients represent, how the degrees of freedom are determined, and how to construct contrast tests.

13.3 Using PROC CATMOD for Weighted Least Squares Analysis

Since the CATMOD procedure is very general, it offers great flexibility in its input. Standard uses that take advantage of defaults may require no more than three or four statements. More statements are required if you take advantage of the facilities for repeated measurements analysis or loglinear model analysis. And the input can be quite rich if you choose to create your own response functions through the specification of the appropriate matrix operations or create your own parameterization by directly inputting your model matrix.

The analysis for the colds data requires minimal input. You need to specify the input data set, the WEIGHT variable if the data are in count form, the response function, and the desired model in a MODEL statement. The MODEL statement is the only required statement for PROC CATMOD.

First, a SAS data set is created for the colds data.

```
data colds;
   input sex $ residence $ periods count @@;
   datalines;
female rural 0 45 female rural 1 64  female rural 2 71
female urban 0 80 female urban 1 104 female urban 2 116
```

```
male rural    0 84 male    rural 1 124 male    rural 2 82
male urban    0 106 male   urban 1 117 male    urban 2 87
;
run;
```

The following set of SAS statements request that a weighted least squares analysis be performed for the mean response, using the saturated model.

```
proc catmod;
   weight count;
   response means;
   model periods = sex residence sex*residence /freq prob;
run;
```

The WEIGHT statement works the same as it does for the FREQ procedure; the WEIGHT variable contains the count of observations that have the values listed in the data line. As with PROC FREQ, you can supply input data in raw form, one observation per data line, or in count form. The RESPONSE statement specifies the response functions. If you leave out this statement, PROC CATMOD models generalized logits with maximum likelihood estimation. Specifying the MEANS keyword requests that mean response functions be constructed for each subpopulation; the default estimation method for functions other than generalized logits is weighted least squares.

The MODEL statement requests that PROC CATMOD fit a model that includes main effects for sex and residence as well as their interaction. The effects specification is similar to that used in the GLM procedure. The effects for sex and residence each have 1 df, and their interaction also has 1 df. Since the model also includes an intercept by default, this model is saturated. There are four parameters for the four response functions.

PROC CATMOD uses the explanatory variables listed in the right-hand side of the MODEL statement to determine the rows of the underlying contingency table. Since the variable SEX has two levels and the variable RESIDENCE has two levels, PROC CATMOD forms a contingency table that has four rows. The columns of the underlying contingency table are determined by the number of values for the response variable on the left-hand side of the MODEL statement. Since there can be 0, 1, or 2 periods with colds, there are three columns in this table.

The FREQ and PROB options in the MODEL statement cause the frequencies and proportions from the underlying contingency table to be printed.

Output 13.1 displays the population and response profiles, which represent the rows and columns of the underlying table, respectively. Output 13.2 displays the underlying frequency table and the corresponding table of proportions. PROC CATMOD labels each group or subpopulation "Sample n"; you often need to refer back to the "Population Profiles" table to interpret other parts of the PROC CATMOD output. You should always check the population and response profiles to ensure that you have defined the underlying frequency table as you intended.

Output 13.1   Population and Response Profiles

```
                      Population Profiles

        Sample    sex        residence     Sample Size
        ------------------------------------------------
           1      female     rural               180
           2      female     urban               300
           3      male       rural               290
           4      male       urban               310


                      Response Profiles

                  Response     periods
                  ------------------------
                     1            0
                     2            1
                     3            2
```

Output 13.2   Table Frequencies and Proportions

```
                     Response Frequencies

                         Response Number
          Sample      1          2          3
          -----------------------------------------
             1        45         64         71
             2        80        104        116
             3        84        124         82
             4       106        117         87


                    Response Probabilities

                          Response Number
          Sample      1          2           3
          ------------------------------------------
             1      0.25000    0.35556    0.39444
             2      0.26667    0.34667    0.38667
             3      0.28966    0.42759    0.28276
             4      0.34194    0.37742    0.28065
```

PROC CATMOD output includes a table of response function values and the model matrix, labeled "Design Matrix" in Output 13.3. The response functions are the mean number of periods with colds for each of the populations.

Output 13.3   Observed Response Functions and Model Matrix

```
               Response Functions and Design Matrix

                   Response          Design Matrix
          Sample   Function      1      2      3      4
          ---------------------------------------------------
             1     1.14444       1      1      1      1
             2     1.12000       1      1     -1     -1
             3     0.99310       1     -1      1     -1
             4     0.93871       1     -1     -1      1
```

Model-fitting results are displayed in Output 13.4 in a table labeled "Analysis of Variance" for its similarity in function to an ANOVA table.

Output 13.4   ANOVA Table

```
                      Analysis of Variance

          Source           DF   Chi-Square   Pr > ChiSq
          ----------------------------------------------
          Intercept         1     1841.13      <.0001
          sex               1       11.57      0.0007
          residence         1        0.65      0.4202
          sex*residence     1        0.09      0.7594

          Residual          0         .           .
```

The effects listed in the right-hand side of the MODEL statement are listed under "Source." Unless otherwise specified, an intercept is included in the model. If there is one response function per subpopulation, the intercept has 1 df. The statistics printed under "Chi-Square" are Wald statistics. Also provided are the degrees of freedom for each effect and corresponding *p*-value.

The last row contains information labeled "Residual." Normally, this line contains a chi-square value that serves as a goodness-of-fit test for the specified model. However, in this case, the model uses four parameters to fit four response functions. The fit must necessarily be perfect, and thus the model explains all the variation among the response functions. The degrees of freedom are zero since the degrees of freedom for Q_W are equal to the difference in the number of response functions and the number of parameters. The SAS System prints out missing values under "Chi-Square" and "Prob" for zero degrees of freedom.

Since the model fits, it is appropriate to examine the chi-square statistics for the individual effects. With a chi-square value of 0.09 and $p = 0.7594$, the SEX*RESIDENCE interaction is clearly nonsignificant. SEX appears to be a strong effect and RESIDENCE a negligible effect, but these are better assessed in the context of the main effects model that remains after the interaction term is deleted, since the estimation of these main effects is better in the absence of the interaction.

The following statements request the main effects model and produce the analysis of variance table displayed in Output 13.5.

```
proc catmod;
   weight count;
   response means;
   model periods = sex residence;
run;
```

Output 13.5   Preliminary Colds Output

```
                   Analysis of Variance

        Source       DF   Chi-Square   Pr > ChiSq
        ------------------------------------------
        Intercept     1     1882.77       <.0001
        sex           1       12.08       0.0005
        residence     1        0.76       0.3839

        Residual      1        0.09       0.7594
```

Look at the goodness-of-fit statistic. $Q_W = 0.09$ with 1 df and $p = 0.7594$. The main effects model adequately fits the data. The smaller the goodness-of-fit chi-square value, and correspondingly the larger the p value, the better the fit. This is different from the model F statistic in the usual linear model setting, where the F value is high for a model that fits the data well in the sense of explaining a large amount of the variation. Strictly speaking, using the usual significance level of $\alpha = 0.05$, any p-value greater than 0.05 supports an adequate model fit. However, many analysts are more comfortable with goodness-of-fit p-values that are greater than 0.15.

The effect for sex is highly significant, $p < 0.001$. However, the effect for residence remains nonsignificant when the interaction is removed from the model, $p = 0.3839$. These results suggest that a model with a single main effect for SEX is appropriate.

Consider the following statements to perform this task. The MODEL statement contains the response variable PERIODS and a single explanatory variable, SEX. This should produce the desired model. However, recall that the variables listed in the right-hand side of the MODEL statement are also used to determine the underlying contingency table structure. This table has its rows determined by both SEX and RESIDENCE. If RESIDENCE is *not* included in the MODEL statement, as shown in the following statements, then PROC CATMOD would create two groups based on SEX instead of four groups based on SEX and RESIDENCE.

```
proc catmod;
   weight count;
   response means;
   model periods = sex;
run;
```

However, you need to maintain the sampling structure of the underlying table. The solution is the addition of the POPULATION statement. When a POPULATION statement is included, the variables listed in it determine the populations, not the variables listed in the MODEL statement. So, you can let the right-hand variables on the MODEL statement determine the populations so long as all the necessary variables are included; if not, you need to use a POPULATION statement. Some analysts use the POPULATION statement for all PROC CATMOD invocations as a precautionary measure.

The following code requests the single main effect model.

```
proc catmod;
   population sex residence;
   weight count;
   response means;
   model periods = sex;
run;
```

The table of population profiles for the invocation using the POPULATION statement is identical to those produced by previous invocations without it, but including both SEX and RESIDENCE as explanatory variables.

Output 13.6   POPULATION Statement Results

```
                        Population Profiles

        Sample     sex        residence    Sample Size
        ---------------------------------------------
           1       female     rural               180
           2       female     urban               300
           3       male       rural               290
           4       male       urban               310
```

The analysis of variance table now includes only one main effect, SEX. The residual goodness-of-fit $Q_W = 0.85$, with 2 df and $p = 0.6531$, indicating an adequate fit.

Output 13.7   Single Main Effect ANOVA Table

```
                    Analysis of Variance

        Source       DF    Chi-Square    Pr > ChiSq
        ---------------------------------------------
        Intercept     1      1899.55       <.0001
        sex           1        11.53       0.0007

        Residual      2         0.85       0.6531
```

Compare this analysis of variance table with that displayed in Output 13.5.

Note that Q_W for the reduced model (0.85) is the sum of Q_W for the two effects model ($Q_W = 0.09$) plus the value of the Wald statistic for the effect for residence (0.76). This is a property of weighted least squares. When you delete a term from a model, the residual chi-square for the goodness of fit for the new model is equal to the old model's residual chi-square value plus the chi-square value for the particular effect. This is also true for maximum likelihood estimation when likelihood ratio tests are used for goodness of fit and for particular effects, but not when the Wald statistic is used with maximum likelihood estimation. Similarly, note that the $Q_W = 0.09$ for the two main effects model of Output 13.5 is equal to the chi-square for the interaction term in the saturated model (Output 13.4).

When an effect is deleted, any variation attributed to that effect is put into the residual variation, which is the variation that the model does not explain; this variation is essentially random for well-fitting models. If the residual variation is low, the residual chi-square will be small, indicating that the model explains the variation in the response fairly well. If the residual variation is high, the residual chi-square will be large, with a correspondingly low *p*-value, indicating that the residual variation is significantly different from zero. The implication is that the model lacks necessary terms.

Finally, note that the degrees of freedom for the goodness of fit for the reduced model are increased by the number of degrees of freedom for the deleted effect, in this case from 1 to 2, since residence had one degree of freedom.

PROC CATMOD also prints out a table containing the parameter estimates. Since the model fits, it is appropriate to examine this table, displayed in Output 13.8.

Output 13.8   Single Main Effect Model

```
                  Analysis of Weighted Least Squares Estimates

                                            Standard       Chi-
         Effect       Parameter   Estimate    Error       Square    Pr > ChiSq
         ------------------------------------------------------------------------
         Intercept        1         1.0477    0.0240      1899.55     <.0001
         sex              2         0.0816    0.0240        11.53     0.0007
```

Listed under "Effect" are the parameters estimated for the model. Since sex is represented by one parameter, only one estimate is listed. Since females are listed first under SEX in the population profile of Output 13.6, the effect for sex is the differential effect for females. If an effect has more than one parameter, each of them is listed, as well as the associated standard error, Wald statistic, and *p*-value. Since sex is represented by only one parameter, the chi-square value listed in the table of WLS estimates is identical to that listed in the analysis of variance table. This won't happen for those effects comprised of more than one parameter, since the effect test listed in the analysis of variance table is the test of whether all the effect's parameters are jointly zero, and the chi-square tests listed in the parameter estimates table are always one degree of freedom tests for each of the individual parameters.

To summarize, the model that most effectively describes these data is a single main effect model where sex is the main effect. Its goodness of fit is satisfactory and the model is parsimonious in the sense of not including factors with essentially no association with the response. Girls reported more colds than boys; the model-predicted mean number of periods with colds for girls is

$$\bar{F}_{\text{girls}} = \alpha + \beta_1 = 1.0477 + 0.0816 = 1.1293$$

and the model-predicted mean number of periods with colds for boys is

$$\bar{F}_{\text{boys}} = \alpha - \beta_1 = 1.0477 - 0.0816 = 0.9661$$

13.4   Analysis of Means: Performing Contrast Tests

Frequently, the underlying contingency table is based on more than two factors. This section discusses how to build models in a multifactor framework, how to specify scores for the response variable, and how to construct contrast tests with the CATMOD procedure. In addition, the interactive use of PROC CATMOD is explained.

Model building for more complicated cross-classification structures follows a similar strategy to that illustrated in the analysis of the colds data set. Consider the following data from a randomized clinical trial of chronic pain. Investigators were interested in comparing an active treatment with a placebo for an aspect of the condition of patients in the study. These patients were obtained from two investigators whose research design included stratified randomization relative to four diagnostic classes.

Table 13.4.   Chronic Pain Clinical Trial

Diagnostic Class	Investigator	Treatment	Patient Status				
			Poor	Fair	Moderate	Good	Excellent
I	A	Active	3	2	2	1	0
I	A	Placebo	7	0	1	1	1
I	B	Active	1	6	1	5	3
I	B	Placebo	5	4	2	3	3
II	A	Active	1	0	1	2	2
II	A	Placebo	1	1	0	1	1
II	B	Active	0	1	1	1	6
II	B	Placebo	3	1	1	5	0
III	A	Active	2	0	3	3	2
III	A	Placebo	5	0	0	8	1
III	B	Active	2	4	1	10	3
III	B	Placebo	2	5	1	4	2
IV	A	Active	8	1	3	4	0
IV	A	Placebo	5	0	3	3	0
IV	B	Active	1	5	2	3	1
IV	B	Placebo	3	4	3	4	2

If you look at the cell sizes in this table, you will see that they are small, ranging from 0 to 10. Such small sample sizes rule out the possibility of modeling multiple response functions per group, such as generalized logits or cumulative logits, as was discussed in previous chapters. However, there is marginally adequate sample size to model one function per group, such as a mean score. If you assign scores to the categories of patient status, such as the integers 1–5 to poor–excellent, respectively, then you can model the mean patient response score with weighted least squares.

The following SAS statements input the data.

```
data cpain;
   input dstatus $ invest $ treat $ status $ count @@;
   datalines;
I   A active poor   3 I   A active   fair 2 I A active moderate 2
I   A active good   1 I   A active   excel 0
I   A placebo poor 7 I   A placebo fair 0 I A placebo moderate 1
I   A placebo good 1 I   A placebo excel 1
I   B active poor   1 I   B active   fair 6 I B active moderate 1
I   B active good   5 I   B active   excel 3
I   B placebo poor 5 I   B placebo fair 4 I B placebo moderate 2
I   B placebo good 3 I   B placebo excel 3
II  A active poor   1 II  A active   fair 0 II A active moderate 1
II  A active good   2 II  A active   excel 2
II  A placebo poor 1 II  A placebo fair 1 II A placebo moderate 0
II  A placebo good 1 II  A placebo excel 1
II  B active poor   0 II  B active   fair  1 II B active moderate 1
II  B active good   1 II  B active   excel 6
II  B placebo poor 3 II  B placebo fair 1 II B placebo moderate 1
II  B placebo good 5 II  B placebo excel 0
III A active poor   2 III A active   fair 0 III A active moderate 3
III A active good   3 III A active   excel 2
III A placebo poor 5 III A placebo fair 0 III A placebo moderate 0
III A placebo good 8 III A placebo excel 1
III B active poor   2 III B active   fair 4 III B active moderate 1
III B active good 10 III B active   excel 3
III B placebo poor 2 III B placebo fair 5 III B placebo moderate 1
III B placebo good 4 III B placebo excel 2
IV  A active poor   8 IV  A active   fair 1 IV A active moderate 3
IV  A active good   4 IV  A active   excel 0
IV  A placebo poor 5 IV  A placebo fair 0 IV A placebo moderate 3
IV  A placebo good 3 IV  A placebo excel 0
IV  B active poor   1 IV  B active   fair 5 IV B active moderate 2
IV  B active good   3 IV  B active   excel 1
IV  B placebo poor 3 IV  B placebo fair 4 IV B placebo moderate 3
IV  B placebo good 4 IV  B placebo excel 2
;
```

The saturated model is fit as the preliminary model. The following PROC CATMOD statements request this analysis. Note the use of the bar notation in the MODEL statement to specify that the model includes all interactions of the specified factors in addition to their main effects.

```
proc catmod order=data;
   weight count;
   response 1 2 3 4 5;
   model status=dstatus|invest|treat;
run;
```

Since the response variable STATUS is character valued, you need to specify scores for its levels. You can do this by specifying numeric values in the RESPONSE statement. This RESPONSE statement causes a mean score to be created based on scoring the first response variable level as 1, the second response variable level as 2, and so on. Specifying the ORDER=DATA option in the PROC CATMOD statement forces the levels of the response variable to be the same as the order in which they appear in the data, that is, poor, fair, moderate, good, and excellent.

Output 13.9 contains the population profiles, and Output 13.10 contains the response profiles.

Output 13.9 Population Profiles

```
                          Population Profiles

        Sample     dstatus     invest     treat      Sample Size
        --------------------------------------------------------
             1        I          A        active          8
             2        I          A        placebo        10
             3        I          B        active         16
             4        I          B        placebo        17
             5       II          A        active          6
             6       II          A        placebo         4
             7       II          B        active          9
             8       II          B        placebo        10
             9      III          A        active         10
            10      III          A        placebo        14
            11      III          B        active         20
            12      III          B        placebo        14
            13       IV          A        active         16
            14       IV          A        placebo        11
            15       IV          B        active         12
            16       IV          B        placebo        16
```

Output 13.10 Response Profiles

```
                   Response Profiles

             Response       status
             --------------------
                  1         poor
                  2         fair
                  3         moderate
                  4         good
                  5         excel
```

Output 13.11 contains the model matrix and the response functions. The functions are the mean patient status scores based on integer scoring.

Output 13.11   Response Functions

```
               Response Functions and Design Matrix

           Response                    Design Matrix
  Sample   Function    1    2    3    4    5    6    7    8
  ------------------------------------------------------------------
    1      2.12500     1    1    0    0    1    1    0    0
    2      1.90000     1    1    0    0    1    1    0    0
    3      3.18750     1    1    0    0   -1   -1    0    0
    4      2.70588     1    1    0    0   -1   -1    0    0
    5      3.66667     1    0    1    0    1    0    1    0
    6      3.00000     1    0    1    0    1    0    1    0
    7      4.33333     1    0    1    0   -1    0   -1    0
    8      2.80000     1    0    1    0   -1    0   -1    0
    9      3.30000     1    0    0    1    1    0    0    1
   10      3.00000     1    0    0    1    1    0    0    1
   11      3.40000     1    0    0    1   -1    0    0   -1
   12      2.92857     1    0    0    1   -1    0    0   -1
   13      2.18750     1   -1   -1   -1    1   -1   -1   -1
   14      2.36364     1   -1   -1   -1    1   -1   -1   -1
   15      2.83333     1   -1   -1   -1   -1    1    1    1
   16      2.87500     1   -1   -1   -1   -1    1    1    1


               Response Functions and Design Matrix

                            Design Matrix
  Sample    9    10    11    12    13    14    15    16
  ------------------------------------------------------------------
    1       1     1     0     0     1     1     0     0
    2      -1    -1     0     0    -1    -1     0     0
    3       1     1     0     0    -1    -1     0     0
    4      -1    -1     0     0     1     1     0     0
    5       1     0     1     0     1     0     1     0
    6      -1     0    -1     0    -1     0    -1     0
    7       1     0     1     0    -1     0    -1     0
    8      -1     0    -1     0     1     0     1     0
    9       1     0     0     1     1     0     0     1
   10      -1     0     0    -1    -1     0     0    -1
   11       1     0     0     1    -1     0     0    -1
   12      -1     0     0    -1     1     0     0     1
   13       1    -1    -1    -1     1    -1    -1    -1
   14      -1     1     1     1    -1     1     1     1
   15       1    -1    -1    -1    -1     1     1     1
   16      -1     1     1     1     1    -1    -1    -1
```

The saturated model results are displayed in Output 13.12.

Output 13.12   ANOVA Table for Saturated Model

```
                 Analysis of Variance

  Source                 DF   Chi-Square   Pr > ChiSq
  ------------------------------------------------------------
  Intercept               1      764.24      <.0001
  dstatus                 3       13.83      0.0031
  invest                  1        4.36      0.0368
  dstatus*invest          3        3.28      0.3506
  treat                   1        4.21      0.0401
  dstatus*treat           3        3.55      0.3143
  invest*treat            1        0.72      0.3966
  dstatus*invest*treat    3        0.34      0.9515

  Residual                0        .           .
```

Since the model is saturated, there is no residual variation and the fit is perfect. The interaction DSTATUS*INVEST*TREAT has a chi-square of 0.34 with 3 df ($p = 0.9515$), which is clearly nonsignificant. Thus, the next stage of modeling is to remove this interaction and examine the two-way interactions to see if they change. Since the reduced model is reduced by this one term, its goodness-of-fit chi-square will be equal to 0.34. With PROC CATMOD, you can enter statements interactively. For example, you can specify that the pairwise interactions model be fit by submitting the following MODEL statement.

```
model status=dstatus|invest|treat@2; run;
```

The procedure will be invoked using the previously submitted statements (from page 379) and substituting the new MODEL statement for the previous one. PROC CATMOD remains in this interactive mode until it encounters a QUIT, PROC, or DATA statement. (Most of the examples in this book contain the full code for completeness).

Output 13.13 contains the ANOVA table for this reduced model.

Output 13.13 ANOVA Table for Pairwise Interactions Model

```
                        Analysis of Variance

        Source             DF   Chi-Square   Pr > ChiSq
        ------------------------------------------------
        Intercept           1     776.58       <.0001
        dstatus             3      13.50        0.0037
        invest              1       4.87        0.0274
        dstatus*invest      3       3.22        0.3589
        treat               1       5.31        0.0212
        dstatus*treat       3       5.33        0.1494
        invest*treat        1       0.47        0.4943

        Residual            3       0.34        0.9515
```

Since the pairwise interactions DSTATUS*INVEST, DSTATUS*TREAT, and INVEST*TREAT are all nonsignificant, with p-values of 0.3589, 0.1494, and 0.4943, respectively, the model excluding these terms is fit next.

```
model status=dstatus invest treat; run;
```

Output 13.14 contains the ANOVA table for the main effects model.

Output 13.14 ANOVA Table for Main Effects Model

```
                        Analysis of Variance

        Source             DF   Chi-Square   Pr > ChiSq
        -----------------------------------------------
        Intercept           1     843.00       <.0001
        dstatus             3      15.41        0.0015
        invest              1       6.74        0.0094
        treat               1       3.71        0.0540

        Residual           10      10.20        0.4229
```

$Q_W = 10.20$, with 10 df and $p = 0.4229$, and is indicative of a satisfactory fit. All of the main effects are significant, although TREAT is on the border for the $\alpha = 0.05$ significance level criterion with $p = 0.0540$. It is kept in the model.

The model matrix for this model is displayed in Output 13.15, and the parameter estimates are displayed in Output 13.16.

Output 13.15   Model Matrix

```
                 Response Functions and Design Matrix

                 Response                  Design Matrix
      Sample     Function     1      2      3      4      5      6
      ------------------------------------------------------------
         1       2.12500      1      1      0      0      1      1
         2       1.90000      1      1      0      0      1     -1
         3       3.18750      1      1      0      0     -1      1
         4       2.70588      1      1      0      0     -1     -1
         5       3.66667      1      0      1      0      1      1
         6       3.00000      1      0      1      0      1     -1
         7       4.33333      1      0      1      0     -1      1
         8       2.80000      1      0      1      0     -1     -1
         9       3.30000      1      0      0      1      1      1
        10       3.00000      1      0      0      1      1     -1
        11       3.40000      1      0      0      1     -1      1
        12       2.92857      1      0      0      1     -1     -1
        13       2.18750      1     -1     -1     -1      1      1
        14       2.36364      1     -1     -1     -1      1     -1
        15       2.83333      1     -1     -1     -1     -1      1
        16       2.87500      1     -1     -1     -1     -1     -1
```

Output 13.16   Parameter Estimates

```
            Analysis of Weighted Least Squares Estimates

                                        Standard    Chi-
      Effect       Parameter   Estimate   Error    Square   Pr > ChiSq
      ------------------------------------------------------------------
      Intercept        1        2.9079    0.1002   843.00     <.0001
      dstatus          2       -0.3904    0.1621     5.80     0.0161
                       3        0.5660    0.1916     8.73     0.0031
                       4        0.1893    0.1569     1.46     0.2275
      invest           5       -0.2511    0.0967     6.74     0.0094
      treat            6        0.1816    0.0942     3.71     0.0540
```

DSTATUS is represented by three parameters, and both INVEST and TREAT are represented by one parameter. Table 13.5 contains the parameter interpretations and illustrates how they relate to the numbered parameters in the PROC CATMOD output. Referring to the order of the variable values in the population profiles of Output 13.9 enables you to determine that the three DSTATUS effects are for diagnostic classes I, II, and III, respectively (the first three levels listed), and that the TREAT effect is for active treatment.

Table 13.5. Parameter Interpretations

CATMOD Parameter	Model Parameter	Interpretation
1	α	intercept
2	β_1	differential effect for diagnosis I
3	β_2	differential effect for diagnosis II
4	β_3	differential effect for diagnosis III
5	β_4	differential effect for investigator A
6	β_5	differential effect for active treatment

13.4.1 Constructing Contrast Tests

Researchers were interested in whether there were differences between the diagnostic classes. If you look at the values for these effects and their standard errors, $\hat{\beta}_1 = -0.3904(0.1621)$, $\hat{\beta}_2 = 0.5660(0.1916)$, $\hat{\beta}_3 = 0.1893(0.1569)$, and the implied effect for diagnostic class IV, which is $(-\hat{\beta}_1 - \hat{\beta}_2 - \hat{\beta}_3) = -0.365$, it seems like there are probably several individual differences. You can formally address questions about the parameters with the use of contrast tests.

The tests of interest, barring a priori considerations, are whether diagnostic class I is different from classes II, III, and IV; whether diagnostic class II is different from classes III and IV; and whether diagnostic class III is different from class IV. You perform the hypothesis test

$$H_0: \beta_1 - \beta_2 = 0$$

to determine whether the differential effect for diagnostic class I is equal to the differential effect for diagnostic class II, and you construct similar hypotheses to test for the differences between classes I and III and classes II and III.

Since the effect for diagnostic class IV is equal to $(-\hat{\beta}_1 - \hat{\beta}_2 - \hat{\beta}_3)$, you test whether there is a difference between diagnostic class I and class IV with the hypothesis:

$$H_0: \beta_1 - (-\beta_1 - \beta_2 - \beta_3) = 2\beta_1 + \beta_2 + \beta_3 = 0$$

You construct similar hypotheses to test whether there is a difference between the other diagnostic classes and class IV.

In Chapter 8, the TEST statement in PROC LOGISTIC and the CONTRAST statement in PROC GENMOD were discussed. These statements serve the same purpose as the CONTRAST statement in PROC CATMOD, testing linear combinations of the parameters, but they all function somewhat differently. With PROC CATMOD, you specify the following CONTRAST statement to request the test of the first hypothesis listed:

```
contrast 'Diag I versus II' dstatus 1 -1 0;
```

You list a character string that labels the contrast, list the effect whose parameters you are interested in, and then supply a coefficient for each of the effect parameters that PROC CATMOD estimates. Remember that since PROC CATMOD uses full rank parameterization, it produces $s - 1$ estimated parameters for an effect that has s levels. Thus, since DSTATUS is represented by parameters β_1, β_2, and β_3, you need to specify three coefficients when you name the variable DSTATUS in the CONTRAST statement. Since the hypothesis being tested is $H_0: \beta_1 - \beta_2 = 0$, you supply the coefficient 1 for β_1, -1 for β_2, and 0 for β_3. The other contrast statements are specified similarly, inserting the coefficients corresponding to the hypothesis being tested.

The following statements produced the desired results. You can specify as many CONTRASTS as you want; the results are placed into a single table.

```
contrast 'I versus II'    dstatus 1 -1  0;
contrast 'I versus III'   dstatus 1  0 -1;
contrast 'I versus IV'    dstatus 2  1  1;
contrast 'II versus III'  dstatus 0  1 -1;
contrast 'II versus IV'   dstatus 1  2  1;
contrast 'III versus IV'  dstatus 1  1  2;
contrast 'dstatus'   dstatus 1 0 0 ,
                     dstatus 0 1 0 ,
                     dstatus 0 0 1 ;
run;
```

The last contrast specified is included to demonstrate that the results in the ANOVA table can be generated with contrast tests. This contrast is testing the hypothesis

$$H_0: \beta_1 = \beta_2 = \beta_3 = 0$$

and is performed with the contrast matrix

$$\mathbf{C} = \begin{bmatrix} 1 & 0 & 0 \\ 0 & 1 & 0 \\ 0 & 0 & 1 \end{bmatrix}$$

The test has 3 df, one for each linearly independent row of the contrast matrix. It requires three sets of coefficients in the CONTRAST statement, each separated by a comma. The variable name DSTATUS needs to be listed in each of the three lines. All of the results are displayed in Output 13.17.

Output 13.17   Contrast Results

```
                          Analysis of Contrasts

            Contrast        DF    Chi-Square    Pr > ChiSq
            ---------------------------------------------------
            I versus II      1        10.30        0.0013
            I versus III     1         5.18        0.0229
            I versus IV      1         0.01        0.9194
            II versus III    1         1.67        0.1966
            II versus IV     1        10.25        0.0014
            III versus IV    1         5.11        0.0238
            dstatus          3        15.41        0.0015
```

First, note that $Q_C = 15.41$ for DSTATUS, with 3 df, which is the same as the chi-square value listed for DSTATUS in Output 13.14 in the ANOVA table. This contrast test is the same as the test automatically produced for the DSTATUS effect in the model.

The contrast test results indicate that diagnostic class I is different from both II and III but not IV; both II and III are different from IV. There are substantial differences in how diagnostic class influences patient response score; diagnostic classes I and IV decrease patient scores, and diagnostic classes II and III increase patient scores. The influences of classes I and IV are not significantly different from each other, and the influences of classes II and III are not significantly different from each other.

13.5   Analysis of Proportions: Occupational Data

The previous section focused on the analysis of means with weighted least squares. Weighted least squares analysis can be performed for many different types of response functions composed from the proportions in a contingency table. The simplest of these is the response proportions themselves, which are analyzed in this section. In addition, the capability of the CATMOD procedure to fit *nested* models is demonstrated.

13.5.1   Occupational Data Example

The data displayed in Table 13.6 are from a cross-sectional prevalence study done in 1973 to investigate textile worker complaints about respiratory symptoms experienced while working in the mills (Higgins and Koch 1977). Investigators were interested in whether occupational environment was related to the prevalence of respiratory ailments associated with the disease byssinosis.

Since this was a cross-sectional study rather than a prospective one, you cannot make inferences to a more general population without making some rather restrictive assumptions. If such assumptions can be made—that the symptoms remained for the duration of a worker's presence in the same work environment, that worker departures were not related to the presence or absence of symptoms, and so on—then you may be able to make inferences to all workers employed in those mills or possibly even to workers engaged in similar work in similar mills. If not, then the results of the analysis apply only to the observed population and serve only to describe the variation found for that observed population. It is always important to clarify the inferential implications of an analysis based on the relevant sampling framework.

Table 13.6. Byssinosis Complaints

Workplace Condition	Years Employment	Smoking	Complaints Yes	Complaints No
Dusty	<10	Yes	30	203
Dusty	<10	No	7	119
Dusty	≥ 10	Yes	57	161
Dusty	≥ 10	No	11	81
Not Dusty	<10	Yes	14	1340
Not Dusty	<10	No	12	1004
Not Dusty	≥ 10	Yes	24	1360
Not Dusty	≥ 10	No	10	986

If you can assume that there is a justifiable target population, then it becomes reasonable to think of these frequencies as coming from some stratified simple random sampling scheme, so that the table is distributed as product multinomial. A logical response function to model for these dichotomous responses is the logit, and logits are usually analyzed with the maximum likelihood estimation of logistic regression. However, you may be interested in modeling the proportion of byssinosis complaints in each classification group. Using the proportion has the interpretative advantage that model parameters have a direct effect on the size of the proportions. Fitting a model to proportions is not so easily performed with maximum likelihood methods; however, it is easily done with weighted least squares methods.

Since there are eight groups in the contingency table formed from the different combinations of workplace condition, years of employment, and smoking status, and two possible responses, respiratory complaints or not, there are sixteen elements in the overall proportion vector. Consider the elements p_{ij} where $i = 1, \ldots, 8$ represents the groups, and $j = 1, 2$ represents yes and no, respectively. Since the proportions in each row add up to 1, only one response per group needs to be included in the analysis. Otherwise, the responses would be linearly dependent and the computations would fail.

The transformation matrix that generates the 'yes' proportions from the proportion vector is straightforward: each row of the matrix picks up the 'yes' proportion from the corresponding row of the underlying contingency table. The matrix formulation required to construct the response functions is

$$F(\mathbf{p}) = \mathbf{Ap} = \begin{bmatrix} 1 & 0 & 0 & 0 & 0 & 0 & 0 & 0 & 0 & 0 & 0 & 0 & 0 & 0 & 0 & 0 \\ 0 & 0 & 1 & 0 & 0 & 0 & 0 & 0 & 0 & 0 & 0 & 0 & 0 & 0 & 0 & 0 \\ 0 & 0 & 0 & 0 & 1 & 0 & 0 & 0 & 0 & 0 & 0 & 0 & 0 & 0 & 0 & 0 \\ 0 & 0 & 0 & 0 & 0 & 0 & 1 & 0 & 0 & 0 & 0 & 0 & 0 & 0 & 0 & 0 \\ 0 & 0 & 0 & 0 & 0 & 0 & 0 & 0 & 1 & 0 & 0 & 0 & 0 & 0 & 0 & 0 \\ 0 & 0 & 0 & 0 & 0 & 0 & 0 & 0 & 0 & 0 & 1 & 0 & 0 & 0 & 0 & 0 \\ 0 & 0 & 0 & 0 & 0 & 0 & 0 & 0 & 0 & 0 & 0 & 0 & 1 & 0 & 0 & 0 \\ 0 & 0 & 0 & 0 & 0 & 0 & 0 & 0 & 0 & 0 & 0 & 0 & 0 & 0 & 1 & 0 \end{bmatrix} \mathbf{p} = \begin{bmatrix} 0.129 \\ 0.056 \\ 0.261 \\ 0.120 \\ 0.010 \\ 0.012 \\ 0.017 \\ 0.010 \end{bmatrix}$$

This is a linear transformation, just as the transformation required to compute means from the previous section's proportion vector was a linear transformation. Thus, the covariance matrix for $F(\mathbf{p})$ is $\mathbf{V_F} = \mathbf{A V_p A'}$.

13.5.2 Fitting a Preliminary Model

The preliminary model for this analysis is the saturated model that includes all interactions. This includes pairwise and three-way interactions.

$$
\begin{bmatrix} F(\mathbf{p}_{11}) \\ F(\mathbf{p}_{21}) \\ F(\mathbf{p}_{31}) \\ F(\mathbf{p}_{41}) \\ F(\mathbf{p}_{51}) \\ F(\mathbf{p}_{61}) \\ F(\mathbf{p}_{71}) \\ F(\mathbf{p}_{81}) \end{bmatrix}
=
\begin{bmatrix}
1 & 1 & 1 & 1 & 1 & 1 & 1 & 1 \\
1 & 1 & 1 & 1 & -1 & -1 & -1 & -1 \\
1 & 1 & -1 & -1 & 1 & 1 & -1 & -1 \\
1 & 1 & -1 & -1 & -1 & -1 & 1 & 1 \\
1 & -1 & 1 & -1 & 1 & -1 & 1 & -1 \\
1 & -1 & 1 & -1 & -1 & 1 & -1 & 1 \\
1 & -1 & -1 & 1 & 1 & -1 & -1 & 1 \\
1 & -1 & -1 & 1 & -1 & 1 & 1 & -1
\end{bmatrix}
\begin{bmatrix} \alpha \\ \beta_1 \\ \beta_2 \\ \beta_3 \\ \beta_4 \\ \beta_5 \\ \beta_6 \\ \beta_7 \end{bmatrix}
$$

In this model, α is the intercept term, and β_1, β_2, and β_4 are parameters for workplace, years employment, and smoking behavior, respectively. β_3 is the parameter for interaction between workplace and years employment, β_5 is the parameter for interaction between workplace and smoking behavior, and β_6 is the parameter for interaction between years of employment and smoking behavior. Finally, β_7 is the parameter for the three-way interaction.

The following statements fit this model with PROC CATMOD. Instead of using the keyword MEANS in the RESPONSE statement to generate the mean of the responses, you use the keyword MARGINALS. This specifies that the response functions are marginal proportions defined across the population profiles. For a dichotomous response outcome, this generates the proportions corresponding to the first response level. For response variables with r outcome levels, the keyword MARGINALS generates proportions for the first $r - 1$ response levels. In Chapter 14, "Modeling Repeated Measurements Data with WLS," you learn about analyses where multiple response variables are used. In these cases, *marginal distributions* are of interest.

The following DATA step creates the data set BYSS.

```
data byss;
   input workplace $ em_years $ smoking $ status $ count @@;
   datalines ;
dusty          <10    yes   yes 30   dusty       <10    yes no  203
dusty          <10    no    yes  7   dusty       <10    no  no  119
dusty          >=10   yes   yes 57   dusty       >=10   yes no  161
dusty          >=10   no    yes 11   dusty       >=10   no  no   81
notdusty       <10    yes   yes 14   notdusty    <10    yes no 1340
notdusty       <10    no    yes 12   notdusty    <10    no  no 1004
notdusty       >=10   yes   yes 24   notdusty    >=10   yes no 1360
notdusty       >=10   no    yes 10   notdusty    >=10   no  no  986
;
run;
```

The CATMOD procedure invocation is very similar to those seen before; the keyword MARGINALS in the RESPONSE statement is the only difference. Note the use of the

ORDER=DATA option in the PROC CATMOD statement. This maintains the response variable levels in the order in which they occur in the input data set. Thus, "yes" is the first ordered value, as desired, and the response functions will be Pr{STATUS=yes}. PROC CATMOD by default orders its response levels alphanumerically. Since the response function is not the default logit, you get weighted least squares estimation.

```
proc catmod order=data;
   weight count;
   response marginals;
   model status = workplace|em_years|smoking;
run;
```

Output 13.18 displays the population and response profiles as well as the response functions and model matrix.

Output 13.18   Populations, Response Profiles, and Response Functions

```
                           Population Profiles

        Sample    workplace    em_years    smoking    Sample Size
        -----------------------------------------------------------
          1       dusty         <10         yes            233
          2       dusty         <10         no             126
          3       dusty         >=10        yes            218
          4       dusty         >=10        no              92
          5       notdusty      <10         yes           1354
          6       notdusty      <10         no            1016
          7       notdusty      >=10        yes           1384
          8       notdusty      >=10        no             996


                           Response Profiles

                         Response    status
                         ------------------
                            1        yes
                            2        no


                 Response Functions and Design Matrix

                  Response                    Design Matrix
        Sample    Function    1    2    3    4    5    6    7    8
        -----------------------------------------------------------
          1       0.12876     1    1    1    1    1    1    1    1
          2       0.05556     1    1    1    1   -1   -1   -1   -1
          3       0.26147     1    1   -1   -1    1    1   -1   -1
          4       0.11957     1    1   -1   -1   -1   -1    1    1
          5       0.01034     1   -1    1   -1    1   -1    1   -1
          6       0.01181     1   -1    1   -1   -1    1   -1    1
          7       0.01734     1   -1   -1    1    1   -1   -1    1
          8       0.01004     1   -1   -1    1   -1    1    1   -1
```

The ANOVA table displayed in Output 13.19 indicates a nonsignificant three-way interaction, with a Wald statistic of $Q_W = 1.21$ and a corresponding $p = 0.2714$. Based on this result, the model including all two-way interactions is fit. Recall that since the only effect being eliminated is the three-way interaction, the residual goodness-of-fit statistic for the reduced model will have the same value as the three-way interaction in the full model. Note that the label for the three-way interaction includes the truncation WORKPLC for WORKPLCE.

Output 13.19   ANOVA Table for Saturated Model

```
                        Analysis of Variance

     Source                    DF    Chi-Square    Pr > ChiSq
     ----------------------------------------------------------
     Intercept                  1       127.33       <.0001
     workplace                  1        89.61       <.0001
     em_years                   1        13.74       0.0002
     workplace*em_years         1        12.35       0.0004
     smoking                    1        16.44       <.0001
     workplace*smoking          1        14.75       0.0001
     em_years*smoking           1         2.02       0.1551
     workpla*em_years*smoking   1         1.21       0.2714

     Residual                   0         .            .
```

The following code requests the model with all pairwise interactions.

```
proc catmod order=data;
    weight count;
    response marginals;
    model status = workplace|em_years workplace|smoking
                   em_years|smoking;
run;
```

The resulting ANOVA table is displayed in Output 13.20.

Output 13.20   ANOVA Table for All Pairwise Interactions

```
                        Analysis of Variance

     Source                 DF    Chi-Square    Pr > ChiSq
     -------------------------------------------------------
     Intercept               1       131.73       <.0001
     workplace               1        92.90       <.0001
     em_years                1        14.33       0.0002
     workplace*em_years      1        12.92       0.0003
     smoking                 1        15.46       <.0001
     workplace*smoking       1        13.70       0.0002
     em_years*smoking        1         2.26       0.1324

     Residual                1         1.21       0.2714
```

Note that the residual $Q_W = 1.21$ for this model is the same as the Q_W for the three-way interaction in the saturated model (Output 13.19). Thus, you could eliminate the step of fitting the saturated model and assess the three-way interaction from the analysis results for the model with all pairwise interactions.

This model fits adequately; however, the years of employment and smoking behavior interaction appears to be an unimportant source of variation, with $Q_W = 2.26$ and $p = 0.1324$. Thus, a further reduced model is fit that includes all main effects and only the workplace $\times$ smoking behavior and workplace $\times$ years of employment interactions. The code required for this PROC CATMOD invocation is not reproduced here; the results are displayed in Output 13.21.

Output 13.21 ANOVA Table for Reduced Model

```
                    Analysis of Variance

     Source            DF    Chi-Square    Pr > ChiSq
     ------------------------------------------------------
     Intercept          1      131.56        <.0001
     workplace          1       93.89        <.0001
     em_years           1       14.47        0.0001
     workplace*em_years 1       12.95        0.0003
     smoking            1       14.82        0.0001
     workplace*smoking  1       13.29        0.0003

     Residual           2        3.47        0.1761
```

This model fits the data with a residual goodness-of-fit statistic of $Q_W = 3.47$ with 2 df. It includes the main effects workplace, years of employment, and smoking, as well as the interactions workplace $\times$ years of employment and workplace $\times$ smoking. Often it is useful to examine the nature of the interactions to determine exactly where the differences are occurring.

13.5.3 Reduced Models Using Nested-By-Value Effects

Pairwise interactions occur when one variable's effect depends on the level of a second variable. A main effect in the absence of an interaction means that the variable's effect has roughly the same influence at all levels of the second variable. One explanation for the occurrence of an interaction is when one variable has a measurable effect at one level of a second variable but virtually no effect at a different level of that variable. Nested-by-value effects coding enables you to determine whether this behavior is occurring and also allows you to fit a reduced model that incorporates this behavior.

To investigate both the workplace $\times$ smoking behavior interaction and the workplace $\times$ years of employment interaction with nested-by-value effects, you replace the interaction term and the main effect for the term that you want nested with the corresponding nested-by-value terms. For example,

`workplace|em_years` and `em_years`

in the MODEL statement are replaced with

`em_years(workplace='dusty')` and `em_years(workplace='notdusty')`

You can see what happens by examining the model matrices that are produced by the two different models in Table 13.7 and Table 13.8. In the first model matrix, the second column corresponds to workplace, the third column to years of employment, and the fourth column represents their interaction. In the second model matrix, the third and fourth columns represent the effect of employment years nested in the dusty workplace and the not dusty workplace, respectively. For each, there are nonzero coefficients only in the rows that correspond to that particular value for workplace. You can think of this as splitting the main effect for employment into two components: one for dusty workplace and one for not dusty workplace. The same principle applies to splitting the main effect for smoking into two such components.

Table 13.7.   Model Matrix for Interactions Model

Group			Columns 1	2	3	4	5	6
dusty	<10	yes	1	1	1	1	1	1
dusty	<10	no	1	1	1	1	−1	−1
dusty	>=10	yes	1	1	−1	−1	1	1
dusty	>=10	no	1	1	−1	−1	−1	−1
notdusty	<10	yes	1	−1	1	−1	1	−1
notdusty	<10	no	1	−1	1	−1	−1	1
notdusty	>=10	yes	1	−1	−1	1	1	−1
notdusty	>=10	no	1	−1	−1	1	−1	1

Table 13.8.   Model Matrix for Nested-by-Value Model

Group			Columns 1	2	3	4	5	6
dusty	<10	yes	1	1	1	0	1	0
dusty	<10	no	1	1	1	0	−1	0
dusty	>=10	yes	1	1	−1	0	1	0
dusty	>=10	no	1	1	−1	0	−1	0
notdusty	<10	yes	1	−1	0	1	0	1
notdusty	<10	no	1	−1	0	1	0	−1
notdusty	>=10	yes	1	−1	0	−1	0	1
notdusty	>=10	no	1	−1	0	−1	0	−1

The following statements fit the nested-by-value model.

```
proc catmod order=data;
    weight count;
    response marginals;
    model status = workplace
                em_years(workplace='dusty')
                em_years(workplace='notdusty')
                smoking(workplace='dusty')
                smoking(workplace='notdusty') ;
run;
```

The model matrix in Output 13.22 is the same as that displayed in Table 13.8.

Output 13.22   Model Matrix for Nested-By-Value Effects Model

```
              Response Functions and Design Matrix

              Response              Design Matrix
    Sample    Function     1     2     3     4     5     6
    ---------------------------------------------------------
       1       0.12876     1     1     1     0     1     0
       2       0.05556     1     1     1     0    -1     0
       3       0.26147     1     1    -1     0     1     0
       4       0.11957     1     1    -1     0    -1     0
       5       0.01034     1    -1     0     1     0     1
       6       0.01181     1    -1     0     1     0    -1
       7       0.01734     1    -1     0    -1     0     1
       8       0.01004     1    -1     0    -1     0    -1
```

The ANOVA table displayed in Output 13.23 lists the nested-by-value effects.

Output 13.23   ANOVA Table for Nested-By-Value Effects Model

```
                    Analysis of Variance

    Source                      DF   Chi-Square   Pr > ChiSq
    ---------------------------------------------------------
    Intercept                    1     131.56       <.0001
    workplace                    1      93.89       <.0001
    em_years(workplac=dusty)     1      13.90       0.0002
    em_yea(workpla=notdusty)     1       0.75       0.3860
    smoking(workplace=dusty)     1      14.27       0.0002
    smokin(workpla=notdusty)     1       0.64       0.4225

    Residual                     2       3.47       0.1761
```

The residual goodness-of-fit test is the same as for the previous reduced model with interactions. The degrees of freedom have not decreased with the nested-by-value model, but they have been used differently with the new parameterization. The statistic for WORKPLCE has also remained the same, since the new coding does not change the meaning of the parameter for WORKPLCE. Neither years of employment nor smoking behavior are significant effects for the notdusty level of WORKPLCE, $p = 0.3860$ and $p = 0.4225$, respectively, while they both appear to be strongly significant for the dusty level of WORKPLCE, $p = 0.0002$ for both effects. Thus, it appears that both interactions can be explained by the interplay of years of employment and smoking behavior in the dusty workplace.

The model excluding the nonsignificant nested-by-value effects is fit next.

```
proc catmod order=data;
   weight count ;
   response marginals;
   model status = workplace
                  em_years(workplace='dusty')
                  smoking(workplace='dusty') / pred;
run;
```

Output 13.24 contains the ANOVA table for the final model, and Output 13.25 contains the parameter estimates.

Output 13.24   ANOVA Table for Final Model

```
                    Analysis of Variance

    Source                          DF   Chi-Square   Pr > ChiSq
    ----------------------------------------------------------------
    Intercept                        1      131.50       <.0001
    workplace                        1       93.98       <.0001
    em_years(workplac=dusty)         1       13.90       0.0002
    smoking(workplace=dusty)         1       14.27       0.0002

    Residual                         4        4.68       0.3215
```

Output 13.25   Parameter Estimates

```
            Analysis of Weighted Least Squares Estimates

                                          Standard    Chi-
    Effect                    Parameter  Estimate  Error    Square  Pr > ChiSq
    ---------------------------------------------------------------------------
    Intercept                     1       0.0776   0.00677  131.50    <.0001
    workplace                     2       0.0656   0.00677   93.98    <.0001
    em_years(workplac=dusty)      3      -0.0503   0.0135    13.90    0.0002
    smoking(workplace=dusty)      4       0.0471   0.0125    14.27    0.0002
```

The model goodness-of-fit chi-square is 4.68 with 4 df, clearly an adequate fit. This model contains a main effect for workplace and effects for years of employment and smoking only in the dusty workplace. See Table 13.9 for a display of parameter interpretations.

Table 13.9.   Parameter Interpretations

CATMOD Parameter	Model Parameter	Interpretation
1	α	intercept
2	β_1	differential effect for dusty workplace
3	β_2	differential effect for < 10 years employment within a dusty workplace
4	β_3	differential effect for smoking within a dusty workplace

A dusty workplace increases the proportion of subjects in this study who had byssinosis complaints. Having less than ten years employment lessens the proportion with byssinosis complaints within a dusty workplace, $\hat{\beta}_2 = -0.0503$, which means that having ten or more years of employment increases the proportion with byssinosis complaints in a dusty workplace. Smoking increases the proportion with byssinosis complaints if the workplace is dusty.

The PRED option in the MODEL statement produces predicted values for the proportions with byssinosis symptoms for each group. In order to interpret this table, compare the "Sample" values with those listed in the "Population Profiles" of Output 13.18. For example, the model-predicted proportion with byssinosis complaints for smokers with less than ten years employment who also worked in a dusty workplace (sample 1) is 0.140 ± 0.020. For smokers with at least ten years employment who worked in a dusty workplace (sample 3), the predicted proportion with byssinosis symptoms is 0.241 ± 0.025.

Output 13.26   Predicted Values for Final Model

```
                    Predicted Values for Response Functions

                                     -----Observed---- ----Predicted----
                            Function          Standard          Standard
workplace em_years smoking  Number  Function     Error Function     Error Residual
------------------------------------------------------------------------------------
dusty     <10      yes         1    0.128755  0.021942 0.140053  0.020057  -0.0113
dusty     <10      no          1    0.055556  0.020406 0.045784    0.0189  0.009771
dusty     >=10     yes         1    0.261468  0.029762 0.240683  0.024856  0.020785
dusty     >=10     no          1    0.119565  0.033827 0.146415  0.026402  -0.02685
notdusty  <10      yes         1    0.01034   0.002749 0.012003   0.00158  -0.00166
notdusty  <10      no          1    0.011811  0.003389 0.012003   0.00158  -0.00019
notdusty  >=10     yes         1    0.017341  0.003509 0.012003   0.00158  0.005338
notdusty  >=10     no          1    0.01004   0.003159 0.012003   0.00158  -0.00196
```

13.6   Obstetrical Pain Data: Advanced Modeling of Means

Sections 13.3 and 13.4 discussed analyses of means. This section is concerned with an advanced application of modeling means, and it includes a lengthy analysis of contrasts to fully investigate various effects and interactions.

The data displayed in Table 13.10 are from a multicenter randomized study of obstetrical-related pain for women who had recently delivered a baby (Koch et al. 1985). Investigators were interested in comparing four treatments: placebo, drug a, drug b, and a combined treatment of drug a and drug b. Each patient was classified as initially having some pain or a lot of pain. Then, a randomly assigned treatment was administered at the beginning of the study period and again at 4 hours. Each patient was observed at hourly intervals for 8 hours and pain status was recorded as little or no pain or some or more pain. The response measure of interest is the average proportion of hours for which the patient reported little or no pain.

The patients for each center × initial status × treatment group can be considered to be representative of some corresponding large target population in a manner that is consistent with stratified simple random sampling. Each of the patient's responses can also be assumed to be independent of other patient responses. Thus, the data in Table 13.10 are distributed as product multinomial. There are many small cell frequencies (less than 5) in this table. This means that the asymptotic requirements necessary for modeling functions such as multiple cell proportions or generalized logits are not met. However, the average proportion of hours with little or no pain is a reasonable response measure, and there is sufficient sample size for modeling means.

Table 13.10. Number of Hours with Little or No Pain for Women Who Recently Delivered a Baby

Center	Initial Pain Status	Treatment	0	1	2	3	4	5	6	7	8	Total
1	lot	placebo	6	1	2	2	2	3	7	3	0	26
1	lot	a	6	3	1	2	4	4	7	1	0	28
1	lot	b	3	1	0	4	2	3	11	4	0	28
1	lot	ba	0	0	0	1	1	7	9	6	2	26
1	some	placebo	1	0	3	0	2	2	4	4	2	18
1	some	a	2	1	0	2	1	2	4	5	1	18
1	some	b	0	0	0	1	0	3	7	6	2	19
1	some	ba	0	0	0	0	1	3	5	4	6	19
2	lot	placebo	7	2	3	2	3	2	3	2	2	26
2	lot	a	3	1	0	0	3	2	9	7	1	26
2	lot	b	0	0	0	1	1	5	8	7	4	26
2	lot	ba	0	1	0	0	1	2	8	9	5	26
2	some	placebo	2	0	2	1	3	1	2	5	4	20
2	some	a	0	0	0	1	1	1	8	1	7	19
2	some	b	0	2	0	1	0	1	4	6	6	20
2	some	ba	0	0	0	1	3	0	4	7	5	20
3	lot	placebo	6	0	2	2	2	6	1	2	1	22
3	lot	a	4	2	1	5	1	1	3	2	3	22
3	lot	b	5	0	2	3	1	0	2	6	7	26
3	lot	ba	3	2	1	0	0	2	5	9	4	26
3	some	placebo	5	0	0	1	3	1	4	4	5	23
3	some	a	1	0	0	1	3	5	3	3	6	22
3	some	b	3	0	1	1	0	0	3	7	11	26
3	some	ba	0	0	0	1	1	4	2	4	13	25
4	lot	placebo	4	0	1	3	2	1	1	2	2	16
4	lot	a	0	1	3	1	1	6	1	3	6	22
4	lot	b	0	0	0	0	2	7	2	2	9	22
4	lot	ba	1	0	3	0	1	2	3	4	8	22
4	some	placebo	1	0	1	1	4	1	1	0	10	19
4	some	a	0	0	0	1	0	2	2	1	13	19
4	some	b	0	0	0	1	1	1	1	5	11	20
4	some	ba	1	0	0	0	0	2	2	2	14	21

The proportion vector for each group i is written

$$\mathbf{p} = (p_{i0}, p_{i1}, p_{i2}, p_{i3}, p_{i4}, p_{i5}, p_{i6}, p_{i7}, p_{i8})'$$

where p_{ij} is the proportion of patients with j hours of pain for the ith group, and $i = 1, \ldots, 32$ is the group corresponding to the ith row of Table 13.10. You compute the average proportion response function by applying the following matrix operation to the proportion vector for each group.

$$\mathbf{F}_i = \mathbf{F}(\mathbf{p}_i) = \mathbf{A}\mathbf{p} = \left[\tfrac{0}{8}, \tfrac{1}{8}, \tfrac{2}{8}, \tfrac{3}{8}, \tfrac{4}{8}, \tfrac{5}{8}, \tfrac{6}{8}, \tfrac{7}{8}, \tfrac{8}{8} \right] \mathbf{p}_i$$

A useful preliminary model is one that includes effects for center, initial pain, treatment, and initial pain × treatment interaction; initial pain and treatment are believed to be similar across centers, so their interactions with center are not included.

13.6.1 Performing the Analysis with PROC CATMOD

The following DATA step creates the SAS data set PAIN. The raw data contains an observation for each line of Table 13.10, although they are in a different order. The ARRAY and OUTPUT statements in the DATA step modify the input data by creating an individual observation for each different response value, or number of hours with little or no pain. It creates the variables NO_HOURS and COUNT. The CATMOD procedure requires that each response value be represented on a different observation. This data set contains 288 observations, or 9 observations for each of the 32 original data lines. The values for the variable INITIAL are 'some' for some pain and 'lot' for a lot of pain.

```
data pain (drop=h0-h8);
   input center initial $ treat $ h0-h8;
   array hours h0-h8;
   do i=1 to 9;
      no_hours=i-1; count=hours(i); output;
   end;
   datalines;
1 some placebo  1 0 3 0 2 2 4 4 2
1 some treat_a  2 1 0 2 1 2 4 5 1
1 some treat_b  0 0 0 1 0 3 7 6 2
1 some treat_ba 0 0 0 0 1 3 5 4 6
1 lot  placebo  6 1 2 2 2 3 7 3 0
1 lot  treat_a  6 3 1 2 4 4 7 1 0
1 lot  treat_b  3 1 0 4 2 3 11 4 0
1 lot  treat_ba 0 0 0 1 1 7 9 6 2
2 some placebo  2 0 2 1 3 1 2 5 4
2 some treat_a  0 0 0 1 1 1 8 1 7
2 some treat_b  0 2 0 1 0 1 4 6 6
2 some treat_ba 0 0 0 1 3 0 4 7 5
2 lot  placebo  7 2 3 2 3 2 3 2 2
2 lot  treat_a  3 1 0 0 3 2 9 7 1
2 lot  treat_b  0 0 0 1 1 5 8 7 4
2 lot  treat_ba 0 1 0 0 1 2 8 9 5
3 some placebo  5 0 0 1 3 1 4 4 5
3 some treat_a  1 0 0 1 3 5 3 3 6
3 some treat_b  3 0 1 1 0 0 3 7 11
3 some treat_ba 0 0 0 1 1 4 2 4 13
3 lot  placebo  6 0 2 2 2 6 1 2 1
3 lot  treat_a  4 2 1 5 1 1 3 2 3
3 lot  treat_b  5 0 2 3 1 0 2 6 7
3 lot  treat_ba 3 2 1 0 0 2 5 9 4
4 some placebo  1 0 1 1 4 1 1 0 10
4 some treat_a  0 0 0 1 0 2 2 1 13
4 some treat_b  0 0 0 1 1 1 1 5 11
4 some treat_ba 1 0 0 0 0 2 2 2 14
```

```
4 lot  placebo  4 0 1 3 2 1 1 2 2
4 lot  treat_a  0 1 3 1 1 6 1 3 6
4 lot  treat_b  0 0 0 0 2 7 2 2 9
4 lot  treat_ba 1 0 3 0 1 2 3 4 8
;
proc print data=pain(obs=9);
run;
```

Output 13.27 displays the observations for the group from center 1 who had some initial pain and received the placebo.

Output 13.27   Partial Listing of Data Set PAIN

OBS	center	initial	treat	i	no_hours	count
1	1	some	placebo	1	0	1
2	1	some	placebo	2	1	0
3	1	some	placebo	3	2	3
4	1	some	placebo	4	3	0
5	1	some	placebo	5	4	2
6	1	some	placebo	6	5	2
7	1	some	placebo	7	6	4
8	1	some	placebo	8	7	4
9	1	some	placebo	9	8	2

The following SAS statements invoke the CATMOD procedure and fit the preliminary model. Note that the RESPONSE statement includes the coefficients required to compute the average proportions per group. Using the MEANS keyword on the RESPONSE statement would compute the mean number of hours with little or no pain, not the average proportion of hours with little or no pain, which is desired here. Actually the results will be the same; the decision is whether you want the parameter estimates to apply to proportions or means.

```
proc catmod;
   weight count;
   response 0 .125 .25 .375 .5 .625 .75 .875 1;
   model no_hours =  center initial treat
                     treat*initial;
run;
```

The population profiles and the response profiles are displayed in Output 13.28 and Output 13.29, respectively.

Output 13.28 Population Profiles

```
                     Population Profiles

  Sample    center    initial     treat      Sample Size
  ------------------------------------------------------------
       1       1       lot        placebo         26
       2       1       lot        treat_a         28
       3       1       lot        treat_b         28
       4       1       lot        treat_ba        26
       5       1       some       placebo         18
       6       1       some       treat_a         18
       7       1       some       treat_b         19
       8       1       some       treat_ba        19
       9       2       lot        placebo         26
      10       2       lot        treat_a         26
      11       2       lot        treat_b         26
      12       2       lot        treat_ba        26
      13       2       some       placebo         20
      14       2       some       treat_a         19
      15       2       some       treat_b         20
      16       2       some       treat_ba        20
      17       3       lot        placebo         22
      18       3       lot        treat_a         22
      19       3       lot        treat_b         26
      20       3       lot        treat_ba        26
      21       3       some       placebo         23
      22       3       some       treat_a         22
      23       3       some       treat_b         26
      24       3       some       treat_ba        25
      25       4       lot        placebo         16
      26       4       lot        treat_a         22
      27       4       lot        treat_b         22
      28       4       lot        treat_ba        22
      29       4       some       placebo         19
      30       4       some       treat_a         19
      31       4       some       treat_b         20
      32       4       some       treat_ba        21
```

Output 13.29 Response Profiles

```
               Response Profiles

       Response       no_hours
       --------------------
              1         0
              2         1
              3         2
              4         3
              5         4
              6         5
              7         6
              8         7
              9         8
```

The response functions listed in Output 13.30 are the average proportions of hours with little or no pain.

Output 13.30   Model Matrix

```
                     Response Functions and Design Matrix

                Response                    Design Matrix
     Sample     Function     1     2     3     4     5     6     7     8
     -----------------------------------------------------------------------
          1     0.46635      1     1     0     0     1     1     0     0
          2     0.42857      1     1     0     0     1     0     1     0
          3     0.58036      1     1     0     0     1     0     0     1
          4     0.74038      1     1     0     0     1    -1    -1    -1
          5     0.63889      1     1     0     0    -1     1     0     0
          6     0.61111      1     1     0     0    -1     0     1     0
          7     0.77632      1     1     0     0    -1     0     0     1
          8     0.82237      1     1     0     0    -1    -1    -1    -1
          9     0.40385      1     0     1     0     1     1     0     0
         10     0.64423      1     0     1     0     1     0     1     0
         11     0.77404      1     0     1     0     1     0     0     1
         12     0.79808      1     0     1     0     1    -1    -1    -1
         13     0.64375      1     0     1     0    -1     1     0     0
         14     0.80921      1     0     1     0    -1     0     1     0
         15     0.77500      1     0     1     0    -1     0     0     1
         16     0.80000      1     0     1     0    -1    -1    -1    -1
         17     0.43182      1     0     0     1     1     1     0     0
         18     0.47727      1     0     0     1     1     0     1     0
         19     0.61058      1     0     0     1     1     0     0     1
         20     0.66827      1     0     0     1     1    -1    -1    -1
         21     0.60870      1     0     0     1    -1     1     0     0
         22     0.72159      1     0     0     1    -1     0     1     0
         23     0.76923      1     0     0     1    -1     0     0     1
         24     0.85500      1     0     0     1    -1    -1    -1    -1
         25     0.46875      1    -1    -1    -1     1     1     0     0
         26     0.67614      1    -1    -1    -1     1     0     1     0
         27     0.80114      1    -1    -1    -1     1     0     0     1
         28     0.73864      1    -1    -1    -1     1    -1    -1    -1
         29     0.73684      1    -1    -1    -1    -1     1     0     0
         30     0.89474      1    -1    -1    -1    -1     0     1     0
         31     0.88125      1    -1    -1    -1    -1     0     0     1
         32     0.88095      1    -1    -1    -1    -1    -1    -1    -1
```

Output 13.31   Model Matrix (continued)

```
             Response Functions and Design Matrix

                          Design Matrix
              Sample     9        10       11
              --------------------------------
                 1       1         0        0
                 2       0         1        0
                 3       0         0        1
                 4      -1        -1       -1
                 5      -1         0        0
                 6       0        -1        0
                 7       0         0       -1
                 8       1         1        1
                 9       1         0        0
                10       0         1        0
                11       0         0        1
                12      -1        -1       -1
                13      -1         0        0
                14       0        -1        0
                15       0         0       -1
                16       1         1        1
                17       1         0        0
                18       0         1        0
                19       0         0        1
                20      -1        -1       -1
                21      -1         0        0
                22       0        -1        0
                23       0         0       -1
                24       1         1        1
                25       1         0        0
                26       0         1        0
                27       0         0        1
                28      -1        -1       -1
                29      -1         0        0
                30       0        -1        0
                31       0         0       -1
                32       1         1        1
```

The goodness-of-fit statistic for this preliminary model is $Q_W = 26.90$ with 21 df, as displayed in Output 13.32. With $p = 0.1743$, this indicates that the model fits the data adequately. All the constituent effects are highly significant, $p < 0.01$.

Output 13.32   Preliminary ANOVA Table

```
                  Analysis of Variance

         Source          DF    Chi-Square    Pr > ChiSq
         ---------------------------------------------------
         Intercept        1      5271.98       <.0001
         center           3        29.02       <.0001
         initial          1        62.65       <.0001
         treat            3        92.15       <.0001
         initial*treat    3        12.63       0.0055

         Residual        21        26.90       0.1743
```

It is useful to examine further the interaction between the treatments and initial pain status. The significant interaction means that some of the treatment effects depend on the level of initial pain status.

Some questions of interest are:

- Which treatment effects depend on the level of initial pain status? Where exactly is the interaction occurring?

- In which levels of initial pain do treatments differ?

To address these questions, it is helpful to fit a differently parameterized model. If you nest the effects of treatment within levels of initial pain, then these questions can be addressed with the use of contrasts. By using the nested effects model, you are trading the 3 df for TREAT and 3 df for TREAT*INITIAL for 6 df for the nested effect TREAT(INITIAL). Accordingly, the TREAT(INITIAL) effect is associated with six parameters, three of which pertain to the effects of treatment within some initial pain and three of which pertain to the effects of treatment within a lot of initial pain.

The following PROC CATMOD statements fit the nested model. The difference between specifying the nested effect TREAT(INITIAL) and the nested-by-values effects (see page 391) TREAT(INITIAL=some) and TREAT(INITIAL=lot) is that the former yields the 6 df test in the ANOVA table that tests whether the six parameters for treatment effects within both some and a lot of initial pain levels are essentially zero; the latter results in two separate 3 df tests in the ANOVA table, one for whether the three treatment parameters for some pain are essentially zero and one for whether the three treatment parameters for a lot of pain are essentially zero.

```
proc catmod;
   weight count;
   response 0 .125 .25 .375 .5 .625 .75 .875 1;
   model no_hours = center initial
                    treat(initial);
run;
```

Submitting these statements produces the results contained in Output 13.33.

Output 13.33   Nested Value ANOVA Table

```
                    Analysis of Variance

       Source            DF    Chi-Square    Pr > ChiSq
       -----------------------------------------------------
       Intercept          1      5271.98       <.0001
       center             3        29.02       <.0001
       initial            1        62.65       <.0001
       treat(initial)     6       102.70       <.0001

       Residual          21        26.90       0.1743
```

The model goodness-of-fit test is the same, $Q_W = 26.90$ with 21 df, since no model reduction was performed. The current model simply redistributes the variation over different degrees of freedom. Geometrically, you can think of the model space as being spanned by a different, but equivalent, vector set. The tests for CENTER effect and INITIAL effect remain the same as well. However, now there is the 6 df nested effect TREAT(INITIAL) in place of a TREAT effect and a TREAT*INITIAL interaction.

The next table that PROC CATMOD produces is the table of parameter estimates. These are listed in the same order as the corresponding explanatory variables in the MODEL statement. Besides the intercept, there are three parameters corresponding to CENTER for the effects of center 1, center 2, and center 3, respectively. As discussed previously, for the full rank parameterization that PROC CATMOD uses, you can determine the estimate of the effect for center 4 by taking the negative of the sum of these three displayed center effect parameters. Since INITIAL has two levels, it is represented by one parameter; this is the effect for a lot of initial pain.

Output 13.34 Nested Value ANOVA Table

			Analysis of Weighted Least Squares Estimates		
Effect	Parameter	Estimate	Standard Error	Chi-Square	Pr > ChiSq
Intercept	1	0.6991	0.00963	5271.98	<.0001
center	2	-0.0484	0.0145	11.24	0.0008
	3	0.0187	0.0145	1.66	0.1982
	4	-0.0415	0.0176	5.56	0.0184
initial	5	-0.0753	0.00951	62.65	<.0001
treat(initial)	6	-0.1739	0.0283	37.81	<.0001
	7	-0.0644	0.0255	6.39	0.0115
	8	0.0952	0.0206	21.45	<.0001
	9	-0.1159	0.0284	16.68	<.0001
	10	0.00740	0.0217	0.12	0.7331
	11	0.0347	0.0206	2.84	0.0921

The last six parameters are those for the nested effect TREAT(INITIAL). It is useful to examine the model matrix to understand what these parameters mean. The model matrix is displayed in Output 13.35–Output 13.36.

The first column of the model matrix corresponds to the intercept, columns 2–4 correspond to the parameters for CENTER, and column 5 corresponds to the parameter for INITIAL. Notice the structure of the next six columns, 6–11. There are nonzero entries in columns 6–8 for those rows that correspond to groups that are from the same initial pain level. You can compare the "Sample" column value to the same values in the "Population Profiles" table previously displayed to determine exactly which group is represented by each row of the model matrix. Columns 6–8 are the effects for placebo, treatment a, and treatment b, nested within the 'a lot of pain' level. This ordering is also determined by the order in which the TREAT values are listed in the "Population Profiles" table. You see a similar pattern in the final three columns of the model matrix, columns 9–11. Effects for placebo, treatment a, and treatment b are nested within the 'some pain' level.

Output 13.35   Model Matrix for Nested Model

```
                  Response Functions and Design Matrix

            Response                     Design Matrix
  Sample    Function    1     2     3     4     5     6     7     8
-------------------------------------------------------------------------
       1    0.46635     1     1     0     0     1     1     0     0
       2    0.42857     1     1     0     0     1     0     1     0
       3    0.58036     1     1     0     0     1     0     0     1
       4    0.74038     1     1     0     0     1    -1    -1    -1
       5    0.63889     1     1     0     0    -1     0     0     0
       6    0.61111     1     1     0     0    -1     0     0     0
       7    0.77632     1     1     0     0    -1     0     0     0
       8    0.82237     1     1     0     0    -1     0     0     0
       9    0.40385     1     0     1     0     1     1     0     0
      10    0.64423     1     0     1     0     1     0     1     0
      11    0.77404     1     0     1     0     1     0     0     1
      12    0.79808     1     0     1     0     1    -1    -1    -1
      13    0.64375     1     0     1     0    -1     0     0     0
      14    0.80921     1     0     1     0    -1     0     0     0
      15    0.77500     1     0     1     0    -1     0     0     0
      16    0.80000     1     0     1     0    -1     0     0     0
      17    0.43182     1     0     0     1     1     1     0     0
      18    0.47727     1     0     0     1     1     0     1     0
      19    0.61058     1     0     0     1     1     0     0     1
      20    0.66827     1     0     0     1     1    -1    -1    -1
      21    0.60870     1     0     0     1    -1     0     0     0
      22    0.72159     1     0     0     1    -1     0     0     0
      23    0.76923     1     0     0     1    -1     0     0     0
      24    0.85500     1     0     0     1    -1     0     0     0
      25    0.46875     1    -1    -1    -1     1     1     0     0
      26    0.67614     1    -1    -1    -1     1     0     1     0
      27    0.80114     1    -1    -1    -1     1     0     0     1
      28    0.73864     1    -1    -1    -1     1    -1    -1    -1
      29    0.73684     1    -1    -1    -1    -1     0     0     0
      30    0.89474     1    -1    -1    -1    -1     0     0     0
      31    0.88125     1    -1    -1    -1    -1     0     0     0
      32    0.88095     1    -1    -1    -1    -1     0     0     0
```

Output 13.36 Model Matrix for Nested Model (continued)

```
                Response Functions and Design Matrix

                            Design Matrix
                  Sample     9        10        11
                  ----------------------------------
                     1       0         0         0
                     2       0         0         0
                     3       0         0         0
                     4       0         0         0
                     5       1         0         0
                     6       0         1         0
                     7       0         0         1
                     8      -1        -1        -1
                     9       0         0         0
                    10       0         0         0
                    11       0         0         0
                    12       0         0         0
                    13       1         0         0
                    14       0         1         0
                    15       0         0         1
                    16      -1        -1        -1
                    17       0         0         0
                    18       0         0         0
                    19       0         0         0
                    20       0         0         0
                    21       1         0         0
                    22       0         1         0
                    23       0         0         1
                    24      -1        -1        -1
                    25       0         0         0
                    26       0         0         0
                    27       0         0         0
                    28       0         0         0
                    29       1         0         0
                    30       0         1         0
                    31       0         0         1
                    32      -1        -1        -1
```

Table 13.11 describes each parameter.

Table 13.11. Parameter Interpretations

CATMOD Parameter	Model Parameter	Interpretation
1	α	intercept
2	β_1	differential effect for center 1
3	β_2	differential effect for center 2
4	β_3	differential effect for center 3
5	β_4	differential effect for a lot of initial pain
6	β_5	differential effect for placebo for a lot of pain
7	β_6	differential effect for treatment a for a lot of pain
8	β_7	differential effect for treatment b for a lot of pain
9	β_8	differential effect for placebo for some pain
10	β_9	differential effect for treatment a for some pain
11	β_{10}	differential effect for treatment b for some pain

Consider testing to see if the effect for treatment a is the same as the effect for placebo for those patients with some pain. The appropriate hypothesis is stated

$$H_0: \beta_9 - \beta_8 = 0$$

Since the implicit effect for treatment ba is written in terms of the other treatment parameters,

$$\beta_{\text{treatment ba}} = -\beta_8 - \beta_9 - \beta_{10}$$

the hypothesis test to see if the effect for treatment ba is the same as the effect for placebo is written

$$H_0: -2\beta_8 - \beta_9 - \beta_{10} = 0$$

The hypotheses of interest and their corresponding contrasts and coefficients are displayed in Table 13.12. The coefficients are required in the CONTRAST statement in PROC CATMOD to perform a particular contrast test.

Table 13.12. Hypothesis Tests

Hypothesis	Initial Pain	Contrast	Coefficients e_5	e_6	e_7	e_8	e_9	e_{10}
treatment a vs. placebo	a lot	$-e_5 + e_6$	-1	1	0	0	0	0
treatment b vs. placebo	a lot	$-e_5 + e_7$	-1	0	1	0	0	0
treatment ba vs. placebo	a lot	$-2e_5 - e_6 - e_7$	-2	-1	-1	0	0	0
treatment ba vs. a	a lot	$-e_5 - 2e_6 - e_7$	-1	-2	-1	0	0	0
treatment ba vs. b	a lot	$-e_5 - e_6 - 2e_7$	-1	-1	-2	0	0	0
treatment a vs. placebo	some	$-e_8 + e_9$	0	0	0	-1	1	0
treatment b vs. placebo	some	$-e_8 + e_{10}$	0	0	0	-1	0	1
treatment ba vs. placebo	some	$-2e_8 - e_9 - e_{10}$	0	0	0	-2	-1	-1
treatment ba vs. a	some	$-e_8 - 2e_9 - e_{10}$	0	0	0	-1	-2	-1
treatment ba vs. b	some	$-e_8 - e_9 - 2e_{10}$	0	0	0	-1	-1	-2

The following CONTRAST statements request that the CATMOD procedure perform the appropriate tests. The statements can be submitted interactively, following the previous nested model invocation, or in batch, included at the end of the nested model invocation. Since you are only interested in the parameters corresponding to the TREAT(INITIAL) effect, you list that effect on the CONTRAST statement and then specify the appropriate six coefficients that pertain to the contrast involving the parameters e_5–e_{10}.

```
contrast 'lot: a-placebo'   treat(initial) -1  1  0  0  0  0 ;
contrast 'lot: b-placebo'   treat(initial) -1  0  1  0  0  0 ;
contrast 'lot: ba-placebo'  treat(initial) -2 -1 -1  0  0  0 ;
contrast 'lot: ba-a'        treat(initial) -1 -2 -1  0  0  0 ;
```

```
contrast 'lot: ba-b'       treat(initial) -1 -1 -2  0  0  0 ;
contrast 'some:a-placebo'  treat(initial)  0  0  0 -1  1  0 ;
contrast 'some:b-placebo'  treat(initial)  0  0  0 -1  0  1 ;
contrast 'some:ba-placebo' treat(initial)  0  0  0 -2 -1 -1 ;
contrast 'some:ba-a'       treat(initial)  0  0  0 -1 -2 -1 ;
contrast 'some:ba-b'       treat(initial)  0  0  0 -1 -1 -2 ;
run;
```

These statements produce the following "Analysis of Contrasts" table. It includes the results for all the individual hypothesis tests.

Output 13.37   Contrast Results

```
                      Analysis of Contrasts

          Contrast          DF    Chi-Square    Pr > ChiSq
          ----------------------------------------------------
          lot: a-placebo     1        5.59        0.0180
          lot: b-placebo     1       42.81        <.0001
          lot: ba-placebo    1       61.48        <.0001
          lot: ba-a          1       32.06        <.0001
          lot: ba-b          1        2.59        0.1076
          some:a-placebo     1        8.19        0.0042
          some:b-placebo     1       12.83        0.0003
          some:ba-placebo    1       21.45        <.0001
          some:ba-a          1        4.37        0.0365
          some:ba-b          1        1.67        0.1964
```

Most of these contrasts are significant, using the $\alpha = 0.05$ criterion; however, it appears that the difference between the ba treatment and the b treatment is marginal for both some initial pain and a lot of initial pain.

Additional contrasts of interest are the individual components of the treatment $\times$ initial pain status interaction, detailed in the first three rows of Table 13.13, as well as the individual components of the overall treatment effect. The interaction components are constructed by taking the difference of the pertinent contrasts for the some pain and a lot of pain contrasts, that is, taking the differences of rows 1 and 6, 2 and 7, and 3 and 8, in Table 13.12. The treatment effect is the joint effect of the various treatments averaged over the some initial pain and a lot of initial pain domains, so the individual components are the sum of the following rows from Table 13.12: 1 and 6, 2 and 7, 3 and 8. (You naturally think of averaging as summing and dividing by the number of summands; however, multiplying the contrast equations by a constant yields equivalent results, and so summing is all you really have to do.)

Table 13.13.   Hypothesis Tests

Hypothesis	Coefficients					
	e_5	e_6	e_7	e_8	e_9	e_{10}
treatment a vs. placebo, some vs. a lot	1	1	0	1	−1	0
treatment b vs. placebo, some vs. a lot	−1	0	1	1	0	−1
treatment ba vs. placebo, some vs. a lot	−2	−1	−1	2	1	1
treatment ba vs. a, some vs. a lot	−1	−2	−1	1	2	1
treatment ba vs. b, some vs. a lot	−1	−1	−2	1	1	2
average treatment a effect	−1	1	0	−1	1	0
average treatment b effect	−1	0	1	−1	0	1
average treatment ba effect	−2	−1	−1	−2	−1	−1
average ba vs. a	−1	−2	−1	−1	−2	−1
average ba vs. b	−1	−1	−2	−1	−1	−2

The next block of CONTRAST statements performs these tests. The last two CONTRAST statements request the 3 df TREAT*INITIAL interaction and the 3 df TREAT effect, respectively. The output is displayed in Output 13.38.

```
contrast 'interact:a-placebo'  treat(initial) -1  1  0  1 -1  0 ;
contrast 'interact:b-placebo'  treat(initial) -1  0  1  1  0 -1 ;
contrast 'interact:ba-placebo' treat(initial) -2 -1 -1  2  1  1 ;
contrast 'average:a-placebo'   treat(initial) -1  1  0 -1  1  0 ;
contrast 'average:b-placebo'   treat(initial) -1  0  1 -1  0  1 ;
contrast 'average:ba-placebo'  treat(initial) -2 -1 -1 -2 -1 -1 ;
contrast 'average:ba-a'        treat(initial) -1 -2 -1 -1 -2 -1 ;
contrast 'average:ba-b'        treat(initial) -1 -1 -2 -1 -1 -2 ;
contrast 'interaction'         treat(initial) -1  1  0  1 -1  0 ,
                               treat(initial) -1  0  1  1  0 -1 ,
                               treat(initial) -2 -1 -1  2  1  1 ;
contrast 'treatment effect'    treat(initial) -1  1  0 -1  1  0 ,
                               treat(initial) -1  0  1 -1  0  1 ,
                               treat(initial) -2 -1 -1 -2 -1 -1 ;
run;
```

Output 13.38   Contrast Results

```
              Analysis of Contrasts

        Contrast           DF    Chi-Square    Pr > ChiSq
        -------------------------------------------------
        interact:a-placebo  1         0.05        0.8266
        interact:b-placebo  1         4.05        0.0441
        interact:ba-placebo 1         4.89        0.0271
        average:a-placebo   1        13.51        0.0002
        average:b-placebo   1        50.93        <.0001
        average:ba-placebo  1        77.60        <.0001
        average:ba-a        1        31.42        <.0001
        average:ba-b        1         4.22        0.0399
        interaction         3        12.63        0.0055
        treatment effect    3        92.15        <.0001
```

These contrasts indicate that the interaction component corresponding to the comparison of treatment a and placebo is nonsignificant; therefore, treatment a has similar effects for an initial pain status of some or a lot. Treatments b and ba do appear to have different effects at the different levels of initial pain when compared to placebo; however, these effects may be quite similar. The interaction component corresponding to the comparison of treatment a and treatment ba is significant; their difference depends on the level of initial pain. This is not the case for treatment b compared to treatment ba; their interaction component is nonsignificant.

All the components of the treatment effect, those tests marked average, are significant.

The 'interaction' and the 'treatment effect' are testing the same thing as the TREAT effect and the TREAT*INITIAL effects listed in the ANOVA table for the preliminary model. Compare the Q_W values, 12.63 and 92.15, to those listed in that table. They are identical.

These results address the pertinent questions of this analysis. Other approaches may include fitting a reduced model that incorporates the results of these hypothesis tests. This is not pursued in this example.

13.7 Analysis of Survey Sample Data

In addition to analyzing data based on an underlying contingency table, the CATMOD procedure provides a convenient way to analyze data that come in the form of a function vector and covariance matrix. Often, such data come from complex surveys, and the covariance matrix has been computed using other software that takes the sampling design into account. If the number of response function estimates and the corresponding covariance matrix is large, then software designed for survey data analysis may be more appropriate.

13.7.1 HANES Data

The following data are from the Health and Nutrition Examination Survey (HANES) that was conducted in the United States from 1971–1974. This survey obtained various information concerning health from over 10,000 households in the United States. One of the measures constructed for analysis of these data was a well-being index, a composite index comprised from the answers to a questionnaire on general well-being. Table 13.14 contains the well-being ordered categorical estimates and standard errors for a cross-classification based on sex and age. The covariance matrix was computed using other software that used balanced repeated replication and took into account the sampling framework of the survey (Koch and Stokes 1979).

Table 13.14.   Well-being Index

Sex	Age	Estimate	S.E.
Female	25–34	7.250	0.105
Female	35–44	7.190	0.153
Female	45–54	7.360	0.103
Female	55–64	7.319	0.152
Female	65–74	7.552	0.139
Male	25–34	7.937	0.086
Male	35–44	7.925	0.108
Male	45–54	7.828	0.102
Male	55–64	7.737	0.116
Male	65–74	8.168	0.120

13.7.2   Direct Input of Response Functions

In the typical PROC CATMOD analysis, you input a contingency table or raw data and specify the response functions, and PROC CATMOD computes the appropriate covariance matrix based on the product multinomial distribution. In this case, your data are the vector of response functions and its covariance matrix. Thus, in order to describe the variation of these functions across various groups, you need to inform the procedure of the structure of your underlying cross-classification. Ordinarily, you would do this with the explanatory variables that define the contingency table. For this case, you need to rely on the FACTOR statement to express the cross-classification relationships.

The following is the DATA step that inputs the response functions and their covariance matrix into the SAS data set WBEING. The first two data lines are the well-being estimates, listed in the same order as they appear in Table 13.14. The following data lines contain the 10×10 covariance matrix corresponding to the estimates; each row takes two lines. The variable _TYPE_ identifies whether each data line corresponds to parameter estimates or covariance estimates. The value 'parms' identifies the lines with parameter estimates, and the value 'cov' identifies the lines with the covariance estimates. The variable _NAME_ identifies the name of the variable that has its covariance elements stored in that data line. Note that the diagonal element in the ith row of the covariance matrix is the variance for the ith well-being estimate. (The square root of the element is the standard error for the estimate.)

```
data wbeing;
    input #1 b1-b5 _type_ $ _name_ $8. #2 b6-b10;
    datalines;
 7.24978   7.18991   7.35960   7.31937   7.55184 parms
 7.93726   7.92509   7.82815   7.73696   8.16791
 0.01110   0.00101   0.00177  -0.00018  -0.00082 cov       b1
 0.00189  -0.00123   0.00434   0.00158  -0.00050
 0.00101   0.02342   0.00144   0.00369   0.25300 cov       b2
 0.00118  -0.00629  -0.00059   0.00212  -0.00098
 0.00177   0.00144   0.01060   0.00157   0.00226 cov       b3
 0.00140  -0.00088  -0.00055   0.00211   0.00239
-0.00018   0.00369   0.00157   0.02298   0.00918 cov       b4
-0.00140  -0.00232   0.00023   0.00066  -0.00010
```

```
 -0.00082    0.00253    0.00226    0.00918    0.01921 cov          b5
  0.00039    0.00034   -0.00013    0.00240    0.00213
  0.00189    0.00118    0.00140   -0.00140    0.00039 cov          b6
  0.00739    0.00019    0.00146   -0.00082    0.00076
 -0.00123   -0.00629   -0.00088   -0.00232    0.00034 cov          b7
  0.00019    0.01172    0.00183    0.00029    0.00083
  0.00434   -0.00059   -0.00055    0.00023   -0.00013 cov          b8
  0.00146    0.00183    0.01050   -0.00173    0.00011
  0.00158    0.00212    0.00211    0.00066    0.00240 cov          b9
 -0.00082    0.00029   -0.00173    0.01335    0.00140
 -0.00050   -0.00098    0.00239   -0.00010    0.00213 cov          b10
  0.00076    0.00083    0.00011    0.00140    0.01430
;
```

13.7.3   The FACTOR Statement

The FACTOR statement is where you define the cross-classification structure of the estimates. You need to specify names for the CATMOD procedure to use internally that correspond to grouping variables. You specify the number of levels for each and whether their values are character. This is done in the first part of the FACTOR statement. The internal variable SEX has two levels, and its values are character, as denoted by the dollar sign; the internal variable AGE has five levels, and its values are also character. The values for these internal variables are listed under the PROFILE option after a slash (/) in the FACTOR statement. The values are listed according to the order of the estimates; thus, they are listed in the same order as they appear in Table 13.14.

Since SEX and AGE are internal variables, not part of the input data set, you cannot refer to them in the MODEL statement. Thus, you use the keyword _RESPONSE_ to specify the desired model effects. In the following code, the saturated model is assigned to the keyword _RESPONSE_ in the FACTOR statement. This keyword is later used on the right-hand side of the MODEL statement. The _RESPONSE_ construction is also used to perform repeated measurements analyses and loglinear model analyses with the CATMOD procedure. Since the response functions are input directly, the keyword _F_ is used to represent them on the left-hand side of the MODEL statement.

The following PROC CATMOD statements invoke the procedure and specify that a saturated model be fit to the data. The keyword READ on the RESPONSE statement tells PROC CATMOD that the response functions and covariance matrix are to be directly input. The variables B1-B10 after the keyword READ specify that ten response functions are involved and thus that the covariance matrix is 10×10.

```
proc catmod data=wbeing;
    response read b1-b10;
    factors sex $ 2, age $  5 /
        _response_ = sex|age
    profile = (female '25-34',
              female '35-44',
              female '45-54',
              female '55-64',
              female '65-74',
              male   '25-34',
```

```
                        male    '35-44',
                        male    '45-54',
                        male    '55-64',
                        male    '65-74');
        model _f_ = _response_;
run;
```

13.7.4   Preliminary Analysis

Since populations and responses are not determined by input data variables, the population
profiles and response profiles are not printed as usual at the beginning of the PROC
CATMOD output. Instead, the first table displayed contains the response functions and the
model matrix, as shown in Output 13.39.

Output 13.39   Directly Input Response Functions

```
          Response Functions Directly Input from Data Set WBEING


                 Response Functions and Design Matrix

        Function   Response                  Design Matrix
Sample  Number     Function    1    2    3    4    5    6    7
------------------------------------------------------------------------
   1       1        7.24978     1    1    1    0    0    0    1
           2        7.18991     1    1    0    1    0    0    0
           3        7.35960     1    1    0    0    1    0    0
           4        7.31937     1    1    0    0    0    1    0
           5        7.55184     1    1   -1   -1   -1   -1   -1
           6        7.93726     1   -1    1    0    0    0   -1
           7        7.92509     1   -1    0    1    0    0    0
           8        7.82815     1   -1    0    0    1    0    0
           9        7.73696     1   -1    0    0    0    1    0
          10        8.16791     1   -1   -1   -1   -1   -1    1

                 Response Functions and Design Matrix

                      Function      Design Matrix
             Sample   Number      8      9      10
             ------------------------------------------
                1        1         0      0      0
                         2         1      0      0
                         3         0      1      0
                         4         0      0      1
                         5        -1     -1     -1
                         6         0      0      0
                         7        -1      0      0
                         8         0     -1      0
                         9         0      0     -1
                        10         1      1      1
```

Next is the analysis of variance table. The internal variables SEX and AGE are listed under
"Source" just as if they were explanatory variables on the input data set. The SEX*AGE
interaction is clearly nonsignificant, with $p = 0.5713$. Thus, the additive model with
effects SEX and AGE has an adequate goodness of fit with $Q_W = 2.92$ and 4 df.

Output 13.40   Saturated Model

```
         Response Functions Directly Input from Data Set WBEING


                         Analysis of Variance

           Source          DF    Chi-Square    Pr > ChiSq
           ----------------------------------------------
           Intercept        1     27117.73       <.0001
           sex              1        47.07       <.0001
           age              4        10.87       0.0281
           sex*age          4         2.92       0.5713


           Residual         0          .            .
```

The additive model is fit next, with contrasts requested to determine whether any of the parameters for age are essentially the same. The following statements fit the additive model.

```
proc catmod data=wbeing;
   response read b1-b10;
   factors sex $ 2, age $ 5 /
      _response_ = sex age
      profile = (male '25-34' ,
                 male '35-44',
                 male '45-54' ,
                 male '55-64',
                 male '65-74' ,
                 female '25-34',
                 female '35-44',
                 female '45-54',
                 female '55-64' ,
                 female '65-74');
   model _f_ = _response_;
```

The contrasts are set up to compare the first parameter for the age effect with each of the others. Recall that the implicit parameter for the last level age effect (ages 65–74) is the negative of the sum of the other parameters. Since the response functions are input directly, coefficients must be supplied for all the effects, including the intercept. Thus, the ALL_PARMS keyword is required. When you specify this keyword, you must supply coefficients for all the model parameters. Here, 0s are supplied on all contrasts for the intercept term and the sex effect term, and the final four coefficients apply to the age effect.

```
contrast '25-34 vs. 35-44' all_parms  0 0 1 -1  0  0;
contrast '25-34 vs. 45,54' all_parms  0 0 1  0 -1  0;
contrast '25-34 vs. 55,64' all_parms  0 0 1  0  0 -1;
contrast '25-34 vs. 65,74' all_parms  0 0 2  1  1  1;
run;
```

When all these statements are submitted, they produce the results displayed in Output 13.41 and Output 13.42. There are six parameters, one for the intercept, one for the sex effect, and four for the age effect. None of the age effect parameters listed appears to be of much importance. However, there does appear to be suggestive variation among age groups, with the *p*-value for the age effect at 0.0561.

Output 13.41 Additive Model

```
          Response Functions Directly Input from Data Set WBEING


                              Analysis of Variance

                 Source        DF    Chi-Square    Pr > ChiSq
                 --------------------------------------------------
                 Intercept      1     28089.07       <.0001
                 sex            1        65.84        <.0001
                 age            4         9.21        0.0561


                 Residual       4         2.92        0.5713


          Analysis of Weighted Least Squares Estimates

                                          Standard      Chi-
           Effect      Parameter  Estimate   Error    Square   Pr > ChiSq
           --------------------------------------------------------------------
           Intercept       1       7.6319    0.0455   28089.07    <.0001
           sex             2      -0.2900    0.0357      65.84    <.0001
           age             3      -0.00780   0.0645       0.01    0.9037
                           4      -0.0465    0.0636       0.54    0.4642
                           5      -0.0343    0.0557       0.38    0.5387
                           6      -0.1098    0.0764       2.07    0.1506
```

The contrasts indicate that the first four age groups act essentially the same and that the oldest age group is responsible for the age effect, $p = 0.0744$; note that its estimate is $-\{-0.008 - 0.046 - 0.034 - 0.110\} = 0.198$.

Output 13.42 Contrasts for Age Effect

```
          Response Functions Directly Input from Data Set WBEING


                            Analysis of Contrasts

             Contrast            DF    Chi-Square    Pr > ChiSq
             --------------------------------------------------------
             25-34 vs. 35-44      1       0.16         0.6937
             25-34 vs. 45,54      1       0.12         0.7288
             25-34 vs. 55,64      1       0.72         0.3954
             25-34 vs. 65,74      1       3.18         0.0744
```

One more contrast is specified to test the joint hypothesis that the lower four age groups are essentially the same. The following CONTRAST statement is submitted. Note that these three sets of coefficients, separated by commas, result in a 3 df test.

```
contrast '25-64 the same'    all_parms 0 0 1 -1  0  0,
                             all_parms 0 0 1  0 -1  0,
                             all_parms 0 0 1  0  0 -1;

run;
```

The result of this test is nonsignificant, $p = 0.8678$.

Output 13.43   Joint Test for Ages 25-64

```
         Response Functions Directly Input from Data Set WBEING


                        Analysis of Contrasts

       Contrast          DF    Chi-Square    Pr > ChiSq
       ---------------------------------------------------
       25-64 the same     3        0.72        0.8678
```

13.7.5 Inputting the Model Matrix Directly

These results suggest that an appropriate model for these data is one that includes the sex effect and an effect for the oldest age group. Since the response functions and their covariance matrix are inputted directly, it isn't possible to create a new explanatory variable in a DATA step that takes the value 0 for ages 25-64 and the value 1 for ages 65–74 and then use it in a DIRECT statement.

However, PROC CATMOD does allow you to specify your model matrix directly, instead of building one based on the explanatory variables or the effects represented by the _RESPONSE_ keyword. This means that you can fit the desired model with a sex effect and an older age effect. The MODEL statement containing such a model matrix specification follows. You write in the coefficients for the model matrix row-wise, separating each row with a comma. The entire matrix is enclosed by parentheses. This is similar to how you would input a matrix in the SAS/IML matrix programming language.

```
model _f_ = ( 1 0 0 ,
              1 0 0 ,
              1 0 0 ,
              1 0 0 ,
              1 0 1 ,
              1 1 0 ,
              1 1 0 ,
              1 1 0 ,
              1 1 0 ,
              1 1 1 );
```

This matrix represents an incremental effects model. The first column of 1s is for the intercept, the second column is for the incremental effect of sex, where the reference level is for females and the incremental effect is for males, and the third column is an incremental effect for age, where the increment is for those aged 65–74.

Note that model matrices can be inputted directly for all applications of the CATMOD procedure, if desired.

The PROC CATMOD statements required to fit this model follow. After the model matrix is listed a set of labels for the effects; the numbers correspond to the columns of the model matrix. Without the _RESPONSE_ keyword in the MODEL statement, the CATMOD procedure has no way of knowing how to divide the model variability into various components. You can request that various column parameters be tested jointly, or singly, as specified here. Refer to the CATMOD procedure chapter in the *SAS/STAT User's Guide, Version 8* for more detail. If you don't specify information concerning the columns, PROC CATMOD performs a joint test for the significance of the model beyond an overall mean, labeling this effect MODEL|MEAN in the ANOVA table.

```
proc catmod data=wbeing;
   response read b1-b10;
   factors sex $ 2, age $ 5 /
      _response_ = sex age
         profile = (female '25-34',
                    female '35-44',
                    female '45-54',
                    female '55-64',
                    female '65-74',
                    male   '25-34',
                    male   '35-44',
                    male   '45-54',
                    male   '55-64',
                    male   '65-74');
   model _f_ = ( 1 0 0 ,
                 1 0 0 ,
                 1 0 0 ,
                 1 0 0 ,
                 1 0 1 ,
                 1 1 0 ,
                 1 1 0 ,
                 1 1 0 ,
                 1 1 0 ,
                 1 1 1 ) (1='Intercept', 2='Sex', 3='65-74')
                      / pred;
```

The resulting tables for the analysis of variance and the parameter estimates are displayed in Output 13.44. The model fits very well, with a Q_W of 3.64 and 7 df, which results in $p = 0.8198$. The effects are listed as specified in the MODEL statement, and each is significant.

Output 13.44 Reduced Model

```
       Response Functions Directly Input from Data Set WBEING


                          Analysis of Variance

              Source        DF    Chi-Square    Pr > ChiSq
              -----------------------------------------------
              Intercept      1     14432.10       <.0001
              Sex            1        72.64       <.0001
              65-74          1         8.49       0.0036

              Residual       7         3.64       0.8198


          Analysis of Weighted Least Squares Estimates

                                       Standard      Chi-
     Effect    Parameter    Estimate    Error       Square    Pr > ChiSq
     ------------------------------------------------------------------------
     Model         1         7.3079     0.0608     14432.10      <.0001
                   2         0.5601     0.0657        72.64      <.0001
                   3         0.2607     0.0895         8.49      0.0036
```

Finally, the predicted values are displayed in Output 13.45. Fitting this model has resulted in estimates of the standard error that are on the order of twice as small as the standard errors for the original data.

Output 13.45 Reduced Model

```
       Response Functions Directly Input from Data Set WBEING



                 Predicted Values for Response Functions

                   ------Observed------    ------Predicted-----
     Function                  Standard                Standard
     Number      Function       Error      Function     Error      Residual
     --------------------------------------------------------------------------
         1        7.24978     0.105357     7.30786     0.060831    -0.05808
         2        7.18991     0.153036     7.30786     0.060831    -0.11795
         3        7.3596      0.102956     7.30786     0.060831     0.05174
         4        7.31937     0.151592     7.30786     0.060831     0.01151
         5        7.55184     0.1386       7.568608    0.098147    -0.01677
         6        7.93726     0.085965     7.867955    0.047737     0.069305
         7        7.92509     0.108259     7.867955    0.047737     0.057135
         8        7.82815     0.10247      7.867955    0.047737    -0.0398
         9        7.73696     0.115542     7.867955    0.047737    -0.13099
        10        8.16791     0.119583     8.128703    0.095929     0.039207
```

13.8   Modeling Rank Measures of Association Statistics

Many studies include outcomes that are ordinal in nature. When the treatment is either dichotomous or ordinal, you can model rank measures of correlation using WLS methods and use that framework to investigate various treatment effects and interactions. Such an analysis can often complement statistical models such as the proportional odds model. Refer to Carr, Hafner, and Koch (1989) for an example of such an analysis applied to Goodman-Kruskal rank correlation coefficients, also known as gamma coefficients.

The Mann-Whitney rank measure of association statistics are useful statistics for assessing the association between an ordinal outcome and a dichotomous explanatory variable. Consider the chronic pain data again. Investigators compared a new treatment with a placebo and assessed the response for a particular condition. Patients were obtained from two investigators whose design included stratification relative to four diagnostic classes. See Table 13.4. You may be interested in computing the Mann-Whitney rank measure of association as a way of assessing the extent to which patients with active treatments are more likely to have better response status than those with placebo. You may then be interested in seeing whether diagnostic status and investigator influence this association through model-fitting. You can perform such modeling by first computing the Mann-Whitney statistics and their standard errors and then using these estimates as input to the CATMOD procedure to perform modeling.

You can compute the Mann-Whitney measures as functions of the Somer's D measures, which are produced by the FREQ procedure.

$$U_i = \frac{\{\text{Somer's D C|R} + 1\}}{2} \text{ and } S_i = \frac{SE}{2}$$

S_i is the standard error of U_i, the Mann-Whitney statistic.

The following statements produce measures of association for the eight 2×5 tables formed for the combination of investigator and treatment.

```
proc freq data=cpain;
   weight count;
   tables dstatus*invest*treat*status/ measures;
run;
```

Output 13.46 displays the table for Diagnostic Status I and Investigator A. Output 13.47 displays the measures of association for that table.

Output 13.46 Frequency Counts

```
                        Table 1 of treat by status
                     Controlling for dstatus=I invest=A

       treat      status

       Frequency|
       Percent  |
       Row Pct  |
       Col Pct  |excel   |fair    |good    |moderate|poor    |  Total
       ---------+--------+--------+--------+--------+--------+
       active   |      0 |      2 |      1 |      2 |      3 |       8
                |   0.00 |  11.11 |   5.56 |  11.11 |  16.67 |  44.44
                |   0.00 |  25.00 |  12.50 |  25.00 |  37.50 |
                |   0.00 | 100.00 |  50.00 |  66.67 |  30.00 |
       ---------+--------+--------+--------+--------+--------+
       placebo  |      1 |      0 |      1 |      1 |      7 |      10
                |   5.56 |   0.00 |   5.56 |   5.56 |  38.89 |  55.56
                |  10.00 |   0.00 |  10.00 |  10.00 |  70.00 |
                | 100.00 |   0.00 |  50.00 |  33.33 |  70.00 |
       ---------+--------+--------+--------+--------+--------+
       Total           1        2        2        3       10      18
                    5.56    11.11    11.11    16.67    55.56   100.00
```

Output 13.47 Measures of Association

```
                  Statistics for Table 1 of treat by status
                     Controlling for dstatus=I invest=A

         Statistic                              Value       ASE
         -----------------------------------------------------
         Gamma                                 0.4286    0.3222
         Kendall's Tau-b                       0.2644    0.2144
         Stuart's Tau-c                        0.2963    0.2388

         Somers' D C|R                         0.3000    0.2412
         Somers' D R|C                         0.2330    0.1922

         Pearson Correlation                   0.2155    0.2392
         Spearman Correlation                  0.2850    0.2300

         Lambda Asymmetric C|R                 0.0000    0.0000
         Lambda Asymmetric R|C                 0.3750    0.2210
         Lambda Symmetric                      0.1875    0.1133

         Uncertainty Coefficient C|R           0.1291    0.0650
         Uncertainty Coefficient R|C           0.2394    0.1312
         Uncertainty Coefficient Symmetric     0.1678    0.0864

                       Sample Size = 18
```

Table 13.15 displays Somer's D values and asymptotic standard errors produced by the FREQ procedure and the calculated values of U_i and S_i.

Table 13.15.   Mann-Whitney Statistics

Diagnostic Class	Researcher	Somer's	ASE	U_i	S_i
I	A	0.2000	0.3515	0.6000	0.1758
I	B	0.2002	0.1915	0.6001	0.0958
II	A	0.2083	0.3622	0.6042	0.1811
II	B	0.6778	0.1834	0.8389	0.0917
III	A	0.0260	0.2271	0.5130	0.1136
III	B	0.1893	0.1923	0.5947	0.0962
IV	A	0.0000	0.2007	0.5000	0.1004
IV	B	−0.0156	0.2116	0.4922	0.1058

You compute the variances and then create a data set that contains the estimates and the covariance matrix. The following DATA step creates the data set MANNWHITNEY.

```
data MannWhitney;
   input b1-b8 _type_ $ _name_ $8.;
   datalines;
.6000   .6011   .6042 .8389 .5130 .5947 .5000 .4922 parms
.03091  .0000   .0000 .0000 .0000 .0000 .0000 .0000 cov b1
.0000   .00918  .0000 .0000 .0000 .0000 .0000 .0000 cov b2
.0000   .0000   .3280 .0000 .0000 .0000 .0000 .0000 cov b3
.0000   .0000   .0000 .0084 .0000 .0000 .0000 .0000 cov b4
.0000   .0000   .0000 .0000 .0129 .0000 .0000 .0000 cov b5
.0000   .0000   .0000 .0000 .0000 .0093 .0000 .0000 cov b6
.0000   .0000   .0000 .0000 .0000 .0000 .0101 .0000 cov b7
.0000   .0000   .0000 .0000 .0000 .0000 .0000 .0112 cov b8
;
```

This data set is then input into the CATMOD procedure. Instead of generating functions from an underlying contingency table, the CATMOD procedure does modeling directly on the input functions using the input covariance matrix as the weights. You define the profiles for each function with the PROFILE option in the FACTORS statement. You also define your factors, or explanatory variable structure, along with the number of levels for each, and describe the effects you want to include in your model with the _RESPONSE_ option.

```
proc catmod data=MannWhitney;
    response read b1-b8;
    factors diagnosis $ 4 , invest $ 2 /
        _response_ = diagnosis invest
    profile = (I    A,
               I    B,
               II   A,
               II   B,
               III  A,
               III  B,
               IV   A,
               IV   B);
    model _f_ = _response_ / cov;
run;
```

The ANOVA table results follow. The residual Wald test is a test of the diagnostic class and investigator interaction on the treatment effect, which is nonsignificant with a p-value of 0.9540. Neither diagnostic class nor investigator appear to explain significant variation, with diagnostic class appearing to be modestly influential with a p-value of 0.0725.

Output 13.48 ANOVA Table

```
Response Functions Directly Input from Data Set MANNWHITNEY


              Analysis of Variance

    Source        DF    Chi-Square    Pr > ChiSq
    ------------------------------------------------
    Intercept      1      193.69        <.0001
    diagnosis      3        6.98        0.0725
    invest         1        0.14        0.7122

    Residual       3        0.33        0.9540
```

By submitting another MODEL statement that specifies the vector of 1s as the model matrix, you can obtain a test of the hypothesis that the measures have the same value for each diagnostic class and investigator combination through the residual Wald test.

```
model _f_ =( 1,
             1,
             1,
             1,
             1,
             1,
             1,
             1 );
```

This is the seven degree test that is labeled 'Residual.' In the "Analysis of Variance" table. Note that there are no degrees of freedom left over for the "Model|Mean" source of variation, which is why the redundant or restricted parameter message appears.

Output 13.49   ANOVA Table

```
        Response Functions Directly Input from Data Set MANNWHITNEY


                          Analysis of Variance

              Source          DF   Chi-Square    Pr > ChiSq
              ------------------------------------------------
              Model|Mean       0*       .             .

              Residual         7       9.41          0.2247

         NOTE: Effects marked with '*' contain one or more
                 redundant or restricted parameters.


               Analysis of Weighted Least Squares Estimates

                                           Standard      Chi-
       Effect    Parameter    Estimate      Error       Square    Pr > ChiSq
       -----------------------------------------------------------------------
       Model         1         0.6027       0.0396      231.30       <.0001
```

The p-value of 0.2247 suggests that this hypothesis is compatible with these data. The estimate of the common Mann-Whitney rank measure for the eight strata is 0.6027 with standard error 0.0396 and $p < 0.0001$. It is interpretable as an estimated probability of 0.6027 for a randomly selected patient with active treatment having better responses than one with placebo.

Appendix A: Statistical Methodology for Weighted Least Squares

Consider the general contingency table displayed in Table 13.16, where s represents the number of rows, or groups, in the table, and r represents the number of responses.

Table 13.16.   Underlying Contingency Table

Group	Response 1	2	$\cdots$	r	Total
1	n_{11}	n_{12}	$\cdots$	n_{1r}	n_{1+}
2	n_{21}	n_{22}	$\cdots$	n_{2r}	n_{2+}
$\cdots$	$\cdots$	$\cdots$	$\cdots$	$\cdots$	$\cdots$
s	n_{s1}	n_{s2}	$\cdots$	n_{sr}	n_{s+}

The proportion of subjects in the ith group who have the jth response is written

$$p_{ij} = n_{ij}/n_{i+}$$

Suppose $\mathbf{n}'_i = (n_{i1}, n_{i2}, \ldots, n_{ir})$ represents the vector of responses for the ith subpopulation. If $\mathbf{n}' = (\mathbf{n}'_1, \mathbf{n}'_2, \ldots, \mathbf{n}'_r)$, then $\mathbf{n}$ follows the product multinomial distribution, given that each group has an independent sample. You can write the likelihood of **n** as

$$\Pr\{\mathbf{n}\} = \prod_{i=1}^{s} n_{i+}! \prod_{j=1}^{r} \pi_{ij}^{n_{ij}} / n_{ij}!$$

where π_{ij} is the probability that a randomly selected subject from the ith group has the jth response profile. The π_{ij} satisfy the natural restrictions

$$\sum_{j=1}^{r} \pi_{ij} = 1 \text{ for } i = 1, 2, \ldots, s$$

Suppose $\mathbf{p}_i = \mathbf{n}_i / n_{i+}$ is the $r \times 1$ vector of observed proportions associated with the ith group and suppose $\mathbf{p}' = (\mathbf{p}_1', \mathbf{p}_2', \ldots, \mathbf{p}_s')$ is the $(sr \times 1)$ compound vector of proportions.

A consistent estimator of the covariance matrix for the proportions in the ith row is

$$\mathbf{V}(\mathbf{p}_i) = \frac{1}{n_i} \begin{bmatrix} p_{i1}(1 - p_{i1}) & -p_{i1}p_{i2} & \cdots & -p_{i1}p_{ir} \\ -p_{i2}p_1 & p_{i2}(1 - p_{i2}) & \cdots & -p_{i2}p_{ir} \\ \vdots & \vdots & \vdots & \vdots \\ -p_{ir}p_{i1} & -p_{ir}p_{i2} & \cdots & p_{ir}(1 - p_{ir}) \end{bmatrix}$$

and the covariance matrix for the vector **p** is

$$\mathbf{V}_{\mathbf{p}} = \begin{bmatrix} \mathbf{V}_1 & \mathbf{0} & \cdots & \mathbf{0} \\ \mathbf{0} & \mathbf{V}_2 & \cdots & \mathbf{0} \\ \vdots & \vdots & \vdots & \vdots \\ \mathbf{0} & \mathbf{0} & \cdots & \mathbf{V}_s \end{bmatrix}$$

where $\mathbf{V}_i$ is the covariance matrix for $\mathbf{p}_i$.

Suppose $\mathbf{F}_1(\mathbf{p}), \mathbf{F}_2(\mathbf{p}), \ldots, \mathbf{F}_u(\mathbf{p})$ is a set of u functions of **p**. Each of the functions is required to have continuous partial derivatives through order two, and **F** must have a nonsingular covariance matrix, which can be written

$$\mathbf{V}_{\mathbf{F}}(\boldsymbol{\pi}) = [\mathbf{H}(\boldsymbol{\pi})][\mathbf{V}(\boldsymbol{\pi})][\mathbf{H}(\boldsymbol{\pi})]'$$

where $\mathbf{H}(\boldsymbol{\pi}) = [\partial \mathbf{F} / \partial \mathbf{z} | \mathbf{z} = \boldsymbol{\pi}]$ is the first derivative matrix of $\mathbf{F}(\mathbf{z})$.

F is a consistent estimator of $\mathbf{F}(\boldsymbol{\pi})$, so you can investigate the variation among the elements of $\mathbf{F}(\boldsymbol{\pi})$ with the linear model

$$E_A\{\mathbf{F}(\mathbf{p})\} = \mathbf{F}(\boldsymbol{\pi}) = \mathbf{X}\boldsymbol{\beta}$$

where $\mathbf{X}$ is a known model matrix with rank $t \leq u$, β is a $t \times 1$ vector of unknown parameters, and E_{As} means asymptotic expectation.

The goodness of fit of the model is assessed with

$$Q(\mathbf{X}, \mathbf{F}) = (\mathbf{WF})'[\mathbf{WV_F W}']^{-1}\mathbf{WF}$$

where $\mathbf{W}$ is any full rank $[(u - t) \times u]$ matrix orthogonal to $\mathbf{X}$. The quantity $Q(\mathbf{X}, \mathbf{F})$ is approximately distributed as chi-square with $(u - t)$ degrees of freedom when the sample sizes n_{i+} are large enough so that the elements of $\mathbf{F}$ have an approximate multivariate normal distribution. Such statistics are known as Wald statistics (Wald 1943).

The following statistic

$$Q_W = (\mathbf{F} - \mathbf{Xb})'\mathbf{V_F}^{-1}(\mathbf{F} - \mathbf{Xb})$$

is identical to $Q(\mathbf{X}, \mathbf{F})$ and is obtained by using weighted least squares to produce an estimate for β.

$$\mathbf{b} = (\mathbf{X}'\mathbf{V_F}^{-1}\mathbf{X})^{-1}\mathbf{X}'\mathbf{V_F}^{-1}\mathbf{F}$$

which is the minimum modified chi-square estimator (Neyman 1949).

A consistent estimator for the covariance matrix of $\mathbf{b}$ is given by

$$V(\mathbf{b}) = (\mathbf{X}'\mathbf{V_F}^{-1}\mathbf{X})^{-1}$$

If the model adequately characterizes the data as indicated by the goodness-of-fit criterion, then linear hypotheses of the form $\mathbf{C}\beta = \mathbf{0}$, where $\mathbf{C}$ is a known $c \times t$ matrix of constants of rank c, can be tested with the Wald statistic

$$Q_C = (\mathbf{Cb})'[\mathbf{C}(\mathbf{X}'\mathbf{V_F}^{-1}\mathbf{X})^{-1}\mathbf{C}']^{-1}(\mathbf{Cb})$$

Q_C is distributed as chi-square with degrees of freedom equal to c.

Predicted values for $\mathbf{F}(\pi)$ can be calculated from

$$\hat{\mathbf{F}} = \mathbf{Xb} = \mathbf{X}(\mathbf{X}'\mathbf{V_F}^{-1}\mathbf{X})^{-1}\mathbf{X}'\mathbf{V_F}^{-1}\mathbf{F}$$

and consistent estimators for the variances of $\hat{\mathbf{F}}$ can be obtained from the diagonal elements of

$$\mathbf{V}_{\hat{\mathbf{F}}} = \mathbf{X}(\mathbf{X}'\mathbf{V_F}^{-1}\mathbf{X})^{-1}\mathbf{X}'$$

While the functions $\mathbf{F(p)}$ can take on a wide range of forms, a few functions are commonly used. In particular, you can fit a strictly linear model

$$\mathbf{F(p)} = \mathbf{Ap}$$

where $\mathbf{A}$ is a matrix of known constants. The covariance matrix of $\mathbf{F}$ is written

$$\mathbf{V_F} = \mathbf{A V_p A'}$$

Another common model is loglinear:

$$\mathbf{F(p)} = \mathbf{A} \log \mathbf{p}$$

where log transforms a vector to the corresponding vector of natural logarithms and $\mathbf{A}$ is orthogonal to $\mathbf{1}$ (vector of 1s), that is, $\mathbf{A1} = \mathbf{0}$. In this case,

$$\mathbf{V_F} = \mathbf{A D_p}^{-1} \mathbf{A'}$$

where $\mathbf{D_p}$ is a diagonal matrix with the elements of $\mathbf{p}$ on the diagonal.

Many other useful functions can be generated as a sequence of linear, logarithmic, and exponential operations on the vector $\mathbf{p}$.

- linear transformations: $\mathbf{F_1(p)} = \mathbf{A_1 p} = \mathbf{a_1}$

- logarithmic: $\mathbf{F_2(p)} = \log(\mathbf{p}) = \mathbf{a_2}$

- exponential: $\mathbf{F_3(p)} = \exp(\mathbf{p}) = \mathbf{a_3}$

The corresponding $\mathbf{H}_k$ matrix operators needed to produce the covariance matrix for $\mathbf{F}$ are

- $\mathbf{H_1} = \mathbf{A_1}$

- $\mathbf{H_2} = \mathbf{D_p}^{-1}$

- $\mathbf{H_3} = \mathbf{D_{a_3}}$

$\mathbf{V_F}$ is estimated by $\mathbf{V_F} = [\mathbf{H(p)}] \mathbf{V_p} [\mathbf{H(p)}]'$ where $\mathbf{H(p)}$ is a product of the first derivative matrices $\mathbf{H}_k(\mathbf{p})$ where k indicates the ith operation in accordance with the chain rule.

Chapter 14
Modeling Repeated Measurements Data with WLS

Chapter Table of Contents

Chapter 14
Modeling Repeated Measurements Data with WLS

14.1 Introduction

Many types of studies have research designs that involve multiple measurements of a response variable. Longitudinal studies, in which repeated measures are obtained over time from each subject, are one important and commonly used type of repeated measures study. In other applications, the response from each experimental unit is measured under multiple conditions rather than at multiple time points. In some settings in which repeated measures data are obtained, the independent experimental units are not individual subjects. For example, in a toxicological study the experimental units might be litters; responses are then obtained from the multiple newborns in each litter. In a genetic study, experimental units might be defined by families; responses are then obtained from the members of each family.

There are two main difficulties in the analysis of data from repeated measures studies. First, the analysis is complicated by the dependence among repeated observations made on the same experimental unit. Second, the investigator often cannot control the circumstances for obtaining measurements, so that the data may be unbalanced or partially incomplete. For example, in a longitudinal study the response from a subject may be missing at one or more of the time points due to factors that are unrelated to the outcome of interest. In toxicology or genetic studies, litter or family sizes are variable rather than fixed; hence, the number of repeated measures is not constant across experimental units.

While many approaches to the analysis of repeated measures data have been studied, most are restricted to the setting in which the response variable is normally distributed and the data are balanced and complete. Although the development of methods for the analysis of repeated measures categorical data has received substantially less attention in the past, this has recently become an important and active area of research. Still, the methodology is not nearly as well developed as for continuous, normally distributed outcomes.

The SAS System provides several useful methodologies for analyzing repeated measures categorical data. These methodologies are applicable when a univariate response variable is measured repeatedly for each independent experimental unit.

One of these approaches, based on Mantel-Haenszel (MH) test statistics, is described as an advanced topic in Chapter 6, "Sets of $s \times r$ Tables." The MH methodology is useful for testing the null hypothesis of no association between the response variable and the repeated time points or conditions within each subject (that is, interchangeability).

Although these randomization model methods require minimal assumptions and the sample size requirements are less stringent than for other methods, they have important limitations. First, the MH methods are restricted to the analysis of data from a single sample; thus, the effects of additional factors (for example, treatment group) cannot be incorporated. In addition, the methods are oriented primarily to hypothesis testing rather than to parameter estimation.

Another approach is to model categorical repeated measurements in terms of a parsimonious number of parameters. Chapter 8, "Logistic Regression I: Dichotomous Response," introduces statistical modeling of categorical data using maximum likelihood to estimate parameters of models for logits, and Chapter 13, "Weighted Least Squares," describes weighted least squares (WLS) methodology for modeling a wide range of types of categorical data outcomes. Both of these chapters, however, focus on statistical modeling of the relationship between a single dependent categorical variable and one or more explanatory variables. When you model repeated measurements data, you are dealing with multiple dependent variables that reflect different times or conditions under which the outcome of interest was measured.

This chapter describes methods for analyzing repeated measurements data with weighted least squares methods. The WLS techniques are a direct extension of the general approach introduced and described in Chapter 13. The WLS methodology is an extremely versatile modeling approach that can be used efficiently for parameter estimation and hypothesis testing. However, the price of this versatility is that large sample sizes are required.

While such methods are still useful in analyzing repeated measurements data, generalized estimating equations (GEE) has become a popular technique for the analysis of repeated categorical measures as well as clustered data. GEE methods handle continuous explanatory variables, missing data, and time-dependent covariates. Chapter 15 discusses the GEE methodology and its application through a series of practical examples.

Many repeated measurements analyses are now undertaken with the GEE strategy. However, the weighted least squares approach is still a very reasonable one for data that meet the sample size requirements and include a minimum number of discrete explanatory variables, complete data, and limited time-dependent explanatory variables. For these situations, weighted least squares offers full efficiency, provides well-defined goodness-of-fit statistics, accounts for all degrees of freedom, and is asymptotically equivalent to maximum likelihood methods. You do lose these properties with the GEE approach, but you gain greater scope in your analysis. In some sense, you can consider the analogy to the classical MANOVA model in the continuous response setting. The weighted least squares and MANOVA methods have very desirable properties but limited scope; GEE methods, like mixed model methods, extend the possible scope of analyses with some reasonable choice of assumptions. Choosing one method or another depends on the data at hand and your analysis objectives.

14.2   Weighted Least Squares

14.2.1   Introduction

Chapter 13 discusses the use of weighted least squares in modeling categorical data. The first step of this methodology is to arrange the data from the experiment or study as an

$s \times r$ contingency table. In this general WLS framework, there are s groups defined by the cross-classification of the factors of interest (explanatory variables) and r response profiles. Chapter 13 considers situations in which there is a single outcome (dependent) variable, so that the response profiles are defined by the r possible levels of the dependent variable. In each group, at most $r - 1$ linearly independent response functions can be analyzed. Thus, in the applications in which the response variable is dichotomous, there is one response function per group. In other applications, the response is polytomous but a single response function, such as a mean score, is computed for each group. In both of these situations, there are s independent response functions (one for each row of the table) and their estimated covariance matrix $\mathbf{V_F}$ is diagonal.

However, the methodology can also be used when there are multiple response functions per group. In these situations, the response functions from the same group are correlated and their covariance matrix $\mathbf{V_F}$ is block diagonal. Since the usual covariance structure based on the multinomial distribution accounts for correlated proportions, it is a natural candidate for handling the correlation structure of repeated measurements.

In repeated measures applications, interest generally focuses on the analysis of the marginal distributions of the response at each time point, that is, regardless of the responses at the other time points. Thus, there are multiple response functions per group, and the correlation structure induced by the repeated measures must be taken into consideration. In the general situation in which a c-category response variable is measured at t time points, the cross-classification of the possible outcomes results in $r = c^t$ response profiles. You will generally consider $t(c - 1)$ correlated marginal proportions, generalized logits, or cumulative logits, or t correlated mean scores (if the response is ordinal), in the analysis.

Provided that the appropriate covariance matrix is computed for these correlated response functions, the WLS computations are no different from those described in Chapter 13. Koch and Reinfurt (1971) and Koch et al. (1977) first described the application of WLS to repeated measures categorical data. Further work is described in Stanish, Gillings, and Koch (1978), Koch et al. (1985), and Koch et al (1989). Stanish (1986), Landis et al. (1988), Agresti (1988, 1989), and Davis (1992) further developed this methodology and also illustrated various aspects of the use of the CATMOD procedure in analyzing categorical repeated measures.

The following sections illustrate several basic types of WLS analyses of repeated measurements data when the outcome is categorical. The examples progress in difficulty and gradually introduce more sophisticated analyses. Section 13.2.2 illustrates the methodology with a basic example.

14.2.2   One Population, Dichotomous Response

Grizzle, Starmer, and Koch (1969) analyze data in which 46 subjects were treated with three drugs (A, B, and C). The response to each drug was recorded as favorable or unfavorable. The null hypothesis of interest is that the marginal probability of a favorable response is the same for all three drugs, that is, the hypothesis of marginal homogeneity (see 6.4.1). Since the same 46 subjects were used in testing each of the three drugs, the estimates for the three marginal probabilities are correlated. In Section 6.4, this null hypothesis was tested using the Mantel-Haenszel general association statistic. The conclusion of this analysis was that there was a statistically significant difference among the three marginal probabilities.

Table 14.1 displays the data from Section 6.4 in the general WLS framework. There is one subpopulation (since there is a single group of subjects) and $r = 2^3 = 8$ response profiles, corresponding to the possible combinations of favorable and unfavorable response for the three drugs. For example, there are 6 subjects who had a favorable response to all three drugs (FFF) and 16 subjects who responded favorably to drugs A and B and unfavorably to drug C (FFU). In the notation of Section 14.2.1, $s = 1$, $c = 2$, $t = 3$, and $r = 2^3 = 8$.

Based on the underlying multinomial distribution of the cell counts, computation of response functions of interest and subsequent analysis using the WLS approach follows the same principles described in Chapter 13. However, the eight response profiles are not defined by the eight levels of a single response but rather by the response combinations resulting from the measurement of three dichotomous variables. From the proportions of these eight profiles, you can construct three correlated marginal proportions that correspond to those subjects who responded favorably to Drug A, Drug B, and Drug C, respectively.

Table 14.1. Drug Response Data

	F=favorable, U=unfavorable								
Drug A response	F	F	F	F	U	U	U	U	
Drug B response	F	F	U	U	F	F	U	U	
Drug C response	F	U	F	U	F	U	F	U	Total
Number of subjects	6	16	2	4	2	4	6	6	46

Suppose that p_i denotes the observed proportion of subjects in the ith response profile (ordered from left to right as displayed in Table 14.1) and let $\mathbf{p} = (p_1, \ldots, p_8)'$. For example, $p_1 = \Pr\{FFF\}$ is the probability of a favorable response to all three drugs. Now let p_A, p_B, and p_C denote the marginal proportions with a favorable response to drugs A, B, and C, respectively. For example, $p_A = \Pr\{FFF \text{ or } FFU \text{ or } FUF \text{ or } FUU\}$. The vector of response functions $\mathbf{F}(\mathbf{p}) = (p_A, p_B, p_C)'$ can be computed by the linear transformation $\mathbf{F}(\mathbf{p}) = \mathbf{Ap}$, where

$$\mathbf{A} = \begin{bmatrix} 1 & 1 & 1 & 1 & 0 & 0 & 0 & 0 \\ 1 & 1 & 0 & 0 & 1 & 1 & 0 & 0 \\ 1 & 0 & 1 & 0 & 1 & 0 & 1 & 0 \end{bmatrix}$$

The first row of $\mathbf{A}$ sums p_1, p_2, p_3, and p_4 to compute the proportion of subjects with a favorable response to drug A. Similarly, the second row of $\mathbf{A}$ sums p_1, p_2, p_5, and p_6 to yield the proportion with a favorable response to drug B. Finally, the corresponding proportion for drug C is computed by summing p_1, p_3, p_5, and p_7. The hypothesis of marginal homogeneity specifies that the marginal proportions with a favorable response to drugs A, B, and C are equal. This hypothesis can be tested by fitting a model of the form $\mathbf{F}(\boldsymbol{\pi}) = \mathbf{X}\boldsymbol{\beta}$, where $\boldsymbol{\pi}$ is the vector of population probabilities estimated by $\mathbf{p}$, $\mathbf{X}$ is a known model matrix, and $\boldsymbol{\beta}$ is a vector of unknown parameters. If the drug effect is significant, then the hypothesis of marginal homogeneity can be rejected.

This analysis is performed with the CATMOD procedure. The following statements create the SAS data set DRUG. The variables DRUGA, DRUGB, and DRUGC contain the responses for drugs A, B, and C, respectively.

```
data drug;
   input druga $ drugb $ drugc $ count;
   datalines;
F F F  6
F F U 16
F U F  2
F U U  4
U F F  2
U F U  4
U U F  6
U U U  6
;
```

The next group of statements requests a repeated measurements analysis that tests the hypothesis of marginal homogeneity.

```
proc catmod;
   weight count;
   response marginals;
   model druga*drugb*drugc=_response_ / oneway cov;
   repeated drug 3 / _response_=drug;
run;
```

A major difference between this PROC CATMOD invocation and those discussed in Chapter 13 is the syntax of the MODEL statement. One function of the MODEL statement is to specify the underlying $s \times r$ contingency table; that is, it defines the r response profiles by the values of the response variable and the s population profiles by the cross-classification of the levels of the explanatory variables. The fundamental distinction of repeated measures analyses is that there are now multiple response variables and they determine both the response functions and the variation to be modeled.

The response variables are crossed (separated by asterisks) on the left-hand side of the MODEL statement, and the r response profiles are defined by the cross-classification of their levels.

```
model druga*drugb*drugc=_response_ / oneway cov;
```

The response profiles are ordered so that the rightmost variable on the left-hand side of the MODEL statement varies fastest and the leftmost variable varies slowest. In this example, the drug C response changes from favorable to unfavorable most rapidly, followed by drug B, with drug A changing the slowest. Look ahead to Output 14.2 to see these response profiles listed in the resulting PROC CATMOD output. Since MARGINALS is specified in the RESPONSE statement, the marginal proportions for Drug A, Drug B, and Drug C are computed as the three response functions, as seen in Output 14.3.

Since the right-hand side of the MODEL statement does not include any explanatory variables, the data are correctly structured as a single population with $r = 8$ response profiles. The keyword _RESPONSE_ specifies that the variation among the dependent variables is to be modeled; by default, PROC CATMOD builds a full factorial _RESPONSE_ effect with respect to the repeated measurement factors. In this case, there is only one repeated factor, drug, so the full factorial includes only the drug main effect.

However, you can specify a different model matrix in the REPEATED statement, which is usually used in repeated measurements analysis. The general purpose of the REPEATED statement is to specify how to incorporate repeated measurement factors into the model. You can specify a name for each repeated measurement factor in the REPEATED statement, as well as specify the type (numeric or character), number of levels, and the label or value of each level. You can also define the model matrix in terms of the repeated measurement factors.

```
repeated drug 3 / _response_=drug;
```

In this example, the REPEATED statement specifies that there is a single repeated measurement factor that has three levels (drugs A, B, C). Although it is convenient to name this factor DRUG, any valid SAS variable name can be used, with the restriction that it cannot be the same as the name of an existing variable in the data set. If there is only one repeated measurements factor and the number of levels is omitted, then the CATMOD procedure assumes that the number of levels is equal to the number of response functions per group. So, in this case, the number 3 could have been omitted from the REPEATED statement.

The _RESPONSE_= option in the REPEATED statement specifies the effects to be included in the model matrix as a result of using the _RESPONSE_ keyword in the MODEL statement. The variables named in the effects must be listed in the REPEATED statement. If this option is omitted, then PROC CATMOD builds a full factorial _RESPONSE_ effect with respect to the repeated measurement factors. In this example, the _RESPONSE_ option specifies that the model matrix include a DRUG main effect. Note that since there is only one repeated measurement factor, you could replace the preceding REPEATED statement with

```
repeated drug;
```

Note that the ONEWAY option in the MODEL statement prints one-way marginal frequency distributions for each response variable in the MODEL statement. This is very useful in verifying that your model is set up as intended. The COV option in the MODEL statement prints the covariance matrix of the vector of response functions $\mathbf{F}(\mathbf{p})$.

Output 14.1 displays the one-way frequency distributions of the variables DRUGA, DRUGB, and DRUGC; they are useful for checking that the response functions are defined as desired. The variables DRUGA, DRUGB, and DRUGC have two levels, so the marginal proportion of subjects with the first level (F) is computed for each variable.

Output 14.1   One-Way Frequency Distributions

```
                    One-Way Frequencies

          Variable    Value    Frequency
          --------------------------------
          druga         F           28
                        U           18

          drugb         F           28
                        U           18

          drugc         F           16
                        U           30
```

Output 14.2 displays the population and response profiles.

Output 14.2   Population and Response Profiles

```
                    Population Profiles

              Sample      Sample Size
              --------------------------
                 1             46


                    Response Profiles

          Response    druga    drugb    drugc
          ------------------------------------
              1         F        F        F
              2         F        F        U
              3         F        U        F
              4         F        U        U
              5         U        F        F
              6         U        F        U
              7         U        U        F
              8         U        U        U
```

Output 14.3 displays the vector of response functions, its covariance matrix, and the model matrix. Compare these three response functions with the one-way distributions in Output 14.1 and verify that they are equal to the marginal proportions with a favorable response to drugs A, B, and C, respectively; for example, $28/(28 + 18) = 0.6087$ for drugs A and B, $16/(16 + 30) = 0.34783$ for drug C. The covariance matrix $\mathbf{V_F} = \mathbf{AV_pA'}$ of the response function vector $\mathbf{F}$ is printed because the COV option was specified in the MODEL statement. While $\mathbf{V_p}$ is the 8×8 covariance matrix of the proportions in the eight response categories, $\mathbf{V_F}$ is the 3×3 covariance matrix of $\mathbf{F}$. Note that the off-diagonal elements of $\mathbf{V_F}$ are nonzero, since the three marginal proportions are correlated. The model matrix has three columns, and the corresponding parameters are an overall intercept, an effect for drug A, and an effect for drug B.

Output 14.3   Response Functions and Model Matrix

```
                   Response Functions and Covariance Matrix

               Function      Response              Covariance Matrix
    Sample      Number       Function         1            2            3
    -------------------------------------------------------------------------
       1           1          0.60870     0.0051779    0.0023424    -0.000822
                   2          0.60870     0.0023424    0.0051779    -0.000822
                   3          0.34783    -0.000822    -0.000822     0.0049314


                              Design Matrix

                         Function         Design Matrix
               Sample     Number       1       2       3
               -----------------------------------------
                  1          1         1       1       0
                             2         1       0       1
                             3         1      -1      -1
```

Output 14.4 displays the analysis of variance (ANOVA) table. The source of variation labeled "drug" tests the null hypothesis that the probability of a favorable response is the same for all three drugs. Since the observed value of the 2 df test statistic is 6.58, the hypothesis of marginal homogeneity is rejected at the 0.05 level of significance ($p = 0.0372$).

Output 14.4   ANOVA Table

```
                       Analysis of Variance

           Source        DF     Chi-Square     Pr > ChiSq
           ------------------------------------------------
           Intercept      1       146.84        <.0001
           drug           2         6.58        0.0372

           Residual       0          .             .
```

From inspection of the marginal proportions of favorable response, it is clear that drug C is inferior to drugs A and B. You can test the equality of drugs A and C using a contrast statement. Since β_2 and β_3 are the parameters for drugs A and B (corresponding to the second and third columns of the model matrix in Output 14.3), the null hypothesis is

$$H_0\!: \beta_2 = -\beta_2 - \beta_3$$

or, equivalently,

$$H_0\!: 2\beta_2 + \beta_3 = 0$$

The corresponding CONTRAST statement is

```
contrast 'A versus C' _response_ 2 1;
```

Note that the keyword _RESPONSE_ is specified in the CONTRAST statement. You could also test this hypothesis using the ALL_PARMS keyword. The CONTRAST statement would be

```
contrast 'A versus C' all_parms 0 2 1;
```

The results in Output 14.5 indicate a significant difference between drugs A and C ($Q_W = 5.79$, 1 df, $p = 0.0161$).

Output 14.5   Contrast Results

```
                    Analysis of Contrasts

    Contrast      DF    Chi-Square    Pr > ChiSq
    -------------------------------------------------
    A versus C    1         5.79         0.0161
```

14.2.3   Two Populations, Dichotomous Response

The previous example involved the analysis of three responses from a single population. This section extends the methodology to situations in which there are multiple groups of subjects.

The Iowa 65+ Rural Health Study (Cornoni-Huntley et al. 1986) followed a cohort of elderly males and females over a six-year period. At each of three surveys, the response to one of the variables of interest, church attendance, was classified as yes if the subject was a regular church attender, and no if the subject was not a regular church attender. Table 14.2 displays the data from the 1311 females and 662 males who responded to all three surveys. Interest focuses on determining if church attendance rates change over time, if the attendance rates differ between females and males, and if the observed patterns of change over time are the same for females and males.

Table 14.2.   Church Attendance Data

Gender	Regular Church Attender at:			Count
	Year 0	Year 3	Year 6	
Female	Yes	Yes	Yes	904
	Yes	Yes	No	88
	Yes	No	Yes	25
	Yes	No	No	51
	No	Yes	Yes	33
	No	Yes	No	22
	No	No	Yes	30
	No	No	No	158
Male	Yes	Yes	Yes	391
	Yes	Yes	No	36
	Yes	No	Yes	12
	Yes	No	No	26
	No	Yes	Yes	15
	No	Yes	No	21
	No	No	Yes	18
	No	No	No	143

When you obtain repeated measures data from multiple populations, you are interested not only in the effect of the repeated measures factor but also in the effect of the explanatory variables defining the multiple populations. In fact, when there are explanatory variables (factors) in a study involving repeated measures, there are three types of variation:

- main effects and interactions of the repeated measurement factors (within subjects variation)

- main effects and interactions of the explanatory variables (between subjects variation)

- interactions between the explanatory variables and the repeated measurement factors

In this example, there are two populations (females, males). Since a dichotomous response variable is measured at each of three time points (the repeated measurement factor), there are $r = 2^3 = 8$ response profiles. The between subjects variation is due to differences between females and males, and the within subjects variation is due to differences among time points. The analysis investigates both sources of variation, as well as the variation due to their interaction.

The following SAS statements read in the counts displayed in Table 14.2 and fit a saturated model with effects due to gender, time, and their interaction.

```
data church;
   input gender $ attend0 $ attend3 $ attend6 $ count;
   datalines;
F Y Y Y 904
```

```
F Y Y N  88
F Y N Y  25
F Y N N  51
F N Y Y  33
F N Y N  22
F N N Y  30
F N N N 158
M Y Y Y 391
M Y Y N  36
M Y N Y  12
M Y N N  26
M N Y Y  15
M N Y N  21
M N N Y  18
M N N N 143
;
proc catmod order=data;
   weight count;
   response marginals;
   model attend0*attend3*attend6=gender|_response_ / oneway;
   repeated year;
run;
```

The ORDER=DATA option in the PROC CATMOD statement keeps the levels of the explanatory and response variables in the same order as in Table 14.2 and ensures that the response functions are the marginal probabilities of attendance, rather than nonattendance. The MODEL statement specifies a saturated model with all potential sources of variation for the marginal probabilities of regular church attendance using the usual vertical bar (|) notation. The REPEATED statement is not necessary here, but it makes the output a little clearer by naming the repeated effect YEAR instead of _RESPONSE_. The populations are determined by the effects listed in the right-hand side of the MODEL statement since a POPULATION statement is not used. Here, two populations based on the values of GENDER will be formed.

Output 14.6 displays the one-way frequency distributions of the variables ATTEND0, ATTEND3, ATTEND6, and GENDER.

Output 14.6   One-Way Frequency Distributions

```
                 One-Way Frequencies

            Variable   Value   Frequency
            -----------------------------
            attend0      Y        1533
                         N         440

            attend3      Y        1510
                         N         463

            attend6      Y        1428
                         N         545

            gender       F        1311
                         M         662
```

When a repeated measures analysis contains explanatory variables, the ONEWAY option of the MODEL statement produces the marginal distributions of the response variables for the total sample of observations. Thus, this option does not provide the marginal distributions of females and males regularly attending church at each of the three surveys. You can obtain the marginal distributions for females and males by using the FREQ procedure as follows:

```
proc freq;
   weight count;
   by gender;
   tables attend0 attend3 attend6;
run;
```

Output 14.7 shows that the proportions of females who regularly attend church at years 0, 3, and 6 are 0.815, 0.799, and 0.757, respectively.

Output 14.7   One-Way Frequency Distributions for Females

```
--------------------------------- gender=F ---------------------------------

                                     Cumulative    Cumulative
         attend0   Frequency   Percent  Frequency   Percent
         -------------------------------------------------------
         N             243      18.54        243      18.54
         Y            1068      81.46       1311     100.00


                                     Cumulative    Cumulative
         attend3   Frequency   Percent  Frequency   Percent
         -------------------------------------------------------
         N             264      20.14        264      20.14
         Y            1047      79.86       1311     100.00


                                     Cumulative    Cumulative
         attend6   Frequency   Percent  Frequency   Percent
         -------------------------------------------------------
         N             319      24.33        319      24.33
         Y             992      75.67       1311     100.00
```

The corresponding proportions of males regularly attending church are 0.702, 0.699, and 0.659.

Output 14.8 displays the population and response profiles produced by the CATMOD procedure. These results verify that there are two groups (populations) and eight response profiles.

Output 14.8 Population and Response Profiles

```
                       Population Profiles

              Sample    gender    Sample Size
              -----------------------------
                 1       F             1311
                 2       M              662


                       Response Profiles

         Response    attend0    attend3    attend6
         ---------------------------------------
            1          Y          Y          Y
            2          Y          Y          N
            3          Y          N          Y
            4          Y          N          N
            5          N          Y          Y
            6          N          Y          N
            7          N          N          Y
            8          N          N          N
```

The response functions displayed in Output 14.9 agree with those from the FREQ
procedure. Refer to the population profiles (Output 14.8) to determine that sample 1
corresponds to females and sample 2 to males. The model has six parameters: an overall
intercept for the probability of regular church attendance, a gender effect, two survey year
effects (corresponding to columns 3 and 4 of the model matrix) and two differential survey
year effects for females and males, that is, the gender $\times$ year interaction (corresponding to
columns 5 and 6 of the model matrix).

Output 14.9 Response Functions and Model Matrix

```
                   Response Functions and Design Matrix

                 Function   Response              Design Matrix
        Sample    Number    Function    1     2     3     4     5     6
        ----------------------------------------------------------------
           1         1       0.81465     1     1     1     0     1     0
                     2       0.79863     1     1     0     1     0     1
                     3       0.75667     1     1    -1    -1    -1    -1

           2         1       0.70242     1    -1     1     0    -1     0
                     2       0.69940     1    -1     0     1     0    -1
                     3       0.65861     1    -1    -1    -1     1     1
```

Output 14.10 displays the ANOVA table. There are clearly significant effects due to
gender ($Q_W = 30.04$, 1 df, $p < 0.001$) and survey year ($Q_W = 34.83$, 2 df, $p < 0.001$),
but the gender $\times$ year interaction is not significant ($Q_W = 0.87$, 2 df, $p = 0.6476$).

Output 14.10   ANOVA Table

```
                  Analysis of Variance

        Source          DF   Chi-Square   Pr > ChiSq
        ------------------------------------------------
        Intercept        1     6154.13      <.0001
        gender           1       30.04      <.0001
        year             2       34.83      <.0001
        gender*year      2        0.87       0.6476

        Residual         0         .           .
```

The following statements fit the model with main effects for gender and survey year.

```
proc catmod order=data;
   weight count;
   response marginals;
   model attend0*attend3*attend6=gender _response_ / noprofile;
   repeated year;
run;
```

The NOPROFILE option of the MODEL statement suppresses the listing of the population and response profiles. You may want to use this option when fitting multiple models to the same data.

Output 14.11 displays the resulting ANOVA table. Note that the residual chi-square statistic from this model is identical to the chi-square statistic for the gender × survey year interaction from the saturated model (Output 14.10).

Output 14.11   ANOVA Table for Main Effects Model

```
                  Analysis of Variance

        Source          DF   Chi-Square   Pr > ChiSq
        ------------------------------------------------
        Intercept        1     6156.36      <.0001
        gender           1       30.63      <.0001
        year             2       40.58      <.0001

        Residual         2        0.87       0.6476
```

The results of this model indicate that the gender effect and the survey year effect are both clearly significant. Moreover, since the interaction was not significant, the observed patterns of change over time are not significantly different for males and females. You may also want to test if the survey year effect departs from linearity. In terms of the default parameterization, the effects at years 0, 3, and 6 are given by β_3, β_4, and $-\beta_3 - \beta_4$, respectively. The null hypothesis of no departure from linearity is thus equivalent to

$$H_0: \beta_4 - \beta_3 = (-\beta_3 - \beta_4) - \beta_4$$

which simplifies to $H_0: \beta_4 = 0$. Although this hypothesis could be tested using a CONTRAST statement, it is also provided in the table of parameter estimates, which are displayed in Output 14.12.

Output 14.12 Parameter Estimates for Main Effects Model

```
           Analysis of Weighted Least Squares Estimates

                                      Standard      Chi-
 Effect         Parameter    Estimate    Error     Square    Pr > ChiSq
 ------------------------------------------------------------------------
 Intercept          1         0.7385    0.00941    6156.36     <.0001
 gender             2         0.0520    0.00940      30.63     <.0001
 year               3         0.0216    0.00460      22.13     <.0001
                    4         0.00978   0.00418       5.47     0.0193
```

In this example, there is significant departure from linearity ($Q_W = 5.47$, 1 df, $p = 0.0193$); the difference between year 6 and year 3 is significantly larger than that between year 3 and baseline.

14.2.4 Two Populations, Polytomous Response

The previous two sections describe repeated measures analyses when the response variable is dichotomous. In these situations, there is a single response function at each time point (level of the repeated measurement factor). This section describes the application of the WLS methodology when the response variable has more than two levels and the repeated measurement factor isn't time.

Table 14.3 * displays unaided distance vision data from 30–39 year old employees of United Kingdom Royal Ordnance factories during the years 1943–1946 (Kendall and Stuart 1961, pp. 564 and 586). Vision was graded in both the right eye and the left eye on a four-point ordinal categorical scale where 1=highest grade and 4=lowest grade. Interest focuses on determining if the marginal vision grade distributions are the same in the right eye as in the left eye, if the marginal distributions differ between females and males, and if differences between right eye and left eye vision are the same for females and males.

Table 14.3. Unaided Distance Vision Data

Gender	Right Eye Grade	Left Eye Grade 1	2	3	4	Total
Female	1	1520	266	124	66	1976
	2	234	1512	432	78	2256
	3	117	362	1772	205	2456
	4	36	82	179	492	789
	Total	1907	2222	2507	841	7477
Male	1	821	112	85	35	1053
	2	116	494	145	27	782
	3	72	151	583	87	893
	4	43	34	106	331	514
	Total	1952	791	919	480	3242

In this example, there are two populations (females, males). Two measurements of an ordered four-category response variable were obtained from each subject. Thus, there are

---

*Reprinted by permission of Edward Arnold.

$r = 4^2 = 16$ response profiles defined by the possible combinations of right-eye and left-eye vision grade. The between subjects variation is due to differences between females and males and the within subjects variation is due to differences between the right eye and the left eye.

The following SAS statements read in the counts displayed in Table 14.3 and create the SAS data set VISION.

```
data vision;
   input gender $ right left count;
   datalines;
F 1 1 1520
F 1 2  266
F 1 3  124
F 1 4   66
F 2 1  234
F 2 2 1512
F 2 3  432
F 2 4   78
F 3 1  117
F 3 2  362
F 3 3 1772
F 3 4  205
F 4 1   36
F 4 2   82
F 4 3  179
F 4 4  492
M 1 1  821
M 1 2  112
M 1 3   85
M 1 4   35
M 2 1  116
M 2 2  494
M 2 3  145
M 2 4   27
M 3 1   72
M 3 2  151
M 3 3  583
M 3 4   87
M 4 1   43
M 4 2   34
M 4 3  106
M 4 4  331
;
```

Since there are two populations, the null hypothesis of marginal homogeneity can be tested separately for females and males. The marginal distribution of vision grade in each eye involves three linearly independent proportions, since the proportions in the four categories sum to one. Thus, the null hypothesis of marginal homogeneity has 3 df for each gender. The following statements produce the analysis.

```
proc catmod;
   weight count;
   response marginals;
   model right*left=gender _response_(gender='F')
                          _response_(gender='M');
   repeated eye 2;
run;
```

The RESPONSE statement computes six correlated marginal proportions in each of the two populations. The first three response functions in each population are the proportions of subjects with right-eye vision grades of 1, 2, and 3, while the next three are the proportions with left-eye vision grades of 1, 2, and 3. For example, the response function for sample 1 (females), function number 1 (right-eye vision grade of 1) is the marginal proportion of subjects in this category:

$$\frac{\text{number of females with right-eye grade 1}}{\text{total number of females}} = \frac{1976}{7477} = 0.26428$$

In this example, you must specify that the repeated measures factor labeled EYE has two levels. If this specification is omitted, PROC CATMOD constructs a model matrix to test the 5 df null hypothesis that the six response functions from each population are equal. It is, however, not necessary to include the option _RESPONSE_=EYE in the REPEATED statement, since there is only one repeated measures factor and the default factorial _RESPONSE_ effect is desired.

Output 14.13 displays the population and response profiles, and Output 14.14 displays the response functions and model matrix.

Output 14.13   Population and Response Profiles

```
                    Population Profiles

        Sample      gender     Sample Size
        -------------------------------
            1         F               7477
            2         M               3242


                    Response Profiles

        Response      right      left
        -------------------------
            1           1         1
            2           1         2
            3           1         3
            4           1         4
            5           2         1
            6           2         2
            7           2         3
            8           2         4
            9           3         1
           10           3         2
           11           3         3
           12           3         4
           13           4         1
           14           4         2
           15           4         3
           16           4         4
```

Output 14.14   Response Functions and Model Matrix

```
                  Response Functions and Design Matrix

          Function    Response                   Design Matrix
  Sample   Number     Function     1     2     3     4     5     6     7
  ---------------------------------------------------------------------------
    1         1        0.26428      1     0     0     1     0     0     1
              2        0.30173      0     1     0     0     1     0     0
              3        0.32847      0     0     1     0     0     1     0
              4        0.25505      1     0     0     1     0     0    -1
              5        0.29718      0     1     0     0     1     0     0
              6        0.33529      0     0     1     0     0     1     0

    2         1        0.32480      1     0     0    -1     0     0     0
              2        0.24121      0     1     0     0    -1     0     0
              3        0.27545      0     0     1     0     0    -1     0
              4        0.32449      1     0     0    -1     0     0     0
              5        0.24399      0     1     0     0    -1     0     0
              6        0.28347      0     0     1     0     0    -1     0


                  Response Functions and Design Matrix

            Function                 Design Matrix
  Sample     Number      8     9     10     11     12
  ----------------------------------------------------------
    1           1         0     0     0      0      0
                2         1     0     0      0      0
                3         0     1     0      0      0
                4         0     0     0      0      0
                5        -1     0     0      0      0
                6         0    -1     0      0      0

    2           1         0     0     1      0      0
                2         0     0     0      1      0
                3         0     0     0      0      1
                4         0     0    -1      0      0
                5         0     0     0     -1      0
                6         0     0     0      0     -1
```

The first three parameters, which correspond to the first three columns of the model matrix, are overall intercepts for the probability of vision grades 1, 2, and 3. Recall that with a dichotomous response, the first column of the model matrix is an overall intercept for the probability of the first level of response. Likewise, with a polytomous response with r levels, there are $r - 1$ columns in the model matrix corresponding to overall intercepts for the probability of the first $r - 1$ levels of response, respectively. The next three parameters compare females to males at vision grades 1, 2, and 3, respectively. Parameters 7–9 (10–12) compare the right eye to the left eye at grades 1, 2, and 3 for females (males).

Output 14.15 displays the resulting ANOVA table. The test of marginal homogeneity is clearly significant in females ($Q_W = 11.98$, 3 df, $p = 0.0075$), but the differences between the right- and left- eye vision grade distributions in males are not statistically significant ($Q_W = 3.68$, 3 df, $p = 0.2984$).

Output 14.15 ANOVA Table for Gender-Specific Tests of Marginal Homogeneity

```
                       Analysis of Variance

          Source              DF   Chi-Square    Pr > ChiSq
          ----------------------------------------------------
          Intercept            3     71753.50      <.0001
          gender               3       142.07      <.0001
          eye(gender=F)        3        11.98       0.0075
          eye(gender=M)        3         3.68       0.2984

          Residual             0          .           .
```

If the differences between right-eye and left-eye vision are the same for females and males, there is no interaction between gender and eye. This hypothesis is tested using the following CONTRAST statement to compare parameters within the EYE(GENDER=F) and EYE(GENDER=M) effects.

```
contrast 'Interaction' all_parms 0 0 0 0 0 0 1 0 0 -1  0  0,
                        all_parms 0 0 0 0 0 0 0 1 0  0 -1  0,
                        all_parms 0 0 0 0 0 0 0 0 1  0  0 -1;
run;
```

The results in Output 14.16 indicate that there is evidence of interaction ($Q_W = 8.27$, 3 df, $p = 0.0407$).

Output 14.16 Test of Interaction

```
                     Analysis of Contrasts

          Contrast      DF    Chi-Square    Pr > ChiSq
          ------------------------------------------------
          Interaction    3        8.27        0.0407
```

You could also test the hypothesis of no interaction between gender and eye by fitting the model

```
model right*left=gender|_response_;
repeated eye 2;
run;
```

and looking at the GENDER*EYE effect in the ANOVA table. Although this model would provide a more straightforward test of no interaction, it would not provide tests of marginal homogeneity in females and males.

Since vision grade is an ordinal dependent variable, an alternative approach is to assign scores to its four levels and test the hypothesis that the average vision scores in the right and left eyes are the same. Using the scores 1, 2, 3, and 4 (the actual vision grades recorded), you can test the hypothesis of homogeneity for females and males by requesting that mean scores be computed as follows:

```
proc catmod;
   weight count;
   response means;
   model right*left=gender _response_(gender='F')
         _response_(gender='M') / noprofile;
   repeated eye;
run;
```

You do not need to specify the number of levels of the repeated measures factor in the
REPEATED statement since there are only two response functions per group and, by
default, the model matrix will be constructed to test their equality.

Output 14.17 displays the response functions and model matrix. Response function 1 in
sample 1 is the average right-eye vision grade for females. This is computed as follows:

$$\frac{1 \times 1976 + 2 \times 2256 + 3 \times 2456 + 4 \times 789}{7477} = 2.27524$$

The model matrix now includes an overall intercept, a gender effect, and two eye effects
(one for females and one for males).

Output 14.17   Response Functions and Model Matrix for Mean Score Model

		Response Functions and Design Matrix				
Sample	Function Number	Response Function	1	Design Matrix 2	3	4
1	1	2.27524	1	1	1	0
	2	2.30520	1	1	-1	0
2	1	2.26774	1	-1	0	1
	2	2.25509	1	-1	0	-1

Output 14.18 displays the resulting ANOVA table. The test of homogeneity is again
clearly significant in females ($Q_W = 11.97$, 1 df, $p = 0.0005$), and the difference between
the right- and left- average vision scores in males is not statistically significant
($Q_W = 0.73$, 1 df, $p = 0.3916$).

Output 14.18   ANOVA Table for Mean Score Model

	Analysis of Variance		
Source	DF	Chi-Square	Pr > ChiSq
Intercept	1	50866.50	<.0001
gender	1	2.04	0.1534
eye(gender=F)	1	11.97	0.0005
eye(gender=M)	1	0.73	0.3916
Residual	0	.	.

The following CONTRAST statement tests the null hypothesis that the mean score differences between right eye and left eye are equal for females and males.

```
contrast 'Interaction' all_parms 0 0 1 -1;
run;
```

The results in Output 14.19 again indicate that there is evidence of interaction ($Q_W = 6.20$, 1 df, $p = 0.0128$).

Output 14.19   Test of Interaction for Mean Score Model

```
                    Analysis of Contrasts

Contrast        DF     Chi-Square     Pr > ChiSq
-------------------------------------------------
Interaction      1          6.20         0.0128
```

Note that the values of the test statistics are affected only by the spacing between scores, not by their values. Thus, the same test statistics would have been obtained using any set of equally spaced scores, for example, vision scores of (1 3 5 7) instead of (1 2 3 4). If it is not reasonable to assume that the vision grades levels are equally-spaced, you may redefine the values of the RIGHT and LEFT variables to a different set of scores in a DATA step prior to invoking PROC CATMOD.

14.2.5   Multiple Repeated Measurement Factors

In each of the previous examples, a single categorical outcome variable was measured on multiple occasions or under multiple conditions. The multiple measurements were defined by the values of a single repeated measures factor, such as time, drug, and so on. In some applications, there may be more than one repeated measurement factor. For example, an outcome variable might be measured at four time points under each of two conditions, resulting in a total of eight repeated measures.

MacMillan et al. (1981) analyze data from a one-population ($s = 1$) observational study involving 793 subjects. For each subject, two diagnostic procedures (standard and test) were carried out at each of two times. The results of the four evaluations were classified as positive or negative. Since a dichotomous response ($c = 2$) was measured at $t = 4$ occasions, there are $r = 2^4 = 16$ response profiles. Table 14.4 displays the resulting data.

In this example, the four repeated measures are obtained from a factorial design involving two factors, each with two levels. Although the hypothesis of marginal homogeneity (with 3 df) could be tested, it is of greater interest to investigate the effects of time and treatment on the probability of a positive test result.

Table 14.4. Diagnostic Test Results for 793 Subjects

Time 1		Time 2		No. of
Standard	Test	Standard	Test	Subjects
Negative	Negative	Negative	Negative	509
Negative	Negative	Negative	Positive	4
Negative	Negative	Positive	Negative	17
Negative	Negative	Positive	Positive	3
Negative	Positive	Negative	Negative	13
Negative	Positive	Negative	Positive	8
Negative	Positive	Positive	Negative	0
Negative	Positive	Positive	Positive	8
Positive	Negative	Negative	Negative	14
Positive	Negative	Negative	Positive	1
Positive	Negative	Positive	Negative	17
Positive	Negative	Positive	Positive	9
Positive	Positive	Negative	Negative	7
Positive	Positive	Negative	Positive	4
Positive	Positive	Positive	Negative	9
Positive	Positive	Positive	Positive	170

The following statements read in the data and fit a model incorporating main effects for time and treatment, as well as the time × treatment interaction.

```
data diagnos;
   input std1 $ test1 $ std2 $ test2 $ count;
   datalines;
Neg Neg Neg Neg 509
Neg Neg Neg Pos   4
Neg Neg Pos Neg  17
Neg Neg Pos Pos   3
Neg Pos Neg Neg  13
Neg Pos Neg Pos   8
Neg Pos Pos Neg   0
Neg Pos Pos Pos   8
Pos Neg Neg Neg  14
Pos Neg Neg Pos   1
Pos Neg Pos Neg  17
Pos Neg Pos Pos   9
Pos Pos Neg Neg   7
Pos Pos Neg Pos   4
Pos Pos Pos Neg   9
Pos Pos Pos Pos 170
;
proc catmod;
   weight count;
   response marginals;
   model std1*test1*std2*test2=_response_ / oneway;
   repeated time 2, trtment 2 / _response_=time|trtment;
run;
```

Output 14.20 displays the one-way frequency distributions for each of the four response variables. Since this is a single population example, these distributions are useful for checking that the response functions are defined as desired.

Output 14.20   One-Way Frequency Distributions

```
                One-Way Frequencies

           Variable   Value   Frequency
           ------------------------------
           std1       Neg          562
                      Pos          231

           test1      Neg          574
                      Pos          219

           std2       Neg          560
                      Pos          233

           test2      Neg          586
                      Pos          207
```

The population and response profiles are displayed in Output 14.21. Note that only 15 of the 16 potential response profiles occur in the data. There were no subjects who were negative for standard at time 1, positive for test at time 1, positive for standard at time 2, and negative for test at time 2.

Output 14.21   Population and Response Profiles

```
                    Population Profiles

               Sample    Sample Size
               ---------------------
                  1           793


                    Response Profiles

     Response    std1    test1    std2    test2
     ---------------------------------------------
         1       Neg     Neg      Neg     Neg
         2       Neg     Neg      Neg     Pos
         3       Neg     Neg      Pos     Neg
         4       Neg     Neg      Pos     Pos
         5       Neg     Pos      Neg     Neg
         6       Neg     Pos      Neg     Pos
         7       Neg     Pos      Pos     Pos
         8       Pos     Neg      Neg     Neg
         9       Pos     Neg      Neg     Pos
        10       Pos     Neg      Pos     Neg
        11       Pos     Neg      Pos     Pos
        12       Pos     Pos      Neg     Neg
        13       Pos     Pos      Neg     Pos
        14       Pos     Pos      Pos     Neg
        15       Pos     Pos      Pos     Pos
```

Output 14.22 displays the response functions and the model matrix. Since the MODEL statement lists the four dependent variables in the order standard at time 1, test at time 1, standard at time 2, and test at time 2, the response functions computed by the keyword MARGINALS on the RESPONSE statement are the corresponding four marginal proportions with a negative response. The marginal probability of a negative response is used since 'Neg' is the first level for each response variable. The REPEATED statement specifies that there are two repeated measures factors, each with two levels, and that the model matrix includes the main effects and interaction of these two factors. Since the MODEL statement groups the results from time 1 together, followed by the results from time 2, the repeated measures factor for time is listed first in the REPEATED statement. Recall that the factor that changes most slowly is listed first.

Output 14.22 Response Functions and Model Matrix

```
                 Response Functions and Design Matrix

                  Function     Response           Design Matrix
       Sample      Number      Function      1      2      3      4
       ------------------------------------------------------------
          1           1         0.70870       1      1      1      1
                      2         0.72383       1      1     -1     -1
                      3         0.70618       1     -1      1     -1
                      4         0.73897       1     -1     -1      1
```

The results of fitting this model (Output 14.23) indicate that the main effect of time ($Q_W = 0.85$, 1 df, $p = 0.3570$) and the time $\times$ treatment interaction ($Q_W = 2.40$, 1 df, $p = 0.1215$) are not significantly different from zero.

Output 14.23 ANOVA Table

```
                    Analysis of Variance

       Source         DF     Chi-Square     Pr > ChiSq
       ----------------------------------------------------
       Intercept       1       2385.34        <.0001
       time            1          0.85        0.3570
       trtment         1          8.20        0.0042
       time*trtment    1          2.40        0.1215

       Residual        0            .             .
```

The preceding results could also have been produced using a different ordering of the response variables and repeated measurement factors on the MODEL and REPEATED statements:

```
model std1*std2*test1*test2=_response_ / oneway;
repeated trtment 2, time 2 / _response_=time|trtment;
```

In this case, the responses from the standard treatment precede those from the test treatment.

A reduced model including only the treatment main effect can be fit by modifying the _RESPONSE_= option of the REPEATED statement, as shown below.

```
proc catmod;
   weight count;
   response marginals;
   model std1*test1*std2*test2=_response_ / noprofile;
   repeated time 2, trtment 2 / _response_=trtment;
run;
```

The results in Output 14.24 indicate that this model provides a good fit to the observed data ($Q_W = 3.51$, 2 df, $p = 0.1731$) and that the treatment effect is clearly significant ($Q_W = 9.55$, 1 df, $p = 0.0020$). Since the parameter for the first treatment (standard) is -0.0128, the parameter estimate for the test treatment is 0.0128. Consequently, it is estimated that the marginal probability of a negative response is $2 \times 0.0128 = 0.0256$ higher for the test treatment than for the standard treatment.

Output 14.24 Results from the Reduced Model

```
                Response Functions and Design Matrix

                   Function      Response     Design Matrix
         Sample     Number       Function        1      2
         -----------------------------------------------------
           1           1         0.70870         1      1
                       2         0.72383         1     -1
                       3         0.70618         1      1
                       4         0.73897         1     -1


                     Analysis of Variance

         Source         DF    Chi-Square    Pr > ChiSq
         -----------------------------------------------
         Intercept       1      2386.97       <.0001
         trtment         1         9.55       0.0020

         Residual        2         3.51       0.1731


         Analysis of Weighted Least Squares Estimates

                                       Standard      Chi-
         Effect      Parameter  Estimate  Error     Square    Pr > ChiSq
         -----------------------------------------------------------------
         Intercept       1       0.7196   0.0147    2386.97      <.0001
         trtment         2      -0.0128   0.00416      9.55      0.0020
```

14.3 Advanced Topic: Further Weighted Least Squares Applications

14.3.1 One Population Regression Analysis of Marginal Proportions and Logits

In a longitudinal study of the health effects of air pollution (Ware, Lipsitz, and Speizer 1988), children were examined annually at ages 9, 10, 11, and 12. At each examination, the response measured was the presence of wheezing. Two questions of interest are

- Does the prevalence of wheezing change with age?
- Is there a quantifiable trend in the age-specific prevalence rates?

Table 14.5 *, from Agresti (1990, p. 408), displays data from 1019 children included in this study. In this single population example, the cross-classification of a dichotomous outcome at four time points defines $r = 2^4 = 16$ response profiles.

Table 14.5.   Breath Test Results at Four Ages for 1019 Children

Wheeze				No. of
Age 9	Age 10	Age 11	Age 12	Children
Present	Present	Present	Present	94
Present	Present	Present	Absent	30
Present	Present	Absent	Present	15
Present	Present	Absent	Absent	28
Present	Absent	Present	Present	14
Present	Absent	Present	Absent	9
Present	Absent	Absent	Present	12
Present	Absent	Absent	Absent	63
Absent	Present	Present	Present	19
Absent	Present	Present	Absent	15
Absent	Present	Absent	Present	10
Absent	Present	Absent	Absent	44
Absent	Absent	Present	Present	17
Absent	Absent	Present	Absent	42
Absent	Absent	Absent	Present	35
Absent	Absent	Absent	Absent	572

The following SAS statements read the observed counts for each of the 16 response profiles and test the hypothesis of marginal homogeneity using PROC CATMOD. The RESPONSE, MODEL, and REPEATED statements are used in the same way as was described in Section 14.2.

```
data wheeze;
    input wheeze9 $ wheeze10 $ wheeze11 $ wheeze12 $ count;
    datalines;
Present Present Present Present   94
Present Present Present Absent    30
Present Present Absent  Present   15
Present Present Absent  Absent    28
Present Absent  Present Present   14
Present Absent  Present Absent     9
Present Absent  Absent  Present   12
Present Absent  Absent  Absent    63
Absent  Present Present Present   19
Absent  Present Present Absent    15
```

```
   Absent    Present  Absent    Present    10
   Absent    Present  Absent    Absent     44
   Absent    Absent   Present   Present    17
   Absent    Absent   Present   Absent     42
   Absent    Absent   Absent    Present    35
   Absent    Absent   Absent    Absent    572
   ;
   proc catmod order=data;
      weight count;
      response marginals;
      model wheeze9*wheeze10*wheeze11*wheeze12=_response_ / oneway;
      repeated age;
   run;
```

Since this is a single population example, the one-way frequency distributions for each of the four response variables (Output 14.25) are useful in checking that the response functions are defined as desired.

Output 14.25 One-Way Frequency Distributions

```
                  One-Way Frequencies

             Variable    Value      Frequency
             ------------------------------------
             wheeze9     Present        265
                         Absent         754

             wheeze10    Present        255
                         Absent         764

             wheeze11    Present        240
                         Absent         779

             wheeze12    Present        216
                         Absent         803
```

The response profile listing in Output 14.26 verifies that all 16 possible response profiles occur in the data.

Output 14.26   Response Profiles

```
                        Response Profiles

    Response    wheeze9    wheeze10    wheeze11    wheeze12
    -------------------------------------------------------------
       1        Present    Present     Present     Present
       2        Present    Present     Present     Absent
       3        Present    Present     Absent      Present
       4        Present    Present     Absent      Absent
       5        Present    Absent      Present     Present
       6        Present    Absent      Present     Absent
       7        Present    Absent      Absent      Present
       8        Present    Absent      Absent      Absent
       9        Absent     Present     Present     Present
      10        Absent     Present     Present     Absent
      11        Absent     Present     Absent      Present
      12        Absent     Present     Absent      Absent
      13        Absent     Absent      Present     Present
      14        Absent     Absent      Present     Absent
      15        Absent     Absent      Absent      Present
      16        Absent     Absent      Absent      Absent
```

Output 14.27 displays the response functions and the model matrix. The results of the ANOVA table in Output 14.28 indicate that the hypothesis of marginal homogeneity is rejected ($Q_W = 12.85$, 3 df, $p = 0.0050$). Thus, the prevalence of wheezing changes with age.

Output 14.27   Response Functions and Model Matrix

```
                Response Functions and Design Matrix

                Function    Response          Design Matrix
    Sample      Number      Function     1      2      3      4
    ---------------------------------------------------------------
       1          1         0.26006      1      1      0      0
                  2         0.25025      1      0      1      0
                  3         0.23553      1      0      0      1
                  4         0.21197      1     -1     -1     -1
```

Output 14.28   ANOVA Table

```
                    Analysis of Variance

           Source       DF    Chi-Square    Pr > ChiSq
           ----------------------------------------------
           Intercept     1       523.63        <.0001
           age           3        12.85        0.0050

           Residual      0          .             .
```

In order to quantify the trend in the age-specific prevalence rates, the proportion of children with wheezing present is modeled as a linear function of age. In this case, it is not possible to use the _RESPONSE_ keyword of the MODEL statement in conjunction with the REPEATED statement, since a repeated measurement factor cannot be specified in a DIRECT statement. Therefore, the model matrix must be specified explicitly, as described in Section 12.8.5 The SAS statements are as follows:

```
proc catmod order=data;
   weight count;
   response marginals;
   model wheeze9*wheeze10*wheeze11*wheeze12=(1   9,
                                             1  10,
                                             1  11,
                                             1  12)
                                            (1='Intercept',
                                             2='Linear Age')
                                            / noprofile;
run;
```

When you input the model matrix directly, you have the option of testing the significance of selected parameters or subsets of parameters. The specification enclosed in parentheses after the model matrix stipulates that the ANOVA table include tests that the first (intercept) and second (linear age) parameters are equal to zero. Note that the label describing the parameter or subset to be tested must be 24 characters or less. The results of this model statement are displayed in Output 14.29 and Output 14.30.

Output 14.29   Response Functions and Model Matrix

```
               Response Functions and Design Matrix

                   Function      Response     Design Matrix
         Sample     Number       Function       1       2
         ----------------------------------------------------
            1          1          0.26006        1       9
                       2          0.25025        1      10
                       3          0.23553        1      11
                       4          0.21197        1      12
```

Output 14.30   ANOVA Table

```
                    Analysis of Variance

         Source       DF    Chi-Square    Pr > ChiSq
         ----------------------------------------------
         Intercept     1       66.70        <.0001
         Linear Age    1       12.31        0.0005

         Residual      2        0.54        0.7620
```

The residual chi-square tests the null hypothesis that the nonlinear (quadratic, cubic) components of the age effect are not significantly different from zero. These data provide little evidence of nonlinearity ($Q_W = 0.54$, 2 df, $p = 0.7620$). If this test had been statistically significant, the quadratic and, if necessary, cubic terms could also have been specified in the model matrix. The test of the linear age effect is clearly significant ($Q_W = 12.31$, 1 df, $p < 0.0005$).

With reference to the estimated parameters displayed in Output 14.31, the resulting model for predicting the effect of age on the prevalence of wheezing is

$$\Pr\{\text{wheezing}\} = 0.4083 - 0.0161 \times \text{age in years}$$

Thus, the prevalence is estimated to decrease by 0.0161 per year of age.

Output 14.31 Parameter Estimates

```
              Analysis of Weighted Least Squares Estimates

                                        Standard        Chi-
        Effect    Parameter    Estimate    Error      Square    Pr > ChiSq
       ---------------------------------------------------------------------
        Model         1         0.4083     0.0500      66.70       <.0001
                      2        -0.0161     0.00460     12.31       0.0005
```

Chapter 8 describes logistic models for dichotomous response variables. As an alternative to modeling the probability of wheezing as a linear function of age, you could choose to model the marginal logit of the probability of wheezing. In this case, the logarithm of the odds is modeled as a linear function of age. Even if there are no substantive grounds for preferring a logit analysis over the analysis on the proportion scale, you may decide to consider both types of models and select the model that provides the simplest interpretation.

Since it is not possible to analyze repeated measurements using the LOGISTIC procedure, maximum likelihood parameter estimates can not be obtained. However, PROC CATMOD can be used to estimate model parameters using weighted least squares.

Suppose L_x denotes the observed log odds of wheezing at age x, for $x = 9, 10, 11, 12$, respectively, that is,

$$L_x = \log\left(\frac{p_x}{1 - p_x}\right)$$

where p_x denotes the marginal probability of wheezing at age x. The following statements fit the regression model

$$L_x = \alpha + \beta x$$

The only change from the previous model is that the keyword MARGINALS on the RESPONSE statement is replaced by the keyword LOGITS.

```
proc catmod order=data;
   weight count;
   response logits;
   model wheeze9*wheeze10*wheeze11*wheeze12=(1  9,
                                             1 10,
                                             1 11,
                                             1 12)
                              (1='Intercept',
                               2='Linear Age') / noprofile;
run;
```

Output 14.32 displays the marginal logit response functions and the model matrix. The ANOVA table in Output 14.33 indicates that the regression model for marginal logits also provides a good fit to the observed data ($Q_W = 0.67$, 2 df, $p = 0.7167$) and that the linear effect of age is clearly significant ($Q_W = 11.77$, 1 df, $p = 0.0006$).

Output 14.32   Response Functions and Model Matrix

```
                 Response Functions and Design Matrix

                  Function      Response      Design Matrix
        Sample     Number       Function       1        2
        --------------------------------------------------------
           1          1         -1.04566        1        9
                      2         -1.09730        1       10
                      3         -1.17737        1       11
                      4         -1.31308        1       12
```

Output 14.33   ANOVA Table

```
                     Analysis of Variance

        Source        DF    Chi-Square    Pr > ChiSq
        ----------------------------------------------
        Intercept      1        0.76         0.3824
        Linear Age     1       11.77         0.0006

        Residual       2        0.67         0.7167
```

The model for predicting the log odds of wheezing (Output 14.34) is

$$\text{logit}[\text{Pr}\{\text{wheezing}\}] = -0.2367 - 0.0879 \times \text{age in years}$$

The parameter estimates are interpreted in the same manner as was described in Chapter 8. For example, the odds of wheezing are estimated to decrease by $e^{-0.0879} = 0.916$ for each one-year increase in age.

Output 14.34   Parameter Estimates

```
           Analysis of Weighted Least Squares Estimates

                                    Standard      Chi-
      Effect    Parameter  Estimate   Error      Square   Pr > ChiSq
      --------------------------------------------------------------
      Model         1       -0.2367   0.2710      0.76      0.3824
                    2       -0.0879   0.0256     11.77      0.0006
```

As described in Section 12.6.1, the logit function is the default response function for the CATMOD procedure, and maximum likelihood is the default estimation method. However, Output 14.34 displays weighted least squares parameter estimates. In a repeated measures analysis, the specification

```
response logits;
```

analyzes marginal logits using weighted least squares. If the RESPONSE statement is omitted in this example, 15 generalized logits would be computed, comparing each of the first 15 response profiles with the last one. Since the model matrix has only four rows, an error message would then be printed.

14.3.2   Analysis of Respiratory Data

The following example describes a repeated measurements analysis of logits for a data set that has more explanatory variables than you can manage with WLS. An appropriate analysis is performed, with the limitations noted. These data are analyzed again with the GEE strategy in Chapter 15 in Section 15.6, which compares the WLS and GEE analyses and then proceeds to discuss the more comprehensive analysis you can perform with the GEE strategy.

A clinical trial compared two treatments for a respiratory illness (Koch et al. 1990). In each of two centers, eligible patients were randomly assigned to active treatment or placebo. During treatment, respiratory status was determined at four visits and recorded on a five-point scale as 0 for terrible to 4 for excellent. Potential explanatory variables were center, sex, and baseline respiratory status (all dichotomous), as well as age (in years) at the time of study entry. There were 111 patients (54 active, 57 placebo), with no missing data for responses or covariates. One direction of analysis was to focus on the dichotomous response of good outcome (response is 3 or 4) versus poor outcome (response is less than 3). Tables 14.6 displays the data from Center 1.

Table 14.6. Respiratory Disorder Data for 56 Subjects from Center 1

Patient	Treatment	Sex	Age	Respiratory Status (0=poor, 1=good)				
				Baseline	Visit 1	Visit 2	Visit 3	Visit 4
1	P	M	46	0	0	0	0	0
2	P	M	28	0	0	0	0	0
3	A	M	23	1	1	1	1	1
4	P	M	44	1	1	1	1	0
5	P	F	13	1	1	1	1	1
6	A	M	34	0	0	0	0	0
7	P	M	43	0	1	0	1	1
8	A	M	28	0	0	0	0	0
9	A	M	31	1	1	1	1	1
10	P	M	37	1	0	1	1	0
11	A	M	30	1	1	1	1	1
12	A	M	14	0	1	1	1	0
13	P	M	23	1	1	0	0	0
14	P	M	30	0	0	0	0	0
15	P	M	20	1	1	1	1	1
16	A	M	22	0	0	0	0	1
17	P	M	25	0	0	0	0	0
18	A	F	47	0	0	1	1	1
19	P	F	31	0	0	0	0	0
20	A	M	20	1	1	0	1	0
21	A	M	26	0	1	0	1	0
22	A	M	46	1	1	1	1	1
23	A	M	32	1	1	1	1	1
24	A	M	48	0	1	0	0	0
25	P	F	35	0	0	0	0	0
26	A	M	26	0	0	0	0	0
27	P	M	23	1	1	0	1	1
28	P	F	36	0	1	1	0	0
29	P	M	19	0	1	1	0	0
30	A	M	28	0	0	0	0	0
31	P	M	37	0	0	0	0	0
32	A	M	23	0	1	1	1	1
33	A	M	30	1	1	1	1	0
34	P	M	15	0	0	1	1	0
35	A	M	26	0	0	0	1	0
36	P	F	45	0	0	0	0	0
37	A	M	31	0	0	1	0	0
38	A	M	50	0	0	0	0	0
39	P	M	28	0	0	0	0	0
40	P	M	26	0	0	0	0	0
41	P	M	14	0	0	0	0	1
42	A	M	31	0	0	1	0	0
43	P	M	13	1	1	1	1	1
44	P	M	27	0	0	0	0	0
45	P	M	26	0	1	0	1	1
46	P	M	49	0	0	0	0	0
47	P	M	63	0	0	0	0	0
48	A	M	57	1	1	1	1	1
49	P	M	27	1	1	1	1	1
50	A	M	22	0	0	1	1	1
51	A	M	15	0	0	1	1	1
52	P	M	43	0	0	0	1	0
53	A	F	32	0	0	0	1	0
54	A	M	11	1	1	1	1	0
55	P	M	24	1	1	1	1	1
56	A	M	25	0	1	1	0	1

Consider performing a WLS analysis on the logit response. With four visits, that means four logits are modeled for each subpopulation. If you want to include visit as an effect, you can handle at most two other dichotomous explanatory variables. Use treatment since it is the evaluation factor and use center since it is part of the study design structure. The two dichotomous covariates define $2 \times 2 = 4$ subpopulations, which, when you have four logits for the four visits, implies that you have 16 degrees of freedom. You are stretching the limits of the preferred 15 plus subjects per response function. However, since these visits act relatively independently, that is, there is not a tremendous amount of correlation, you could view the sample size requirements as equivalent to those for two response functions instead of four. With about 25 subjects in each subpopulation, you are close to the preferred sample sizes.

The following DATA step inputs these data. The variables VISIT1-VISIT4 are created to be the dichotomous response variables.

```
data resp;
   input center id treatment $ sex $ age baseline visit1-visit4 @@;
   datalines;
1  1 P M 46 0 0 0 0 0   2  1 P F 39 0 0 0 0 0
1  2 P M 28 0 0 0 0 0   2  2 A M 25 0 0 1 1 1
1  3 A M 23 1 1 1 1 1   2  3 A M 58 1 1 1 1 1
1  4 P M 44 1 1 1 1 0   2  4 P F 51 1 1 0 1 1
1  5 P F 13 1 1 1 1 1   2  5 P F 32 1 0 0 1 1
1  6 A M 34 0 0 0 0 0   2  6 P M 45 1 1 0 0 0
1  7 P M 43 0 1 0 1 1   2  7 P F 44 1 1 1 1 1
1  8 A M 28 0 0 0 0 0   2  8 P F 48 0 0 0 0 0
1  9 A M 31 1 1 1 1 1   2  9 A M 26 0 1 1 1 1
1 10 P M 37 1 0 1 1 0   2 10 A M 14 0 1 1 1 1
1 11 A M 30 1 1 1 1 1   2 11 P F 48 0 0 0 0 0
1 12 A M 14 0 1 1 1 0   2 12 A M 13 1 1 1 1 1
1 13 P M 23 1 1 0 0 0   2 13 P M 20 0 1 1 1 1
1 14 P M 30 0 0 0 0 0   2 14 A M 37 1 1 0 0 1
1 15 P M 20 1 1 1 1 1   2 15 A M 25 1 1 1 1 1
1 16 A M 22 0 0 0 0 1   2 16 A M 20 0 0 0 0 0
1 17 P M 25 0 0 0 0 0   2 17 P F 58 0 1 0 0 0
1 18 A F 47 0 0 1 1 1   2 18 P M 38 1 1 0 0 0
1 19 P F 31 0 0 0 0 0   2 19 A M 55 1 1 1 1 1
1 20 A M 20 1 1 0 1 0   2 20 A M 24 1 1 1 1 1
1 21 A M 26 0 1 0 1 0   2 21 P F 36 1 1 0 0 1
1 22 A M 46 1 1 1 1 1   2 22 P M 36 0 1 1 1 1
1 23 A M 32 1 1 1 1 1   2 23 A F 60 1 1 1 1 1
1 24 A M 48 0 1 0 0 0   2 24 P M 15 1 0 0 1 1
1 25 P F 35 0 0 0 0 0   2 25 A M 25 1 1 1 1 0
1 26 A M 26 0 0 0 0 0   2 26 A M 35 1 1 1 1 1
1 27 P M 23 1 1 0 1 1   2 27 A M 19 1 1 0 1 1
1 28 P F 36 0 1 1 0 0   2 28 P F 31 1 1 1 1 1
1 29 P M 19 0 1 1 0 0   2 29 A M 21 1 1 1 1 1
1 30 A M 28 0 0 0 0 0   2 30 A F 37 0 1 1 1 1
1 31 P M 37 0 0 0 0 0   2 31 P M 52 0 1 1 1 1
1 32 A M 23 0 1 1 1 1   2 32 A M 55 0 0 1 1 0
1 33 A M 30 1 1 1 1 0   2 33 P M 19 1 0 0 1 1
1 34 P M 15 0 0 1 1 0   2 34 P M 20 1 0 1 1 1
1 35 A M 26 0 0 0 1 0   2 35 P M 42 1 0 0 0 0
```

```
1 36 P F 45 0 0 0 0 0    2 36 A M 41 1 1 1 1 1
1 37 A M 31 0 0 1 0 0    2 37 A M 52 0 0 0 0 0
1 38 A M 50 0 0 0 0 0    2 38 P F 47 0 1 1 0 1
1 39 P M 28 0 0 0 0 0    2 39 P M 11 1 1 1 1 1
1 40 P M 26 0 0 0 0 0    2 40 P M 14 0 0 0 1 0
1 41 P M 14 0 0 0 0 1    2 41 P M 15 1 1 1 1 1
1 42 A M 31 0 0 1 0 0    2 42 P M 66 1 1 1 1 1
1 43 P M 13 1 1 1 1 1    2 43 A M 34 0 1 1 0 1
1 44 P M 27 0 0 0 0 0    2 44 P M 43 0 0 0 0 0
1 45 P M 26 0 1 0 1 1    2 45 P M 33 1 1 1 0 1
1 46 P M 49 0 0 0 0 0    2 46 P M 48 1 1 0 0 0
1 47 P M 63 0 0 0 0 0    2 47 A M 20 0 1 1 1 1
1 48 A M 57 1 1 1 1 1    2 48 P F 39 1 0 1 0 0
1 49 P M 27 1 1 1 1 1    2 49 A M 28 0 1 0 0 0
1 50 A M 22 0 0 1 1 1    2 50 P F 38 0 0 0 0 0
1 51 A M 15 0 0 1 1 1    2 51 A M 43 1 1 1 1 0
1 52 P M 43 0 0 0 1 0    2 52 A F 39 0 1 1 1 1
1 53 A F 32 0 0 0 1 0    2 53 A M 68 0 1 1 1 1
1 54 A M 11 1 1 1 1 0    2 54 A F 63 1 1 1 1 1
1 55 P M 24 1 1 1 1 1    2 55 A M 31 1 1 1 1 1
1 56 A M 25 0 1 1 0 1
;
```

The following PROC CATMOD statements produce the desired repeated measurements analysis. Since an incremental effects analysis is desired in order to make some comparisons with analyses performed in Chapter 15, direct input of the model matrix is required. As described in Section 13.7.5 you write the coefficients row-wise, separating each row with a comma. This model includes main effects for treatment, center, and visits, and all pairwise interactions. The numbers and labels in the parentheses after the model matrix identify the parameter that each column of the model matrix represents.

```
proc catmod data=resp;
   population treatment center;
   response logits;
   model visit1*visit2*visit3*visit4 =
    ( 1 1 1 1 0 0 1 0 0 1 0 0 1,
      1 1 1 0 1 0 0 1 0 0 1 0 1,
      1 1 1 0 0 1 0 0 1 0 0 1 1,
      1 1 1 0 0 0 0 0 0 0 0 0 1,
      1 1 0 1 0 0 1 0 0 0 0 0 0,
      1 1 0 0 1 0 0 1 0 0 0 0 0,
      1 1 0 0 0 1 0 0 1 0 0 0 0,
      1 1 0 0 0 0 0 0 0 0 0 0 0,
      1 0 1 1 0 0 0 0 0 1 0 0 0,
      1 0 1 0 1 0 0 0 0 0 1 0 0,
      1 0 1 0 0 1 0 0 0 0 0 1 0,
      1 0 1 0 0 0 0 0 0 0 0 0 0,
      1 0 0 1 0 0 0 0 0 0 0 0 0,
      1 0 0 0 1 0 0 0 0 0 0 0 0,
      1 0 0 0 0 1 0 0 0 0 0 0 0,
      1 0 0 0 0 0 0 0 0 0 0 0 0)
      (1='Intercept', 2='treatment', 3='center',
```

```
          4='visit1', 5='visit2', 6='visit3', 7='tr*visit1',
          8='tr*visit2', 9='tr*visit3', 10='ct*visit1',
          11='ct*visit2', 12='ct*visit3', 13='trt*ct');
```

The ANOVA table for this MODEL statement will only provide the single degree of freedom tests. To produce the multidegree of freedom tests required, you need to supply the appropriate CONTRAST statements. The following CONTRAST statements produce a contrast table with tests for all of the desired effects. Note that you have to use the ALL_PARMS keyword and supply coefficients for each column of the model matrix. The following CONTRAST statements come after the MODEL statement in the PROC CATMOD invocation.

```
          contrast 'treatment'
          all_parms  0 1 0 0 0 0 .25 .25 .25 0 0 0 .5;
          contrast 'center'
          all_parms  0 0 1 0 0 0 0 0 0 .25 .25 .25 .5;
          contrast 'visit' all_parms 0 0 0 1 0 0 .5 0 0  .5 0  0 0,
                           all_parms 0 0 0 0 1 0 0 .5 0  0 .5  0 0,
                           all_parms 0 0 0 0 0 1 0 0 .5 0  0 .5 0;
          contrast 'trt*visit' all_parms 0 0 0 0 0 0 1 0 0 0 0 0 0,
                               all_parms 0 0 0 0 0 0 0 1 0 0 0 0 0,
                               all_parms 0 0 0 0 0 0 0 0 1 0 0 0 0;
          contrast 'ct*visit'  all_parms 0 0 0 0 0 0 0 0 0 1 0 0 0,
                               all_parms 0 0 0 0 0 0 0 0 0 0 1 0 0,
                               all_parms 0 0 0 0 0 0 0 0 0 0 0 1 0;
          contrast 'trt*ct'    all_parms 0 0 0 0 0 0 0 0 0 0 0 0 1;
        run;
```

In the resulting output, the population profiles tell you that the underlying contingency table being analyzed is based on the cross-classification of treatment and center.

Output 14.35   Population Profiles

```
                          Population Profiles

             Sample      treatment      center    Sample Size
             -----------------------------------------------
                1          A             1              27
                2          A             2              27
                3          P             1              29
                4          P             2              28
```

There are 25 plus persons per subpopulation; as discussed, this sample size is usually considered to be inadequate for the analysis of four response functions, the logits for each visit. This analysis is stretching the limits by making the assumption that the expected low correlation among the visits makes the sample size requirements for four logits to be more like the sample size requirements for four logits.

The ANOVA table displays a one degree of freedom test for each parameter, which is of limited value at this point. However, it also provides the residual goodness-of-fit test, which has the value 0.59 with 3 df and a p-value of 0.8994. Note that the analysis is a weighted least squares analysis; you get a WLS analysis, not a ML analysis, when you specify logits and you have multiple responses. With the adequate fit, you can proceed with evaluating the contrasts.

Output 14.36   ANOVA Table

```
                    Analysis of Variance

     Source        DF    Chi-Square    Pr > ChiSq
     -------------------------------------------------
     Intercept     1        0.33         0.5675
     treatment     1        4.20         0.0405
     center        1        3.44         0.0635
     visit1        1        0.19         0.6635
     visit2        1        2.76         0.0967
     visit3        1        1.26         0.2613
     tr*visit1     1        0.10         0.7464
     tr*visit2     1        2.31         0.1284
     tr*visit3     1        1.32         0.2507
     ct*visit1     1        0.23         0.6291
     ct*visit2     1        1.75         0.1855
     ct*visit3     1        3.67         0.0554
     trt*ct        1        1.18         0.2770

     Residual      3        0.59         0.8994
```

The following table contains the contrasts requested, which provides the multiple degree of freedom effects. All of the pairwise interactions are not significant, with p-values ranging from 0.4338 for the treatment $\times$ visit interaction to 0.2770 for the treatment $\times$ center interaction. Since the visit effect is represented by 3 parameters, and the effects for treatment and center are represented by 1 each, the corresponding visit interaction effects are represented by 3 parameters each, respectively.

Output 14.37   Analysis of Contrasts

```
                    Analysis of Contrasts

     Contrast      DF    Chi-Square    Pr > ChiSq
     -------------------------------------------------
     treatment     1       10.92         0.0010
     center        1        8.37         0.0038
     visit         3        3.86         0.2769
     trt*visit     3        2.74         0.4338
     ct*visit      3        3.82         0.2821
     trt*ct        1        1.18         0.2770
```

The analysis continues with the main effects model including treatment, center, and visit. Note that, if the interactions were significant, it would be problematic. With the too-limited sample size, it would not be clear whether the interactions effects were real or possibly an artifact of that small sample size.

The following PROC CATMOD statements produce the main effects model.

```
proc catmod data=resp;
   population treatment center;
   response logits;
   model visit1*visit2*visit3*visit4 =
    ( 1 1 1 1 0 0 ,
      1 1 1 0 1 0 ,
      1 1 1 0 0 1 ,
      1 1 1 0 0 0 ,
      1 1 0 1 0 0 ,
      1 1 0 0 1 0 ,
      1 1 0 0 0 1 ,
      1 1 0 0 0 0 ,
      1 0 1 1 0 0 ,
      1 0 1 0 1 0 ,
      1 0 1 0 0 1 ,
      1 0 1 0 0 0 ,
      1 0 0 1 0 0 ,
      1 0 0 0 1 0 ,
      1 0 0 0 0 1 ,
      1 0 0 0 0 0 )
     (1='Intercept', 2='treatment', 3='center',
      4='visit1', 5='visit2', 6='visit3' );
   contrast 'treatment' all_parms  0 1 0 0 0 0 ;
   contrast 'center'    all_parms  0 0 1 0 0 0 ;
   contrast 'visit'     all_parms  0 0 0 1 0 0,
                        all_parms  0 0 0 0 1 0,
                        all_parms  0 0 0 0 0 1;
run;
```

The ANOVA table for this analysis is displayed next and shows a residual chi-square with a value of 13.04 for 10 df and a p-value of 0.2212. This supports adequate fit.

Output 14.38 ANOVA

Analysis of Variance			
Source	DF	Chi-Square	Pr > ChiSq
Intercept	1	0.00	0.9538
treatment	1	10.61	0.0011
center	1	7.62	0.0058
visit2	1	1.39	0.2392
visit2	1	0.01	0.9158
visit3	1	1.16	0.2813
Residual	10	13.04	0.2212

Note that if you subtracted the residual chi-square for the main effects model from the chi-square for the model with interactions, $13.04 - 0.59 = 12.45$ (10 df minus 3 df equals 7 df), you have a chi-square test for the joint test of the three pairwise interactions. This test is nonsignificant.

The analysis of contrasts displayed in Output 14.39 shows that the treatment and center effects are highly significant with p-values of 0.0011 and 0.0058, respectively. The joint test for the visit effect has a chi-square value of 2.76, and a p-value of 0.4303 for 3 df. The visit effect does not appear to be influential.

Output 14.39   Analysis of Contrasts

```
                     Analysis of Contrasts

         Contrast     DF    Chi-Square   Pr > ChiSq
         ------------------------------------------
         treatment    1        10.61       0.0011
         center       1         7.62       0.0058
         visit        3         2.76       0.4303
```

The parameter estimates are displayed in Output 14.40.

Output 14.40   Parameter Estimates

```
              Analysis of Weighted Least Squares Estimates

                                         Standard      Chi-
         Effect    Parameter   Estimate    Error      Square   Pr > ChiSq
         ----------------------------------------------------------------
         Model         1        -0.0168    0.2901       0.00      0.9538
                       2        -1.0434    0.3203      10.61      0.0011
                       3         0.8803    0.3188       7.62      0.0058
                       4        -0.2216    0.1883       1.39      0.2392
                       5        -0.0201    0.1896       0.01      0.9158
                       6        -0.1811    0.1681       1.16      0.2813
```

No further model reductions are attempted because visit is considered part of the analysis structure. This analysis demonstrates that weighted least squares analysis of logits can be a reasonable strategy for the evaluation of a small number of main effects. However, when your data contains additional explanatory variables such as baseline, sex, and age, which were ignored in this analysis, the WLS approach is not appropriate. While more sample size would eventually support the inclusion of the sex variable, it would still not support the inclusion of the continuous variable age. Chapter 15 discusses the generalized estimating equations approach for analyzing repeated categorical responses, and this method does handle continuous explanatory variables as well as time-dependent explanatory variables. See Section 15.6 for the GEE analysis of the same data set and a comparison of the GEE and WLS analyses.

Chapter 15
Generalized Estimating Equations

Chapter Table of Contents

Chapter 15
Generalized Estimating Equations

15.1 Introduction

The weighted least squares methodology described in Chapter 12 is a useful approach to the analysis of repeated binary and ordered categorical outcome variables. However, it can only accommodate categorical explanatory variables and can't easily handle missing values. In addition, the WLS methodology requires sufficient sample size for the marginal response functions at each time in each subpopulation to have an approximately multivariate normal distribution. This requirement can be very restrictive.

In recent years, researchers have begun to use a new method for the repeated measurements analysis of categorical outcomes. The generalized estimating equation (GEE) approach (Liang and Zeger 1986) is an extension of generalized linear models that provides a semiparametric approach to longitudinal data analysis with univariate outcomes for which the quasi-likelihood formulation is sensible, for example, normal, Poisson, binomial, and gamma response variables. This approach encompasses a broad range of data situations, including missing observations, continuous explanatory variables, and time-dependent explanatory variables.

The scope for the GEE strategy is useful for many situations, including the following:

- a two-period crossover study in which researchers study the effects of two treatments and a placebo

- a longitudinal study on the efficacy of a new drug designed to prevent fractures in the elderly. The outcome of interest is the number of fractures that occur.

- a large study on the effects of air pollution on children in which measurements on respiratory symptoms are taken every year for three years. Many children have one or two measurements missing.

In this chapter, the generalized estimating equations approach for the analysis of repeated measurements is discussed and illustrated with a series of examples using the GENMOD procedure. In addition, the use of GEE methods for the analysis of some univariate response outcomes is also discussed.

15.2 Methodology

15.2.1 Motivation

Correlated data come from many sources: longitudinal studies on health care outcomes, crossover studies concerned with drug comparisons, split plot experiments in agriculture, and clinical trials investigating new treatments with baseline and follow-up visits. You may have multiple measurements taken at the same time, such as in a psychometric study. You may also have clusters of correlated measurements: one example results from group randomization, such as randomizations of litters of animals to experimental conditions. Another example is sample selection of physician practices and the assessment of all of the patients in each practice, or cluster. Often, particularly with longitudinal studies, missing data are common.

An important consideration in each of these situations is how to account for the correlated measurements in the analysis. Within-subject factors (visit, time) are likely to have correlated measurements, while between-subject factors (age, gender) are likely to have independent measurements. The correlation must be taken into account, because, if you don't, you may produce incorrect standard errors. In the presence of positive correlations, you would underestimate the standard errors of the between-subject effects and overestimate the standard errors of the within-subject effects, resulting in inefficient estimation.

As discussed in Chapter 14, weighted least squares provides a reasonable strategy for repeated categorical outcomes when you have all of the following:

- complete data

- small number of discrete explanatory variables

- samples that are large enough to support approximately normal distributions

However, when you have continuous explanatory variables, a larger number of categorical variables, missing response values, and/or time-dependent covariates, the WLS approach does not apply. The GEE strategy, however, can handle these situations.

When you have continuous outcomes, the general linear multivariate model for normally distributed errors is often appropriate. It requires complete data for all outcomes and requires the covariates to be measured at the cluster level. If you can assume that the covariances have a spherical structure (compound symmetry), then repeated measures ANOVA applies for univariate tests about the within-subject effects. However, if you have time-dependent covariates, missing data, or non-normality, then that approach may not be adequate. You might consider the mixed model, which handles these issues, but that requires certain covariance matrix assumptions. If these are not met, the GEE method provides an alternative strategy.

GEEs were introduced by Liang and Zeger (1986) as a way of handling correlated data that, except for the correlation among responses, can be modeled with a generalized linear model (GLM). They are ideal for discrete response data such as binary outcomes and Poisson counts. They work for longitudinal studies data and cluster sampling data. You model these data using the same link functions and linear predictor set-up as you do in the GLM for the independent case. The difference between the GLM and GEE methods is that, with the GEE method, you account for the structure of the covariances of the response outcomes through its specification in the estimating process, much like you specify the covariance structure in mixed model analysis, but there is robustness to it. See Liang and Zeger (1986), Zeger and Liang (1986), Wei and Stram (1988), Stram, Wei, and Ware (1988), Moulton and Zeger (1989), and Zhao and Prentice (1990) for more detail.

The focus of this chapter is the analysis of categorical repeated measurements; however, as mentioned above, the GEE methodology also applies to continuous outcomes and often is used as an adjunct to other types of analyses.

15.2.2   Generalized Linear Models

The GEE method is an extension of generalized linear models (GLM), which are an extension of traditional linear models (Nelder and Wedderburn 1972). The GLM relates a mean response to a vector of explanatory variables through a link function:

$$g(E(y_i)) = g(\mu_i) = \mathbf{x}_i'\boldsymbol{\beta}$$

where y_i is a response variable $(i = 1, \ldots, n)$, $\mu_i = E(y_i)$, g is a link function, $\mathbf{x}_i$ is a vector of independent variables, and $\boldsymbol{\beta}$ is a vector of regression parameters to be estimated.

Additionally,

- The variance of y_i is $v_i = v_i(\mu_i)$ and is a specified function of its mean μ_i.
- The y_i are from the exponential family. This includes the binomial, Poisson, normal, gamma, and inverse Gaussian distributions. When you assume the normal distribution and specify the identity link function $g(\mu_i) = \mu_i$, you are fitting the same model as the general linear model.

For logistic regression, the link and variance functions are

$$g(\mu) = \log\left\{\frac{\mu}{1 - \mu}\right\} \text{ and } v(\mu) = \mu(1 - \mu)$$

For Poisson regression, the link and variance functions are

$$g(\mu) = \log(\mu) \text{ and } v(\mu) = \mu$$

You obtain the maximum likelihood estimator $\hat{\boldsymbol{\beta}}$ of the $p \times 1$ parameter vector $\boldsymbol{\beta}$ by solving the estimating equations, which are the score equations shown below. These estimators also maximize the log likelihood.

$$\sum_{i=1}^{n} \frac{\partial \mu_i'}{\partial \boldsymbol{\beta}} v_i^{-1} (y_i - \mu_i(\boldsymbol{\beta})) = \mathbf{0}$$

Generally, these are a set of nonlinear equations with no closed form solution, so you must solve them iteratively. The Newton-Raphson or Fisher scoring methods are often used; the fitting algorithm in the GENMOD procedure begins with a few Fisher scoring steps and then switches to a ridge-stabilized Newton Raphson method.

15.2.3   Generalized Estimating Equations Methodology

Generalized estimating equations are an extension of GLMs to accommodate correlated data; they are an extension of quasi-score equations. The GEE methodology models a known function of the marginal expectation of the dependent variable as a linear function of one or more explanatory variables. With quasi-likelihood methods, you can pursue statistical models by making assumptions about the link function and the relationship between the first two moments, but without fully specifying the complete distribution of the response. With GEEs, you describe the random component of the model for each marginal response with a common link and variance function, similar to what you do with a GLM model. However, unlike GLMs, you have to account for the covariance structure of the correlated measures, although there is robustness to how this is done.

The GEE methodology provides consistent estimators of the regression coefficients and their variances under weak assumptions about the actual correlation among a subject's observations. This approach avoids the need for multivariate distributions by assuming only a functional form for the marginal distribution at each time point or condition. The covariance structure across time or conditions is managed as a nuisance parameter. The method relies on the independence across subjects to consistently estimate the variance of the proposed estimators even when the assumed working correlation structure is incorrect. Zeger (1988), Zeger, Liang, and Albert (1988), and Liang, Zeger, and Qaqish (1992) provide further detail on the GEE methodology.

Data Structure

Suppose repeated measurements are obtained at t_i time points, $1 \le t_i \le t$ from each of n subjects. (Note that if the number and spacing of the repeated measurements are fixed and do not vary among subjects, t_i is equal to the total number of distinct measurement times.) Although this notation is most natural for longitudinal studies, it also applies to the general case of correlated responses. For example, t might instead denote the number of conditions under which dependent measurements are obtained, or there might be n clusters with at most t experimental units per cluster.

Now, suppose y_{ij} denotes the response from subject i at time or condition j, for $i = 1, \ldots, n$ and $j = 1, \ldots, t_i$. These y_{ij} may be binary outcomes or Poisson counts, for example. Also, suppose $x_{ij} = (x_{ij1}, \ldots, x_{ijp})'$ denote a $p \times 1$ vector of explanatory variables (covariates) associated with y_{ij}. If all covariates are time independent, then $x_{i1} = x_{i2} = \cdots = x_{it}$. Note that y_{ij} and x_{ij} are missing if observations are not obtained at time j.

Generalized Estimating Equations

Assume that you have chosen a model that relates a marginal mean to the linear predictor $\mathbf{x}_i'\boldsymbol{\beta}$ through a link function. The generalized estimating equations for estimating $\boldsymbol{\beta}$, an extension of the GLM estimating equation, follow:

$$\sum_{i=1}^{n} \frac{\partial \mu'}{\partial \beta} \mathbf{V}_i^{-1} (\mathbf{Y}_i - \mu_i(\beta)) = \mathbf{0}$$

where μ_i is the corresponding vector of means $\mu_i = (\mu_{i1}, \ldots, \mu_{it_i})'$, $\mathbf{Y}_i = (y_{i1}, y_{i2}, \ldots, y_{it_i})$, and $\mathbf{V}_i$ is an estimator of the covariance matrix of $\mathbf{Y}_i$. These equations are similar to the GLM estimating equations except that, since you have multiple outcomes, they include a vector of means instead of a single mean and a covariance matrix instead of a scalar variance. The covariance matrix of $\mathbf{Y}_i$ is specified as the estimator

$$\mathbf{V}_i = \phi \mathbf{A}_i^{\frac{1}{2}} \mathbf{R}_i(\alpha) \mathbf{A}_i^{\frac{1}{2}}$$

where $\mathbf{A}_i$ is a $t_i \times t_i$ diagonal matrix with $v(\mu_{ij})$ as the jth diagonal element. Note that $\mathbf{V}_i$ can be different from subject to subject, but generally you use a specification that approximates the average dependence among repeated observations over time. Note that the GEE facilities in the GENMOD procedure only allow you to specify the same form of $\mathbf{V}_i$ for all subjects.

$\mathbf{R}_i(\alpha)$ is the *working correlation matrix*. The (j, j') element of $\mathbf{R}_i(\alpha)$ is the known, hypothesized, or estimated correlation between y_{ij} and $y_{ij'}$. This working correlation matrix may depend on a vector of unknown parameters α, which is the same for all subjects. You assume that $\mathbf{R}_i(\alpha)$ is known except for a fixed number of parameters α that must be estimated from the data.

Choosing the Working Correlation Matrix

Several possibilities for the working correlation structure have been suggested (Liang and Zeger 1986). First, when the number of subjects is large relative to the number of observations per subject, the influence of correlation is often small enough so that the GLM regression coefficients are nearly efficient. The correlations among repeated measures, however, may have a substantial effect on the estimated variances of the regression coefficients and hence must be taken into account to make correct inferences.

The following are some choices for $\mathbf{R}$ with matrix formulations for $t = 4$.

Independence: $\mathbf{R} = \mathbf{R}_0 = \mathbf{I}$.

$$\mathbf{R} = \begin{bmatrix} 1 & 0 & 0 & 0 \\ 0 & 1 & 0 & 0 \\ 0 & 0 & 1 & 0 \\ 0 & 0 & 0 & 1 \end{bmatrix}$$

The independence model adopts the working assumption that repeated observations for a subject are independent. In this case, solving the GEE is the same as fitting the usual regression models for independent data and the resulting parameter estimates are the same. However, their standard errors are different. You are choosing not to specify the correlation explicitly but the GEE method still accounts for that correlation by operating at the cluster level. However, the estimation is done with estimation of β only at each step, and not α, so it doesn't improve the precision of the parameter estimates with additional iterations. In this case, the GEE simplifies to the GLM estimating equations.

Fixed: $\mathbf{R} = \mathbf{R}_0$.

Fixed correlation matrices arise when you have determined the form from a previous analysis. You simply input your covariance matrix directly.

Exchangeable:

$$\mathrm{Corr}(y_{ij}, y_{i,j'}) = \left\{ \begin{array}{ll} 1 & j = j' \\ \alpha & j \neq j' \end{array} \right\}$$

$$\mathbf{R} = \begin{bmatrix} 1 & \alpha & \alpha & \alpha \\ \alpha & 1 & \alpha & \alpha \\ \alpha & \alpha & 1 & \alpha \\ \alpha & \alpha & \alpha & 1 \end{bmatrix}$$

The *exchangeable* working correlation specification makes constant the correlations between any two measurements within a subject, that is, $R_{jj'} = \alpha$, for $j \neq j'$. This is the correlation structure assumed in a random effects model with a random intercept and is also known as *compound symmetry* in the repeated measures ANOVA literature. Although the specification of constant correlation between any two repeated measurements may not be justified in a longitudinal study, it is often reasonable in situations in which the repeated measures are not obtained over time. It is probably reasonable when there are a few repeated measurements. An arbitrary number of observations per subject is permissible with both the independence and exchangeable working correlation structures. This structure is commonly used and is relatively easy to explain to investigators. The exchangeable structure is also appropriate when cluster sampling is involved, such as studies in which physician practices are selected as clusters and measurements are obtained for the patients in those practices.

Unstructured:

$$\mathrm{Corr}(y_{ij}, y_{i,j'}) = \left\{ \begin{array}{ll} 1 & j = j' \\ \alpha_{jk} & j \neq j' \end{array} \right\}$$

$$\mathbf{R} = \begin{bmatrix} 1 & \alpha_{21} & \alpha_{31} & \alpha_{41} \\ \alpha_{21} & 1 & \alpha_{32} & \alpha_{42} \\ \alpha_{31} & \alpha_{32} & 1 & \alpha_{43} \\ \alpha_{41} & \alpha_{42} & \alpha_{43} & 1 \end{bmatrix}$$

When the correlation matrix is completely unspecified, there are $t_i(t_i - 1)/2$ parameters to be estimated. This provides the most efficient estimator for β but is useful only when there are relatively few observation times or conditions. In addition, when there are missing data and/or varying numbers of observations per subject, estimation of the complete correlation structure may result in a nonpositive definite matrix and parameter estimation may not proceed.

m-**dependent**:

$$\mathrm{Corr}(y_{ij}, y_{i,j+s}) = \left\{ \begin{array}{ll} 1 & s = 0 \\ \alpha_s & s = 1, 2, \ldots, m \\ 0 & s > m \end{array} \right]$$

$$\mathbf{R} = \begin{bmatrix} 1 & \alpha_1 & \alpha_2 & 0 \\ \alpha_1 & 1 & \alpha_1 & \alpha_2 \\ \alpha_2 & \alpha_1 & 1 & \alpha_1 \\ 0 & \alpha_2 & \alpha_1 & 1 \end{bmatrix}$$

With the m-dependent structure, the correlations depend on the distances between measures; eventually, they diminish to zero for $s \geq m$.

Auto-regressive (AR-1):

$$\mathrm{Corr}(y_{ij}, y_{i,j+s}) = \alpha^s \quad s = 0, 1, 2, \ldots, t_i - j$$

$$\mathbf{R} = \begin{bmatrix} 1 & \alpha & \alpha^2 & \alpha^3 \\ \alpha & 1 & \alpha & \alpha^2 \\ \alpha^2 & \alpha & 1 & \alpha \\ \alpha^3 & \alpha^2 & \alpha & 1 \end{bmatrix}$$

With an auto-regressive correlation structure, the correlations also depend on the distance between the measures; they diminish with increasing distance.

See the PROC GENMOD documentation for specific estimators of the $\mathbf{R}_i(\boldsymbol{\alpha})$ parameters for each of the working correlation matrix types; they involve using the current value of β to compute functions of the Pearson residual

$$r_{ij} = \frac{y_{ij} - \hat{\mu}_{ij}}{\sqrt{v(\hat{\mu}_{ij})}}$$

$\mathbf{R}$ is called a working correlation matrix because, for non-normal data, the actual values may depend on the mean value and on $\mathbf{x}_i'\beta$. See Appendix A at the end of this chapter for more detail on the steps in the GEE solution.

Estimating the Covariance of the Parameter Estimates

The model-based estimator of the covariance matrix for $\hat{\beta}$ is the inverse of the observed information matrix

$$\boldsymbol{\Sigma}_m(\hat{\beta}) = \mathbf{I}_0^{-1}$$

where

$$\mathbf{I}_0 = \sum_{i=1}^{K} \frac{\partial \boldsymbol{\mu}_i}{\partial \beta}' \mathbf{V}_i^{-1} \frac{\partial \boldsymbol{\mu}_i}{\partial \beta}$$

This is a consistent estimator if the model and working correlation matrix are correctly specified. Its use may be preferable in those situations where you have a moderate number of large clusters (Albert and McShane 1995).

The empirical sandwich (robust) estimator of $\mathrm{Cov}(\hat{\beta})$ is given by

$$\boldsymbol{\Sigma}_e = \mathbf{I}_0^{-1} \mathbf{I}_1 \mathbf{I}_0^{-1}$$

where

$$\mathbf{I}_1 = \sum_{i=1}^{K} \frac{\partial \boldsymbol{\mu}_i}{\partial \boldsymbol{\beta}}' \mathbf{V}_i^{-1} \mathrm{Cov}(\mathbf{Y}_i) \mathbf{V}_i^{-1} \frac{\partial \boldsymbol{\mu}_i}{\partial \boldsymbol{\beta}}$$

$\mathrm{Cov}(\mathbf{Y}_i)$ is estimated by

$$(\mathbf{Y}_i - \boldsymbol{\mu}_i(\hat{\boldsymbol{\beta}}))(\mathbf{Y}_i - \boldsymbol{\mu}_i(\hat{\boldsymbol{\beta}}))'$$

This is a consistent estimator even when $\mathrm{Var}(y_{ij}) \neq v(\mu_{ij})$ or when $\mathbf{R}_i(\boldsymbol{\alpha})$ is not the correlation matrix of $\mathbf{Y}_i$ or when the true correlation varies across clusters. You lose efficiency with the misspecification, but if the working correlation structure is approximately correct, the asymptotic efficiency is expected to be relatively high.

You can test linear hypotheses of the form H_0: $\mathbf{C}\boldsymbol{\beta} = \mathbf{0}$, where $\mathbf{C}$ is a known $c \times p$ matrix of constants of rank c, with the Wald statistic

$$Q_C = (\mathbf{C}\hat{\boldsymbol{\beta}})'[\mathbf{C}\mathbf{V}_{\boldsymbol{\beta}}\mathbf{C}']^{-1}(\mathbf{C}\hat{\boldsymbol{\beta}})$$

The statistic Q_C is approximately distributed as chi-square under H_0 with degrees of freedom equal to c. These are also known as tests for contrasts.

15.3   Summary of the GEE Methodology

The GEE method is a practical strategy for the analysis of repeated measurements, particularly categorical repeated measurements. It provides a way to handle continuous explanatory variables, a moderate number of explanatory categorical variables, and time-dependent explanatory variables. It handles missing values, that is, the number of measurements in each cluster can vary from 1 to t.

The following are the important properties of the GEE method:

- GEEs reduce to GLM estimating equations for $t_i = 1$.

- GEEs are the maximum likelihood score equations for multivariate Gaussian data when you specify unstructured correlation.

- The regression parameter estimates are consistent as the number of clusters become large, even if you have misspecified the working correlation matrix, as long as the model for the mean is correct.

- The empirical sandwich estimator of the covariance matrix of $\hat{\boldsymbol{\beta}}$ is also consistent relative to the number of clusters becoming large, even if you have misspecified the working correlation matrix, as long as the model for the mean is correct.

While the GEE method handles missing values, it is important to note that the method requires the missing data to be missing completely at random (MCAR), which roughly means that the missing values may depend only on the explanatory variables that appear in the model. This requirement is more restrictive than the missing at random (MAR) assumption, which is the assumption for likelihood-based inference.

The GEE method depends on asymptotic theory; the number of clusters needs to be large for the method to produce consistent estimates. By that is meant that the sample size is large enough to support the properties of consistency and approximate normality for the estimates from the method. Note that the number of clusters determines adequate sample size, not the number of measurements per cluster or the total number of measurements. The desired number of clusters depends on other factors: if you have a very small number of continuous or dichotomous explanatory variables, 25 clusters may be minimally enough so that you aren't badly misled by your results. If you have 5–12 explanatory variables, you need at least 100 clusters. If you want to be reasonably confident, you probably need 200 clusters. Note that if the correlations are relatively small, you may be able to handle more time-dependent explanatory variables within a subject than if you have a high degree of correlation.

The Z statistics and Wald statistics (with the former being the square root of the latter) presently used in the GENMOD procedure to assess parameter significance and Type 3 contrasts require around 200 clusters to provide a great deal of confidence concerning assessments of statistical significance at the 0.05 confidence level or smaller; the score statistics produced in the Type 3 analyses of the model effects procedure have similar properties (Boos 1992, and Rotnitzky and Jewell 1990) although they are often more conservative in the presence of small numbers of clusters. As the number of degrees of freedom of the contrast for the hypothesis test approaches the number of clusters, these tests are likely to become less reliable. Several simulation studies (for example, Hendricks et al. 1996) show that the Type I errors associated with the robust variance estimators can be inflated. Researchers are investigating adjustments to the Wald statistic based on the number of clusters in order to produce statistics with better properties for moderate sample sizes. Shah, Holt, and Folsom (1977) discuss such strategies in the context of sample survey data analysis. It's likely that the GENMOD procedure will offer some different statistics to assess the significance of the effects in the future. See Section 15.7 for an example of the use of one of these adjusted Wald statistics and the availability of a SAS macro to compute it.

GEE methods are robust to an assigned correlation structure; you can misspecify that correlation structure and still obtain consistent parameter estimates. However, note that the closer the working correlation matrix is to the true structure, the more efficient your estimates will be. You can compare this property to the mixed model, which heavily leverages the correlation assumption; this means that if you have misspecified the correlation structure, you may obtain biased estimates.

The previous discussion did not include goodness-of-fit criteria for the GEE model. Since the GEE method is quasi-likelihood based, there are not readily defined analogs to the fit statistics for maximum likelihood estimation. This is an area of active research, and the current GENMOD procedure does not include any measures for the assessment of fit. Barnhart and Williamson (1998) describe an empirical procedure for GEE fit based on the Hosmer and Lemeshow approach for logistic regression (1989). Preisser and Qaqish (1996) describe diagnostics for GEE that are extensions of Cook's D and DBETA for linear regression.

Marginal Model

The robustness of the GEE method is due to the fact that the GEE method produces a *marginal model*. It models a known function of the marginal expectation of the dependent variable as a linear function of the explanatory variables. The resulting parameter estimates are population-averaged, or estimates "on the average." You can also think of the GEE model as a variational model in which you use estimation to describe the variation among a set of population parameters (Koch, Gillings, and Stokes 1980). You are relying on the independence across clusters to consistently estimate the variance; the covariance matrix parameters are effectively managed as nuisance parameters.

Compare the marginal model to the subject-specific model fit with the conditional logistic regression method described in Chapter 10 or with mixed models. In those analyses, you characterize behavior as a process for individuals. The predictions you produce are individual-based, rather than predictions that apply on average. Your choice of strategy often depends on the goals of your analysis—whether you want to make population statements about your results, on average, or whether you want to produce a model that permits individual prediction. Note that in the standard linear model there is no distinction between the marginal and subject-specific model. Refer to Diggle, Liang, and Zeger (1994) and Zeger, Liang, and Albert (1988) for more discussion of marginal models in longitudinal data analysis.

15.4   Passive Smoking Example

The following data are from a hypothetical study of the effects of air pollution on children. Researchers followed 25 children and recorded whether they were exhibiting wheezing symptoms during the periods of evaluation at ages 8, 9, 10, and 11. The response is recorded as 1 for symptoms and 0 for no symptoms. Explanatory variables included age, city, and a passive smoking index with values 0, 1, and 2 that reflected the degree of smoking in the home.

Note that age and the passive smoking index are time-dependent explanatory variables; their values depend on the period of measurement. This example provides a basic introduction to fitting GEE models with the GENMOD procedure. The dichotomous outcome is modeled with a logistic regression analysis; while four response times may be pushing the limits for the exchangeable structure, the small number of clusters makes a single-parameter covariance structure a more reasonable choice. Since there are only 25 experimental units, or clusters, only a few explanatory variables can be included in the analysis.

Table 15.1. Pollution Study Data

ID	City	Age 8		Age 9		Age 10		Age 11	
		Smoke	Symp	Smoke	Symp	Smoke	Symp	Smoke	Symp
1	steelcity	0	1	0	1	0	1	0	0
2	steelcity	2	1	2	1	2	1	1	0
3	steelcity	2	1	2	0	1	0	0	0
4	greenhills	0	0	1	1	1	1	0	0
5	steelcity	0	0	1	0	1	0	1	0
6	greenhills	0	1	0	0	0	0	0	1
7	steelcity	1	1	1	1	0	1	0	0
8	greenhills	1	0	1	0	1	0	2	0
9	greenhills	2	1	2	0	1	1	1	0
10	steelcity	0	0	0	0	0	0	1	0
11	steelcity	1	1	0	0	0	0	0	1
12	greenhills	0	0	0	0	0	0	0	0
13	steelcity	2	1	2	1	1	0	0	1
14	greenhills	0	1	0	1	0	0	0	0
15	steelcity	2	0	0	0	0	0	2	1
16	greenhills	1	0	1	0	0	0	1	0
17	greenhills	0	0	0	1	0	1	1	1
18	steelcity	1	1	2	1	0	0	1	0
19	steelcity	2	1	1	0	0	1	0	0
20	greenhills	0	0	0	1	0	1	0	0
21	steelcity	1	0	1	0	1	0	2	1
22	greenhills	0	1	0	1	0	0	0	0
23	steelcity	1	1	1	0	0	1	0	0
24	greenhills	1	0	1	1	1	1	2	1
25	greenhills	0	1	0	0	0	0	0	0

The following DATA step inputs the collected measures into the SAS data set named
CHILDREN. Note that the data are stored with all of a particular child's measurements on
a single data line. However, the GENMOD procedure requires that each repeated measure
be managed as a separate observation. So, the DO loop included in the DATA step
statements inputs each measure, age, and the passive smoking index and outputs them,
along with the variable CITY, to the CHILDREN data set. You often need to rearrange
data in this manner when you are dealing with repeated measurements data.

```
data children;
   input id city$ @@;
   do i=1 to 4;
      input age smoke symptom @@;
      output;
   end;
   datalines;
 1 steelcity  8 0 1   9 0 1   10 0 1   11 0 0
 2 steelcity  8 2 1   9 2 1   10 2 1   11 1 0
 3 steelcity  8 2 1   9 2 0   10 1 0   11 0 0
```

```
 4 greenhills 8 0 0   9 1 1   10 1 1   11 0 0
 5 steelcity  8 0 0   9 1 0   10 1 0   11 1 0
 6 greenhills 8 0 1   9 0 0   10 0 0   11 0 1
 7 steelcity  8 1 1   9 1 1   10 0 1   11 0 0
 8 greenhills 8 1 0   9 1 0   10 1 0   11 2 0
 9 greenhills 8 2 1   9 2 0   10 1 1   11 1 0
10 steelcity  8 0 0   9 0 0   10 0 0   11 1 0
11 steelcity  8 1 1   9 0 0   10 0 0   11 0 1
12 greenhills 8 0 0   9 0 0   10 0 0   11 0 0
13 steelcity  8 2 1   9 2 1   10 1 0   11 0 1
14 greenhills 8 0 1   9 0 1   10 0 0   11 0 0
15 steelcity  8 2 0   9 0 0   10 0 0   11 2 1
16 greenhills 8 1 0   9 1 0   10 0 0   11 1 0
17 greenhills 8 0 0   9 0 1   10 0 1   11 1 1
18 steelcity  8 1 1   9 2 1   10 0 0   11 1 0
19 steelcity  8 2 1   9 1 0   10 0 1   11 0 0
20 greenhills 8 0 0   9 0 1   10 0 1   11 0 0
21 steelcity  8 1 0   9 1 0   10 1 0   11 2 1
22 greenhills 8 0 1   9 0 1   10 0 0   11 0 0
23 steelcity  8 1 1   9 1 0   10 0 1   11 0 0
24 greenhills 8 1 0   9 1 1   10 1 1   11 2 1
25 greenhills 8 0 1   9 0 0   10 0 0   11 0 0
;
```

The PROC GENMOD invocation includes the usual MODEL statement as well as the REPEATED statement. You use the MODEL statement to request the logit link function, binomial distribution, and a Type 3 analysis by specifying LINK=LOGIT, DIST=BIN, and TYPE3, respectively. So far, this specification is the same as for any logistic regression using PROC GENMOD. The DESCENDING option in the PROC statement specifies that the model is based on the probability of the largest value of the response variable, which is 1.

```
proc genmod data=children descending;
   class id city;
   model symptom = city age smoke  /
                   link=logit dist=bin type3;
   repeated subject=id / type=exch covb corrw;
run;
```

You request a GEE analysis with the REPEATED statement. The SUBJECT=ID identifies the clustering variable. The SUBJECT= variable must be listed in the CLASS statement and needs to have a unique value for each cluster. Specifying TYPE=EXCH requests the exchangeable working correlation structure. The COVB option requests that the parameter estimate covariance matrix be printed, and the CORRW option specifies that the final working correlation matrix be printed.

Output 15.1 displays the "Model Information" table which provides information about the model specifications, including the specified distribution and link function. In addition, the table describes on which level of the outcome variable the model is based. By default, beginning with Release 8.1, the default is to model the lower alphanumeric ordered response values for dichotomous outcomes. (Note that this is a change from previous versions of the GENMOD procedure.) Thus, by using the DESCENDING option here, you are basing your model on the favorable outcome (SYMPTOM=1).

Output 15.1 Basic Model Information

```
                        Model Information

            Data Set                  WORK.CHILDREN
            Distribution                   Binomial
            Link Function                     Logit
            Dependent Variable              symptom
            Observations Used                   100
            Probability Modeled     Pr( symptom = 1 )
```

Output 15.2 displays the class levels and response profiles, respectively. Since the DESCENDING option is used, the first ordered value is 1 in the "Response Profile" table.

Output 15.2 Class Levels and Response Profiles

```
                        Class Level Information

        Class      Levels    Values

        id           25      1 2 3 4 5 6 7 8 9 10 11 12 13 14 15 16 17 18 19 20
                             21 22 23 24 25
        city          2      greenhil steelcit


                            Response Profile

                        Ordered    Ordered
                         Level      Value        Count

                           1          1            42
                           2          0            58
```

Output 15.3 displays information concerning the parameters, including which parameter pertains to which level of the CLASS variables.

Output 15.3 Information About Parameters

```
                        Parameter Information

            Parameter        Effect        city

            Prm1             Intercept
            Prm2             city          greenhil
            Prm3             city          steelcit
            Prm4             age
            Prm5             smoke
```

Output 15.4 contains the initial parameter estimates. To generate a starting solution, the GENMOD procedure first treats all of the measurements as independent and fits a generalized linear model. These parameter estimates are then used as the starting values for the GEE solution.

Output 15.4 Initial Parameter Estimates

```
                    Analysis Of Initial Parameter Estimates

                                    Standard   Wald 95% Confidence    Chi-
Parameter            DF    Estimate   Error         Limits          Square

Intercept             1     2.4161   1.8673   -1.2438     6.0760      1.67
city     greenhil     1     0.0017   0.4350   -0.8508     0.8543      0.00
city     steelcit     0     0.0000   0.0000    0.0000     0.0000       .
age                   1    -0.3283   0.1914   -0.7035     0.0468      2.94
smoke                 1     0.5598   0.2952   -0.0188     1.1385      3.60
Scale                 0     1.0000   0.0000    1.0000     1.0000

                           Analysis Of Initial
                           Parameter Estimates

                    Parameter            Pr > ChiSq

                    Intercept              0.1957
                    city     greenhil      0.9968
                    city     steelcit        .
                    age                    0.0863
                    smoke                  0.0579
                    Scale

NOTE: The scale parameter was held fixed.
```

The beginning of the output produced by the GEE analysis is the general model information that is displayed in Output 15.5. Since there are 25 subjects with repeated measures, there are 25 clusters. Each subject has 4 measures, and the data are complete. Thus, the minimum and maximum cluster size is 4.

Output 15.5 General GEE Model Information

```
                       GEE Model Information

          Correlation Structure            Exchangeable
          Subject Effect                   id (25 levels)
          Number of Clusters                         25
          Correlation Matrix Dimension                4
          Maximum Cluster Size                        4
          Minimum Cluster Size                        4
```

Output 15.6 contains the Type 3 analysis results for the model effects.

The results indicate that city is not a factor in wheezing status. However, smoking exposure has a nearly significant association ($p = 0.0583$). Age is marginally influential ($p = 0.0981$).

Output 15.6   Type 3 Analysis

```
                Score Statistics For Type 3 GEE Analysis

                                    Chi-
            Source         DF      Square     Pr > ChiSq

            city            1        0.01        0.9388
            age             1        2.74        0.0981
            smoke           1        3.59        0.0583
```

Output 15.7 contains the parameter estimates produced by the GEE analysis. The table also supplies standard errors, confidence intervals, Z statistics, and p-values. The empirical standard errors are produced by default. Since the effects reported in the Type 3 analysis are single degree of freedom effects, the score statistics in that table are assessing the same hypotheses as the Z statistics in this table. Note that the p-value for the Z for smoking is 0.0211, compared to the 0.0583 reported with the score statistic in the "Type 3" table. In a strict testing situation, you would assess the null hypothesis with the score statistic. The Z and Wald statistic generally produce more liberal p-values than the score statistic. Particularly for small sample sizes, you would want to report the more conservative value.

Output 15.7   GEE Parameter Estimates

```
                   Analysis Of GEE Parameter Estimates
                   Empirical Standard Error Estimates

                             Standard    95% Confidence
     Parameter       Estimate   Error        Limits          Z  Pr > |Z|

     Intercept         2.2615  2.0243  -1.7060   6.2290    1.12    0.2639
     city   greenhil    0.0418  0.5435  -1.0234   1.1070    0.08    0.9387
     city   steelcit    0.0000  0.0000   0.0000   0.0000      .       .
     age               -0.3201  0.1884  -0.6894   0.0492   -1.70    0.0893
     smoke              0.6506  0.2821   0.0978   1.2035    2.31    0.0211
```

The procedure prints both the empirical and model-based covariance matrix of the parameter estimates; these are displayed in Output 15.8. Note that their values are often similar, especially for large samples. If these matrices are very similar, you may have some confidence that you have correctly specified the correlation structure and the estimates are relatively efficient. However, recall that, even if you have misspecified the correlation structure, both the parameter estimates and their empirical standard errors are consistent, provided that the specification is correct for the explanatory variables.

Output 15.8   Covariance Matrix Estimates

	Covariance Matrix (Model-Based)			
	Prm1	Prm2	Prm4	Prm5
Prm1	3.26069	-0.16313	-0.32274	-0.12257
Prm2	-0.16313	0.24015	0.002520	0.03422
Prm4	-0.32274	0.002520	0.03379	0.004471
Prm5	-0.12257	0.03422	0.004471	0.09533

	Covariance Matrix (Empirical)			
	Prm1	Prm2	Prm4	Prm5
Prm1	4.09770	-0.55261	-0.37280	-0.29397
Prm2	-0.55261	0.29538	0.03719	0.09143
Prm4	-0.37280	0.03719	0.03550	0.02064
Prm5	-0.29397	0.09143	0.02064	0.07957

Finally, the exchangeable working correlation matrix is also printed. The estimated correlation is fairly low at 0.0883.

Output 15.9   Working Correlation Matrix

	Working Correlation Matrix			
	Col1	Col2	Col3	Col4
Row1	1.0000	0.0883	0.0883	0.0883
Row2	0.0883	1.0000	0.0883	0.0883
Row3	0.0883	0.0883	1.0000	0.0883
Row4	0.0883	0.0883	0.0883	1.0000

Since this is a logistic regression based on reference cell coding, you can exponentiate the parameter estimates to obtain estimates of odds ratios for various explanatory factors. Since the parameter estimate for smoking exposure is 0.6506, the odds of symptoms for those with one higher category of smoking exposure are $e^{0.6506}=1.9$ times the odds of symptoms for those children with the lower exposure.

The GENMOD procedure can produce the odds ratio estimate via the ESTIMATE statement, along with 95% confidence limits. You can produce such results for any estimable linear combination of the parameters from the GEE analysis.

Since the smoking effect is represented by a single parameter, you place the coefficient 1 after listing a label and the SMOKE variable. The EXP option requests that the estimate be exponentiated, which, in the case of reference parameterization, produces the odds ratio estimate.

```
ods select Estimates;
proc genmod data=children descending;
   class id city;
```

```
model symptom = city age smoke  /
                link=logit dist=bin type3;
repeated subject=id / type=exch covb corrw;
estimate 'smoking' smoke 1 / exp;
run;
```

Output 15.10 displays the results from the ESTIMATE statement. The point estimate for the odds ratio is 1.9 with 95% confidence limits of (1.1027, 3.3318) for the extent of increased odds of symptoms per category of increase in passive smoking exposure. The confidence limits are based on the Wald statistic.

Output 15.10   ESTIMATE Results

```
                      Contrast Estimate Results

                      Standard                             Chi-
Label        Estimate    Error   Alpha  Confidence Limits  Square  Pr > ChiSq

smoking        0.6506   0.2821   0.05    0.0978   1.2035   5.32      0.0211
Exp(smoking)   1.9168   0.5407   0.05    1.1027   3.3318
```

Note that specifying the coefficients in the ESTIMATE statement can be more involved with CLASS variables, due to the less than full rank parameterization that is used by PROC GENMOD. Consider a two-level treatment variable, say TREATMENT, with levels A and B and a default reference level of B. The ESTIMATE statement required to produce the estimated odds ratio for A compared to B would be

```
estimate 'treatment' treatment 1 -1 /exp;
```

Refer to the *SAS/STAT User's Guide, Version 8* for more information regarding the ESTIMATE statement and parameterization.

15.5   Crossover Example

Crossover designs provide another form of repeated measurements. In a crossover design, subjects serve as their own controls and receive two or more treatments or conditions in two or more consecutive periods. You can use the GEE method to analyze such data, managing the subjects as clusters and managing the treatment as a time-varying covariate.

The following data are from a two-period crossover study investigating three treatments. These data were analyzed with conditional logistic regression in Chapter 10.

Table 15.2.   Crossover Design Data

Age	Sequence	Response Profiles				Total
		FF	FU	UF	UU	
older	A:B	12	12	6	20	50
older	B:P	8	5	6	31	50
older	P:A	5	3	22	20	50
younger	B:A	19	3	25	3	50
younger	A:P	25	6	6	13	50
younger	P:B	13	5	21	11	50

As described in Chapter 10, this is a two-period crossover design where patients have been stratified to two age groups and, within age group, have been assigned to one of three treatment sequences. These data can be modeled with parameters for period effect, effects for Drug A and Drug B relative to the placebo (P), carryover effects for Drug A and Drug B, and interactions of period with age and drug with age.

The following DATA step enters the data into SAS data set CROSS. The variable AGE contains information on whether the subject is older or younger, and the variable SEQUENCE contains two letters describing the sequence of treatments for that group. For example, the value AB means that treatment A was received in the first period and treatment B was received in the second, and the value BP means that treatment B was received in the first period and the placebo was received in the second. The variables TIME1 and TIME2 have the values F and U depending on whether the treatment produced a favorable or unfavorable response. The data are frequency counts, and the variable COUNT contains the frequency for each response profile for each sequence and age combination. The following DATA step creates an observation for each subject.

```
data cross (drop=count);
   input age $ sequence $ time1 $ time2 $ count;
   do i=1 to count;
      output;
   end;
   datalines;
older AB F F 12
older AB F U 12
older AB U F 6
older AB U U 20
older BP F F 8
older BP F U 5
older BP U F 6
older BP U U 31
older PA F F 5
older PA F U 3
older PA U F 22
older PA U U 20
younger BA F F 19
younger BA F U 3
younger BA U F  25
younger BA U U  3
younger AP F F 25
younger AP F U 6
younger AP U F 6
younger AP U U 13
younger PB F F 13
younger PB F U 5
younger PB U F 21
younger PB U U 11
;
```

The next DATA step creates an observation for each response in each period so that the data are in the correct data structure for the GEE analysis. The variable PERIOD is an indicator variable for whether the observation is from the first period. The RESPONSE variable contains the value 1 if the response was favorable and 0 if it was not.

```
data cross2;
   set cross;
   subject=_n_;
      period=1;
         drug = substr(sequence, 1, 1);
         carry='N';
         response = time1;
         output;
      period=0;
         drug  = substr(sequence, 2, 1);
         carry = substr(sequence, 1, 1);
         if carry='P' then carry='N';
         response = time2;
         output;
   run;
   proc print data=cross2(obs=15);
   run;
```

The variable CARRY takes the value N (no) if the observation is from the first period; it takes the value A or B if it comes from the second period and the treatment in the first period is A or B, respectively. If the subject received the placebo in the first period, the value of CARRY is also set to N for the observations in the second period.

Output 15.11 displays the first 15 observations of SAS data set CROSS2.

Output 15.11   First 15 Observations of Data Set CROSS2

Obs	age	sequence	time1	time2	i	subject	period	drug	carry	response
1	older	AB	F	F	1	1	1	A	N	F
2	older	AB	F	F	1	1	0	B	A	F
3	older	AB	F	F	2	2	1	A	N	F
4	older	AB	F	F	2	2	0	B	A	F
5	older	AB	F	F	3	3	1	A	N	F
6	older	AB	F	F	3	3	0	B	A	F
7	older	AB	F	F	4	4	1	A	N	F
8	older	AB	F	F	4	4	0	B	A	F
9	older	AB	F	F	5	5	1	A	N	F
10	older	AB	F	F	5	5	0	B	A	F
11	older	AB	F	F	6	6	1	A	N	F
12	older	AB	F	F	6	6	0	B	A	F
13	older	AB	F	F	7	7	1	A	N	F
14	older	AB	F	F	7	7	0	B	A	F
15	older	AB	F	F	8	8	1	A	N	F

The following PROC GENMOD statements fit the GEE model. Since there are 300 subjects in the crossover study, there are 300 clusters or experimental units in the GEE analysis. With responses for both periods, the cluster size is two. There are no missing values, so both the minimum and maximum cluster size is two. A logistic regression analysis is appropriate for these data so DIST=BIN is specified in the MODEL statement.

The logit link is used by default. Both SUBJECT and AGE are specified in the CLASS statement, since AGE reflects a classification into two groups. The model includes main effects for period, age, drug, and carryover effects and interactions for period and age and drug and age. The option TYPE=UNSTR specifies the unstructured correlation structure. Since there are only two measurements per subject, this is the same as the exchangeable structure.

```
proc genmod data=cross2;
   class subject age drug carry;
   model response = period age drug
                    period*age carry
                    drug*age / dist=bin type3;
   repeated subject=subject/type=unstr;
run;
```

The "Class Level Information" table lists the variables treated as classification variables and their values.

Output 15.12 Class Level Information

```
                     Class Level Information

      Class        Levels    Values

      subject        300      1  2  3  4  5  6  7  8  9  10 11 12 13 14 15 16 17 18 19 20
                              21 22 23 24 25 26 27 28 29 30 31 32 33 34 35 36 37
                              38 39 40 41 42 43 44 45 46 47 48 49 50 51 52 53 54
                              55 56 57 58 59 60 61 62 63 64 65 66 67 68 69 70 71
                              72 73 74 75 76 77 78 79 80 81 82 83 84 85 86 87
                              ...
      age              2      older younger
      drug             3      A B P
      carry            3      A B N


                         Response Profile

               Ordered      Ordered
               Level        Value        Count

                  1         F            284
                  2         U            316
```

The "Parameter Information" table lists the 18 parameters and tells you what they represent.

Output 15.13 Parameter Information

```
                        Parameter Information

       Parameter       Effect          age       drug    carry

       Prm1            Intercept
       Prm2            period
       Prm3            age             older
       Prm4            age             younger
       Prm5            drug                      A
       Prm6            drug                      B
       Prm7            drug                      P
       Prm8            period*age      older
       Prm9            period*age      younger
       Prm10           carry                             A
       Prm11           carry                             B
       Prm12           carry                             N
       Prm13           age*drug        older     A
       Prm14           age*drug        older     B
       Prm15           age*drug        older     P
       Prm16           age*drug        younger   A
       Prm17           age*drug        younger   B
       Prm18           age*drug        younger   P
```

After the initial estimates are printed, the "GEE Model Information" table is displayed and confirms that you have 300 clusters, each containing two responses.

Output 15.14 GEE Model Information

```
                    GEE Model Information

       Correlation Structure                 Unstructured
       Subject Effect                subject (300 levels)
       Number of Clusters                             300
       Correlation Matrix Dimension                     2
       Maximum Cluster Size                             2
       Minimum Cluster Size                             2
```

The Type 3 table contains the results of effect tests for all the terms specified in the MODEL statement. When you are conducting an analysis of crossover data, you hope that there are no carryover effects. Having such effects greatly complicates the model and interpretation. In this analysis, the carryover effect is not significant. The score statistic for the two-level CARRY variable is 1.15 with p-value equal to 0.5626. In addition, the age $\times$ drug interaction appears to be unimportant, with a score chi-square statistic of 0.72 for 2 df ($p = 0.6981$).

Output 15.15   Type 3 Table

```
             Score Statistics For Type 3 GEE Analysis

                                      Chi-
                Source        DF     Square    Pr > ChiSq

                period         1       4.61       0.0318
                age            1      36.03      <.0001
                drug           2      27.66      <.0001
                period*age     1       4.69       0.0303
                carry          2       1.15       0.5626
                age*drug       2       0.72       0.6981
```

You can use the CONTRAST statement to obtain the joint test for CARRY and the
AGE*DRUG interaction. You submit the following statements. The ODS SELECT
statement restricts the output to the test results. The contrast labeled 'joint' is the joint test
for both the CARRY and AGE*DRUG effects. The contrasts labeled 'carry' and 'inter' are
the main effects tests and should match the results displayed in the Type 3 analysis for
those effects.

```
ods select Contrasts;
proc genmod data=cross2;
   class subject age drug carry;
   model response = period age drug
                    period*age carry
                    drug*age / dist=bin type3;
   repeated subject=subject/type=unstr;
   contrast 'carry' carry 1 0 -1,
                    carry 0 1 -1;
   contrast 'inter' age*drug 1 0 -1 -1 0  1 ,
                    age*drug 0 1 -1  0 -1 1 ;
   contrast 'joint' carry 1 0 -1,
                    carry 0 1 -1,
                    age*drug 1 0 -1 -1 0  1 ,
                    age*drug 0 1 -1  0 -1 1 ;
run;
```

Output 15.16 contains the results of these tests. The joint test is definitely nonsignificant,
with a chi-square value of 1.31 for 4 df and a p-value of 0.8595.

Output 15.16   Type 3 Table

```
             Contrast Results for GEE Analysis

                                   Chi-
             Contrast      DF     Square    Pr > ChiSq    Type

             carry          2       1.15      0.5626      Score
             inter          2       0.72      0.6981      Score
             joint          4       1.31      0.8595      Score
```

A reduced model was then specified, with main effects for period, age, and drug, as well as the period × age interaction. The terms of the reduced model are listed in the MODEL statement, and the CORRW option requests that the estimate of the working correlation matrix be printed. Since interest now lies in the Type 3 effects, the parameter estimates, and the working correlation matrix, the ODS SELECT statement is used to specify that only those tables be produced.

```
ods select GEEEmpPEst Type3 GEEWCorr;
proc genmod data=cross2;
   class subject age drug;
   model response = period age drug
                    period*age
      / dist=bin type3;
   repeated subject=subject/type=unstr corrw;
run;
```

The Type 3 tests indicate that period, age, and drug are highly significant. With a p-value of 0.0240, the period × age interaction cannot be dismissed.

Output 15.17   Type III Table

```
              Score Statistics For Type 3 GEE Analysis

                                     Chi-
           Source            DF    Square    Pr > ChiSq

           period             1     24.98      <.0001
           age                1     35.53      <.0001
           drug               2     39.31      <.0001
           period*age         1      5.10      0.0240
```

Output 15.18   Parameter Estimates

```
                    Analysis Of GEE Parameter Estimates
                    Empirical Standard Error Estimates

                               Standard    95% Confidence
      Parameter        Estimate   Error        Limits          Z  Pr > |Z|

      Intercept         0.5127   0.2063    0.1084   0.9170    2.49   0.0129
      period           -1.1553   0.2304   -1.6069  -0.7037   -5.01   <.0001
      age     older    -1.4994   0.2583   -2.0056  -0.9931   -5.80   <.0001
      age     younger   0.0000   0.0000    0.0000   0.0000     .       .
      drug    A          1.2542   0.2010    0.8602   1.6483    6.24   <.0001
      drug    B          0.3404   0.2016   -0.0546   0.7355    1.69   0.0912
      drug    P          0.0000   0.0000    0.0000   0.0000     .       .
      period*age older   0.7088   0.3131    0.0951   1.3224    2.26   0.0236
      period*age younger 0.0000   0.0000    0.0000   0.0000     .       .
```

Finally, the working correlation matrix is also printed. As discussed, the unstructured correlation structure is the same as the exchangeable correlation structure when you have two responses per cluster. The correlation is estimated to be 0.2274.

Output 15.19 Working Correlation Matrix

```
                      Working Correlation Matrix

                               Col1          Col2

                  Row1        1.0000        0.2274
                  Row2        0.2274        1.0000
```

Note that these results are similar to those presented for the conditional logistic analysis in Chapter 10. Most of the time, the general conclusions for a GEE analysis and the corresponding conditional logistic regression are the same; the p-values are similar, but estimates may be somewhat different. The conditional logistic model is a subject-specific model, producing odds ratio estimates for the individual, while the GEE model is a marginal model, producing odds ratios "on the average."

You may be interested in comparing the two drugs, A and B. This is done with a CONTRAST statement. You test to see if the difference of the two parameters for drugs A and B is equal to zero. The following statements request a contrast test for drug A versus drug B. The ODS SELECT statement restricts the output to just the CONTRAST statement results.

```
ods select Contrasts;
proc genmod data=cross2;
   class subject age drug;
   model response = period age drug
                    period*age
                  / dist=bin type3;
   repeated subject=subject/type=unstr;
   contrast 'A versus B' drug 1 -1 0;
run;
```

Output 15.20 Contrast Results

```
                 Contrast Results for GEE Analysis

                                Chi-
          Contrast        DF    Square    Pr > ChiSq    Type

          A versus B       1    19.15       <.0001      Score
```

This is a single degree of freedom test, and the chi-square value of 19.15 for the score test is highly significant. If you want the Wald statistic, instead of the score statistic, you specify the WALD option in the CONTRAST statement.

15.6 Respiratory Data

In Chapter 14, a repeated measurements analysis of the respiratory data was performed. The response function modeled was the logit, and the response was whether the outcome was good or excellent versus all other responses such as fair, poor, and so on. Explanatory variables included treatment, center, and visit. These data also included other explanatory variables, such as age, sex, and baseline, that couldn't be handled in the WLS repeated measurements setting because there wasn't adequate sample size. The GEE method enables you to take such explanatory variables into account.

First, the same model that resulted from the WLS is fit with the GEE method. The WLS strategy is also a marginal method that provides a robust covariance estimate, and, given large sample sizes, the WLS approach and GEE approach produce very similar estimates. Refer to Miller, Davis, and Landis (1993) for more detail on the relationship between the WLS and GEE methods.

The following SAS DATA step inputs the respiratory data and creates an observation for each response. The baseline and follow-up responses are actually measured on a five-point scale, from terrible to excellent, and this ordinal response is analyzed later in the chapter. For this analysis, the dichotomous outcome of whether the patient experienced good or excellent response is analyzed with a logistic regression. The second DATA step creates the SAS data set RESP2 and computes response variable DICHOT and dichotomous baseline variable DI_BASE. Note that the baseline variable, which was recorded on a five-point scale, could be managed as either ordinal or dichotomous.

```
data resp;
   input center id treatment $ sex $ age baseline
   visit1-visit4 @@;
   visit=1;  outcome=visit1;  output;
   visit=2;  outcome=visit2;  output;
   visit=3;  outcome=visit3;  output;
   visit=4;  outcome=visit4;  output;
   datalines;
1  53 A F  32 1  2 2 4 2   2  30 A F  37 1  3 4 4 4
1  18 A F  47 2  2 3 4 4   2  52 A F  39 2  3 4 4 4
1  54 A M  11 4  4 4 4 2   2  23 A F  60 4  4 3 3 4
1  12 A M  14 2  3 3 3 2   2  54 A F  63 4  4 4 4 4
1  51 A M  15 0  2 3 3 3   2  12 A M  13 4  4 4 4 4
1  20 A M  20 3  3 2 3 1   2  10 A M  14 1  4 4 4 4
1  16 A M  22 1  2 2 2 3   2  27 A M  19 3  3 2 3 3
1  50 A M  22 2  1 3 4 4   2  16 A M  20 2  4 4 4 3
1   3 A M  23 3  3 4 4 3   2  47 A M  20 2  1 1 0 0
1  32 A M  23 2  3 4 4 4   2  29 A M  21 3  3 4 4 4
1  56 A M  25 2  3 3 2 3   2  20 A M  24 4  4 4 4 4
1  35 A M  26 1  2 2 3 2   2   2 A M  25 3  4 3 3 1
1  26 A M  26 2  2 2 2 2   2  15 A M  25 3  4 4 3 3
1  21 A M  26 2  4 1 4 2   2  25 A M  25 2  2 4 4 4
1   8 A M  28 1  2 2 1 2   2   9 A M  26 2  3 4 4 4
1  30 A M  28 0  0 1 2 1   2  49 A M  28 2  3 2 2 1
1  33 A M  30 3  3 4 4 2   2  55 A M  31 4  4 4 4 4
1  11 A M  30 3  4 4 4 3   2  43 A M  34 2  4 4 2 4
1  42 A M  31 1  2 3 1 1   2  26 A M  35 4  4 4 4 4
1   9 A M  31 3  3 4 4 4   2  14 A M  37 4  3 2 2 4
1  37 A M  31 0  2 3 2 1   2  36 A M  41 3  4 4 3 4
1  23 A M  32 3  4 4 3 3   2  51 A M  43 3  3 4 4 2
1   6 A M  34 1  1 2 1 1   2  37 A M  52 1  2 1 2 2
```

```
1   22  A  M   46  4   3  4  3  4   2    19  A  M   55  4   4  4  4  4
1   24  A  M   48  2   3  2  0  2   2    32  A  M   55  2   2  3  3  1
1   38  A  M   50  2   2  2  2  2   2     3  A  M   58  4   4  4  4  4
1   48  A  M   57  3   3  4  3  4   2    53  A  M   68  2   3  3  3  4
1    5  P  F   13  4   4  4  4  4   2    28  P  F   31  3   4  4  4  4
1   19  P  F   31  2   1  0  2  2   2     5  P  F   32  3   2  2  3  4
1   25  P  F   35  1   0  0  0  0   2    21  P  F   36  3   3  2  1  3
1   28  P  F   36  2   3  3  2  2   2    50  P  F   38  1   2  0  0  0
1   36  P  F   45  2   2  2  2  1   2     1  P  F   39  1   2  1  1  2
1   43  P  M   13  3   4  4  4  4   2    48  P  F   39  3   2  3  0  0
1   41  P  M   14  2   2  1  2  3   2     7  P  F   44  3   4  4  4  4
1   34  P  M   15  2   2  3  3  2   2    38  P  F   47  2   3  3  2  3
1   29  P  M   19  2   3  3  0  0   2     8  P  F   48  2   2  1  0  0
1   15  P  M   20  4   4  4  4  4   2    11  P  F   48  2   2  2  2  2
1   13  P  M   23  3   3  1  1  1   2     4  P  F   51  3   4  2  4  4
1   27  P  M   23  4   4  2  4  4   2    17  P  F   58  1   4  2  2  0
1   55  P  M   24  3   4  4  4  3   2    39  P  M   11  3   4  4  4  4
1   17  P  M   25  1   1  2  2  2   2    40  P  M   14  2   1  2  3  2
1   45  P  M   26  2   4  2  4  3   2    24  P  M   15  3   2  2  3  3
1   40  P  M   26  1   2  1  2  2   2    41  P  M   15  4   3  3  3  4
1   44  P  M   27  1   2  2  1  2   2    33  P  M   19  4   2  2  3  3
1   49  P  M   27  3   3  4  3  3   2    13  P  M   20  1   4  4  4  4
1   39  P  M   28  2   1  1  1  1   2    34  P  M   20  3   2  4  4  4
1    2  P  M   28  2   0  0  0  0   2    45  P  M   33  3   3  3  2  3
1   14  P  M   30  1   0  0  0  0   2    22  P  M   36  2   4  3  3  4
1   10  P  M   37  3   2  3  3  2   2    18  P  M   38  4   3  0  0  0
1   31  P  M   37  1   0  0  0  0   2    35  P  M   42  3   2  2  2  2
1    7  P  M   43  2   3  2  4  4   2    44  P  M   43  2   1  0  0  0
1   52  P  M   43  1   1  1  3  2   2     6  P  M   45  3   4  2  1  2
1    4  P  M   44  3   4  3  4  2   2    46  P  M   48  4   4  0  0  0
1    1  P  M   46  2   2  2  2  2   2    31  P  M   52  2   3  4  3  4
1   46  P  M   49  2   2  2  2  2   2    42  P  M   66  3   3  3  4  4
1   47  P  M   63  2   2  2  2  2
;
data resp2; set resp;
   dichot=(outcome=3 or outcome=4);
   di_base = (baseline=3 or baseline=4);
run;
```

Since the final WLS model included terms for treatment, center, and visit, these main effects are included in this PROC GENMOD invocation. The options DIST=BIN and LINK=LOGIT request logistic regression. Since unique identification of the clusters requires both variables ID and CENTER, their crossing is specified in the SUBJECT= option. The WLS method computes a fully specified covariance matrix, so, for comparison purposes, the unstructured working correlation matrix is specified for the GEE analysis with TYPE=UNSTR. The DESCENDING option in the PROC GENMOD statement specifies that the highest ordered response level, 1, is to be modeled.

```
proc genmod descending;
   class id center treatment visit;
   model dichot = center treatment visit /
                  link=logit dist=bin type3;
   repeated subject=id*center / type=unstr;
run;
```

Output 15.21 displays the general model information.

Output 15.21 Model Information

```
                  Model Information

        Data Set                 WORK.RESP2
        Distribution               Binomial
        Link Function                 Logit
        Dependent Variable           dichot
        Observations Used               444
        Probability Modeled   Pr( dichot = 1 )
```

The following table of the response profiles also demonstrates that the model is based on the outcome of good or excellent (DICHOT=1).

Output 15.22 Response Profiles

```
                  Response Profile

        Ordered    Ordered
        Level      Value        Count

           1         1           248
           2         0           196
```

In the "GEE Model Information" table, you can see that there are 111 clusters in the analysis, with all clusters having responses for each of the four visits.

Output 15.23 GEE Model Information

```
              GEE Model Information

   Correlation Structure              Unstructured
   Subject Effect            id*center (111 levels)
   Number of Clusters                          111
   Correlation Matrix Dimension                  4
   Maximum Cluster Size                          4
   Minimum Cluster Size                          4
```

The Type 3 analysis displayed in Output 15.24 finds the treatment and center effects to be influential but not the visit effect.

Output 15.24   Type 3 Analysis

```
              Score Statistics For Type 3 GEE Analysis

                                    Chi-
              Source        DF     Square    Pr > ChiSq

              center         1       8.01      0.0047
              treatment      1       9.89      0.0017
              visit          3       3.47      0.3251
```

Output 15.25 displays the parameter estimates. These are very similar to those obtained from the WLS solution, and so are the standard errors.

Output 15.25   Parameter Estimates

```
                    Analysis Of GEE Parameter Estimates
                     Empirical Standard Error Estimates

                              Standard    95% Confidence
           Parameter   Estimate   Error       Limits          Z  Pr > |Z|

           Intercept     0.0732   0.2946  -0.5042   0.6506   0.25   0.8039
           center    1  -0.9168   0.3157  -1.5355  -0.2982  -2.90   0.0037
           center    2   0.0000   0.0000   0.0000   0.0000    .       .
           treatment A   1.0145   0.3165   0.3941   1.6349   3.21   0.0013
           treatment P   0.0000   0.0000   0.0000   0.0000    .       .
           visit     1   0.2835   0.2094  -0.1269   0.6939   1.35   0.1757
           visit     2   0.0804   0.2053  -0.3220   0.4829   0.39   0.6953
           visit     3   0.2840   0.1932  -0.0946   0.6627   1.47   0.1415
           visit     4   0.0000   0.0000   0.0000   0.0000    .       .
```

Table 15.3 displays these estimates side-by-side for a closer comparison.

Table 15.3.   Comparison of WLS and GEE Estimates

Parameter	WLS Estimate	Standard Error	GEE Estimate	Standard Error
Intercept	0.0168	0.2901	0.0732	0.2946
Treatment A	1.0434	0.3203	1.0145	0.3165
Visit 1	0.2216	0.1883	0.2835	0.2094
Visit 2	0.0201	0.1896	0.0804	0.2053
Visit 3	0.1811	0.1681	0.2840	0.1932
Center	−0.8803	0.3188	−0.9168	0.3157

If you were to compare WLS and GEE analyses with unstructured correlation matrix for data sets with cluster sizes of 400 plus, you would find that the estimates and standard errors were reasonably similar.

Recall that Chapter 14 indicated that fitting treatment, center, and visit for these data and considering their interactions was on the verge of not being defensible in terms of sample size. The data set includes other explanatory variables such as age, a continuous variable that is not generally handled by a WLS analysis, and sex and baseline. The GEE analysis can handle these additional variables reasonably well with 111 clusters.

The next GEE analysis includes the additional explanatory variables in the model and also includes the visit × treatment interaction and the treatment × center interaction. The exchangeable working correlation structure is thought to be a reasonable choice so it is specified with the TYPE=EXCH option in the REPEATED statement.

```
proc genmod descending;
   class id center sex treatment visit;
   model dichot = treatment sex age center di_base
                  visit visit*treatment treatment*center/
                  link=logit dist=bin type3;
   repeated subject=id*center / type=exch;
run;
```

Output 15.26 displays the general model information. Because the DESCENDING option is specified, the probability that the response DICHOT is 1 is modeled.

Output 15.26   Model Information

```
                    Model Information

          Data Set                    WORK.RESP2
          Distribution                  Binomial
          Link Function                    Logit
          Dependent Variable              dichot
          Observations Used                  444
          Probability Modeled    Pr( dichot = 1 )
```

In the "GEE Model Information" table, you can see that there are 111 clusters in the analysis, with all clusters having responses for each of the four visits.

Output 15.27   GEE Model Information

```
                  GEE Model Information

       Correlation Structure              Exchangeable
       Subject Effect             id*center (111 levels)
       Number of Clusters                          111
       Correlation Matrix Dimension                  4
       Maximum Cluster Size                          4
       Minimum Cluster Size                          4
```

The Type 3 analysis displayed in Output 15.28 indicates that the two interaction terms are nonsignificant. The TREAT*VISIT interaction has a score test statistic of 3.10 and a p-value of 0.3760 with 3 df. The CENTER*TREATMENT interaction has a score test statistic value of 2.46 with a p-value of 0.1169 and 1 df.

Output 15.28   Type 3 Tests for Model with Interactions

```
              Score Statistics For Type 3 GEE Analysis

                                      Chi-
              Source            DF    Square    Pr > ChiSq

              treatment         1     12.85       0.0003
              sex               1      0.24       0.6247
              age               1      2.23       0.1351
              center            1      3.32       0.0683
              di_base           1     23.06      <.0001
              visit             3      3.33       0.3429
              treatment*visit   3      3.10       0.3760
              center*treatment  1      2.46       0.1169
```

The same analysis was repeated with the following MODEL statement inserted. Both interactions have been dropped from the model.

```
model dichot = center sex treatment age di_base
               visit / link=logit dist=bin type3;
```

Output 15.29 displays the resulting Type 3 analysis. Visit does not appear to be influential ($p = 0.3251$), and neither does sex ($p = 0.7565$) nor age ($p = 0.1345$).

Output 15.29   Type 3 Tests for Reduced Model

```
              Score Statistics For Type 3 GEE Analysis

                                      Chi-
              Source            DF    Square    Pr > ChiSq

              center            1      3.24       0.0720
              sex               1      0.10       0.7565
              treatment         1     12.11       0.0005
              age               1      2.24       0.1345
              di_base           1     22.53      <.0001
              visit             3      3.47       0.3251
```

The further reduced model has all of the main effects except for visit. Since sex and age were identified as covariates for this analysis ahead of time, they remain in the analysis. The following statements produce the desired GEE analysis:

```
proc genmod descending;
   class id center sex treatment visit;
   model dichot = center sex treatment age di_base
               / link=logit dist=bin type3;
   repeated subject=id*center / type=exch corrw;
run;
```

Output 15.30 displays the Type 3 tests for the final model.

Output 15.30 Type 3 Tests for Final Model

```
            Score Statistics For Type 3 GEE Analysis

                                    Chi-
            Source          DF    Square    Pr > ChiSq

            center           1      3.11      0.0780
            sex              1      0.10      0.7562
            treatment        1     12.52      0.0004
            age              1      2.28      0.1312
            di_base          1     22.97      <.0001
```

There is a very significant treatment effect. As seen in the parameter estimates table in Output 15.31, active treatment increases the odds of a good or excellent response. Baseline is also very influential, with a p-value of less than 0.0001. Sex and age remain nonsignificant, and center is marginally influential with a p-value of 0.0780.

Using the parameter estimates displayed in Output 15.31 you see that those patients on active treatment, have, on the average, $e^{1.2654} = 3.5$ times greater odds of a good or excellent response as those patients on placebo, adjusted for the other effects in the model.

Output 15.31 Parameter Estimates

```
                Analysis Of GEE Parameter Estimates
                Empirical Standard Error Estimates

                       Standard   95% Confidence
        Parameter   Estimate  Error      Limits         Z  Pr > |Z|

        Intercept    -0.2066  0.5776  -1.3388   0.9255  -0.36   0.7206
        center    1  -0.6495  0.3532  -1.3418   0.0428  -1.84   0.0660
        center    2   0.0000  0.0000   0.0000   0.0000    .       .
        sex       F   0.1368  0.4402  -0.7261   0.9996   0.31   0.7560
        sex       M   0.0000  0.0000   0.0000   0.0000    .       .
        treatment A   1.2654  0.3467   0.5859   1.9448   3.65   0.0003
        treatment P   0.0000  0.0000   0.0000   0.0000    .       .
        age          -0.0188  0.0130  -0.0442   0.0067  -1.45   0.1480
        di_base       1.8457  0.3460   1.1676   2.5238   5.33   <.0001
```

Note that visit is usually considered part of the design configuration and generally would be kept in the model, particularly in a clinical trials type of analysis. The design is balanced, and you would not gain that much precision by deleting effects such as visit, age, and sex. However, in the case of an observational study, in which the design was not planned, you will probably encounter collinearity in the predictors and may need to simplify your model to some extent in order to reduce the "noise" and make very real gains in precision. However, such simplification should not be excessive in order to avoid potential bias from overfitting.

The estimated exchangeable working correlation matrix is displayed in Output 15.32.

Output 15.32   Working Correlation Matrix

| | Working Correlation Matrix | | | |
	Col1	Col2	Col3	Col4
Row1	1.0000	0.3270	0.3270	0.3270
Row2	0.3270	1.0000	0.3270	0.3270
Row3	0.3270	0.3270	1.0000	0.3270
Row4	0.3270	0.3270	0.3270	1.0000

There may be some interest in considering the unstructured working correlation matrix, since there are 4 visits per subject. This requires the estimation of more parameters, but that might be appropriate given that the model only contains five terms. The following PROC GENMOD invocation fits the same model but specifies the unstructured working correlation matrix. Note that using the unstructured correlation here, for four responses, requires you to ensure that your responses are in a consistent order, that is, the first observation in a cluster contains the first response, followed by the observation containing the second response, and so on. The DATA step used to create data set RESP on page 495 creates the proper ordering. However, if your data are not ordered correctly, then you need to create a variable that can be used by the GENMOD procedure to identify the correct sequence of responses. You use the WITHINSUBJECT option to specify that variable in the REPEATED statement.

```
proc genmod descending;
   class id center sex treatment visit;
   model dichot = center sex treatment age di_base
                 / link=logit dist=bin type3;
   repeated subject=id*center / type=unstr corrw;
run;
```

Output 15.33 displays the estimated correlation matrix. You can see that there is reasonable homogeneity in the various visit-wise correlations.

Output 15.33   Unstructured Working Correlation Matrix

| | Working Correlation Matrix | | | |
	Col1	Col2	Col3	Col4
Row1	1.0000	0.3351	0.2140	0.2953
Row2	0.3351	1.0000	0.4429	0.3581
Row3	0.2140	0.4429	1.0000	0.3964
Row4	0.2953	0.3581	0.3964	1.0000

Output 15.34 displays the parameter estimates that result from this model.

Output 15.34 Parameter Estimates for Unstructured Working Structure

```
              Analysis Of GEE Parameter Estimates
               Empirical Standard Error Estimates

                            Standard   95% Confidence
   Parameter     Estimate     Error        Limits          Z Pr > |Z|

   Intercept      -0.2324    0.5763   -1.3620    0.8972   -0.40  0.6868
   center    1    -0.6558    0.3512   -1.3442    0.0326   -1.87  0.0619
   center    2     0.0000    0.0000    0.0000    0.0000     .      .
   sex       F     0.1128    0.4408   -0.7512    0.9768    0.26  0.7981
   sex       M     0.0000    0.0000    0.0000    0.0000     .      .
   treatment A     1.2442    0.3455    0.5669    1.9214    3.60  0.0003
   treatment P     0.0000    0.0000    0.0000    0.0000     .      .
   age            -0.0175    0.0129   -0.0427    0.0077   -1.36  0.1728
   di_base         1.8981    0.3441    1.2237    2.5725    5.52  <.0001
```

Compare these estimates to those in Output 15.31. The parameter estimates themselves are quite similar, and, while most of the standard errors are a little smaller for the unstructured correlation model, there really is very little gain in efficiency. Your choice of working correlation structures depends on what you believe is most realistic for your particular data.

If you have no idea of what to specify for your correlation structure, you might want to consider the independent working correlation matrix for these data. Many analysts regularly use the independent working structure with GEE analysis and don't attempt to postulate a correlation structure. They rely on the GEE properties that both the parameter estimates and their standard errors are consistent even if the correlation structure has not been correctly specified. They are not that concerned about the potential loss of efficiency.

Output 15.35 displays the parameter estimates and standard errors that result when you repeat this analysis with the independent working correlation matrix. Again, the parameter estimates are very similar to those obtained by specifying the exchangeable and unstructured correlation structures, respectively.

Output 15.35 Parameter Estimates for Independent Working Structure

```
              Analysis Of GEE Parameter Estimates
               Empirical Standard Error Estimates

                            Standard   95% Confidence
   Parameter     Estimate     Error        Limits          Z Pr > |Z|

   Intercept      -0.2066    0.5776   -1.3388    0.9255   -0.36  0.7206
   center    1    -0.6495    0.3532   -1.3418    0.0428   -1.84  0.0660
   center    2     0.0000    0.0000    0.0000    0.0000     .      .
   sex       F     0.1368    0.4402   -0.7261    0.9996    0.31  0.7560
   sex       M     0.0000    0.0000    0.0000    0.0000     .      .
   treatment A     1.2654    0.3467    0.5859    1.9448    3.65  0.0003
   treatment P     0.0000    0.0000    0.0000    0.0000     .      .
   age            -0.0188    0.0130   -0.0442    0.0067   -1.45  0.1480
   di_base         1.8457    0.3460    1.1676    2.5238    5.33  <.0001
```

15.7   Using a Modified Wald Statistic to Assess Model Effects

Section 15.3 discusses the possibility of using an adjusted Wald statistic to evaluate model effects in the GEE approach. Shah, Holt, and Folsom (1977) describe a modification of the Wald statistic based on a Hotelling T^2 type of transformation of Q_C:

$$\frac{(d-c+1)Q_C}{dc} \quad \text{is distributed as} \quad F_{c,d-c+1}$$

The quantity d is equal to the number of clusters minus 1 and c is equal to the number of rows of the contrast. Thus, for tests concerning effects of explanatory factors, it is equal to the corresponding number of df. This test is more conservative than the Wald test; LaVange, Koch, and Schwartz (2000) suggest that you can use the Wald and F-transform statistic p-values as the lower and upper bounds for judging the robustness of the actual p-value. As the number of clusters becomes very large, these statistics produce very similar conclusions.

The Output Delivery System makes it relatively easy to write a macro that produces these F-transform statistics and append a table containing them to the end of the PROC GENMOD output. Appendix B of this chapter contains the SAS statements making up the macro GEEF that performs this task. It first produces output data sets from the GENMOD invocation that contain the number of clusters, the appropriate Wald statistics, and df for the Type 3 analysis. Then, it does some DATA step manipulations to produce a data set that includes the desired computations. The TEMPLATE procedure is then used to create a template for the new type of table containing the F statistics, and a DATA _NULL_ step is used to print it.

The following example illustrates how to use this macro with the first reduced model fit for the respiratory data in the preceding section. The statements assume that the macro GEEF has been included in a file named MACROS.SAS that is stored in the directory in which you are running your SAS program. The ODS OUTPUT statement puts the GEE model information, including the number of clusters, into a SAS data set named CLUSTOUT, and it puts the Type 3 analysis results, which include the Wald chi-square values and their df, into a SAS data set named SCOREOUT. Because the GEEF macro expects them, the names CLUSTOUT and SCOREOUT must be used. The PROC GENMOD invocation is exactly the same as before except that the WALD option is specified in the MODEL statement to produce Wald statistics in the Type 3 analysis.

```
%include 'macros.sas';
ods output GEEModInfo=clustout
           Type3=scoreout;
proc genmod descending data=resp2;
   class id center sex treatment visit;
   model dichot = treatment sex center age di_base
                  visit / link=logit dist=bin type3 wald;
   repeated subject=id*center / type=exch;
run;
%geef;
```

The statements produce the usual PROC GENMOD output, including the model information, initial parameter estimates, the GEE information, the GEE parameter estimates, and the GEE Type 3 analysis. The created GEE Type 3 table with the F statistics is printed next. Both Type 3 tables are displayed in Output 15.36.

Output 15.36   Type 3 Analyses

```
            Wald Statistics For Type 3 GEE Analysis

                              Chi-
        Source         DF    Square    Pr > ChiSq

        treatment       1    12.95       0.0003
        sex             1     0.10       0.7551
        center          1     3.52       0.0605
        age             1     2.06       0.1516
        di_base         1    27.90      <.0001
        visit           3     3.63       0.3047



            F-Statistics for Type 3 GEE Analysis

        Source         DF    F Value    Pr > F

        treatment       1    12.95       0.0005
        sex             1     0.10       0.7557
        center          1     3.52       0.0631
        age             1     2.06       0.1545
        di_base         1    27.90      <.0001
        visit           3     1.19       0.3182
```

Note that the values of the F statistics are the same as the values for the Wald statistics for the single degree of freedom tests. However, all of the p-values are more conservative. You might choose to generate the transformed F statistics when you are dealing with a small number of clusters, especially when you have marginal significance. Note that the default score statistics for this Type 3 analysis provide the most conservative p-values for all of these effects except the age effect, in which case the score statistic results in the most liberal p-value (0.1345 compared to 0.1516 for the Wald test and 0.1545 for the F statistic).

15.8   Diagnostic Data

The diagnostic data analyzed in Chapter 10 and Chapter 14 are now analyzed with the GEE method. Recall that subjects received test and standard procedures at two times, and researchers recorded response as positive or negative. Besides analyzing these data with conditional logistic regression and repeated measures WLS, you can also analyze these data with the GEE method. There are 793 clusters (corresponding to the number of subjects) and four measurements per subject (corresponding to the two types of tests at two times).

The following DATA steps input the diagnosis data and create an observation for each measurement so that the GEE facilities in the GENMOD procedure can be used. In addition, an indicator variables is created for time. The variable PROCEDURE takes the values of the standard or test procedures.

```
data diagnos;
   input std1 $ test1 $ std2 $ test2 $ count;
   do i=1 to count;
    output;
   end;
   datalines;
Neg Neg Neg Neg 509
Neg Neg Neg Pos   4
Neg Neg Pos Neg  17
Neg Neg Pos Pos   3
Neg Pos Neg Neg  13
Neg Pos Neg Pos   8
Neg Pos Pos Neg   0
Neg Pos Pos Pos   8
Pos Neg Neg Neg  14
Pos Neg Neg Pos   1
Pos Neg Pos Neg  17
Pos Neg Pos Pos   9
Pos Pos Neg Neg   7
Pos Pos Neg Pos   4
Pos Pos Pos Neg   9
Pos Pos Pos Pos 170
;
data diagnos2;
   set diagnos;
    drop std1 test1 std2 test2;
   subject=_n_;
   time=1; procedure='standard';
   response=std1; output;
    time=1; procedure='test';
   response=test1; output;
   time=2; procedure='standard';
    response=std2; output;
    time=2; procedure='test';
   response=test2; output;
run;
```

The model consists of time and procedure main effects as well as their interaction. The exchangeable working correlation structure is specified with the TYPE=EXCH option. Logistic regression is requested with the LINK=LOGIT and DIST=BIN options in the MODEL statement. The model is based on the probability of the positive response since the DESCENDING option is used in the PROC statement.

```
proc genmod descending;
   class subject time procedure;
   model response = time procedure time*procedure /
                    link=logit dist=bin type3;
   repeated subject=subject /type=exch;
run;
```

Output 15.37 displays the model information. Note that 'Pos' is the first ordered response value.

Output 15.37   Model Information

```
                        Model Information

           Data Set                        WORK.DIAGNOS2
           Distribution                         Binomial
           Link Function                           Logit
           Dependent Variable                   response
           Observations Used                        3172
           Probability Modeled     Pr( response = Pos )
```

Output 15.38 defines the parameters.

Output 15.38   Parameter Information

```
                      Parameter Information

         Parameter      Effect          time    procedure

         Prm1           Intercept
         Prm2           time             1
         Prm3           time             2
         Prm4           procedure                standard
         Prm5           procedure                test
         Prm6           time*procedure   1        standard
         Prm7           time*procedure   1        test
         Prm8           time*procedure   2        standard
         Prm9           time*procedure   2        test
```

The model information indicates that 793 clusters are analyzed with a cluster size of four and no missing data; the exchangeable working correlation structure is requested.

Output 15.39   Model Information

```
                     GEE Model Information

       Correlation Structure              Exchangeable
       Subject Effect             subject (793 levels)
       Number of Clusters                          793
       Correlation Matrix Dimension                  4
       Maximum Cluster Size                          4
       Minimum Cluster Size                          4
```

The score statistics in the Type 3 analysis indicate that the time $\times$ procedure interaction is not significant, using an $\alpha = 0.05$ criterion.

Output 15.40   Type 3 Test Results

```
            Score Statistics For Type 3 GEE Analysis

                                    Chi-
            Source         DF      Square     Pr > ChiSq

            time            1        0.91        0.3390
            procedure       1        8.17        0.0043
            time*procedure  1        2.49        0.1142
```

The reduced model fit consists of the main effects only.

```
proc genmod descending;
   class subject time procedure;
   model response = time procedure /
                    link=logit dist=bin type3;
   repeated subject=subject / type=exch corrw;
run;
```

The Type 3 analysis for the reduced model finds procedure significant with a chi-square value of 8.11 and 1 df ($p = 0.0044$).

Output 15.41   Reduced Model

Score Statistics For Type 3 GEE Analysis			
Source	DF	Chi-Square	Pr > ChiSq
time	1	0.85	0.3573
procedure	1	8.11	0.0044

Output 15.42 displays the parameter estimates and their test statistics. The first procedure, the standard, is associated with a higher odds of getting the positive response as compared to the test treatment. The odds of the positive response with the standard procedure are $e^{0.1188}$ or 1.13 times higher than the odds for the test procedure.

Output 15.42   Parameter Estimates

Analysis Of GEE Parameter Estimates Empirical Standard Error Estimates						
Parameter	Estimate	Standard Error	95% Confidence Limits		Z	Pr > |Z|
Intercept	-1.0173	0.0792	-1.1726	-0.8621	-12.84	<.0001
time 1	0.0313	0.0340	-0.0353	0.0978	0.92	0.3573
time 2	0.0000	0.0000	0.0000	0.0000	.	.
procedure standard	0.1188	0.0415	0.0373	0.2002	2.86	0.0042
procedure test	0.0000	0.0000	0.0000	0.0000	.	.

Output 15.43 contains the estimated exchangeable working correlation matrix. Note that with over 700 cluster and four measurements, you may want to specify the unstructured correlation matrix for a possible gain in precision of the estimates.

Output 15.43 Estimated Exchangeable Correlation

```
                      Working Correlation Matrix

                  Col1        Col2        Col3        Col4

        Row1    1.0000      0.8041      0.8041      0.8041
        Row2    0.8041      1.0000      0.8041      0.8041
        Row3    0.8041      0.8041      1.0000      0.8041
        Row4    0.8041      0.8041      0.8041      1.0000
```

When you re-submit the preceding PROC GENMOD statements with the following REPEATED statement

```
repeated subject=subject /type=unstr corrw;
```

you obtain the following parameter estimates. These have minimally smaller standard errors than those for the exchangeable structure.

Output 15.44 Parameter Estimates for Unstructured

```
                     Analysis Of GEE Parameter Estimates
                     Empirical Standard Error Estimates

                                Standard   95% Confidence
        Parameter     Estimate    Error       Limits          Z Pr > |Z|

        Intercept     -1.0208    0.0793   -1.1762   -0.8654  -12.88   <.0001
        time       1   0.0344    0.0339   -0.0321    0.1009    1.01   0.3103
        time       2   0.0000    0.0000    0.0000    0.0000     .        .
        procedure standard  0.1240 0.0414  0.0429    0.2052    3.00   0.0027
        procedure test 0.0000    0.0000    0.0000    0.0000     .        .
```

When you compare the estimated unstructured correlation matrix with the estimated exchangeable correlation matrix, you can see that they are fairly similar.

Output 15.45 Estimated Unstructured Correlation

```
                      Working Correlation Matrix

                  Col1        Col2        Col3        Col4

        Row1    1.0000      0.7855      0.8369      0.7763
        Row2    0.7855      1.0000      0.7691      0.8560
        Row3    0.8369      0.7691      1.0000      0.8163
        Row4    0.7763      0.8560      0.8163      1.0000
```

Note that the results of the analyses for these data—a GEE analysis, the WLS repeated analysis, and the conditional logistic regression analysis—all provided similar conclusions. The WLS analysis focused on the marginal proportion of the negative response, the GEE analysis was a marginal analysis of the logit function, and the conditional logistic analysis was a subject-specific analysis of the logit function. The conditional analysis, through its subject-specific focus, found the standard procedure effect to be stronger, with the odds of positive response for the standard procedure being nearly twice the odds of positive response for the test procedure. However, note that the conditional analysis produces a subject-specific odds ratio, whereas the GEE odds ratio is a population-averaged odds ratio. Your choice of strategy depends on the overall objectives of the study analysis.

15.9 Using GEE for Count Data

Sometimes, categorical data come in the form of count data. For example, you may record the number of acute pain episodes in a time interval in a clinical trial evaluating treatments. Other examples might be the number of insurance claims registered during the year or the number of unscheduled medical visits made during a study of a new protocol for asthma medication. Often, Poisson regression is the appropriate strategy for analyzing such data. See Chapter 12,"Poisson Regression," for further discussion of Poisson regression.

Since Poisson regression is an application of the generalized linear model with the Poisson distribution and the log link function, you can fit models for clustered or repeated data with GEE methods. In this example, researchers evaluated a new drug to treat osteoporosis in women past menopause. In a double-blind study, a group of women were assigned the treatment and a group of women were assigned the placebo. Both groups of women were provided with calcium supplements, given nutritional counseling, and encouraged to be physically active through the availability of exercise programs.

The study ran for three years, and the number of fractures occurring in each of those years was recorded. The length of each of the years, the corresponding risk periods, is 12 months. However, there were a few drop-outs in the third year, and those risk periods were set at 6 months. The offset variable is log of months at risk, as contained in the variable LMONTHS.

The following DATA step inputs the fracture data.

```
data fracture;
   input ID age center $ treatment $ year1 year2 year3 @@;
   total=year1+year2+year3;
   lmonths=log(12);
   datalines;
 1  56 A p 0 0 0    2 71 A p 1 0 0    3 60 A p 0 0 1    4 71 A p 0 1 0
 5  78 A p 0 0 0    6 67 A p 0 0 0    7 49 A p 0 0 0
 9  75 A p 1 0 0    8 68 A p 0 0 0   11 82 A p 0 0 0
13  56 A p 0 0 0   12 71 A p 0 0 0   15 66 A p 1 0 0
17  78 A p 0 0 0   16 63 A p 0 2 0   19 61 A p 0 0 0
21  75 A p 1 0 0   20 68 A p 0 0 0   23 63 A p 1 1 1
25  54 A p 0 0 0   24 65 A p 0 0 0   27 71 A p 0 0 0
29  56 A p 0 0 0   28 64 A p 0 0 0   31 78 A p 0 0 2
33  76 A p 0 0 0   32 61 A p 0 0 0   35 76 A p 0 0 0
37  74 A p 0 0 0   36 56 A p 0 0 0   39 62 A p 0 0 0
41  56 A p 0 0 0   40 72 A p 0 0 1   43 76 A p 0 0 0
45  75 A p 0 0 0   44 77 A p 2 2 0   47 78 A p 0 0 0
49  71 A p 0 0 0   48 68 A p 0 0 0   51 74 A p 0 0 0
```

```
 53   69 A p 0 0 0    52 78 A p 1 0 0    55 81 A p 2 0 1
 57   68 A p 0 0 0    56 77 A p 0 0 0    59 77 A p 0 0 0
 61   75 A p 0 0 0    60 83 A p 0 0 0    63 72 A p 0 0 0    64 88 A p 0 0 0
 65   69 A p 0 0 0    66 55 A p 0 0 0    67 76 A p 0 0 0    68 55 A p 0 0 0
 69   63 A t 0 0 2    70 52 A t 0 0 0    71 56 A t 0 0 0    72 52 A t 0 0 0
 73   74 A t 0 0 0    74 61 A t 0 0 0    75 69 A t 0 0 0    76 61 A t 0 0 0
 77   84 A t 0 0 0    78 76 A t 0 1 0    79 59 A t 0 0 1    80 76 A t 0 0 0
 81   66 A t 0 0 1    82 78 A t 0 0 1    83 77 A t 0 0 0    84 75 A t 1 0 0
 85   75 A t 0 0 0    86 62 A t 0 0 0    87 67 A t 0 0 0    88 62 A t 0 0 0
 89   71 A t 0 0 0    90 63 A t 0 0 0                       92 68 A t 0 0 0
 93   69 A t 0 0 0    94 61 A t 0 0 0                       96 61 A t 0 0 0
 97   67 A t 0 0 0    98 77 A t 0 0 0 91   70 A t 0 0 1 102 81 A t 0 0 0
 95   49 A t 0 0 0   106 55 A t 0 0 0
 99   63 A t 2 1 0   100 52 A t 0 0 0   101 48 A t 0 0 0
103   71 A t 0 0 0   104 61 A t 0 0 0   105 74 A t 0 0 0
107   67 A t 0 0 0   108 56 A t 0 0 0   109 54 A t 0 0 0
111   56 A t 0 0 0   112 77 A t 1 0 0   113 65 A t 0 0 0
115   66 A t 0 0 0   116 71 A t 0 0 0   117 71 A t 0 0 0   128 71 A t 0 0 0
119   86 A t 1 0 0   120 81 A t 0 0 0   121 64 A t 0 0 0   132 76 A t 0 0 0
123   71 A t 0 0 0   124 76 A t 0 0 0   125 66 A t 0 0 0   136 76 A t 0 0 0
  1   68 B p 0 0 0     2 63 B p 0 0 0     3 66 B p 0 0 0     4 63 B p 0 0 0
  5   70 B p 0 1 0     6 62 B p 0 0 0     7 54 B p 1 0 0     8 66 B p 0 0 0
  9   71 B p 0 0 0    10 76 B p 0 0 0    11 72 B p 0 0 1    12 65 B p 0 1 0
 13   55 B p 0 1 0    14 59 B p 0 0 2    15 61 B p 1 0 0    16 56 B p 0 1 0
 17   54 B p 0 0 0    18 68 B p 0 0 0    19 68 B p 0 0 0    20 81 B p 0 0 0
 21   81 B p 1 0 0    22 61 B p 2 0 1    23 72 B p 1 0 0    24 67 B p 0 0 0
 25   56 B p 0 0 0    26 66 B p 0 0 0    27 71 B p 0 1 0    28 75 B p 0 1 0
 29   76 B p 0 0 0    30 73 B p 2 0 0    31 56 B p 0 0 0    32 89 B p 0 0 0
 33   56 B p 0 0 0    34 78 B p 0 0 0    35 55 B p 0 0 0    36 73 B p 0 0 1
 37   71 B p 0 0 0    38 56 B p 0 0 0    39 69 B p 0 0 0    40 77 B p 0 0 0
 41   89 B p 0 0 0    42 63 B p 0 0 0    43 67 B p 0 0 0    44 73 B p 0 0 0
 45   60 B p 0 0 0    46 67 B p 0 0 0    47 56 B p 0 0 0    48 78 B p 0 0 0
 49   73 B t 1 0 0    50 76 B t 0 0 0    51 61 B t 0 0 0    52 81 B t 0 0 0
 53   55 B t 0 0 0    54 82 B t 0 0 0    55 78 B t 0 0 0    56 60 B t 0 0 0
 57   56 B t 0 0 0    58 83 B t 0 0 0    59 55 B t 0 0 0    60 60 B t 0 0 0
 61   80 B t 0 0 0    62 78 B t 0 0 0    63 67 B t 0 0 0    64 67 B t 0 0 0
 65   56 B t 0 0 0    66 72 B t 0 0 0    67 71 B t 0 0 0    68 83 B t 0 0 0
 69   66 B t 0 0 0    70 71 B t 0 0 1    71 78 B t 1 0 2    72 61 B t 0 0 0
 73   56 B t 0 0 0    74 61 B t 0 0 0    75 55 B t 0 0 0    76 69 B t 1 1 0
 77   71 B t 0 0 0    78 76 B t 0 0 0    79 56 B t 0 0 0    80 75 B t 0 0 0
 81   89 B t 0 0 0    82 77 B t 0 0 0    83 77 B t 1 0 0    84 73 B t 0 0 0
 85   60 B t 0 0 0    86 61 B t 0 0 0    87 79 B t 0 0 0    88 71 B t 0 0 0
 89   61 B t 0 0 0    90 79 B t 0 0 0    91 87 B t 1 0 0    92 55 B t 0 0 0
 93   55 B t 0 0 0    94 79 B t 0 0 0    95 66 B t 0 0 0    96 49 B t 0 0 0
 97   56 B t 0 0 0    98 64 B t 0 0 0    99 88 B t 0 0 0   100 62 B t 1 0 0
101   80 B t 0 0 1   102 65 B t 0 0 0   103 57 B t 0 0 1   104 85 B t 0 0 0
  ;
```

The next DATA step creates one observation per year. It also sets the variable LMONTHS to a value of log(6) for the known drop-outs in the third year. Counts were recorded for that year for each subject, for the time they were still in the study.

```
data fracture2;
   set fracture;
   drop year1-year3;
   year=1; fractures=year1; output;
   year=2; fractures=year2; output;
   do; if center = A then do;
     if (ID=85 or ID=66 or ID=124 or ID=51) then lmonths=log(6); end;
     if center = B then do;
     if (ID=29 or ID=45 or ID=55) then lmonths=log(6); end;
   end;
   year=3; fractures=year3; output;
run;
```

The specification for the Poisson GEE analysis is straightforward. You specify the link function, LINK=LOG, and also specify the distribution, DIST=POISSON. The response variable is FRACTURES and the variables in the model include CENTER, TREATMENT, AGE, YEAR, the TREATMENT*CENTER interaction, and the TREATMENT*YEAR interaction. In addition, in Poisson regression you usually specify an offset variable. In this situation, this is the variable LMONTHS which is the log length of time at risk in each year. The offset is specified with the OFFSET=LMONTHS option in the MODEL statement. Since there is not a unique subject identifier, you use the crossing of ID and CENTER with the SUBJECT= option to create unique values that determine the experimental units.

```
proc genmod;
   class id treatment center year;
   model fractures = center treatment age year treatment*center
                     treatment*year/
                     dist=poisson type3 offset=lmonths;
   repeated subject=id*center / type=exch corrw;
run;
```

Output 15.46 displays the general information about the model being fit: the Poisson distribution is requested, the offset variable is LMONTHS, and the response variable is FRACTURES.

Output 15.46   Model Information

```
                    Model Information

           Data Set              WORK.FRACTURE2
           Distribution                 Poisson
           Link Function                    Log
           Dependent Variable         fractures
           Offset Variable              lmonths
           Observations Used                642
```

In Output 15.47, you can see that there are 214 clusters in the analysis, with all clusters having responses for each of the three years (even though some of the responses for the third year were for a reduced risk period).

Output 15.47   GEE Model Information

```
                      GEE Model Information

      Correlation Structure                      Exchangeable
      Subject Effect              ID*center (214 levels)
      Number of Clusters                                214
      Correlation Matrix Dimension                        3
      Maximum Cluster Size                                3
      Minimum Cluster Size                                3
```

The Type 3 analysis displayed in Output 15.48 finds the two interaction terms to be nonsignificant. The treatment × center interaction has a score test statistic of 0.04 and a p-value of 0.8364 with 1 df. The treatment × year interaction has a score test statistic value of 3.15 with a p-value of 0.2074 and 2 df.

Output 15.48   Type 3 Tests for Model with Interactions

```
            Score Statistics For Type 3 GEE Analysis

                                   Chi-
          Source            DF    Square    Pr > ChiSq

          center             1     0.02       0.8750
          treatment          1     4.69       0.0303
          age                1     2.44       0.1180
          year               2     7.64       0.0220
          treatment*center   1     0.04       0.8364
          treatment*year     2     3.15       0.2074
```

The same analysis was repeated with just the main effects. Note that the nesting of ID and CENTER in the SUBJECT= option is just another way to specify a unique set of values with which to identify the individual experimental units. The CORRW option requests that the working correlation matrix be printed.

```
proc genmod;
   class id treatment center year;
   model fractures = center treatment age year /
                   dist=poisson type3 offset=lmonths;
   repeated subject=id(center) / type=exch corrw;
run;
```

Output 15.49 displays the resulting Type 3 analysis.

Output 15.49   Type 3 Tests for Reduced Model

```
            Score Statistics For Type 3 GEE Analysis

                               Chi-
          Source        DF    Square    Pr > ChiSq

          center         1     0.02       0.8930
          treatment      1     3.41       0.0647
          age            1     2.22       0.1359
          year           2     4.71       0.0948
```

Treatment is nearly significant here, with a p-value of 0.0647. Year also has some modest influence, with a p-value of 0.0948.

Output 15.50 contains the parameter estimates.

Output 15.50   Parameter Estimates

```
                  Analysis Of GEE Parameter Estimates
                  Empirical Standard Error Estimates

                        Standard   95% Confidence
        Parameter     Estimate    Error        Limits          Z  Pr > |Z|

        Intercept      -6.6379    1.1201   -8.8333   -4.4424  -5.93   <.0001
        center    A     0.0400    0.2968   -0.5416    0.6216   0.13   0.8928
        center    B     0.0000    0.0000    0.0000    0.0000     .       .
        treatment p     0.5715    0.3042   -0.0248    1.1678   1.88   0.0603
        treatment t     0.0000    0.0000    0.0000    0.0000     .       .
        age             0.0223    0.0147   -0.0065    0.0512   1.52   0.1294
        year      1     0.2763    0.2940   -0.2999    0.8524   0.94   0.3473
        year      2    -0.3830    0.3747   -1.1173    0.3513  -1.02   0.3067
        year      3     0.0000    0.0000    0.0000    0.0000     .       .
```

The placebo increases the log fracture rate by 0.5715; the test treatment lowers the log fracture rate by -0.5715.

Output 15.51 contains the estimate of the working correlation matrix. It indicates small, but not ignorable, correlations among the respective years.

Output 15.51   Working Correlation Matrix

```
                     Working Correlation Matrix

                        Col1       Col2       Col3

              Row1     1.0000     0.1049     0.1049
              Row2     0.1049     1.0000     0.1049
              Row3     0.1049     0.1049     1.0000
```

15.10   Fitting the Proportional Odds Model

Recall that the respiratory data analyzed in Section 15.6 contained an ordinal response that ranged from 0 for poor to 4 for excellent. (The responses were dichotomized in the previous analyses.) The proportional odds model provides a strategy that takes into account the ordinality of the data. See Chapter 9, "Logistic Regression II: Polytomous Response," for a discussion of the proportional odds model in the univariate case, and refer to Lipsitz, Kim, and Zhao (1994) and Miller, Davis, and Landis (1993) for discussions on fitting the proportional odds model with GEE.

The following statements request a proportional odds model to be fit with the GEE method. The SAS data set RESP is the same as created in Section 15.6.

```
proc genmod data=resp2 descending;
   class id center sex treatment visit;
   model outcome = treatment sex center age baseline
                   visit visit*treatment /
                   link=clogit dist=mult type3;
   repeated subject=id*center / type=ind;
run;
```

The variable OUTCOME has five levels, ranging from 0 to 4, for poor to excellent. Since interest lies in assessing how much better the subjects receiving the active treatment were, you form the cumulative logits that focus on the comparison of better to poorer outcomes. By default, the GENMOD procedure forms the cumulative logits based on the ratio of the probability of the lower ordered response values to the probability of the higher ordered response values. In this case, this would be poorer outcomes compared to better outcomes. To reverse this ordering, you simply specify the DESCENDING option in the PROC statement.

You specify the LINK=CLOGIT option to request the cumulative logit link and the DIST=MULT option to request the multinomial distribution. Together, these options specify the proportional odds model. The preliminary model includes TREATMENT, SEX, CENTER, AGE, BASELINE, VISIT, and the VISIT*TREATMENT interaction as the explanatory variables. Note that the BASELINE variable also lies on a 0–4 scale.

Since a unique patient identification requires the ID value and the CENTER value, you specify the SUBJECT=ID*CENTER option in the REPEATED statement. The TYPE=IND option specifies the independent working correlation matrix, which is currently the only correlation structure available with the ordinal response model. Output 15.52 displays the class level and response profile information. Since the ordered values are listed in descending order, the cumulative logits are modeling the better outcomes compared to the poorer outcomes.

Output 15.52   Class Level and Response Information

```
                        Class Level Information

   Class          Levels    Values

   id               56       1  2  3  4  5  6  7  8  9  10  11  12  13  14  15  16  17  18  19  20
                            21 22 23 24 25 26 27 28 29 30 31 32 33 34 35 36 37
                            38 39 40 41 42 43 44 45 46 47 48 49 50 51 52 53 54
                            55 56
   center            2       1 2
   sex               2       F M
   treatment         2       A P
   visit             4       1 2 3 4


                           Response Profile

                    Ordered      Ordered
                    Level        Value         Count

                        1          4            152
                        2          3             96
                        3          2            116
                        4          1             40
                        5          0             40
```

The parameter information is displayed in Output 15.53.

Output 15.53   Parameter Information

```
                         Parameter Information

      Parameter        Effect            center    sex    treatment    visit

      Prm1             treatment                            A
      Prm2             treatment                            P
      Prm3             sex                          F
      Prm4             sex                          M
      Prm5             center              1
      Prm6             center              2
      Prm7             age
      Prm8             baseline
      Prm9             visit                                             1
      Prm10            visit                                             2
      Prm11            visit                                             3
      Prm12            visit                                             4
      Prm13            treatment*visit                      A            1
      Prm14            treatment*visit                      A            2
      Prm15            treatment*visit                      A            3
      Prm16            treatment*visit                      A            4
      Prm17            treatment*visit                      P            1
      Prm18            treatment*visit                      P            2
      Prm19            treatment*visit                      P            3
      Prm20            treatment*visit                      P            4
```

From the GEE Model Information table, you can see that there are 111 clusters, with four visit outcomes in each of the clusters.

Output 15.54 Model Information

```
                      GEE Model Information

         Correlation Structure                   Independent
         Subject Effect                 id*center (111 levels)
         Number of Clusters                              111
         Correlation Matrix Dimension                      4
         Maximum Cluster Size                              4
         Minimum Cluster Size                              4
```

The Type 3 analysis indicates that the treatment × visit interaction is significant, at least at the $\alpha = 0.05$ level. Gender doesn't appear to be an important factor, and neither does age. Center also appears to be non-influential, but as a pre-stated design covariate, it stays in the model regardless.

Baseline and treatment (for visit 4) have strongly significant effects.

Output 15.55 Type 3 Test Results

```
              Score Statistics For Type 3 GEE Analysis

                                      Chi-
         Source             DF       Square     Pr > ChiSq

         treatment           1        15.33        <.0001
         sex                 1         0.53        0.4664
         center              1         1.33        0.2482
         age                 1         2.68        0.1016
         baseline            1        21.60        <.0001
         visit               3         0.66        0.8837
         treatment*visit     3        10.47        0.0150
```

The next PROC GENMOD invocation simplifies the model by excluding the AGE and SEX terms. Since CENTER is part of the study design, it remains in the model. The VISIT*TREATMENT term also stays.

```
proc genmod data=resp2 descending;
   class id center sex treatment visit;
   model outcome = treatment center baseline
                   visit visit*treatment /
                   link=clogit dist=mult type3;
   repeated subject=id*center / type=ind;
run;
```

The treatment × visit interaction remains important in this simplified model, as indicated in Output 15.56. Note that the analysis of the dichotomous outcome did not find the treatment × visit interaction to be noteworthy.

Output 15.56   Reduced Model

```
                Score Statistics For Type 3 GEE Analysis

                                            Chi-
                Source              DF     Square    Pr > ChiSq

                treatment           1      16.40      <.0001
                center              1       1.25      0.2636
                baseline            1      21.27      <.0001
                visit               3       0.54      0.9106
                treatment*visit     3      10.50      0.0148
```

Baseline and treatment (for visit 4) remain extremely significant in this reduced model, with p-values less than 0.001.

Output 15.57 contains the parameter estimates.

Output 15.57   Parameter Estimates

```
                     Analysis Of GEE Parameter Estimates
                      Empirical Standard Error Estimates

                                    Standard    95% Confidence
          Parameter        Estimate   Error        Limits          Z  Pr > |Z|

          Intercept1        -3.3645   0.5766   -4.4945   -2.2345  -5.84   <.0001
          Intercept2        -2.2049   0.5412   -3.2657   -1.1441  -4.07   <.0001
          Intercept3        -0.6060   0.5193   -1.6239    0.4119  -1.17   0.2433
          Intercept4         0.2929   0.5643   -0.8131    1.3988   0.52   0.6037
          treatment    A     0.9995   0.3625    0.2891    1.7100   2.76   0.0058
          treatment    P     0.0000   0.0000    0.0000    0.0000    .       .
          center       1    -0.3491   0.3023   -0.9415    0.2434  -1.15   0.2482
          center       2     0.0000   0.0000    0.0000    0.0000    .       .
          baseline           0.8993   0.1670    0.5719    1.2266   5.38   <.0001
          visit        1     0.2581   0.2501   -0.2321    0.7484   1.03   0.3021
          visit        2    -0.2505   0.2303   -0.7019    0.2010  -1.09   0.2768
          visit        3    -0.0360   0.1615   -0.3525    0.2806  -0.22   0.8238
          visit        4     0.0000   0.0000    0.0000    0.0000    .       .
          treatment*visit A 1 -0.3049  0.3927   -1.0746    0.4648  -0.78   0.4375
          treatment*visit A 2  0.7247  0.3547    0.0296    1.4198   2.04   0.0410
          treatment*visit A 3  0.2990  0.3321   -0.3519    0.9500   0.90   0.3679
          treatment*visit A 4  0.0000  0.0000    0.0000    0.0000    .       .
          treatment*visit P 1  0.0000  0.0000    0.0000    0.0000    .       .
          treatment*visit P 2  0.0000  0.0000    0.0000    0.0000    .       .
          treatment*visit P 3  0.0000  0.0000    0.0000    0.0000    .       .
          treatment*visit P 4  0.0000  0.0000    0.0000    0.0000    .       .
```

15.11   GEE Analyses for Data with Missing Values

One of the main advantages of the GEE method is that it addresses the possibility of missing values. The number of responses per subject, or cluster, can vary; recall that you can have t_i responses per subject, where t_i depends on the ith subject. While the data sets analyzed in previous sections were complete, or balanced, you are faced with missing data in many situations, especially for observational data that are longitudinal. Loss to follow-up is a common problem for planned studies that involve repeated visits. The GEE method works nicely for many of these data situations. Note however, that the GEE method does assume that the missing values are missing completely at random, or MCAR.

Crossover Study with Missing Data

Consider a two-period crossover study on treatments for a skin disorder where patients were given sequences of the standard drug A, a new drug B, and a placebo. Investigators introduced a skin irritant and then applied topical treatments. Subjects were stratified by gender. 300 patients participated at the first session, but 50 patients failed to attend the second session one week later. Investigators determined that none of the losses to follow-up was actually due to the failure of the treatments, but rather due to the usual attrition plus a breakdown in the communication to emphasize the importance of the return visit. For this reason, although much more missing data occurred than was expected, the analysis proceeded.

You can analyze these data in a similar manner to the way in which the crossover data were analyzed in Section 15.5. The design is exactly the same; the only difference is that the 50 subjects with only one measurement have a cluster size of 1. Note the number of missing values for the second period. These data are input into SAS data set SKINCROSS.

```
data skincross;
   input subject gender $ sequence $ Time1 $ Time2 $ @@;
   datalines;
 1  m AB  Y  Y  101 m  PA  Y  Y  201 f  AP  Y  Y
 2  m AB  Y  .  102 m  PA  Y  Y  202 f  AP  Y  Y
 3  m AB  Y  Y  103 m  PA  Y  Y  203 f  AP  Y  Y
 4  m AB  Y  .  104 m  PA  Y  Y  204 f  AP  Y  Y
 5  m AB  Y  Y  105 m  PA  Y  Y  205 f  AP  Y  Y
 6  m AB  Y  .  106 m  PA  Y  N  206 f  AP  Y  Y
 7  m AB  Y  .  107 m  PA  Y  .  207 f  AP  Y  Y
 8  m AB  Y  Y  108 m  PA  Y  N  208 f  AP  Y  Y
 9  m AB  Y  Y  109 m  PA  N  .  209 f  AP  Y  Y
10  m AB  Y  Y  110 m  PA  N  Y  210 f  AP  Y  Y
11  m AB  Y  .  111 m  PA  N  Y  211 f  AP  Y  Y
12  m AB  Y  Y  112 m  PA  N  Y  212 f  AP  Y  Y
13  m AB  Y  N  113 m  PA  N  .  213 f  AP  Y  Y
14  m AB  Y  N  114 m  PA  N  .  214 f  AP  Y  .
15  m AB  Y  N  115 m  PA  N  Y  215 f  AP  Y  .
16  m AB  Y  N  116 m  PA  N  Y  216 f  AP  Y  .
17  m AB  Y  N  117 m  PA  N  Y  217 f  AP  Y  Y
18  m AB  Y  N  118 m  PA  N  Y  218 f  AP  Y  Y
19  m AB  Y  .  119 m  PA  N  Y  219 f  AP  Y  Y
20  m AB  Y  N  120 m  PA  N  Y  220 f  AP  Y  Y
21  m AB  Y  N  121 m  PA  N  Y  221 f  AP  Y  .
22  m AB  Y  N  122 m  PA  N  Y  222 f  AP  Y  Y
23  m AB  Y  .  123 m  PA  N  Y  223 f  AP  Y  Y
24  m AB  Y  N  124 m  PA  N  Y  224 f  AP  Y  Y
25  m AB  N  Y  125 m  PA  N  Y  225 f  AP  Y  Y
26  m AB  N  .  126 m  PA  N  Y  226 f  AP  Y  N
27  m AB  N  .  127 m  PA  N  Y  227 f  AP  Y  N
28  m AB  N  .  128 m  PA  N  Y  228 f  AP  Y  N
29  m AB  N  Y  129 m  PA  N  Y  229 f  AP  Y  .
30  m AB  N  Y  130 m  PA  N  Y  230 f  AP  Y  N
31  m AB  N  .  131 m  PA  N  .  231 f  AP  Y  N
32  m AB  N  N  132 m  PA  N  N  232 f  AP  N  Y
33  m AB  N  N  133 m  PA  N  N  233 f  AP  N  Y
```

34	m	AB	N	N	134	m	PA	N	N	234	f	AP	N	Y
35	m	AB	N	N	135	m	PA	N	N	235	f	AP	N	Y
36	m	AB	N	N	136	m	PA	N	.	236	f	AP	N	Y
37	m	AB	N	N	137	m	PA	N	N	237	f	AP	N	Y
38	m	AB	N	N	138	m	PA	N	N	238	f	AP	N	N
39	m	AB	N	N	139	m	PA	N	.	239	f	AP	N	N
40	m	AB	N	N	140	m	PA	N	N	240	f	AP	N	N
41	m	AB	N	.	141	m	PA	N	N	241	f	AP	N	N
42	m	AB	N	N	142	m	PA	N	N	242	f	AP	N	N
43	m	AB	N	N	143	m	PA	N	N	243	f	AP	N	N
44	m	AB	N	N	144	m	PA	N	N	244	f	AP	N	N
45	m	AB	N	.	145	m	PA	N	N	245	f	AP	N	N
46	m	AB	N	N	146	m	PA	N	N	246	f	AP	N	N
47	m	AB	N	N	147	m	PA	N	N	247	f	AP	N	N
48	m	AB	N	N	148	m	PA	N	N	248	f	AP	N	N
49	m	AB	N	N	149	m	PA	N	N	249	f	AP	N	N
50	m	AB	N	N	150	m	PA	N	.	250	f	AP	N	N
51	m	BP	Y	Y	151	f	BA	Y	Y	251	f	PB	Y	.
52	m	BP	Y	Y	152	f	BA	Y	Y	252	f	PB	Y	Y
53	m	BP	Y	Y	153	f	BA	Y	Y	253	f	PB	Y	Y
54	m	BP	Y	Y	154	f	BA	Y	.	254	f	PB	Y	Y
55	m	BP	Y	Y	155	f	BA	Y	Y	255	f	PB	Y	Y
56	m	BP	Y	Y	156	f	BA	Y	Y	256	f	PB	Y	.
57	m	BP	Y	Y	157	f	BA	Y	Y	257	f	PB	Y	Y
58	m	BP	Y	Y	158	f	BA	Y	Y	258	f	PB	Y	.
59	m	BP	Y	N	159	f	BA	Y	Y	259	f	PB	Y	Y
60	m	BP	Y	.	160	f	BA	Y	Y	260	f	PB	Y	Y
61	m	BP	Y	N	161	f	BA	Y	.	261	f	PB	Y	Y
62	m	BP	Y	.	162	f	BA	Y	.	262	f	PB	Y	.
63	m	BP	Y	N	163	f	BA	Y	Y	263	f	PB	Y	.
64	m	BP	N	Y	164	f	BA	Y	Y	264	f	PB	Y	N
65	m	BP	N	Y	165	f	BA	Y	Y	265	f	PB	Y	N
66	m	BP	N	Y	166	f	BA	Y	Y	266	f	PB	Y	N
67	m	BP	N	Y	167	f	BA	Y	Y	267	f	PB	Y	N
68	m	BP	N	Y	168	f	BA	Y	Y	268	f	PB	Y	N
69	m	BP	N	Y	169	f	BA	Y	Y	269	f	PB	N	Y
70	m	BP	N	.	170	f	BA	Y	.	270	f	PB	N	Y
71	m	BP	N	N	171	f	BA	Y	N	271	f	PB	N	Y
72	m	BP	N	N	172	f	BA	Y	N	272	f	PB	N	.
73	m	BP	N	N	173	f	BA	N	Y	273	f	PB	N	.
74	m	BP	N	N	174	f	BA	N	Y	274	f	PB	N	Y
75	m	BP	N	N	175	f	BA	N	.	275	f	PB	N	Y
76	m	BP	N	N	176	f	BA	N	Y	276	f	PB	N	Y
77	m	BP	N	N	177	f	BA	N	Y	277	f	PB	N	.
78	m	BP	N	N	178	f	BA	N	Y	278	f	PB	N	Y
79	m	BP	N	N	179	f	BA	N	.	279	f	PB	N	Y
80	m	BP	N	N	180	f	BA	N	.	280	f	PB	N	Y
81	m	BP	N	.	181	f	BA	N	Y	281	f	PB	N	Y
82	m	BP	N	N	182	f	BA	N	Y	282	f	PB	N	Y
83	m	BP	N	.	183	f	BA	N	Y	283	f	PB	N	Y
84	m	BP	N	N	184	f	BA	N	Y	284	f	PB	N	Y
85	m	BP	N	.	185	f	BA	N	Y	285	f	PB	N	Y
86	m	BP	N	N	186	f	BA	N	Y	286	f	PB	N	Y
87	m	BP	N	N	187	f	BA	N	Y	287	f	PB	N	Y

```
 88  m BP  N  N  188 f  BA  N  Y   288 f  PB  N  Y
 89  m BP  N  N  189 f  BA  N  Y   289 f  PB  N  Y
 90  m BP  N  N  190 f  BA  N  Y   290 f  PB  N  N
 91  m BP  N  N  191 f  BA  N  Y   291 f  PB  N  N
 92  m BP  N  .  192 f  BA  N  Y   292 f  PB  N  .
 93  m BP  N  N  193 f  BA  N  Y   293 f  PB  N  N
 94  m BP  N  N  194 f  BA  N  Y   294 f  PB  N  N
 95  m BP  N  N  195 f  BA  N  Y   295 f  PB  N  N
 96  m BP  N  N  196 f  BA  N  Y   296 f  PB  N  N
 97  m BP  N  N  197 f  BA  N  Y   297 f  PB  N  N
 98  m BP  N  N  198 f  BA  N  N   298 f  PB  N  N
 99  m BP  N  N  199 f  BA  N  N   299 f  PB  N  N
100  m BP  N  .  200 f  BA  N  N   300 f  PB  N  N
;
```

The next step manipulates the data the same as in Section 15.5. The DATA step creates observations for each period and creates indicator variables for the carryover effects.

```
data skincross2;
   set skincross;
   period=1;
   treatment=substr(sequence, 1, 1);
   carryA=0;
   carryB=0;
   response=Time1;
   output;
   period=2;
   Treatment=substr(sequence, 2, 1);
   carrya=(substr(sequence, 1, 1)='A');
   carryb=(substr(sequence, 1, 1)='B');
   response=Time2;
   output;
run;
```

The following PROC GENMOD invocation requests a GEE analysis for a model including effects for treatment, period, gender, carryover, and the period $\times$ gender interaction. The DESCENDING option specifies that the probability of a 'yes' response is to be modeled. Logistic regression is used along with the exchangeable working correlation structure.

```
proc genmod data=skincross2 descending;
   class subject treatment period gender;
   model response = treatment period carrya carryb
                    gender gender*period /type3
                    dist=bin link=logit;
   repeated subject=subject / type=exch;
run;
```

The "GEE Model Information" table shows that there are 300 clusters total and 50 clusters with missing data; 250 clusters have two measurements and 50 clusters have only one measurement corresponding to the first period.

Output 15.58 GEE Model Information

```
                    GEE Model Information

      Correlation Structure                  Exchangeable
      Subject Effect                subject (300 levels)
      Number of Clusters                            300
      Clusters With Missing Values                   50
      Correlation Matrix Dimension                    2
      Maximum Cluster Size                            2
      Minimum Cluster Size                            1
```

The Type 3 analysis displayed in Output 15.58 suggests that the carryover effects are not influential.

Output 15.59 Type 3 Analysis

```
           Score Statistics For Type 3 GEE Analysis

                                   Chi-
           Source          DF    Square    Pr > ChiSq

           Treatment        2     29.38       <.0001
           period           1      7.11       0.0077
           carryA           1      0.02       0.8870
           carryB           1      0.68       0.4088
           gender           1     29.94       <.0001
           period*gender    1      4.21       0.0401
```

The next model fit includes the treatment, period, and gender main effects as well as the gender $\times$ period interaction. The ESTIMATE specifies that odds ratio estimates be computed to compare the effect of drug A to the placebo effect, the effect of drug B to the placebo effect, and the effect of drug A to the effect of drug B.

```
proc genmod data=skincross2 descending;
   class subject treatment period gender;
   model response = treatment period gender*period
                    gender /type3
                    dist=bin link=logit;
   repeated subject=subject / type=exch;
   estimate 'OR:A-B' treatment 1 -1 0 /exp;
   estimate 'OR:A-P' treatment 1 0 -1 / exp;
   estimate 'OR:B-P' treatment 0 1 -1 / exp;
run;
```

The main effects and gender $\times$ period interaction remain important, as indicated in Output 15.60.

Output 15.60 Type 3 Analysis

```
          Score Statistics For Type 3 GEE Analysis

                                 Chi-
          Source         DF     Square    Pr > ChiSq

          Treatment       2      40.02      <.0001
          period          1      20.97      <.0001
          period*gender   1       3.93      0.0474
          gender          1      28.89      <.0001
```

The parameter estimates are displayed in Output 15.61.

Output 15.61 Parameter Estimates

```
                    Analysis Of GEE Parameter Estimates
                    Empirical Standard Error Estimates

                             Standard  95% Confidence
    Parameter         Estimate  Error       Limits        Z Pr > |Z|

    Intercept          -0.9287  0.2249  -1.3696  -0.4879  -4.13  <.0001
    Treatment     A     1.2622  0.2079   0.8548   1.6696   6.07  <.0001
    Treatment     B     0.1722  0.2141  -0.2473   0.5918   0.80  0.4210
    Treatment     P     0.0000  0.0000   0.0000   0.0000    .      .
    period        1    -0.4520  0.2257  -0.8944  -0.0095  -2.00  0.0453
    period        2     0.0000  0.0000   0.0000   0.0000    .      .
    period*gender 1 f   0.7938  0.2531   0.2978   1.2898   3.14  0.0017
    period*gender 1 m   0.0000  0.0000   0.0000   0.0000    .      .
    period*gender 2 f   1.4443  0.2816   0.8925   1.9961   5.13  <.0001
    period*gender 2 m   0.0000  0.0000   0.0000   0.0000    .      .
    gender        f     0.0000  0.0000   0.0000   0.0000    .      .
    gender        m     0.0000  0.0000   0.0000   0.0000    .      .
```

The odds ratio estimates displayed in Output 15.62 show that those subjects receiving drug A had almost 3.5 higher odds of improvement as those on the placebo. Those subjects receiving drug A have almost three times higher odds of improvement as those subjects receiving drug B. The odds ratio estimate comparing the odds of drug B and placebo is 1.1880 and its confidence limits contain the value 1. Subjects on the new drug B did no better than the placebo.

Output 15.62 Odds Ratio Estimates

```
                       Contrast Estimate Results

                     Standard                         Chi-
    Label     Estimate  Error  Alpha  Confidence Limits  Square  Pr > ChiSq

    OR:A-B      1.0899  0.2193  0.05   0.6601   1.5198   24.70    <.0001
    Exp(OR:A-B) 2.9741  0.6523  0.05   1.9350   4.5713
    OR:A-P      1.2622  0.2079  0.05   0.8548   1.6696   36.87    <.0001
    Exp(OR:A-P) 3.5331  0.7344  0.05   2.3508   5.3099
    OR:B-P      0.1722  0.2141  0.05  -0.2473   0.5918    0.65    0.4210
    Exp(OR:B-P) 1.1880  0.2543  0.05   0.7809   1.8072
```

Observational Study with Missing Data

Missing data in a longitudinal study is almost a certainty, particularly if the study has an observational nature. Environmental researchers in the 1970s studied various aspects of the air pollution impacting various cities under investigation, including the potential effect of air pollution on children. In one part of the overall study, researchers collected information on colds that children experienced during each of three years. The following DATA step includes information on over 5000 children and whether they experienced substantial colds during three years in the middle 1970s. More than 3000 children had missing data for at least one of the three years (Stokes 1986).

While there is an excessive amount of missing data, its extent is not unusual. You can still apply the GEE method with this much missing data, but you would want to be cautious about your findings. Often, you would complement your analysis with a complete data analysis for those children with all three years of data. Researchers felt that the missing data occurred at random so the MCAR assumption required for GEE analysis was considered to be realistic.

The following DATA step creates SAS data set COLDS. These data represent two areas, 1 and 2; gender of the subject, f for female and m for male; and whether the response was yes, no, or missing, y for yes, n for no, and a '.' for missing. These characteristics are assigned to the variables AREA, GENDER, and YEAR1 through YEAR3. The variable COUNT contains the number of children who had the various possible profiles based on the possible combinations of y, n, and '.' for the three years. The variable PATTERN is created to contain type of missing data pattern for each set of responses; these can be single year missing, two combinations of two years missing, or no missing data. None of the subjects had all missing data.

```
data colds;
   input area gender $ year1 $ year2 $ year3 $ count @@;
    if year1 =' ' and year2 =' ' and year3=' ' then pattern ='mmm';
      else if year1=' ' and year2=' ' then pattern='mmh';
        else if year1=' ' and year3=' ' then pattern= 'mhm';
          else if year2=' ' and year3 =' ' then pattern ='hmm';
            else if year1=' ' then pattern= 'mhh';
              else if year2=' ' then pattern='hmh';
                else if year3=' ' then pattern='hhm';
                  else pattern= 'hhh';
      do i=1 to count; output; end;
    datalines;
1 m y y y  80  1 m y y n  46 1 m y n y 38 1 m y n n   61
1 m n y y  57  1 m n y n  60 1 m n n y 59 1 m n n n  121
1 m y y .  20  1 m y n .  14 1 m n y . 14 1 m n n .   39
1 m y . y  16  1 m y . n   5 1 m n . y 15 1 m n . n   13
1 m . y y  47  1 m . y n  32 1 m . n y 32 1 m . n n   50
1 m y . . 141  1 m n . . 191 1 m . y . 87 1 m . n .   83
1 m . . y 156  1 m . . n 173
1 f y y y 109  1 f y y n  48 1 f y n y 39 1 f y n n   47
1 f n y y  45  1 f n y n  43 1 f n n y 47 1 f n n n   79
1 f y y .  34  1 f y n .  10 1 f n y . 19 1 f n n .   28
1 f y . y  13  1 f y . n   8 1 f n . y 14 1 f n . n    9
1 f . y y  60  1 f . y n  15 1 f . n y 30 1 f . n n   39
```

```
1 f y . . 170 1 f n . .  155 1 f . y . 91 1 f . n .  84
1 f . . y 173 1 f . . n  152
2 m y y y  59 2 m y y n   31 2 m y n y 22 2 m y n n  30
2 m n y y  35 2 m n y n   15 2 m n n y 41 2 m n n n  55
2 m y y .  44 2 m y n .   23 2 m n y . 28 2 m n n .  41
2 m y . y   7 2 m y . n    4 2 m n . y 10 2 m n . n  16
2 m . y y  26 2 m . y n   26 2 m . n y 23 2 m . n n  22
2 m y . . 129 2 m n . .  140 2 m . y . 65 2 m . n .  88
2 m . . y 129 2 m . . n  167
2 f y y y  94 2 f y y n   31 2 f y n y 11 2 f y n n  32
2 f n y y  28 2 f n y n   21 2 f n n y 30 2 f n n n  45
2 f y y .  34 2 f y n .   17 2 f n y . 10 2 f n n .  28
2 f y . y   9 2 f y . n    4 2 f n . y 6 2 f n . n   6
2 f . y y  23 2 f . y n   11 2 f . n y 11 2 f . n n  7
2 f y . . 133 2 f n . .   91 2 f . y . 85 2 f . n .  51
2 f . . y 116 2 f . . n  113
;
```

The next DATA step assigns a subject number to each subject and creates an observation for each of the three response periods to produce SAS data set COLDS2.

```
data colds2; set colds;
   drop year1 year2 year3;
   subject=_n_;
   resp= year1; year=1; output;
   resp=year2; year=2; output;
   resp=year3; year=3; output;
run;
```

The GEE analysis begins with the main effects and all pairwise interactions. The exchangeable working correlation structure is reasonable for three time points. The variable PATTERN is included to assess the relationship of the missing data pattern to the response variable and to see if it interacts with explanatory variables GENDER and AREA. If such effects are nonsignificant, this can be interpreted as evidence supporting the assumption of MCAR, or the missing completely at random assumption. The visit $\times$ pattern interaction is not included although it could be; the relationship between these two variables makes interpretation of any interaction less than straightforward.

```
proc genmod;
   class subject area gender year pattern;
   model resp = pattern area gender pattern*area pattern*gender
               year area*gender area*year gender*year/
               dist=bin link=logit type3;
   repeated subject=subject / type=exch;
run;
```

The "GEE Model Information" table shows that 5534 children are included in the analysis, and 3975 children had missing responses for at least one year.

Output 15.63   GEE Model Information

```
                        GEE Model Information

            Correlation Structure                  Exchangeable
            Subject Effect              subject (5534 levels)
            Number of Clusters                             5534
            Clusters With Missing Values                   3975
            Correlation Matrix Dimension                      3
            Maximum Cluster Size                              3
            Minimum Cluster Size                              1
```

Output 15.64 displays the Type 3 analysis. All of the interaction terms are nonsignificant.
Also, neither the missing pattern effect or its interactions with AREA or GENDER are
significant. This can be interpreted as evidence for the reasonableness of the MCAR
assumption.

Output 15.64   Type 3 Analysis

```
            Score Statistics For Type 3 GEE Analysis

                                       Chi-
            Source              DF    Square    Pr > ChiSq

            pattern              6      9.91        0.1285
            area                 1      1.56        0.2111
            gender               1     33.40       <.0001
            area*pattern         6      9.62        0.1416
            gender*pattern       6      0.92        0.9885
            year                 2      2.70        0.2597
            area*gender          1      0.39        0.5345
            area*year            2      0.34        0.8451
            gender*year          2      1.21        0.5471
```

The next PROC GENMOD invocation includes only the main effects. Note that, to
proceed with more certainty, you could first perform a joint test of the pattern effect plus
its interactions, and then perform a joint test of the pattern-related effects and the other
interactions.

```
proc genmod;
class subject area gender year ;
    model resp = area gender year/
              dist=bin link=logit type3;
    repeated subject=subject / type=exch;
run;
```

Output 15.65 contains the Type 3 analysis for the main effects model. Both the area and
gender effects are very significant; year does not have an effect on whether colds were
reported.

Output 15.65   Type 3 Analysis Results for Final Model

```
          Score Statistics For Type 3 GEE Analysis

                                   Chi-
          Source           DF     Square    Pr > ChiSq

          area              1      7.42       0.0064
          gender            1     53.75      <.0001
          year              2      3.95       0.1386
```

Finally, Output 15.66 contains the parameter estimates.

Output 15.66   Parameter Estimates for Final Model

```
                   Analysis Of GEE Parameter Estimates
                   Empirical Standard Error Estimates

                             Standard   95% Confidence
        Parameter   Estimate   Error        Limits          Z Pr > |Z|

        Intercept      0.0364   0.0483  -0.0582   0.1311   0.75   0.4507
        area      1    0.1256   0.0461   0.0353   0.2159   2.73   0.0064
        area      2    0.0000   0.0000   0.0000   0.0000    .       .
        gender    f   -0.3330   0.0454  -0.4220  -0.2440  -7.33  <.0001
        gender    m    0.0000   0.0000   0.0000   0.0000    .       .
        year      1    0.0743   0.0473  -0.0185   0.1670   1.57   0.1166
        year      2   -0.0106   0.0459  -0.1006   0.0794  -0.23   0.8179
        year      3    0.0000   0.0000   0.0000   0.0000    .       .
```

15.12   Alternating Logistic Regression

There are some limitations of the correlation approach to fit models to binary data. The data influences the range of the correlation since the estimates of r_{jk} are constrained by the means, $\mu_{jk} = \Pr\{y_{ij} = 1\}$. Consider:

$$\mathrm{Corr}(Y_{ij}, Y_{ik}) = r_{jk} = \frac{Pr(Y_{ij} = 1, Y_{ik} = 1) - \mu_{ij}\mu_{ik}}{\sqrt{\mu_{ij}(1 - \mu_{ij})\mu_{ik}(1 - \mu_{ik})}}$$

The odds ratio appears to be a more natural choice for modeling the association in binary data as they are not constrained by the means.

$$\mathrm{OR}(Y_{ij}, Y_{ik}) = \frac{Pr(Y_{ij} = 1, Y_{ik} = 1)Pr(Y_{ij} = 0, Y_{ik} = 0)}{Pr(Y_{ij} = 1, Y_{ik} = 0)Pr(Y_{ij} = 0, Y_{ik} = 1)}$$

In GEE, the correlations are treated as nuisance parameters, and the use of correlations versus odds ratios usually has little influence on inference on β, the regression parameters for the marginal mean model.

In some applications, you may want your analysis to focus both on regressing the outcome on the explanatory variables and describing the association between the outcomes. The generalized estimating equations discussed in this chapter are known as the first-order estimating equations, and they are efficient for the estimation of β but not necessarily efficient in the estimation of the association parameters, which are the correlations estimated in PROC GENMOD with the method of moments. Prentice (1988) describes second-order estimating equations and the simultaneous modeling of the responses and all pairwise products as a method of producing more efficient estimation of the association parameters. However, as the number of clusters grows large, this method can become computationally infeasible.

Carey, Zeger, and Diggle (1993) describe the alternating logistic regression (ALR) algorithm, which provides a means of both fitting the first-order GEE model and simultaneously modeling the association in a manner that produces relatively efficient estimators. With this method, you obtain $\hat{\beta}$ and also obtain estimates of the association parameters that relate to log odds ratios, as well as their standard errors and confidence intervals.

The ALR algorithm models the log of the odds ratio as

$$\psi_{ijk} = \mathbf{z}'\boldsymbol{\alpha}$$

where $\psi_{ijk} = \log(OR(y_{ij}, y_{ik}))$, $\boldsymbol{\alpha}$ is a $q \times 1$ vector of regression parameters, and $\mathbf{z}_{ijk}$ is a fixed vector of coefficients. The method switches between the first-order GEE estimation of the β and a modified (with offset) logistic regression estimate of the $\boldsymbol{\alpha}$ until convergence, updating the GEE with product-moments from the newly estimated OR, and then updating the offsets in the association model with the new $\hat{\beta}$s. Thus, you are applying alternating logistic regressions, one for α and one for β.

There are numerous choices for modeling the log odds ratio: you can choose to specify the log odds ratio as a constant across clusters; for pairs (j, k), you can specify that the log odds ratio is a constant within different levels of a blocking factor such as clinics; and you can specify fully parameterized clusters in which each cluster is parameterized the same way. There are numerous other possibilities for model structures for the log odds ratio. For more information on the motivation and the details of the ALR approach, refer to Carey, Zeger, and Diggle (1993), Lipsitz, Laird, and Harrington (1991), and Firth (1992).

Beginning with Version 8, the GENMOD procedure produces the ALR algorithm for binary data. The following log odds ratio structures are available:

- exchangeable (constant over all clusters)

- covariate (block effect)

- fully parameterized within cluster (parameter for each pair)

- nested (one parameter for pairs within same subcluster, one for between subclusters)

- user-specified **Z**-matrix

The ALR algorithm provides a reasonable approach when the focus of your analysis is estimating association as much as modeling the response; this method provides estimates of association with more efficiency than the usual GEE method. The resulting parameter estimates β are consistent (Pickles 1998) and so is the estimated covariance matrix of β. That is, you retain the robustness properties of the first-order GEE even if the association structure is misspecified. A possible limitation of ALR is that no covariance estimator has been suggested that is analogous to the model-based estimator of GEE.

15.12.1 Respiratory Data

You may recall the respiratory data analyzed in previous sections of this book. In this example, the ALR algorithm is applied in modeling the dichotomous outcome of whether respiratory symptoms were good or excellent at the four visits. The first analysis models the log odds ratios as exchangeable: in this case, α is the common log odds ratio.

$$\log(\text{OR}(Y_{ij}, Y_{ik})) = \alpha \ \text{ for all } \ i, j \neq k$$

The following SAS statements produce this analysis. The DATA step creating RESP2 is listed on page 495. To specify the ALR algorithm, you include the LOGOR option in the REPEATED statement. Here, the exchangeable structure for the log odds ratio is requested.

```
proc genmod data=resp2 descending;
   class id treatment sex center visit;
   model dichot = center sex treatment age di_base visit
                  / dist=bin type3 link=logit;
   repeated subject=id*center / logor=exch;
run;
```

Output 15.67 contains information about the GEE modeling; it tells you that the log OR structure is exchangeable.

Output 15.67   GEE Model Information

```
                   GEE Model Information

   Log Odds Ratio Structure                Exchangeable
   Subject Effect                  id*center (111 levels)
   Number of Clusters                               111
   Correlation Matrix Dimension                       4
   Maximum Cluster Size                               4
   Minimum Cluster Size                               4
```

Output 15.68 contains Type 3 tests; the results are very similar to those obtained in the first-order GEE based on the exchangeable structure defined with the Pearson correlations. See Output 15.29.

Output 15.68   Type 3 Analysis

```
                  Score Statistics For Type 3 GEE Analysis

                                    Chi-
                Source        DF   Square   Pr > ChiSq

                center         1     3.13       0.0767
                sex            1     0.09       0.7642
                treatment      1    12.56       0.0004
                age            1     2.03       0.1542
                di_base        1    22.48      <.0001
                visit          3     2.99       0.3932
```

Output 15.69 displays the parameter estimates for the elements of β. The estimates are also similar to those obtained in the standard GEE model; the parameter labeled 'Alpha 1' is the estimate of the common log odds ratio and has the value 1.7524. Note that you need to interpret the 'Alpha' estimates somewhat cautiously since they assume their model specification is correct (as compared to the estimates of β, which are robust.)

Output 15.69   Parameter Estimates

```
                     Analysis Of GEE Parameter Estimates
                     Empirical Standard Error Estimates

                          Standard   95% Confidence
          Parameter      Estimate    Error      Limits              Z Pr > |Z|

          Intercept      -0.4137    0.5760   -1.5428    0.7153   -0.72   0.4726
          center     1   -0.6590    0.3517   -1.3483    0.0303   -1.87   0.0610
          center     2    0.0000    0.0000    0.0000    0.0000      .       .
          sex        F    0.1329    0.4365   -0.7226    0.9884    0.30   0.7608
          sex        M    0.0000    0.0000    0.0000    0.0000      .       .
          treatment  A    1.2696    0.3432    0.5969    1.9423    3.70   0.0002
          treatment  P    0.0000    0.0000    0.0000    0.0000      .       .
          age            -0.0180    0.0127   -0.0429    0.0068   -1.42   0.1552
          di_base         1.8381    0.3439    1.1642    2.5121    5.35   <.0001
          visit      1    0.3138    0.2494   -0.1751    0.8027    1.26   0.2084
          visit      2    0.1065    0.2409   -0.3657    0.5786    0.44   0.6585
          visit      3    0.3269    0.2314   -0.1266    0.7804    1.41   0.1577
          visit      4    0.0000    0.0000    0.0000    0.0000      .       .
          Alpha1          1.7524    0.2767    1.2102    2.2947    6.33   <.0001
```

Another approach with the ALR strategy is to estimate a separate log odds ratio for each center. The following PROC GENMOD statements produce that analysis. The LOGOR=LOGORVAR(CENTER) specifies that each center has its own log odds ratio.

```
proc genmod data=resp2 descending;
   class id treatment sex center visit;
   model dichot = center sex treatment age di_base visit
                  / dist=bin type3 link=logit;
   repeated subject=id*center / logor=logorvar(center) corrw;
run;
```

Output 15.70 indicates that, this time, the exchangeable structure for the log OR is based on the CENTER variable in this specification.

Output 15.70 GEE Model Information

```
                    GEE Model Information

Log Odds Ratio Covariate                          center
Subject Effect                      id*center (111 levels)
Number of Clusters                                   111
Correlation Matrix Dimension                           4
Maximum Cluster Size                                   4
Minimum Cluster Size                                   4
```

Output 15.71 is produced when the log OR structure has more than one α parameter; this table lists the group(center) levels associated with the log OR parameters. In this case, there is a common log OR for all clusters in Center 1 and a common log OR for all clusters in Center 2.

Output 15.71 Log Odds Ratio Parameter Information

```
            Log Odds Ratio Parameter
                  Information

        Parameter        Group

        Alpha1           1
        Alpha2           2
```

Output 15.72 contains Type 3 tests; the results are still similar to those obtained in the usual GEE model.

Output 15.72 Type 3 Analysis

```
          Score Statistics For Type 3 GEE Analysis

                                   Chi-
          Source        DF       Square    Pr > ChiSq

          center        1          2.96       0.0852
          sex           1          0.12       0.7325
          treatment     1         12.31       0.0004
          age           1          2.32       0.1276
          di_base       1         22.53      <.0001
          visit         3          2.77       0.4292
```

Output 15.73 displays the parameter estimates. The parameter labeled 'Alpha 1' is the common log odds ratio for the Center 1 subjects; the parameter labeled 'Alpha 2' is the common log odds ratio for the Center 2 subjects. In Center 1, the common log odds ratio is 1.3677 with a 95% confidence interval of (0.7423, 1.9930). For Center 2, the common log odds ratio is 2.0886 with a 95% confidence interval of (1.2362, 2.9410). The association would appear to be slightly stronger in Center 2.

Output 15.73   Parameter Estimates

```
                  Analysis Of GEE Parameter Estimates
                   Empirical Standard Error Estimates

                          Standard   95% Confidence
       Parameter   Estimate   Error       Limits           Z Pr > |Z|

       Intercept    -0.3729   0.5693  -1.4887   0.7430   -0.65   0.5125
       center    1  -0.6423   0.3527  -1.3335   0.0489   -1.82   0.0686
       center    2   0.0000   0.0000   0.0000   0.0000     .       .
       sex       F   0.1506   0.4318  -0.6958   0.9969    0.35   0.7273
       sex       M   0.0000   0.0000   0.0000   0.0000     .       .
       treatment A   1.2392   0.3396   0.5735   1.9048    3.65   0.0003
       treatment P   0.0000   0.0000   0.0000   0.0000     .       .
       age          -0.0194   0.0126  -0.0440   0.0052   -1.54   0.1226
       di_base       1.9210   0.3396   1.2555   2.5866    5.66   <.0001
       visit     1   0.3017   0.2511  -0.1904   0.7937    1.20   0.2295
       visit     2   0.0640   0.2398  -0.4060   0.5341    0.27   0.7895
       visit     3   0.2782   0.2310  -0.1745   0.7310    1.20   0.2284
       visit     4   0.0000   0.0000   0.0000   0.0000     .       .
       Alpha1        1.3677   0.3191   0.7423   1.9930    4.29   <.0001
       Alpha2        2.0886   0.4349   1.2362   2.9410    4.80   <.0001
```

Note that inserting the following statement in the previous PROC GENMOD invocation requests a fully parameterized cluster model for the log odds ratio parameters:

```
repeated subject=id*center / logor=fullclust;
```

Information about what the parameters mean is presented in Output 15.74.

Output 15.74   Log OR Parameter Information

```
             Log Odds Ratio Parameter
                   Information

             Parameter        Group

             Alpha1          (1,  2)
             Alpha2          (1,  3)
             Alpha3          (1,  4)
             Alpha4          (2,  3)
             Alpha5          (2,  4)
             Alpha6          (3,  4)
```

The estimated parameters are presented in Output 15.75.

Output 15.75   Parameter Estimates

```
                Analysis Of GEE Parameter Estimates
                 Empirical Standard Error Estimates

                           Standard   95% Confidence
    Parameter    Estimate    Error       Limits           Z Pr > |Z|

    Intercept     -0.0871   0.6174   -1.2972    1.1230   -0.14   0.8878
    center    1   -0.6343   0.3487   -1.3177    0.0491   -1.82   0.0689
    center    2    0.0000   0.0000    0.0000    0.0000     .       .
    sex       F    0.1110   0.4369   -0.7454    0.9674    0.25   0.7995
    sex       M    0.0000   0.0000    0.0000    0.0000     .       .
    treatment A    1.2581   0.3405    0.5906    1.9255    3.69   0.0002
    treatment P    0.0000   0.0000    0.0000    0.0000     .       .
    age           -0.0164   0.0126   -0.0410    0.0082   -1.31   0.1916
    di_base        1.8983   0.3407    1.2306    2.5660    5.57   <.0001
    visit         -0.0839   0.0805   -0.2417    0.0740   -1.04   0.2978
    Alpha1         1.6242   0.4911    0.6617    2.5867    3.31   0.0009
    Alpha2         1.0553   0.4862    0.1023    2.0083    2.17   0.0300
    Alpha3         1.6305   0.4818    0.6863    2.5747    3.38   0.0007
    Alpha4         2.0924   0.5016    1.1093    3.0756    4.17   <.0001
    Alpha5         1.8811   0.4694    0.9610    2.8011    4.01   <.0001
    Alpha6         2.1411   0.4943    1.1724    3.1099    4.33   <.0001
```

The relative magnitudes of these six estimated log odds ratios have a similar pattern as the unstructured working correlation estimates presented on page 503, and this is what you expect. This pattern seems consistent with exchangeable structure.

15.13   Using GEE to Fit a Partial Proportional Odds Model: Univariate Outcome

Chapter 9, "Logistic Regression II: Polytomous Response," describes the use of the proportional odds model for response outcomes that are ordinal. Instead of modeling logits as in logistic regression for a dichotomous response, you model cumulative logits. However, sometimes you have data with an ordinal outcome for which the proportional odds assumption doesn't apply. You can use the GEE approach to fit a *partial proportional odds* model in which you assume proportional odds for some of the explanatory variables but not others. You form multiple response outcomes from your univariate outcome by forming logits corresponding to the different cutpoints of the ordinal values. For example, if your response variable has the values 1, 2, and 3, you would form two logits: the first logit would compare 1 versus 2 and 3, and the second logit would compare 1 and 2 versus 3. Then, you consider the logits to be multiple response functions for the same subject and perform a GEE analysis with a model that includes interactions between the explanatory variables and different types of logit. If some interactions are significant, there is a relationship between those explanatory variables and type of logit, and proportional odds doesn't hold for those explanatory variables. If an interaction is nonsignificant, then you do have proportional odds for that explanatory variable and you can remove the interaction terms.

The following data come from a study on dental pain for a new analgesic (Gansky, Koch, and Wilson 1994). Patients were administered the treatment and followed up at one hour, two hours, and three hours. The outcome at the first hour is of interest here. The response was measured on a five-point scale from 0 to 4, where 0 represents no relief and 4 represents maximum relief. Also recorded was the baseline severity of the pain, 1 for substantial and 0 for not substantial. There were five types of treatment, representing various levels of the dosages of the treatment drug. The study was conducted at two dental research centers.

The following DATA step creates SAS data set DENT:

```
data dent;
   input patient center trt $ baseline ldose resp @@;
   datalines;
  2  1  ACL  0  5.29832  0 131  1  TL   0  3.91202  2
  1  1  ACH  1  5.99146  1   3  1  TH   0  4.60517  1
130  1  P    0  0.00000  0 132  1  P    0  0.00000  0
  4  1  P    0  0.00000  0 133  1  P    0  0.00000  0
  5  1  P    0  0.00000  0 134  2  ACH  0  5.99146  4
  6  1  TL   1  3.91202  2 135  2  ACL  0  5.29832  4
  7  1  ACH  0  5.99146  1 136  2  TH   0  4.60517  3
  8  1  ACL  0  5.29832  0 137  2  ACL  0  5.29832  4
  9  1  TL   1  3.91202  0 138  2  TL   0  3.91202  3
 10  1  TL   1  3.91202  4 139  2  P    0  0.00000  4
 11  1  ACL  0  5.29832  2 140  2  TL   0  3.91202  3
 12  1  ACH  0  5.99146  0 141  2  TL   0  3.91202  3
 13  1  P    0  0.00000  0 142  2  ACL  1  5.29832  3
 14  1  TL   0  3.91202  0 143  2  ACH  0  5.99146  1
 15  1  P    0  0.00000  0 144  2  ACH  0  5.99146  3
 16  1  TH   1  4.60517  4 145  2  P    0  0.00000  1
 17  1  TH   0  4.60517  2 146  2  P    0  0.00000  0
 18  1  ACH  0  5.99146  1 147  2  ACH  0  5.99146  4
 19  1  ACL  0  5.29832  0 148  2  TL   0  3.91202  2
 20  1  TH   0  4.60517  3 149  2  TH   0  4.60517  3
 21  1  P    1  0.00000  1 150  2  P    0  0.00000  0
 22  1  TH   1  4.60517  0 151  2  TH   0  4.60517  2
 23  1  TL   0  3.91202  2 152  2  ACL  1  5.29832  3
 24  1  ACL  1  5.29832  0 153  2  TH   0  4.60517  2
 25  1  P    0  0.00000  0 154  2  ACH  0  5.99146  3
 26  1  ACH  0  5.99146  0 155  2  ACL  0  5.29832  1
 27  1  ACL  0  5.29832  0 156  2  ACL  0  5.29832  0
 28  1  P    0  0.00000  0 157  2  ACL  0  5.29832  3
 29  1  ACH  0  5.99146  2 158  2  TH   0  4.60517  3
 30  1  TL   0  3.91202  0 159  2  ACL  0  5.29832  1
 31  1  P    0  0.00000  0 160  2  TL   0  3.91202  3
 32  1  ACH  0  5.99146  1 161  2  P    0  0.00000  2
 33  1  TL   0  3.91202  0 162  2  TH   0  4.60517  3
 34  1  TH   1  4.60517  4 163  2  TH   0  4.60517  4
 35  1  TL   0  3.91202  2 164  2  ACH  0  5.99146  3
 36  1  ACH  0  5.99146  0 165  2  TH   0  4.60517  2
 37  1  ACL  1  5.29832  1 166  2  P    0  0.00000  3
 38  1  ACL  0  5.29832  3 167  2  ACH  0  5.99146  1
 39  1  TH   0  4.60517  2 168  2  P    0  0.00000  2
 40  1  TH   0  4.60517  0 169  2  TL   1  3.91202  2
 41  1  ACL  0  5.29832  2 170  2  P    0  0.00000  0
 42  1  TL   0  3.91202  3 171  2  TL   0  3.91202  0
 43  1  ACL  0  5.29832  2 172  2  TL   0  3.91202  3
 44  1  ACL  0  5.29832  0 173  2  ACH  1  5.99146  2
```

45	1	TH	0	4.60517	2	174	2	ACL	0	5.29832	3
46	1	ACH	0	5.99146	0	175	2	P	0	0.00000	4
47	1	P	0	0.00000	0	176	2	TL	1	3.91202	0
48	1	ACL	0	5.29832	0	177	2	ACH	0	5.99146	1
49	1	TL	0	3.91202	0	178	2	ACH	1	5.99146	2
50	1	TL	0	3.91202	2	179	2	ACL	0	5.29832	2
51	1	TL	0	3.91202	0	180	2	ACL	0	5.29832	2
52	1	ACH	0	5.99146	4	181	2	TH	0	4.60517	2
53	1	TH	0	4.60517	4	182	2	TL	0	3.91202	3
54	1	TH	0	4.60517	0	183	2	TH	0	4.60517	3
55	1	P	0	0.00000	0	184	2	ACH	0	5.99146	1
56	1	ACH	0	5.99146	3	185	2	P	0	0.00000	1
57	1	ACH	0	5.99146	2	186	2	ACL	0	5.29832	1
58	1	P	0	0.00000	0	187	2	TH	0	4.60517	2
59	1	TH	0	4.60517	0	188	2	ACH	1	5.99146	3
60	1	P	0	0.00000	0	189	2	TH	0	4.60517	2
61	1	TL	0	3.91202	1	190	2	TL	0	3.91202	3
62	1	P	0	0.00000	0	191	2	P	0	0.00000	1
63	1	TH	1	4.60517	1	192	2	TL	0	3.91202	3
64	1	TL	0	3.91202	0	193	2	P	0	0.00000	0
65	1	ACH	0	5.99146	2	194	2	TH	0	4.60517	2
66	1	ACL	0	5.29832	2	195	2	ACH	0	5.99146	4
67	1	P	0	0.00000	2	196	2	ACH	0	5.99146	2
68	1	TH	0	4.60517	1	197	2	ACL	0	5.29832	3
69	1	ACH	0	5.99146	0	198	2	P	0	0.00000	0
70	1	P	0	0.00000	0	199	2	P	0	0.00000	3
71	1	TL	0	3.91202	0	200	2	ACL	0	5.29832	0
72	1	ACH	0	5.99146	2	201	2	ACL	1	5.29832	4
73	1	P	0	0.00000	0	202	2	TH	0	4.60517	3
74	1	TL	0	3.91202	2	203	2	P	0	0.00000	1
75	1	TH	0	4.60517	2	204	2	TH	0	4.60517	1
76	1	ACL	1	5.29832	0	205	2	TH	0	4.60517	3
77	1	TH	1	4.60517	0	206	2	TL	0	3.91202	3
78	1	ACL	0	5.29832	0	207	2	TL	0	3.91202	3
79	1	ACL	1	5.29832	3	208	2	TL	0	3.91202	3
80	1	ACH	0	5.99146	2	209	2	ACL	0	5.29832	2
81	1	ACL	0	5.29832	0	210	2	ACH	0	5.99146	3
82	1	P	0	0.00000	0	211	2	TL	1	3.91202	1
83	1	TH	0	4.60517	0	212	2	ACH	0	5.99146	3
84	1	ACH	0	5.99146	1	213	2	P	0	0.00000	2
85	1	TL	0	3.91202	0	214	2	P	0	0.00000	0
86	1	TH	0	4.60517	3	215	2	TL	0	3.91202	0
87	1	ACH	0	5.99146	0	216	2	TH	0	4.60517	4
88	1	P	0	0.00000	0	217	2	ACH	0	5.99146	2
89	1	ACH	0	5.99146	1	218	2	P	0	0.00000	0
90	1	TL	0	3.91202	0	219	2	TH	0	4.60517	2
91	1	ACL	0	5.29832	1	220	2	TL	0	3.91202	1
92	1	TH	0	4.60517	0	221	2	ACH	0	5.99146	2
93	1	ACL	0	5.29832	1	222	2	TL	0	3.91202	4
94	1	TL	1	3.91202	1	223	2	TH	0	4.60517	2
95	1	TL	1	3.91202	3	224	2	TH	1	4.60517	1
96	1	P	0	0.00000	0	225	2	ACH	0	5.99146	1
97	1	TH	0	4.60517	0	226	2	ACL	0	5.29832	4
98	1	ACL	0	5.29832	0	227	2	P	1	0.00000	3

```
 99  1  P    1  0.00000  0 228  2  ACL  0  5.29832  2
100  1  ACH  0  5.99146  2 229  2  TL   0  3.91202  0
101  1  TH   0  4.60517  0 230  2  ACL  0  5.29832  1
102  1  TL   0  3.91202  0 231  2  ACH  0  5.99146  3
103  1  ACL  0  5.29832  1 232  2  ACL  0  5.29832  3
104  1  TL   0  3.91202  0 233  2  P    0  0.00000  1
105  1  P    0  0.00000  1 234  2  ACL  0  5.29832  4
106  1  ACL  0  5.29832  0 235  2  ACH  0  5.99146  1
107  1  TH   1  4.60517  2 236  2  TH   0  4.60517  1
108  1  P    0  0.00000  0 237  2  ACL  0  5.29832  0
109  1  ACH  0  5.99146  1 238  2  ACL  1  5.29832  4
110  1  TH   1  4.60517  1 239  2  ACL  0  5.29832  3
111  1  TL   0  3.91202  0 240  2  P    0  0.00000  0
112  1  ACH  0  5.99146  0 241  2  P    1  0.00000  3
113  1  TL   0  3.91202  0 242  2  TL   0  3.91202  2
114  1  ACH  0  5.99146  1 243  2  P    0  0.00000  0
115  1  P    0  0.00000  0 244  2  TH   0  4.60517  3
116  1  ACL  0  5.29832  0 245  2  TL   1  3.91202  4
117  1  P    0  0.00000  0 246  2  ACH  1  5.99146  1
118  1  ACH  0  5.99146  3 247  2  P    0  0.00000  1
119  1  TH   0  4.60517  3 248  2  TH   0  4.60517  4
120  1  ACL  0  5.29832  2 249  2  TL   0  3.91202  0
121  1  TH   0  4.60517  2 250  2  TL   0  3.91202  3
122  1  TH   0  4.60517  1 251  2  ACH  0  5.99146  3
123  1  TL   0  3.91202  0 252  2  TH   0  4.60517  3
124  1  ACH  0  5.99146  0 253  2  ACH  0  5.99146  1
125  1  ACL  0  5.29832  0 254  2  TH   0  4.60517  3
126  1  TH   1  4.60517  0 255  2  ACL  0  5.29832  0
127  1  ACL  0  5.29832  0 256  2  TL   0  3.91202  3
128  1  ACH  0  5.99146  1 257  2  P    0  0.00000  1
129  1  ACL  0  5.29832  3 258  2  ACH  0  5.99146  3
;
```

First, PROC LOGISTIC is employed to see whether the proportional odds model fits these data. The following SAS statements invoke the LOGISTIC procedure and specify a main effects model. Since the response variable RESP has five values, the LOGISTIC procedure fits a proportional odds model: it forms the cumulative logits and performs the test of the proportional odds assumption.

```
proc logistic data=dent descending;
   class patient center baseline trt;
   model resp = center baseline trt;
run;
```

The results are displayed in Output 15.76 and indicate that the proportional odds assumption does not hold.

Output 15.76   Test for Proportional Odds Assumption

```
           Score Test for the Proportional Odds Assumption

              Chi-Square        DF       Pr > ChiSq

                35.2185         18        0.0089
```

To proceed with the partial proportional odds model, you first assess whether you have adequate sample size. When you form two-way cross-classifications of the explanatory variables with the response variable, the counts for each cell need to be at least 5. The following PROC FREQ statements request these crosstabulations.

```
proc freq data=dent;
    tables center*resp baseline*resp trt*resp /
        nocol norow nopct;
run;
```

Output 15.77, Output 15.78, and Output 15.79 contain the resulting tables: there appears to be adequate sample size. The table for treatment × response is a five × five table with a few counts less than 5; however, most of these counts are 4, so the sample size is likely to be adequate.

Output 15.77   CENTER*RESP

```
                          Table of center by resp

          center      resp

          Frequency|      0|      1|      2|      3|      4|  Total
          ---------+--------+--------+--------+--------+--------+
                 1 |    72 |    22 |    24 |    10 |     5 |    133
          ---------+--------+--------+--------+--------+--------+
                 2 |    18 |    24 |    24 |    43 |    16 |    125
          ---------+--------+--------+--------+--------+--------+
          Total          90      46      48      53      21      258
```

Output 15.78   BASELINE*RESP

```
                          Table of baseline by resp

          baseline    resp

          Frequency|      0|      1|      2|      3|      4|  Total
          ---------+--------+--------+--------+--------+--------+
                 0 |    82 |    37 |    43 |    46 |    15 |    223
          ---------+--------+--------+--------+--------+--------+
                 1 |     8 |     9 |     5 |     7 |     6 |     35
          ---------+--------+--------+--------+--------+--------+
          Total          90      46      48      53      21      258
```

Output 15.79   TRT*RESP

```
                              Table of trt by resp

          trt          resp

          Frequency|        0|        1|        2|        3|        4|  Total
          ---------+--------+--------+--------+--------+--------+
          ACH      |     8 |    17 |    11 |    11 |     4 |     51
          ---------+--------+--------+--------+--------+--------+
          ACL      |    19 |     8 |     9 |    10 |     6 |     52
          ---------+--------+--------+--------+--------+--------+
          P        |    33 |     9 |     4 |     4 |     2 |     52
          ---------+--------+--------+--------+--------+--------+
          TH       |    10 |     8 |    15 |    13 |     6 |     52
          ---------+--------+--------+--------+--------+--------+
          TL       |    20 |     4 |     9 |    15 |     3 |     51
          ---------+--------+--------+--------+--------+--------+
          Total          90       46       48       53       21      258
```

The following DATA step creates four logits for each observation. They compare levels 4 versus levels 3, 2, 1, and 0; levels 4 and 3 versus 2, 1, 0; levels 4, 3, 2 versus levels 1 and 0; and levels 4, 3, 2, and 1 versus 0. The type of logit is contained in the variable LOGTYPE with the values 4, 3, 2, or 1 to represent the cutpoint, and the new response variable PRESP is assigned a 1 if that observation's value meets that cutpoint criteria and is 0 otherwise. Then, the response is output. Four observations are created from each individual observation, one for each type of logit.

```
data dent2; set dent;
   do;  if resp=4 then presp =1;
   else presp=0; logtype=4; output; end;
   do; if resp=4 or resp=3 then presp=1;
   else presp=0; logtype=3 ; output; end;
   do; if resp=4 or resp=3 or resp=2 then presp=1;
   else presp=0; logtype=2; output; end;
   do; if resp=4 or resp=3 or resp=2 or resp=1 then presp=1;
   else presp=0; logtype=1; output; end;
run;
```

The following PROC GENMOD statements request the GEE method. Each patient is now considered a cluster in the GEE machinery, with the four types of logits comprising the multiple responses. The main effects are included in the model along with interactions of each explanatory variable with the variable LOGTYPE. The unstructured working correlation matrix is specified. In order to make the placebo the reference level for treatment, the ORDER=DATA option is used in the PROC GENMOD statement. This creates an ordering of the classification variables that is determined by the order in which the variable values appear in the data. (The careful reader may have noticed that the first three data lines in the DATA step were adjusted in order to produce the desired order.)

```
proc genmod descending order=data;
   class logtype patient center baseline trt;
   model presp = center baseline trt logtype
      logtype*center logtype*baseline logtype*trt /
      link=logit dist=bin type3;
   repeated subject=patient / type=unstr;
run;
```

The "Class Level Information" table in Output 15.80 displays the classification levels. The placebo is the reference level for variable TRT, and logit type 1 is the reference level for variable LOGTYPE.

Output 15.80 Class Level Information

```
                    Class Level Information

   Class        Levels    Values

   logtype          4     4 3 2 1
   patient        258     2 131 1 3 130 132 4 133 5 134 6 135 7 136 8 137 9
                          138 10 139 11 140 12 141 13 142 14 143 15 144 16
                          145 17 146 18 147 19 148 20 149 21 150 22 151 23
                          152 24 153 25 154 26 155 27 156 28 157 29 158 30
                          159 31 160 32 161 33 162 34 163 35 164 36 165 37
                          166 38 ...
   center           2     1 2
   baseline         2     0 1
   trt              5     ACL TL ACH TH P
```

The information in the "GEE Model Information" table contained in Output 15.81 indicates that the analysis includes 258 clusters with four responses for each cluster; there are four types of logits.

Output 15.81 GEE Model Information

```
              GEE Model Information

   Correlation Structure              Unstructured
   Subject Effect             patient (258 levels)
   Number of Clusters                          258
   Correlation Matrix Dimension                  4
   Maximum Cluster Size                          4
   Minimum Cluster Size                          4
```

The score statistics for the effects of the model are displayed in Output 15.82. Only the BASELINE*LOGTYPE interaction is definitely nonsignificant.

Output 15.82   Type 3 Analysis

```
              Score Statistics For Type 3 GEE Analysis

                                      Chi-
           Source               DF   Square    Pr > ChiSq

           center                1    37.94      <.0001
           baseline              1     2.76      0.0964
           trt                   4    23.45      0.0001
           logtype               3    36.28      <.0001
           logtype*center        3     7.01      0.0715
           logtype*baseline      3     5.73      0.1256
           logtype*trt          12    21.49      0.0437
```

Variable BASELINE is kept in the model since it was part of the original study design.
The next model fit assumes proportional odds for the baseline effect but not for center and
treatment.

```
proc genmod descending order=data;
    class logtype patient center baseline trt;
    model presp = center trt baseline logtype
                  logtype*center logtype*trt /
                  link=logit dist=bin type3;
    repeated subject=patient / type=unstr;
run;
```

Output 15.83 contains the results of the Type 3 analysis for the reduced model. Both center
and treatment for reference logit 1 are strongly significant. The treatment interaction with
type of logit is significant at the 0.05 level of significance, and the center interaction with
type of logit is not; however, no further model reduction is performed.

Output 15.83   Type 3 Analysis

```
              Score Statistics For Type 3 GEE Analysis

                                      Chi-
           Source               DF   Square    Pr > ChiSq

           center                1    37.39      <.0001
           trt                   4    23.19      0.0001
           baseline              1     1.73      0.1884
           logtype               3   120.68      <.0001
           logtype*center        3     6.41      0.0934
           logtype*trt          12    19.96      0.0679
```

Finally, Output 15.84 contains the final parameter estimates. The main effects pertain to
effects of corresponding factors for logit type 1, and interactions are the increments to the
main effects to obtain the effects of the corresponding factors for logit types 2, 3, and 4.
For example, 1.4333 is the estimated log odds ratio for ACL versus placebo for logit type
1, and 1.4333 plus 0.1326 is the estimated log odds ratio for ACL versus placebo for logit

type 2. Thus, $e^{1.4333} = 4.19$ means that those patients receiving ACL had 4.19 higher odds of having relief than those patients receiving the placebo (responses 1–4 represented varying levels of relief and 0 represented no relief). Since $e^{1.4333+0.1326} = 4.78$, those patients receiving ACL had 4.78 times higher odds of the three highest levels of relief (4, 3, 2) than those patients receiving the placebo.

Output 15.84   Parameter Estimates

```
                     Analysis Of GEE Parameter Estimates
                     Empirical Standard Error Estimates


                                   Standard  95% Confidence
    Parameter                 Estimate   Error       Limits          Z  Pr > |Z|

    Intercept                  1.0397  0.4936   0.0723   2.0072    2.11   0.0352
    center          1         -2.2838  0.3288  -2.9282  -1.6394   -6.95  <.0001
    center          2          0.0000  0.0000   0.0000   0.0000     .       .
    trt             ACL        1.4333  0.4746   0.5031   2.3634    3.02   0.0025
    trt             TL         1.2548  0.4829   0.3084   2.2013    2.60   0.0094
    trt             ACH        2.5795  0.4830   1.6328   3.5261    5.34  <.0001
    trt             TH         2.3633  0.4674   1.4472   3.2793    5.06  <.0001
    trt             P          0.0000  0.0000   0.0000   0.0000     .       .
    baseline        0         -0.5977  0.3733  -1.3294   0.1339   -1.60   0.1093
    baseline        1          0.0000  0.0000   0.0000   0.0000     .       .
    logtype         4         -3.2324  0.7778  -4.7569  -1.7078   -4.16  <.0001
    logtype         3         -1.8885  0.4387  -2.7483  -1.0286   -4.30  <.0001
    logtype         2         -1.2480  0.3330  -1.9008  -0.5953   -3.75   0.0002
    logtype         1          0.0000  0.0000   0.0000   0.0000     .       .
    logtype*center  4    1     0.8837  0.5770  -0.2471   2.0146    1.53   0.1256
    logtype*center  4    2     0.0000  0.0000   0.0000   0.0000     .       .
    logtype*center  3    1     0.2325  0.3788  -0.5099   0.9749    0.61   0.5393
    logtype*center  3    2     0.0000  0.0000   0.0000   0.0000     .       .
    logtype*center  2    1     0.5866  0.2705   0.0565   1.1167    2.17   0.0301
    logtype*center  2    2     0.0000  0.0000   0.0000   0.0000     .       .
    logtype*center  1    1     0.0000  0.0000   0.0000   0.0000     .       .
    logtype*center  1    2     0.0000  0.0000   0.0000   0.0000     .       .
    logtype*trt     4    ACL  -0.2481  0.8977  -2.0075   1.5113   -0.28   0.7822
    logtype*trt     4    TL   -0.7605  0.9620  -2.6461   1.1250   -0.79   0.4292
    logtype*trt     4    ACH  -1.8277  0.9663  -3.7217   0.0663   -1.89   0.0586
    logtype*trt     4    TH   -1.1877  0.9167  -2.9844   0.6091   -1.30   0.1951
    logtype*trt     4    P     0.0000  0.0000   0.0000   0.0000     .       .
    logtype*trt     3    ACL   0.0092  0.5554  -1.0793   1.0978    0.02   0.9867
    logtype*trt     3    TL    0.4377  0.5366  -0.6140   1.4894    0.82   0.4147
    logtype*trt     3    ACH  -1.3127  0.6102  -2.5087  -0.1167   -2.15   0.0315
    logtype*trt     3    TH   -0.7568  0.5923  -1.9178   0.4041   -1.28   0.2013
    logtype*trt     3    P     0.0000  0.0000   0.0000   0.0000     .       .
    logtype*trt     2    ACL   0.1326  0.4194  -0.6895   0.9546    0.32   0.7520
    logtype*trt     2    TL    0.5389  0.3823  -0.2105   1.2883    1.41   0.1587
    logtype*trt     2    ACH  -1.0781  0.4885  -2.0355  -0.1208   -2.21   0.0273
    logtype*trt     2    TH   -0.2013  0.4284  -1.0408   0.6383   -0.47   0.6384
    logtype*trt     2    P     0.0000  0.0000   0.0000   0.0000     .       .
    logtype*trt     1    ACL   0.0000  0.0000   0.0000   0.0000     .       .
    logtype*trt     1    TL    0.0000  0.0000   0.0000   0.0000     .       .
    logtype*trt     1    ACH   0.0000  0.0000   0.0000   0.0000     .       .
    logtype*trt     1    TH    0.0000  0.0000   0.0000   0.0000     .       .
    logtype*trt     1    P     0.0000  0.0000   0.0000   0.0000     .       .
```

Note that using the unstructured working correlation matrix instead of the independent working correlation matrix provides a more powerful assessment of the logit type interactions; it produces smaller standard errors for within-subject effects. However, if you used the independent working correlation matrix, you would produce estimates that are most similar to those obtained by performing separate analyses of these logit functions.

15.14   Using GEE to Account for Overdispersion: Univariate Outcome

Section 8.2.7 mentions overdispersion in the case of logistic regression. Overdispersion occurs when the observed variance is larger than the nominal variance for a particular distribution. It occurs with some regularity in the analysis of proportions and discrete counts. This is not surprising for the assumed distributions (binomial and Poisson, respectively) because their respective variances are fixed by a single parameter, the mean. Overdispersion can have a major impact on inference so it needs to be taken into account. Underdispersion also occurs. See McCullagh and Nelder (1989) and Dean (1998) for more detail on overdispersion.

One way to manage overdispersion is to assume a more flexible distribution, such as the negative binomial in the case of overdispersed Poisson data (DIST=NEGBIN in the GENMOD procedure.) You can also adjust the covariance matrix of a Poisson-based analysis with a scaling factor, which is the method PROC GENMOD uses with the SCALE= option in the MODEL statement. Then the covariance matrix is pre-multiplied by the scaling factor ϕ, and the scaled deviance and the log likelihood ratio tests are divided by ϕ, as are the confidence limits based on the profile likelihood. This type of analysis is performed in Section 12.5 of Chapter 12.

Another way of managing the overdispersion is to take the generalized estimating approach. Recall that the robust, or empirical, covariance matrix estimated by the GEE method is robust to the misspecification of the covariance structure, and misspecification is occurring in the case of overdispersion. The variance is not "acting" as it should; it does not take the form for data from a Poisson distribution. With GEE estimation, you are using a subject-to-subject measure for variance estimation instead of a model-based one. The robustness comes from the fact that the variance estimation process involves aggregates at the cluster level. While GEE was devised for the analysis of correlated data with more than one response per subject, you can also use it for the analysis of single outcomes and derive the benefits of the robust standard errors. This section describes the use of GEE for adjusting for overdispersion in the univariate case.

Researchers studying the incidence of lower respiratory illness in infants took repeated observations of infants over one year. They studied 284 children and examined them every two weeks. Explanatory variables evaluated included passive smoking (one or more smokers in the household), socioeconomic status, and crowding. See LaVange et al. (1994) for more information on the study and a discussion of the analysis of incidence densities. One outcome of interest was the total number of times, or counts, of lower respiratory infection recorded for the year. The strategy was to model these counts with Poisson regression. However, it is reasonable to expect some overdispersion since the children that have an infection are more likely to have other infections.

The following DATA step inputs the data into a SAS data set named LRI. The variable COUNT is the total number of infections that year, and the variable RISK is the number of weeks during that year for which the child is considered at risk (when a lower respiratory infection is ongoing, the child is not considered to be at risk for a new one). The variable CROWDING is an indicator variable for whether crowded conditions occur in the household, and SES is an indicator variable for whether the family's socioeconomic status

was considered low (0), medium (1), or high (2). The variable RACE is an indicator
variable for whether the child was white (1) or not (0), and the variable PASSIVE is an
indicator variable for whether the child was exposed to cigarette smoking. Finally, the
AGEGROUP variable takes the values 1, 2, and 3 for under four, four to six, or more than
six months.

```
data lri;
    input id count risk passive crowding ses agegroup race @@;
    logrisk =log(risk/52);
    datalines;
 1 0 42 1 0 2 2 0    96 1 41 1 0 1 2 0    191 0 44 1 0 0 2 0
 2 0 43 1 0 0 2 0    97 1 26 1 1 2 2 0    192 0 45 0 0 0 2 1
 3 0 41 1 0 1 2 0    98 0 36 0 0 0 2 0    193 0 42 0 0 0 2 0
 4 1 36 0 1 0 2 0    99 0 34 0 0 0 2 0    194 1 31 0 0 0 2 1
 5 1 31 0 0 0 2 0   100 1  3 1 1 2 3 1    195 0 35 0 0 0 2 0
 6 0 43 1 0 0 2 0   101 0 45 1 0 0 2 0    196 1 35 1 0 0 2 0
 7 0 45 0 0 0 2 0   102 0 38 0 0 1 2 0    197 1 27 1 0 1 2 0
 8 0 42 0 0 0 2 1   103 0 41 1 1 1 2 1    198 1 33 0 0 0 2 0
 9 0 45 0 0 0 2 1   104 1 37 0 1 0 2 0    199 0 39 1 0 1 2 0
10 0 35 1 1 0 2 0   105 0 40 0 0 0 2 0    200 3 40 0 1 2 2 0
11 0 43 0 0 0 2 0   106 1 35 1 0 0 2 0    201 4 26 1 0 1 2 0
12 2 38 0 0 0 2 0   107 0 28 0 1 2 2 0    202 0 14 1 1 1 1 1
13 0 41 0 0 0 2 0   108 3 33 0 1 2 2 0    203 0 39 0 1 2 2 0
14 0 12 1 1 0 1 0   109 0 38 0 0 0 2 0    204 0  4 1 1 1 3 0
15 0  6 0 0 0 3 0   110 0 42 1 1 2 2 1    205 1 27 1 1 1 2 1
16 0 43 0 0 0 2 0   111 0 40 1 1 2 2 0    206 0 36 1 0 0 2 1
17 2 39 1 0 1 2 0   112 0 38 0 0 0 2 0    207 0 30 1 0 2 2 1
18 0 43 0 1 0 2 0   113 2 37 0 1 1 2 0    208 0 34 0 1 0 2 0
19 2 37 0 0 0 2 1   114 1 42 0 1 0 2 0    209 1 40 1 1 1 2 0
20 0 31 1 1 1 2 0   115 5 37 1 1 1 2 1    210 0  6 1 0 1 1 1
21 0 45 0 1 0 2 0   116 0 38 0 0 0 2 0    211 1 40 1 1 1 2 0
22 1 29 1 1 1 2 1   117 0  4 0 0 0 3 0    212 2 43 0 1 0 2 0
23 1 35 1 1 1 2 0   118 2 37 1 1 1 2 0    213 0 36 1 1 1 2 0
24 3 20 1 1 2 2 0   119 0 39 1 0 1 2 0    214 0 35 1 1 1 2 1
25 1 23 1 1 1 2 0   120 0 42 1 1 0 2 0    215 1 35 1 1 2 2 0
26 1 37 1 0 0 2 0   121 0 40 1 0 0 2 0    216 0 43 1 0 1 2 0
27 0 49 0 0 0 2 0   122 0 36 1 0 0 2 0    217 0 33 1 1 2 2 0
28 0 35 0 0 0 2 0   123 1 42 0 1 1 2 0    218 0 36 0 1 1 2 1
29 3 44 1 1 1 2 0   124 1 39 0 0 0 2 0    219 1 41 0 0 0 2 0
30 0 37 1 0 0 2 0   125 2 29 0 0 0 2 0    220 0 41 1 1 0 2 1
31 2 39 0 1 1 2 0   126 3 37 1 1 2 2 1    221 1 42 0 0 0 2 1
32 0 41 0 0 0 2 0   127 0 40 1 0 0 2 0    222 0 33 0 1 2 2 1
33 1 46 1 1 2 2 0   128 0 40 0 0 0 2 0    223 0 40 1 1 2 2 0
34 0  5 1 1 2 3 1   129 0 39 0 0 0 2 0    224 0 40 1 1 1 2 1
35 1 29 0 0 0 2 0   130 0 40 1 0 1 2 0    225 0 40 0 0 2 2 0
36 0 31 0 1 0 2 0   131 1 32 0 0 0 2 0    226 0 28 1 0 1 2 0
37 0 22 1 1 2 2 0   132 0 46 1 0 1 2 0    227 0 47 0 0 0 2 1
38 1 22 1 1 2 2 1   133 4 39 1 1 0 2 0    228 0 18 1 1 2 2 1
39 0 47 0 0 0 2 0   134 0 37 0 0 0 2 0    229 0 45 1 0 0 2 0
40 1 46 1 1 2 2 1   135 0 51 0 0 1 2 0    230 0 35 0 0 0 2 0
41 0 37 0 0 0 2 0   136 1 39 1 1 0 2 0    231 1 17 1 0 1 1 1
42 1 39 0 0 0 2 0   137 1 34 1 1 0 2 0    232 0 40 0 0 0 2 0
43 0 33 0 1 1 2 1   138 1 14 0 1 0 1 0    233 0 29 1 1 2 2 0
```

```
44 0 34 1 0 1 2 0   139 2 15 1 0 0 2 0   234 1 35 1 1 1 2 0
45 3 32 1 1 1 2 0   140 1 34 1 1 0 2 1   235 0 40 0 0 2 2 0
46 3 22 0 0 0 2 0   141 0 43 0 1 0 2 0   236 1 22 1 1 1 2 0
47 1  6 1 0 2 3 0   142 1 33 0 0 0 2 0   237 0 42 0 0 0 2 0
48 0 38 0 0 0 2 0   143 3 34 1 0 0 2 1   238 0 34 1 1 1 2 1
49 1 43 0 1 0 2 0   144 0 48 0 0 0 2 0   239 6 38 1 0 1 2 0
50 2 36 0 1 0 2 0   145 4 26 1 1 0 2 0   240 0 25 0 0 1 2 1
51 0 43 0 0 0 2 0   146 0 30 0 1 2 2 1   241 0 39 0 1 0 2 0
52 0 24 1 0 0 2 0   147 0 41 1 1 1 2 0   242 1 35 0 1 2 2 1
53 0 25 1 0 1 2 1   148 0 34 0 1 1 2 0   243 1 36 1 1 1 2 1
54 0 41 0 0 0 2 0   149 0 43 0 1 0 2 0   244 0 23 1 0 0 2 0
55 0 43 0 0 0 2 0   150 1 31 1 0 1 2 0   245 4 30 1 1 1 2 0
56 2 31 0 1 1 2 0   151 0 26 1 0 1 2 0   246 1 41 1 1 1 2 1
57 3 28 1 1 1 2 0   152 0 37 0 0 0 2 0   247 0 37 0 1 1 2 0
58 1 22 0 0 1 2 1   153 0 44 0 0 0 2 0   248 0 46 1 1 0 2 0
59 1 11 1 1 1 1 0   154 0 40 1 0 0 2 0   249 0 45 1 1 0 2 1
60 3 41 0 1 1 2 0   155 0  8 1 1 1 3 1   250 1 38 1 1 1 2 0
61 0 31 0 0 1 2 0   156 0 40 1 1 1 2 1   251 0 10 1 1 1 1 0
62 0 11 0 0 1 1 1   157 1 45 0 0 0 2 0   252 0 30 1 1 2 2 0
63 0 44 0 1 0 2 0   158 0  4 0 0 2 3 0   253 0 32 0 1 2 2 0
64 0  9 1 0 0 3 1   159 1 36 0 1 0 2 0   254 0 46 1 0 0 2 0
65 0 36 1 1 1 2 0   160 3 37 1 1 1 2 0   255 5 35 1 1 2 2 1
66 0 29 1 0 0 2 0   161 0 15 1 0 0 1 0   256 0 44 0 0 0 2 0
67 0 27 0 1 0 2 1   162 1 27 1 0 1 2 1   257 0 41 0 1 1 2 0
68 0 36 0 1 0 2 0   163 2 31 0 1 0 2 0   258 2 36 1 0 1 2 0
69 1 33 1 0 0 2 0   164 0 42 0 0 0 2 0   259 0 34 1 1 1 2 1
70 2 13 1 1 2 1 1   165 0 42 1 0 0 2 0   260 1 30 0 1 0 2 1
71 0 38 0 0 0 2 0   166 1 38 0 0 0 2 0   261 1 27 1 0 0 2 0
72 0 41 0 0 0 2 1   167 0 44 1 0 0 2 0   262 0 48 1 0 0 2 0
73 0 41 1 0 2 2 0   168 0 45 0 0 0 2 0   263 1  6 0 1 2 3 1
74 0 35 0 0 1 2 0   169 0 34 0 1 0 2 0   264 0 38 1 1 0 2 1
75 0 45 0 0 0 2 0   170 2 41 0 0 0 2 0   265 0 29 1 1 1 2 1
76 4 38 1 0 2 2 1   171 2 30 1 1 1 2 0   266 1 43 0 1 2 2 1
77 1 42 1 0 0 2 1   172 0 44 0 0 0 2 0   267 0 43 0 1 0 2 0
78 1 42 1 1 2 2 1   173 0 40 1 0 0 2 0   268 0 37 1 0 2 2 0
79 6 36 1 1 0 2 0   174 2 31 0 0 0 2 0   269 1 23 1 1 0 2 1
80 2 23 1 1 1 2 1   175 0 41 1 0 0 2 0   270 0 44 0 0 1 2 0
81 1 32 0 0 1 2 0   176 0 41 0 0 0 2 0   271 0  5 0 1 1 3 1
82 0 41 0 1 0 2 0   177 0 39 1 0 0 2 0   272 0 25 1 0 2 2 0
83 0 50 0 0 0 2 0   178 0 40 1 0 0 2 0   273 0 25 1 0 1 2 0
84 0 42 1 1 1 2 1   179 2 35 1 0 2 2 0   274 1 28 1 1 1 2 1
85 1 30 0 0 0 2 0   180 1 43 1 0 0 2 0   275 0  7 0 1 0 3 1
86 2 47 0 1 0 2 0   181 2 39 0 0 0 2 0   276 0 32 0 0 0 2 0
87 1 35 1 1 2 2 0   182 0 35 1 1 0 2 0   277 0 41 0 0 0 2 0
88 1 38 1 0 1 2 1   183 0 37 0 0 0 2 0   278 1 33 1 1 2 2 1
89 1 38 1 1 1 2 1   184 3 37 0 0 0 2 0   279 2 36 1 1 2 2 0
90 1 38 1 1 1 2 1   185 0 43 0 0 0 2 0   280 0 31 0 0 0 2 0
91 0 32 1 1 1 2 0   186 0 42 0 0 0 2 0   281 0 18 0 0 0 2 0
92 1  3 1 0 1 3 1   187 0 42 0 0 0 2 0   282 1 32 1 0 2 2 0
93 0 26 1 0 0 2 1   188 0 38 0 0 0 2 0   283 0 22 1 1 2 2 1
94 0 35 1 0 0 2 0   189 0 36 1 0 0 2 0   284 0 35 0 0 0 2 1
95 3 37 1 0 0 2 0   190 0 39 0 1 0 2 0
 ;
```

The following SAS statements request the analysis. To produce the Poisson regression, options LINK=LOG and DIST=POISSON are specified. The variable LOGRISK is the offset, and the main effects model is requested.

```
proc genmod data=lri;
    class ses id race agegroup;
    model count = passive crowding ses race agegroup/
                  dist=poisson link=log offset=logrisk type3;
run;
```

Output 15.85 contains the general model information.

Output 15.85   Model Information

```
                   Model Information

         Data Set              WORK.LRI
         Distribution           Poisson
         Link Function              Log
         Dependent Variable       count
         Offset Variable        logrisk
         Observations Used          284
```

Output 15.86 contains the goodness-of-fit statistics, along with the ratios of their values to their degrees of freedom. With values of 1.4788 for the Deviance/df and 1.7951 for Pearson/df, there is evidence of overdispersion. The model-based estimates of standard errors may not be appropriate and therefore any inference is questionable. (When this ratio is close to 1, you conclude that little evidence of over- or under- dispersion exists). The next step is to account for this overdispersion with the GEE-generated robust covariances.

Output 15.86   Goodness-of-Fit Statistics

```
            Criteria For Assessing Goodness Of Fit

    Criterion              DF        Value       Value/DF

    Deviance              276     408.1549        1.4788
    Scaled Deviance       276     408.1549        1.4788
    Pearson Chi-Square    276     495.4494        1.7951
    Scaled Pearson X2     276     495.4494        1.7951
    Log Likelihood               -260.4117
```

The following statements produce the desired GEE analysis. In this case, the subject is the cluster and there is only one measurement per cluster. The working independent correlation structure with the TYPE=IND option although, with a cluster size of 1, the estimates will be the same if you specify exchangeable or unstructured. Otherwise, the model specification is the same as in the previous analysis.

```
proc genmod data=lri;
   class id ses race agegroup;
   model count = passive crowding ses race agegroup /
                 dist=poisson link=log offset=logrisk type3;
   repeated subject=id / type=ind;
run;
```

Output 15.87 reports that the GEE analysis involves one measurement per subject, and that there are 284 subjects, or clusters, in the analysis.

Output 15.87   GEE Model Information

```
                     GEE Model Information

          Correlation Structure             Independent
          Subject Effect               id (284 levels)
          Number of Clusters                       284
          Correlation Matrix Dimension               1
          Maximum Cluster Size                       1
          Minimum Cluster Size                       1
```

Output 15.88 contains the parameter estimates. They are the same as displayed for the unadjusted GLM analysis in Chapter 12, as you would expect, but the standard errors are different. They are larger than the corresponding standard errors in the GLM analysis; this is also what you would expect because overdispersion means that the data are exhibiting additional variance.

Output 15.88   GEE Parameter Estimates

```
                  Analysis Of GEE Parameter Estimates
                  Empirical Standard Error Estimates

                              Standard   95% Confidence
          Parameter  Estimate    Error       Limits         Z Pr > |Z|

          Intercept    0.6047   0.5564  -0.4858   1.6952   1.09   0.2771
          passive      0.4310   0.2105   0.0184   0.8436   2.05   0.0406
          crowding     0.5199   0.2367   0.0559   0.9839   2.20   0.0281
          ses       0 -0.3970   0.2977  -0.9805   0.1865  -1.33   0.1824
          ses       1 -0.0681   0.2520  -0.5619   0.4258  -0.27   0.7871
          ses       2  0.0000   0.0000   0.0000   0.0000    .       .
          race      0  0.1402   0.2211  -0.2931   0.5736   0.63   0.5259
          race      1  0.0000   0.0000   0.0000   0.0000    .       .
          agegroup  1 -0.4792   0.6033  -1.6617   0.7033  -0.79   0.4270
          agegroup  2 -0.9919   0.4675  -1.9082  -0.0756  -2.12   0.0339
          agegroup  3  0.0000   0.0000   0.0000   0.0000    .       .
```

Output 15.89 contains the Type 3 analysis. SES, race, and age group are non-influential. Crowding and smoking exposure are significant at the $\alpha = 0.05$ level of significance.

Output 15.89 Type 3 Analysis

```
         Score Statistics For Type 3 GEE Analysis

                              Chi-
         Source        DF    Square    Pr > ChiSq

         passive        1     3.90       0.0484
         crowding       1     4.72       0.0298
         ses            2     2.11       0.3478
         race           1     0.42       0.5176
         agegroup       2     2.79       0.2484
```

Thus, this section provides an alternative strategy for adjusting for overdispersion to the scaling factor adjustment discussed in Section 12.5 in Chapter 12. Using the GEE method, you are using a measure of variability based on the data to do the adjustment, rather than a single parameter (scaling factor) applied to the covariance matrix. With the GEE method, using the robust variances, you are providing a measure of variability for each parameter you estimate together with the corresponding covariances, all based on your data. In many situations, this strategy may be a practical approach to handling overdispersion.

Appendix A: Steps to Find the GEE Solution

Finding the GEE solution requires a number of steps, including specifying the marginal model for the first moment, specifying the variance function for the relationship between the first and second moments, choosing a working correlation matrix, computing an initial estimate of β, and then using this estimate in an iterative estimation process. In detail:

The **first step** of the GEE method is to relate the marginal response $\mu_{ij} = \mathrm{E}(y_{ij})$ to a linear combination of the covariates: $g(\mu_{ij}) = x'_{ij}\beta$, where $\beta = (\beta_1, \ldots, \beta_p)'$ is a $p \times 1$ vector of unknown parameters and g is a known link function. Common link functions are the logit function $g(x) = \log(x/(1-x))$ for binary responses and the log function $g(x) = \log(x)$ for Poisson counts. The $p \times 1$ parameter vector β characterizes how the cross-sectional response distribution depends on the explanatory variables.

The **second step** is to describe the variance of y_{ij} as a function of the mean: $\mathrm{Var}(y_{ij}) = v(\mu_{ij})\,\phi$, where v is a known variance function and ϕ is a possibly unknown scale parameter. For binary responses, $v(\mu_{ij}) = \mu_{ij}(1-\mu_{ij})$; and for Poisson responses, $v(\mu_{ij}) = \mu_{ij}$. For these two types of response variables, $\phi = 1$. Overdispersion $(\phi > 1)$ may exist for binomial-like or count data, but use of the empirical covariance matrix for the GEE procedure is robust to this overdispersion.

The **third step** is to choose the form of a $t_i \times t_i$ working correlation matrix $\mathbf{R}_i(\alpha)$ for each $\mathbf{y}_i = (y_{i1}, \ldots, y_{it})'$. The (j, j') element of $\mathbf{R}_i(\alpha)$ is the known, hypothesized, or estimated correlation between y_{ij} and $y_{ij'}$. This working correlation matrix may depend on a vector of unknown parameters α, which is the same for all subjects. You assume that $\mathbf{R}_i(\alpha)$ is known except for a fixed number of parameters α that must be estimated from the data. Although this correlation matrix can differ from subject to subject, you commonly use a working correlation matrix $\mathbf{R}(\alpha)$ that approximates the average dependence among repeated observations over subjects.

The GEE method yields consistent estimates of the regression coefficients and their variances, even with misspecification of the structure of the covariance matrix. In addition, the loss of efficiency from an incorrect choice of $\mathbf{R}$ is usually not consequential when the number of subjects is large.

The **fourth step** of the GEE method is to estimate the parameter vector β and its covariance matrix. First, let $\mathbf{A}_i$ be the $t_i \times t_i$ diagonal matrix with $v(\mu_{ij})$ as the jth diagonal element. The working covariance matrix for y_i is $\mathbf{V}_i(\alpha) = \phi \, \mathbf{A}_i^{1/2} \, \mathbf{R}_i(\alpha) \, \mathbf{A}_i^{1/2}$. The GEE estimate of β is the solution of the estimating equation

$$U(\boldsymbol{\beta}) = \sum_{i=1}^{n} \left(\frac{\partial \boldsymbol{\mu}_i}{\partial \boldsymbol{\beta}} \right)' [V_i(\widehat{\alpha})]^{-1} (\mathbf{y}_i - \boldsymbol{\mu}_i) = 0_p$$

where $\boldsymbol{\mu}_i = (\mu_{i1}, \dots, \mu_{it_i})'$, $\mathbf{0}_p$ is the $p \times 1$ vector $(0, \dots, 0)'$, and $\widehat{\boldsymbol{\alpha}}$ is a consistent estimate of $\boldsymbol{\alpha}$.

The estimating equation is solved by iterating between quasi-likelihood methods for estimating β and method of moments estimation of $\boldsymbol{\alpha}$ as a function of β, as follows:

1. Compute an initial estimate of β, using a GLM model or some other method.

2. Compute the standardized Pearson residuals

$$r_{ij} = \frac{y_{ij} - \widehat{\mu}_{ij}}{\sqrt{v(\widehat{\mu}_{ij})}}$$

   and obtain the estimates for the nuisance parameters ϕ and $\boldsymbol{\alpha}$ using moment estimation.

3. Update $\hat{\beta}$ with

$$\hat{\boldsymbol{\beta}} - \left[\sum_{i=1}^{K} \frac{\partial \boldsymbol{\mu}_i}{\partial \boldsymbol{\beta}}' \mathbf{V}_i^{-1} \frac{\partial \boldsymbol{\mu}_i}{\partial \boldsymbol{\beta}} \right]^{-1} \left[\sum_{i=1}^{K} \frac{\partial \boldsymbol{\mu}_i}{\partial \boldsymbol{\beta}}' \mathbf{V}_i^{-1} (\mathbf{Y}_i - \boldsymbol{\mu}_i) \right]$$

4. Iterate until convergence.

Appendix B: Macro for Adjusted Wald Statistic

The following macro is used in Section 15.7.

```
%macro geef;
data temp1;
   set clustout;
   drop Label1 cvalue1;
   if Label1='Number of Clusters';
run;
data temp2;
   set scoreout;
```

```
       drop ProbChiSq;
    run;
    data temp3;
       merge temp1 temp2;
    run;
    data temp4; set temp3;
      retain nclusters; drop nvalue1;
      if _n_=1 then nclusters=nvalue1;
    run;
    data temp5;
       set temp4;
       drop ChiSq nclusters d;
       d=nclusters-1;
       NewF= ((d-df+1)*ChiSq)/(d*df);
       ProbF=1-cdf('F', NewF,df,d-df+1);
    run;

/* Set the ODS path to include your store first (this
   sets the search path order so that ODS looks in your
   store first, followed by the default store */

ods path sasuser.templat (update)
         sashelp.tmplmst (read);

/* Print the path to the log to make sure you will get
   what you expect */

*ods path show;

/* Define your table, and store it */
    proc template;
       define table GEEType3F;
       parent=Stat.Genmod.Type3GEESc;
       header "#F-Statistics for Type 3 GEE Analysis##";
       column Source DF i NewF ProbF;
       define NewF;
       parent = Common.ANOVA.FValue;
     end;
    end;
    run;
    title1;
    data _null_;
       set temp5;
       file print ods=(template='GEEType3F');
       put _ods_;
    run;
  ;
%mend geef;
```

Chapter 16
Loglinear Models

Chapter Table of Contents

Chapter 16
Loglinear Models

16.1 Introduction

Chapters 2–6 discuss methods for testing hypotheses of no association in two-way and stratified two-way contingency tables. These approaches focus on hypothesis testing rather than on model fitting and parameter estimation. In contrast, Chapters 8–12 describe logistic regression and weighted least squares methods for modeling a categorical response variable as a function of one or more categorical and/or continuous explanatory variables. These methods, which are analogous to ANOVA and regression techniques for normally distributed response variables, are appropriate when there is a clearly defined response variable of interest and you want to model how the response is affected by a set of explanatory variables or design factors. In such situations, you are most interested in estimating the parameters of a statistical model and in testing hypotheses concerning model parameters.

Loglinear models are another important tool for the analysis of categorical data. This methodology was primarily developed during the 1960s. Although many investigators made significant contributions, Leo Goodman was a particularly influential researcher who popularized the method in the social sciences. Two of his key papers (Goodman 1968, 1970) summarize much of the earlier work. Bishop, Fienberg, and Holland (1975) first comprehensively described the methodology for the general statistical community.

Loglinear model methodology is most appropriate when there is no clear distinction between response and explanatory variables, for example, when all of the variables are observed simultaneously. The loglinear model point of view treats all variables as response variables, and the focus is on statistical independence and dependence. Loglinear modeling of categorical data is analogous to correlation analysis for normally distributed response variables and is useful in assessing patterns of statistical dependence among subsets of variables.

You perform loglinear model analysis in the SAS System by using the CATMOD procedure, even though the structure and syntax of PROC CATMOD was designed originally for regression analyses of categorical response variables. You can also use the GENMOD procedure to fit loglinear models, although the parameterization of PROC GENMOD (the less than full rank parameterization of PROC GLM) makes it more complicated to interpret lower-order effects in the presence of higher-order effects, although simply testing for only higher order associations is very straightforward and convenient with PROC GENMOD. Other software designed specifically for loglinear modeling may be more convenient to use than PROC CATMOD in carrying out certain routine analyses, such as fitting all possible hierarchical loglinear models to a given data set. On the other hand, the CATMOD procedure permits the fitting of complicated types of loglinear models, some of which can not be fit conveniently using other programs.

Section 16.2 describes the loglinear model for a two-way contingency table and introduces the use of PROC CATMOD for loglinear modeling. Although the simplest application of loglinear models is in testing statistical independence between two categorical variables, the methodology is most useful in situations in which there are several variables. Section 16.3 considers the loglinear model for three-way contingency tables and illustrates how to use PROC GENMOD for fitting loglinear models. Section 16.4 demonstrates loglinear modeling for higher-order tables, and Section 16.5 describes the correspondence between logistic models and loglinear models.

16.2   Two-Way Contingency Tables

16.2.1   Loglinear Model for the 2×2 Table

Table 16.1 displays the 2×2 table of frequencies resulting from the cross-classification of a row variable X and a column variable Y, each with two levels. Chapter 2 discusses tests and estimators of association for 2×2 contingency tables arising from several different sampling frameworks described in Section 2.1. In this chapter, attention focuses on tables representing a simple random sample from one population. Therefore, the cross-classification of the two binary responses X and Y yields a single multinomial distribution with total sample size n and cell probabilities π_{ij} displayed in Table 16.2.

Table 16.1.   Cell Counts in a 2×2 Contingency Table

Level of X	Level of Y		Total
	1	2	
1	n_{11}	n_{12}	n_{1+}
2	n_{21}	n_{22}	n_{2+}
Total	n_{+1}	n_{+2}	n

Table 16.2.   Cell Probabilities in a 2×2 Contingency Table

Level of X	Level of Y		Total
	1	2	
1	π_{11}	π_{12}	π_{1+}
2	π_{21}	π_{22}	π_{2+}
Total	π_{+1}	π_{+2}	1

The motivation for the use of loglinear models is that statistical independence can be expressed in terms of a linear combination of the logarithms of the cell probabilities. In particular, if the variables X and Y in a 2×2 table are statistically independent, then the probability of individuals being in the first row (level 1 of X) among those in the first column (level 1 of Y) would be the same as the probability for the first row among those in the second column (level 2 of Y). Therefore,

$$\frac{\pi_{11}}{\pi_{+1}} = \frac{\pi_{12}}{\pi_{+2}} = \pi_{1+}$$

and $\pi_{11} = \pi_{1+}\pi_{+1}$. Similar arguments lead to the general result that if the row and column variables are independent, then $\pi_{ij} = \pi_{i+}\pi_{+j}$, for $i, j = 1, 2$.

You can then express independence as a general relation involving all four cell probabilities. First, if X and Y are statistically independent

$$\frac{\pi_{11}}{\pi_{+1}} = \frac{\pi_{12}}{\pi_{+2}}$$

Since $\pi_{+1} = \pi_{11} + \pi_{21}$ and $\pi_{+2} = \pi_{12} + \pi_{22}$, the relationship is

$$\frac{\pi_{11}}{\pi_{11} + \pi_{21}} = \frac{\pi_{12}}{\pi_{12} + \pi_{22}}$$

so that $\pi_{11}(\pi_{12} + \pi_{22}) = \pi_{12}(\pi_{11} + \pi_{21})$. This simplifies to $\pi_{11}\pi_{22} = \pi_{12}\pi_{21}$. Therefore, the row and column variables are independent if

$$\Psi = \frac{\pi_{11}\pi_{22}}{\pi_{12}\pi_{21}} = 1$$

where Ψ is called the *cross-product ratio*, or the odds ratio. Taking logarithms of both sides expresses statistical independence as a linear combination of the logarithms of the cell probabilities:

$$\log \Psi = \log \pi_{11} - \log \pi_{12} - \log \pi_{21} + \log \pi_{22} = 0$$

Loglinear models for 2×2 contingency tables involve the logarithm of the cross-product ratio in a special way. The *saturated loglinear model* for a 2×2 table is

$$\log(m_{ij}) = \mu + \lambda_i^X + \lambda_j^Y + \lambda_{ij}^{XY} \qquad i, j = 1, 2$$

where $m_{ij} = n\pi_{ij}$ is the expected frequency in the (i, j) cell. This model is similar to the two-way analysis of variance model for a continuous response y:

$$E(y_{ij}) = \mu + \alpha_i + \beta_j + (\alpha\beta)_{ij}$$

with overall mean μ, main effects α_i and β_j, and interaction effects $(\alpha\beta)_{ij}$. The use of the terms λ_i^X, λ_j^Y, and λ_{ij}^{XY} instead of α_i, β_j, and $(\alpha\beta)_{ij}$ is common loglinear model notation and is especially convenient when considering tables of higher dimensions.

Since there are $1 + 2 + 2 + 4 = 9$ parameters in the saturated loglinear model, but only four observations, the model is overparameterized. Imposing the usual sum-to-zero constraints

$$\sum_{i=1}^{2} \lambda_i^X = 0 \qquad \sum_{j=1}^{2} \lambda_j^Y = 0 \qquad \sum_{i=1}^{2} \lambda_{ij}^{XY} = \sum_{j=1}^{2} \lambda_{ij}^{XY} = 0$$

yields three nonredundant λ parameters (λ_1^X, λ_1^Y, λ_{11}^{XY}). The fourth parameter, μ, is fixed by the total sample size n. Table 16.3 displays the expected cell frequencies m_{ij} in terms of the model parameters μ, λ_1^X, λ_1^Y, and λ_{11}^{XY}.

Table 16.3. Loglinear Model Expected Cell Counts

Level of X	Level of Y	
	1	2
1	$\exp(\mu + \lambda_1^X + \lambda_1^Y + \lambda_{11}^{XY})$	$\exp(\mu + \lambda_1^X - \lambda_1^Y - \lambda_{11}^{XY})$
2	$\exp(\mu - \lambda_1^X + \lambda_1^Y - \lambda_{11}^{XY})$	$\exp(\mu - \lambda_1^X - \lambda_1^Y + \lambda_{11}^{XY})$

The odds ratio can also be expressed as a function of the expected frequencies:

$$\Psi = \frac{m_{11} m_{22}}{m_{12} m_{21}}$$

so that

$$\log \Psi = \log m_{11} - \log m_{12} - \log m_{21} + \log m_{22} = 4\lambda_{11}^{XY}$$

Therefore, the hypothesis of independence of X and Y is equivalent to $H_0: \lambda_{11}^{XY} = 0$. The corresponding *independence loglinear model* is given by

$$\log(m_{ij}) = \mu + \lambda_i^X + \lambda_j^Y \qquad i, j = 1, 2$$

This model has one degree of freedom for testing lack of fit.

Chapter 2 discusses the Pearson chi-square test of independence for a 2×2 contingency table. An alternative approach is to test $H_0: \lambda_{11}^{XY} = 0$ using the likelihood ratio test to compare the fit of the independence and saturated loglinear models.

The likelihood ratio test of independence can be derived directly from the multinomial likelihood

$$f(n_{11}, n_{12}, n_{21}, n_{22}) = \frac{n!}{n_{11}! \, n_{12}! \, n_{21}! \, n_{22}!} \pi_{11}^{n_{11}} \pi_{12}^{n_{12}} \pi_{21}^{n_{21}} \pi_{22}^{n_{22}}$$

The unrestricted maximum likelihood estimates (MLEs) of the π_{ij} values are given by $p_{ij} = n_{ij}/n$. The maximized likelihood is then

$$\max L = \frac{n!}{n_{11}! \, n_{12}! \, n_{21}! \, n_{22}!} \prod_{i=1}^{2} \prod_{j=1}^{2} \left(\frac{n_{ij}}{n}\right)^{n_{ij}}$$

Under the independence hypothesis $H_0: \pi_{ij} = \pi_{i+}\pi_{+j}$, the likelihood is

$$L_0 = \frac{n!}{n_{11}! \, n_{12}! \, n_{21}! \, n_{22}!} \pi_{1+}^{n_{1+}} \pi_{2+}^{n_{2+}} \pi_{+1}^{n_{+1}} \pi_{+2}^{n_{+2}}$$

The MLEs for the π_{ij} under this model are $p_{ij} = n_{i+}n_{+j}/n^2$ and the maximized log likelihood is

$$\max L_0 = \frac{n!}{n_{11}! \, n_{12}! \, n_{21}! \, n_{22}!} \prod_{i=1}^{2} \prod_{j=1}^{2} \left(\frac{n_{i+}n_{+j}}{n^2}\right)^{n_{ij}}$$

The likelihood ratio is

$$\lambda = \frac{\max L_0}{\max L} = \prod_{i=1}^{2} \prod_{j=1}^{2} \left(\frac{\widehat{m}_{ij}}{n_{ij}}\right)^{n_{ij}}$$

where $\widehat{m}_{ij} = n_{i+}n_{+j}/n$, and the likelihood ratio statistic is

$$G^2 = -2\log\lambda = 2\sum_{i=1}^{2}\sum_{j=1}^{2} n_{ij}\log\left(\frac{n_{ij}}{\widehat{m}_{ij}}\right)$$

The statistic G^2 has an asymptotic chi-square distribution with 1 df if H_0 is true, and it is asymptotically equivalent to the Pearson chi-square statistic Q_P discussed in Chapter 2.

16.2.2 Bicycle Example

Table 16.4 displays the cross-classification of type of bicycle (categorized as mountain or other) and safety helmet usage for a sample of 100 bicycle riders. Under the assumption that the variables bicycle type and helmet usage are observed for a sample of 100 riders, both are response variables.

Table 16.4. Bicycle Data

Bicycle Type	Wearing Helmet		Total
	Yes	No	
Mountain	34	32	66
Other	10	24	34
Total	44	56	100

The following statements create a SAS data set containing the cell counts.

```
data bicycle;
   input type $ helmet $ count;
   datalines;
Mountain Yes  34
Mountain No   32
Other    Yes  10
Other    No   24
;
run;
```

Suppose X denotes the row variable (bicycle type) and Y denotes the column variable (helmet usage). The saturated model

$$\log(m_{ij}) = \mu + \lambda_i^X + \lambda_j^Y + \lambda_{ij}^{XY} \qquad i, j = 1, 2$$

is requested by the following PROC CATMOD invocation.

```
proc catmod;
   weight count;
   model type*helmet=_response_ / noresponse noiter noparm;
   loglin type|helmet;
run;
```

The response variables TYPE and HELMET are both listed on the left-hand side of the MODEL statement (separated by an asterisk). This usage of the MODEL statement is similar to that for repeated measures analyses (Chapter 13). Since PROC CATMOD allows only independent variables on the right-hand side of the MODEL statement, you can't specify a loglinear model directly in the MODEL statement. Instead, you use the special keyword _RESPONSE_ on the right-hand side and specify the loglinear model effects in the LOGLIN statement. In this example, the saturated model includes the TYPE and HELMET main effects, as well as the TYPE × HELMET interaction through the TYPE|HELMET specification.

The three options specified in the MODEL statement suppress printed output that may not always be necessary in loglinear model analysis. The NORESPONSE option suppresses printing of the loglinear model design matrix, the NOITER option suppresses printing of the parameter estimates and other information at each iteration of the maximum likelihood procedure, and the NOPARM option suppresses printing of the estimated parameters.

Output 16.1 displays the population and response profiles. There is one population, and the four response profiles are defined by the cross-classification of the response variables TYPE and HELMET.

Output 16.1   Population and Response Profiles

```
                       Population Profiles

                 Sample      Sample Size
                 ---------------------
                    1               100


                       Response Profiles

             Response      type          helmet
             -----------------------------------
                 1         Mountain      No
                 2         Mountain      Yes
                 3         Other         No
                 4         Other         Yes
```

Output 16.2 displays the analysis of variance table. Since the four multinomial cell probabilities sum to one, there are three linearly independent expected frequencies m_{ij}. Since there are also three parameters, the model is saturated and the expected counts m_{ij} are equal to the observed counts n_{ij}. Thus, the likelihood ratio statistic G^2 is equal to zero. Although the model was fit using maximum likelihood, the test statistics in the analysis of variance table are Wald tests.

Output 16.2   Analysis of Variance Table for Saturated Loglinear Model

```
           Maximum Likelihood Analysis of Variance

    Source            DF    Chi-Square    Pr > ChiSq
    ------------------------------------------------------
    type              1       11.29         0.0008
    helmet            1        3.28         0.0700
    type*helmet       1        4.33         0.0374

    Likelihood Ratio  0         .             .
```

The next PROC CATMOD invocation fits the independence loglinear model

$$\log(m_{ij}) = \mu + \lambda_i^X + \lambda_j^Y \qquad i, j = 1, 2$$

which is specified by excluding the TYPE × HELMET term from the LOGLIN statement.

```
proc catmod;
   weight count;
   model type*helmet=_response_ /  noprofile noresponse noiter noparm;
   loglin type helmet;
run;
```

As shown in Output 16.3, the likelihood ratio statistic for testing the null hypothesis of independence of HELMET and TYPE is $G^2 = 4.56$. Therefore, there is clear evidence that the two variables are not independent. Helmet usage is more associated with mountain bikes than other bikes. The main effect TYPE tests the null hypothesis that the subjects are distributed evenly over the two levels of this variable. The strongly significant results of this test ($Q_W = 9.87$, 1 df, $p = 0.0017$) reflect the fact that 66% of the cyclists were riding mountain bikes and only 34% were riding other types of bicycles (Table 16.4). The subjects were relatively evenly distributed over the levels of the HELMET variable (44% wore helmets, 56% did not); this is reflected in the nonsignificant HELMET main effect ($Q_W = 1.43$). However, since there is evidence of interaction between HELMET and TYPE, the main effects should be interpreted with caution.

Output 16.3   Analysis of Variance Table for Independence Loglinear Model

```
           Maximum Likelihood Analysis of Variance

    Source            DF    Chi-Square    Pr > ChiSq
    ------------------------------------------------------
    type              1        9.87         0.0017
    helmet            1        1.43         0.2313

    Likelihood Ratio  1        4.56         0.0328
```

For comparison, the FREQ procedure can also be used to compute the likelihood ratio test of independence. The statements

```
proc freq order=data;
   weight count;
   tables type*helmet / nopercent norow chisq;
run;
```

produce the results shown in Output 16.4. The statistic G^2 is labeled "Likelihood Ratio Chi-Square."

Output 16.4 Likelihood Ratio Test of Independence Using PROC FREQ

```
                    Table of type by helmet

            type        helmet

            Frequency|
            Col Pct  |Yes     |No      |  Total
            ---------+--------+--------+
            Mountain |    34  |    32  |    66
                     | 77.27  | 57.14  |
            ---------+--------+--------+
            Other    |    10  |    24  |    34
                     | 22.73  | 42.86  |
            ---------+--------+--------+
            Total         44       56      100


               Statistics for Table of type by helmet

        Statistic                     DF       Value      Prob
        ------------------------------------------------------
        Chi-Square                     1       4.4494     0.0349
        Likelihood Ratio Chi-Square    1       4.5569     0.0328
        Continuity Adj. Chi-Square     1       3.5975     0.0579
        Mantel-Haenszel Chi-Square     1       4.4049     0.0358
        Phi Coefficient                        0.2109
        Contingency Coefficient                0.2064
        Cramer's V                             0.2109


                      Fisher's Exact Test
            ----------------------------------------
            Cell (1,1) Frequency (F)          34
            Left-sided Pr <= F             0.9906
            Right-sided Pr >= F            0.0280

            Table Probability (P)          0.0186
            Two-sided Pr <= P              0.0549

                   Sample Size = 100
```

16.2.3 Loglinear Model for the $s \times r$ Table

When a sample of n observations is classified with respect to two categorical variables, one having s levels and the other having r levels, the resulting frequencies can be displayed in an $s \times r$ contingency table, as shown in Table 16.5. The corresponding cell probabilities are π_{ij}, with row and column marginal probabilities $\{\pi_{i+}\}$ and $\{\pi_{+j}\}$, respectively.

Table 16.5. Cell Counts in an $s \times r$ Contingency Table

Level of X	Level of Y 1	2	$\cdots$	r	Total
1	n_{11}	n_{12}	$\cdots$	n_{1r}	n_{1+}
2	n_{21}	n_{22}	$\cdots$	n_{2r}	n_{2+}
$\vdots$	$\vdots$	$\vdots$		$\vdots$	$\vdots$
s	n_{s1}	n_{s2}	$\cdots$	n_{sr}	n_{s+}
Total	n_{+1}	n_{+2}	$\cdots$	n_{+r}	n

The generalization of the loglinear model from the 2×2 table to the $s \times r$ table is straightforward. The saturated model is

$$\log(m_{ij}) = \mu + \lambda_i^X + \lambda_j^Y + \lambda_{ij}^{XY} \qquad i = 1, \ldots, s, j = 1, \ldots, r$$

where $m_{ij} = n\pi_{ij}$ is the expected frequency in the (i, j) cell. The parameter μ is fixed by the sample size n and the model has $s + r + sr$ parameters λ_i^X, λ_j^Y, and λ_{ij}^{XY}. The sum-to-zero constraints

$$\sum_{i=1}^{s} \lambda_i^X = 0 \qquad \sum_{j=1}^{r} \lambda_j^Y = 0 \qquad \sum_{i=1}^{s} \lambda_{ij}^{XY} = \sum_{j=1}^{r} \lambda_{ij}^{XY} = 0$$

implies $(s-1) + (r-1) + (s-1)(r-1) = sr - 1$ parameters and zero df for testing lack of fit. Letting $\widehat{m}_{ij} = n_{i+}n_{+j}/n$, the likelihood ratio statistic

$$G^2 = 2 \sum_{i=1}^{s} \sum_{j=1}^{r} n_{ij} \log\left(n_{ij}/\widehat{m}_{ij}\right)$$

tests the null hypothesis $H_0: \lambda_{ij}^{XY} = 0$, for $i = 1, \ldots, s-1, j = 1, \ldots, r-1$. Under the null hypothesis of independence, G^2 has an approximate chi-square distribution with $(s-1)(r-1)$ df.

If H_0 is true, the reduced model $\log(m_{ij}) = \mu + \lambda_i^X + \lambda_j^Y$ is the model of independence of X and Y. This model has $(s-1) + (r-1)$ linearly independent λ parameters and $(s-1)(r-1)$ df for testing lack of fit.

16.2.4 Malignant Melanoma Example

Table 16.6 displays data from a cross-sectional study of 400 patients with malignant melanoma (Roberts et al. 1981). For each patient, the site of the tumor and its histological type were recorded. The following statements create a SAS data set containing the cell frequencies for this 4×3 contingency table.

Table 16.6. Malignant Melanoma Data

Tumor Type	Tumor Site Head and Neck	Trunk	Extremities	Total
Hutchinson's melanotic freckle	22	2	10	34
Superficial spreading melanoma	16	54	115	185
Nodular	19	33	73	125
Indeterminate	11	17	28	56
Total	68	106	226	400

```
data melanoma;
   input type $ site $ count;
   datalines;
Hutchinson's  Head&Neck    22
Hutchinson's  Trunk         2
Hutchinson's  Extremities  10
Superficial   Head&Neck    16
Superficial   Trunk        54
Superficial   Extremities 115
Nodular       Head&Neck    19
Nodular       Trunk        33
Nodular       Extremities  73
Indeterminate Head&Neck    11
Indeterminate Trunk        17
Indeterminate Extremities  28
;
run;
```

The following PROC CATMOD invocation fits the independence loglinear model

$$\log(m_{ij}) = \mu + \lambda_i^X + \lambda_j^Y \qquad i = 1, \ldots, 4;, j = 1, \ldots, 3$$

```
proc catmod;
   weight count;
   model type*site=_response_ / noresponse noiter noparm;
   loglin type site;
run; quit;
```

The analysis of variance table in Output 16.5 provides strong evidence that tumor type and tumor site are not independent ($G^2 = 51.80$, 6 df, $p < 0.0001$). Hutchinson's tumor type is more associated with head and neck, and other types are more associated with extremities.

Output 16.5   Analysis of Variance Table for Independence Loglinear Model

```
              Maximum Likelihood Analysis of Variance

          Source            DF    Chi-Square    Pr > ChiSq
          ----------------------------------------------------
          type               3       121.48        <.0001
          site               2        93.30        <.0001

          Likelihood Ratio   6        51.80        <.0001
```

Fitting the Loglinear Model with PROC GENMOD

To perform loglinear modeling with PROC GENMOD, you actually fit a Poisson regression model. Proportionality of the respective likelihoods implies that the maximum likelihood estimates for the parameters in the Poisson regression model are identical to the corresponding maximum likelihood estimates for the parameters in the loglinear model. See Appendix A in this chapter for more detail concerning this comparability.

The following PROC GENMOD invocation fits the saturated loglinear model:

```
ods select Type3;
proc genmod;
   class type site;
   model count=type|site / link=log dist=poisson type3;
run;
```

You specify the LINK=LOG option and the DIST=POISSON option in the MODEL statement. The ODS SELECT statement restricts the output produced to the table of likelihood ratio statistics. The table of likelihood ratio statistics in Output 16.6 includes $G^2 = 51.80$ (6 df, $p < 0.0001$), which is the same statistic displayed in Output 16.5 as the likelihood ratio test. However, the tests for the single effects are different because PROC GENMOD is producing likelihood ratio tests and PROC CATMOD is producing Wald statistics.

Output 16.6   Likelihood Ratio Statistics from Saturated Loglinear Model

```
              LR Statistics For Type 3 Analysis

                                Chi-
        Source          DF     Square     Pr > ChiSq

        type            3      85.07        <.0001
        site            2      33.34        <.0001
        type*site       6      51.80        <.0001
```

As mentioned in the "Introduction," the GENMOD procedure is a good way to evaluate the association in a loglinear model and to evaluate higher order effects in the process of determining a set of lower order effects that describe the association adequately. However, when it comes to the interpretation of lower order effects in the presence of higher order effects, the parameterization of the CATMOD procedure has advantages. Since the usual constraints of the loglinear model that effects add to zero maps into the deviation-from-the-mean parameterization of the CATMOD procedure, these lower order effects can usually be interpreted as effects averaged over the levels of a variable with which it has an interaction. For the reference cell parameterization that PROC GENMOD uses, such effects are nested within the reference levels of the interacting variables, a much more complicated scenario.

16.2.5   Hierarchical and Nonhierarchical Loglinear Models

Hierarchical loglinear models are defined to be members of the family of models such that if any λ-term is set equal to zero, all effects at the same or higher order with the subscripted λ-terms contained in them are also set equal to zero (Bishop, Fienberg, and Holland 1975, p. 34). Thus, whenever a model contains higher-order effects, it also must contain the corresponding lower-order effects. For two-way tables, the saturated model

$$\log(m_{ij}) = \mu + \lambda_i^X + \lambda_j^Y + \lambda_{ij}^{XY} \qquad i = 1, \ldots, s, j = 1, \ldots, r$$

and the independence model

$$\log(m_{ij}) = \mu + \lambda_i^X + \lambda_j^Y \qquad i = 1, \ldots, s, j = 1, \ldots, r$$

are the only hierarchical loglinear models that involve both variables. The other possible hierarchical models are

$$\log(m_{ij}) = \mu + \lambda_i^X$$
$$\log(m_{ij}) = \mu + \lambda_j^Y$$
$$\log(m_{ij}) = \mu$$

An example of a nonhierarchical model would be

$$\log(m_{ij}) = \mu + \lambda_i^X + \lambda_{ij}^{XY}$$

This model is nonhierarchical since it contains the higher-order term λ_{ij}^{XY} but not the lower-order effect λ_j^Y.

For two-way tables, closed-form estimates of the cell frequencies can be obtained for hierarchical models: $\widehat{m}_{ij} = n_{ij}$ for the saturated model and $\widehat{m}_{ij} = n_{i+}n_{+j}/n$ for the independence model. In multiway tables, explicit estimates are not usually available. Historically, the restriction to consideration of hierarchical loglinear models was at least partially due to the fact that the more readily accessible methods of obtaining MLEs of the cell frequencies were primarily applicable to hierarchical models.

16.3   Three-Way Contingency Tables

16.3.1   Mutual, Joint, Marginal, and Conditional Independence

Consider a three-dimensional table containing the cross-classification of variables X, Y, and Z. The distributions of X, Y cell counts at different levels of Z can be displayed using cross-sections of the three-way table. These cross-sections are called *partial tables*. In the partial tables, the value of Z is held constant.

For example, Section 3.2.2 of Chapter 3 discusses health policy opinion data with variables X=stress, Y=opinion, and Z=residence. Table 16.7 displays the two partial tables of the stress × opinion cross-classification for subjects from urban and rural residences.

Table 16.7.   Partial Tables for Health Policy Opinion Data

Residence	Stress	Opinion Favorable	Opinion Unfavorable	Total
Urban	Low	48	12	60
	High	96	94	190
	Total	144	106	250
Rural	Low	55	135	190
	High	7	53	60
	Total	62	188	250

Alternatively, the two-way contingency table obtained by adding the cell counts in the partial tables is called the X, Y *marginal table*. This table ignores the variable Z.

Table 16.8 displays the marginal stress × opinion cross-classification ignoring the variable residence.

Table 16.8. Marginal Table for Health Policy Opinion Data

Stress	Opinion Favorable	Unfavorable	Total
Low	103	147	250
High	103	147	250
Total	206	294	500

Partial tables can exhibit quite different associations than marginal tables, as was described in Section 3.2.2 for the health policy opinion data. In fact, it can be quite misleading to analyze only the marginal tables of a multiway contingency table. Simpson's Paradox, the result that a pair of variables can have marginal association of different strength from their partial associations, is discussed in Section 3.2.2.

Before describing some of the various types of loglinear models, it is important to consider the four types of independence for cell probabilities in the three-way cross-classification of variables X, Y, and Z. Denote the cell probabilities by π_{ijk}, for $i = 1, \ldots, I$, $j = 1, \ldots, J$, and $k = 1, \ldots, K$, where I, J, and K denote the number of levels of variables X, Y, and Z, respectively.

The three variables are *mutually independent* when

$$\pi_{ijk} = \pi_{i++} \, \pi_{+j+} \, \pi_{++k} \qquad i = 1, \ldots, I, j = 1, \ldots, J, k = 1, \ldots, K$$

Variable Y is *jointly independent* of X and Z when

$$\pi_{ijk} = \pi_{i+k} \, \pi_{+j+} \qquad i = 1, \ldots, I, j = 1, \ldots, J, k = 1, \ldots, K$$

This is ordinary two-way independence between Y and a new variable composed of the IK combinations of the levels of X and Z. Similar definitions apply for X to be jointly independent of Y and Z, and for Z to be jointly independent of X and Y. Note that mutual independence implies joint independence of any one variable from the others.

Variables X and Y are *marginally independent* if

$$\pi_{ij+} = \pi_{i++} \, \pi_{+j+} \qquad i = 1, \ldots, I, \; j = 1, \ldots, J$$

In general, two variables are marginally independent if they are independent in the two-way table obtained by collapsing over the levels of the remaining variables. If Y is jointly independent of X and Z, then X and Y, as well as Y and Z, are marginally independent. Thus, joint independence implies marginal independence.

Next consider the relationship between any pair of variables, controlling for the levels of the third variable. For example, if X and Y are independent in the partial table for the kth category of Z, then X and Y are said to be *conditionally independent at level k of Z*. Suppose

$$\pi_{ij|k} = \pi_{ijk}/\pi_{++k} \quad i = 1, \ldots, I, j = 1, \ldots, J$$

denotes the joint distribution of X and Y at level k of Z. Then conditional independence at level k of Z is

$$\pi_{ijk} = \pi_{i+|k}\,\pi_{+j|k} \qquad i = 1, \ldots, I, j = 1, \ldots, J$$

More generally, the variables X and Y are *conditionally independent given Z* when they are conditionally independent at every level of Z, or when

$$\pi_{ijk} = \pi_{i+k}\,\pi_{+jk}/\pi_{++k} \qquad i = 1, \ldots, I, j = 1, \ldots, J, k = 1, \ldots, K$$

Suppose that Y is jointly independent of X and Z. Then X and Y are conditionally independent, as are Y and Z.

In summary, two variables (say X and Y) are conditionally independent and marginally independent when X, Y, and Z are mutually independent, or when Y is jointly independent of X and Z. However, conditional independence of X and Y, given Z, does not imply that X and Y are marginally independent.

16.3.2   Hierarchical Loglinear Models

The saturated loglinear model for a three-way table is

$$\log(m_{ijk}) = \mu + \lambda_i^X + \lambda_j^Y + \lambda_k^Z + \lambda_{ij}^{XY} + \lambda_{ik}^{XZ} + \lambda_{jk}^{YZ} + \lambda_{ijk}^{XYZ}$$

This model has

$$1 + (I-1) + (J-1) + (K-1) + (I-1)(J-1) + (I-1)(K-1)$$
$$+ (J-1)(K-1) + (I-1)(J-1)(K-1) = IJK$$

parameters and zero df for testing lack of fit. The saturated model allows for three-way interaction, that is, each pair of variables may be conditionally dependent, and an odds ratio for any pair of variables may vary across levels of the third variable.

The reduced model

$$\log(m_{ijk}) = \mu + \lambda_i^X + \lambda_j^Y + \lambda_k^Z + \lambda_{ij}^{XY} + \lambda_{ik}^{XZ} + \lambda_{jk}^{YZ}$$

is called the loglinear model of *no three-factor interaction*. In this model, no pair of variables is conditionally independent. Thus, for each pair of variables, marginal odds ratios may differ from partial odds ratios. The "no three-factor interaction" model implies that the conditional odds ratios between any two variables are identical at each level of the third variable. Except in special cases, closed form expressions for the expected cell frequencies do not exist.

There are three hierarchical models in which only one pair of variables is conditionally independent. For example, if X and Y are conditionally independent, given Z, the corresponding loglinear model is

$$\log(m_{ijk}) = \mu + \lambda_i^X + \lambda_j^Y + \lambda_k^Z + \lambda_{ik}^{XZ} + \lambda_{jk}^{YZ}$$

The parameters $\{\lambda_{ik}^{XZ}\}$ and $\{\lambda_{jk}^{YZ}\}$ pertain to the X, Z and Y, Z partial associations. There are also three models in which only one pair of variables is conditionally dependent. For example, if Y is jointly independent of X and Z, the corresponding model is

$$\log(m_{ijk}) = \mu + \lambda_i^X + \lambda_j^Y + \lambda_k^Z + \lambda_{ik}^{XZ}$$

In this model, the parameters $\{\lambda_{ik}^{XZ}\}$ pertain to the dependence between X and Z.

Finally, the loglinear model corresponding to mutual independence is

$$\log(m_{ijk}) = \mu + \lambda_i^X + \lambda_j^Y + \lambda_k^Z$$

In this model, each pair of variables is also conditionally and marginally independent.

16.3.3 Fitting Loglinear Models

After selecting a loglinear model, the observed data are used to estimate model parameters, cell probabilities, and expected frequencies. Although alternative methods of estimation are sometimes useful, the maximum likelihood (ML) method offers several advantages. First of all, the MLEs for hierarchical loglinear models are relatively easy to compute, since the estimates satisfy certain intuitive marginal constraints. In addition, the ML method can be used when data are sparse, that is, when there are several observed cell counts of zero. (Note that marginal totals, however, cannot be equal to zero.) Although beyond the scope of this book, the ML method also has some theoretical advantages over other approaches (Rao 1961, 1962).

Birch (1963) showed that the MLEs are the same for simple multinomial sampling, independent Poisson sampling, and product multinomial sampling. For hierarchical loglinear models, Birch's (1963) results also enable the derivation of estimates of the expected cell counts without first going through the intermediate step of estimating the λ-terms. For some models, the cell estimates are explicit closed-form functions of the marginal totals. For example, the expected cell frequencies for the independence loglinear model in a two-way table are functions of the row and column marginal totals; specifically $\widehat{m}_{ij} = n_{i+}n_{+j}/n$ (Section 16.2.3). However, many loglinear models do not have direct ML estimates. As one example, direct estimates do not exist for unsaturated models containing all two-factor interactions. When direct estimates do not exist, iterative procedures must be used.

The iterative proportional fitting (IPF) algorithm, originally presented by Deming and Stephan (1940), is a simple method for calculating MLEs of cell frequencies for hierarchical loglinear models. Since the estimated cell counts depend only on the marginal totals, no special provision need be made for sporadic cells with no observations. Any set of starting values may be chosen that conforms to the model being fit; for example, all expected cell counts can initially be set equal to one. If direct estimates exist, the procedure yields these estimates in one cycle. IPF is used by many computer programs, since it is a simple method not requiring matrix inversion or complicated calculations.

The Newton-Raphson method can also be used to fit loglinear models. This method is more complex, since each step requires solving a system of equations. When the contingency table has several dimensions and the parameter vector is large, the Newton-Raphson method may not be feasible. However, since Newton-Raphson is a general purpose method that can solve more complex systems of likelihood equations, restriction to the class of hierarchical loglinear models is not necessary. In addition, Newton-Raphson is more efficient numerically, since the rate of convergence is quadratic (compared to linear for IPF). Of course, this is partially counterbalanced by the fact that each cycle takes less time with IPF. Another advantage of the Newton-Raphson method is that the estimated covariance matrix of the parameter estimates is automatically produced as a by-product. The CATMOD procedure uses the Newton-Raphson method to fit loglinear models.

16.3.4 Testing Goodness of Fit

The goodness of fit of a loglinear model can be assessed by comparing the fitted cell counts to the observed cell counts. The general form of the likelihood ratio chi-square statistic is $G^2 = 2 \sum n \log(n/\widehat{m})$, where n and $\widehat{m}$ denote the observed and fitted cell frequencies. The corresponding Pearson chi-square statistic is equal to $Q_P = \sum (n - \widehat{m})^2/\widehat{m}$. When the model holds, both statistics have asymptotic chi-square distributions with degrees of freedom equal to the number of cells in the table minus the number of linearly independent parameters.

The likelihood ratio statistic G^2 has two important properties not possessed by Q_P. First, it is the statistic that is minimized by the MLEs. In addition, suppose you want to compare two models M_1 and M_2, where M_2 is a special case of M_1. In terms of loglinear model parameters, M_2 contains only a subset of the λ-terms contained in M_1. In this case, the simpler model M_2 is said to be nested within M_1.

Suppose $G^2(M_1)$ and $G^2(M_2)$ denote the goodness-of-fit statistics for models M_1 and M_2, and suppose v_1 and v_2 denote the corresponding df. Since M_2 is simpler than M_1, $v_1 < v_2$ and $G^2(M_1) \leq G^2(M_2)$. Assuming model M_1 holds, the likelihood ratio approach for testing that M_2 holds uses the statistic

$$G^2(M_2 \mid M_1) = G^2(M_2) - G^2(M_1)$$

which has an asymptotic chi-square distribution with $(v_2 - v_1)$ df when model M_2 holds. A comparable decomposition for the Pearson chi-square statistic Q_P does not correspondingly apply.

16.3.5 Job Satisfaction Example

Table 16.9 displays the three-way cross-classification of quality of management, supervisor's job satisfaction, and worker's job satisfaction for a random sample of 715 workers selected from Danish industry (Andersen 1991, p. 155).* Quality of management was categorized from an external evaluation of each factory, while the job satisfaction ratings were based on questionnaires completed by each worker and his or her supervisor. Since all three variables are response variables, the use of loglinear models to investigate the patterns of association among management quality, supervisor's job satisfaction, and worker's job satisfaction seems appropriate.

---

*Reprinted by permission of Springer-Verlag.

Table 16.9. Job Satisfaction Data

Quality of Management	Supervisor's Job Satisfaction	Worker's Job Satisfaction		Total
		Low	High	
Bad	Low	103	87	190
	High	32	42	74
Good	Low	59	109	168
	High	78	205	283

Let X, Y, and Z denote quality of management, supervisor's job satisfaction, and worker's job satisfaction, respectively, and let π_{ijk} denote the corresponding multinomial cell probabilities for $i = 1, 2$, $j = 1, 2$, and $k = 1, 2$. The following statements read in the cell counts and fit the saturated loglinear model

$$\log(m_{ijk}) = \mu + \lambda_i^X + \lambda_j^Y + \lambda_k^Z + \lambda_{ij}^{XY} + \lambda_{ik}^{XZ} + \lambda_{jk}^{YZ} + \lambda_{ijk}^{XYZ}$$

which is expressed using the vertical bar (|) notation in the LOGLIN statement.

```
data satisfac;
   input managmnt $ supervis $ worker $ count;
   datalines;
Bad  Low  Low  103
Bad  Low  High  87
Bad  High Low   32
Bad  High High  42
Good Low  Low   59
Good Low  High 109
Good High Low   78
Good High High 205
;
proc catmod order=data;
   weight count;
   model managmnt*supervis*worker=_response_
         / noresponse noiter noparm;
   loglin managmnt|supervis|worker;
run;
```

Output 16.7 displays the population and response profiles. There is a single multinomial sample with eight categories of response. Since the model is saturated, the likelihood ratio test of fit is equal to zero (see Output 16.8). The Wald test of the three-factor interaction is nonsignificant ($Q_W = 0.06$, 1 df, $p = 0.7989$).

Output 16.7   Population and Response Profiles

```
                    Population Profiles

                Sample      Sample Size
                --------------------
                   1             715



                  Response Profiles

      Response    managmnt    supervis    worker
      ------------------------------------------
         1         Bad         Low         Low
         2         Bad         Low         High
         3         Bad         High        Low
         4         Bad         High        High
         5         Good        Low         Low
         6         Good        Low         High
         7         Good        High        Low
         8         Good        High        High
```

Output 16.8   Analysis of Variance Table for Saturated Model

```
              Maximum Likelihood Analysis of Variance

      Source                    DF    Chi-Square    Pr > ChiSq
      ----------------------------------------------------------
      managmnt                   1       38.30         <.0001
      supervis                   1        8.10         0.0044
      managmnt*supervis          1       65.67         <.0001
      worker                     1       23.59         <.0001
      managmnt*worker            1       18.17         <.0001
      supervis*worker            1        5.24         0.0221
      managmnt*supervis*worker   1        0.06         0.7989

      Likelihood Ratio           0         .             .
```

The second model includes only the main effects and two-factor interactions.

```
proc catmod order=data;
   weight count;
   model managmnt*supervis*worker=_response_
         / noprofile noresponse noiter p=freq;
   loglin managmnt|supervis managmnt|worker supervis|worker;
run;
```

The likelihood ratio test in the analysis of variance table (Output 16.9) compares this model to the saturated model and thus tests the null hypothesis of no three-factor interaction. In this example, the G^2 statistic of 0.06 is the same as the Wald statistic from the saturated model. Although the two statistics are asymptotically equivalent, they are not identical in general.

Output 16.9 Analysis of Variance Table for Model with No Three-Factor Interaction

```
              Maximum Likelihood Analysis of Variance

      Source              DF   Chi-Square   Pr > ChiSq
      ---------------------------------------------------
      managmnt             1      38.37       <.0001
      supervis             1       8.32       0.0039
      managmnt*supervis    1      67.06       <.0001
      worker               1      25.96       <.0001
      managmnt*worker      1      19.57       <.0001
      supervis*worker      1       5.33       0.0210

      Likelihood Ratio     1       0.06       0.7989
```

The Wald tests of the two-factor interactions and main effects are all significant. This indicates that a more parsimonious model for the data may not be justified. However, you may wish to fit each of the three models containing only two of the two-factor interactions and compare these models to the model with no three-factor interaction using likelihood ratio tests. The SAS statements are as follows:

```
proc catmod order=data;
   weight count;
   model managmnt*supervis*worker=_response_
         / noprofile noresponse noiter noparm;
   loglin managmnt|supervis managmnt|worker;
proc catmod order=data;
   weight count;
   model managmnt*supervis*worker=_response_
         / noprofile noresponse noiter noparm;
   loglin managmnt|supervis supervis|worker;
proc catmod order=data;
   weight count;
   model managmnt*supervis*worker=_response_
         / noprofile noresponse noiter noparm;
   loglin managmnt|worker supervis|worker;
run;
```

The corresponding likelihood ratio statistics for goodness of fit (output not shown) are $G^2 = 5.39$, 19.71, and 71.90, all with 2 df. The 1 df likelihood ratio statistics comparing each of these three models to the model with no three-factor interaction are $5.39 - 0.06 = 5.33$, $19.71 - 0.06 = 19.65$, and $71.90 - 0.06 = 71.84$, respectively. Relative to the chi-square distribution with 1 df, all indicate a significant lack of fit.

The model with no three-factor interaction provides a good fit to the observed data. Thus, no pair of variables is conditionally independent. In this model, the conditional odds ratios between any two variables are identical at each level of the third variable. For example, the odds ratio for the association between the employee's job satisfaction and the supervisor's job satisfaction is the same at each level of management quality. You can compute the estimated odds ratios from the table of maximum likelihood estimates (Output 16.10).

Output 16.10   Parameter Estimates from Model with No Three-Factor Interaction

```
                    Analysis of Maximum Likelihood Estimates

                                          Standard      Chi-
Effect              Parameter   Estimate    Error     Square    Pr > ChiSq
-------------------------------------------------------------------------
managmnt                1       -0.2672     0.0431     38.37      <.0001
supervis                2        0.1243     0.0431      8.32       0.0039
managmnt*supervis       3        0.3491     0.0426     67.06      <.0001
worker                  4       -0.2065     0.0405     25.96      <.0001
managmnt*worker         5        0.1870     0.0423     19.57      <.0001
supervis*worker         6        0.0962     0.0417      5.33       0.0210
```

From the model with no three-factor interaction, the log odds of low job satisfaction for employees, at fixed levels of management quality and supervisor's job satisfaction, is

$$
\begin{aligned}
\log(m_{ij1}/m_{ij2}) &= \log(m_{ij1}) - \log(m_{ij2}) \\
&= \lambda_1^Z + \lambda_{i1}^{XZ} + \lambda_{j1}^{YZ} - (\lambda_2^Z + \lambda_{i2}^{XZ} + \lambda_{j2}^{YZ}) \\
&= 2\lambda_1^Z + 2\lambda_{i1}^{XZ} + 2\lambda_{j1}^{YZ}
\end{aligned}
$$

since $\lambda_1^Z + \lambda_2^Z = 0$, $\lambda_{i1}^{XZ} + \lambda_{i2}^{XZ} = 0$, and $\lambda_{j1}^{YZ} + \lambda_{j2}^{YZ} = 0$. Thus, at a fixed level of management quality, the logarithm of the odds ratio at low and high levels of supervisor satisfaction is

$$
\begin{aligned}
\log(m_{i11}/m_{i12}) - \log(m_{i21}/m_{i22}) &= (2\lambda_1^Z + 2\lambda_{i1}^{XZ} + 2\lambda_{11}^{YZ}) - (2\lambda_1^Z + 2\lambda_{i1}^{XZ} + 2\lambda_{21}^{YZ}) \\
&= 2\lambda_{11}^{YZ} - 2\lambda_{21}^{YZ} \\
&= 4\lambda_{11}^{YZ}
\end{aligned}
$$

Since the estimate of λ_{11}^{YZ} from Output 16.10 is 0.0962, the odds of low worker job satisfaction are estimated to be $\exp(4 \times 0.0962) = 1.47$ times higher when the supervisor's job satisfaction is low than when the supervisor's job satisfaction is high. Note that this estimate of the odds ratio is the same for factories with bad and good management quality. Using the observed counts from Table 16.9, the observed odds ratios are

$$
\frac{103 \times 42}{87 \times 32} = 1.55
$$

in factories where the external evaluation of management quality was bad and

$$
\frac{59 \times 205}{109 \times 78} = 1.42
$$

in factories where the quality of management was good.

You can estimate additional odds ratios using the parameter estimates listed in Output 16.10. For a fixed level of supervisor job satisfaction, the odds of low worker satisfaction are estimated to be $\exp(4 \times 0.1870) = 2.1$ times higher when the quality of management is bad than when the management quality is good. This value is in between the corresponding observed odds ratios of

$$\frac{103 \times 109}{87 \times 59} = 2.19$$

when supervisor job satisfaction is low and

$$\frac{32 \times 205}{42 \times 78} = 2.00$$

when supervisor job satisfaction is high. Similarly, for a fixed level of worker job satisfaction, the odds of low supervisor job satisfaction are estimated to be $\exp(4 \times 0.3491) = 4.0$ times higher when the quality of management is bad than when the management quality is good. This value is in between the corresponding observed odds ratios of

$$\frac{103 \times 78}{32 \times 59} = 4.26$$

when worker job satisfaction is low and

$$\frac{87 \times 205}{42 \times 109} = 3.90$$

when worker job satisfaction is high.

These results show that the odds of low worker job satisfaction are somewhat more affected by the quality of management than by the supervisor's job satisfaction. In addition, bad quality management has a greater effect on the job satisfaction of supervisors than on worker job satisfaction.

The P=FREQ option of the MODEL statement prints predicted cell frequencies. Output 16.11 displays the resulting output from the model with no three-factor interaction. The first seven rows are the observed and predicted response functions, given by $\log(m_{ijk}) - \log(m_{222})$. The next eight rows, labeled F1–F8, are the observed and predicted cell counts. Instead of using the parameter estimates from Output 16.10, you could compute the estimated odds ratios using the predicted cell frequencies.

Output 16.11   Predicted Cell Counts

```
          Maximum Likelihood Predicted Values for Response Functions

                   ------Observed------      ------Predicted-----
         Function              Standard                 Standard
         Number    Function      Error     Function       Error    Residual
         -----------------------------------------------------------------------
            1      -0.68828    0.120776    -0.69904     0.113356    0.010757
            2      -0.8571     0.127954    -0.85226     0.1261     -0.00484
            3      -1.85727    0.190074    -1.83812     0.172903   -0.01916
            4      -1.58534    0.169374    -1.60661     0.148406    0.021266
            5      -1.24547    0.14774     -1.23666     0.142911   -0.00881
            6      -0.63166    0.118543    -0.64202     0.111524    0.01036
            7      -0.9663     0.133036    -0.97937     0.12311     0.013066




            Maximum Likelihood Predicted Values for Frequencies

                            ------Observed-----   -----Predicted-----
                                      Standard              Standard
managmnt  supervis  worker  Frequency    Error   Frequency    Error   Residual
-------------------------------------------------------------------------------
Bad       Low       Low        103     9.389475  102.2639   8.904231   0.736105
Bad       Low       High        87     8.741509   87.73611  8.283433  -0.73611
Bad       High      Low         32     5.528818   32.73611  4.783695  -0.73611
Bad       High      High        42     6.287517   41.26389  5.525364   0.736105
Good      Low       Low         59     7.357409   59.73611  6.811457  -0.73611
Good      Low       High       109     9.611619  108.2639   9.138839   0.736105
Good      High      Low         78     8.336121   77.26389  7.782186   0.736105
Good      High      High       205    12.0923    205.7361  11.75535   -0.73611
```

16.4   Higher-Order Contingency Tables

16.4.1   Dyke-Patterson Cancer Knowledge Data

As the number of dimensions of a contingency table increases, there are some complicating factors. One difficulty is the tremendous increase in the number of possible interaction parameters. Another problem is caused by the dramatic increase in the number of cells. Unless the sample size is very large, there may be many observed cell counts equal to zero. There may even be marginal totals equal to zero.

Table 16.10 displays data obtained from a sample of 1729 individuals cross-classified according to five dichotomous variables (Dyke and Patterson 1952). The purpose of the study was to investigate the relationship between cancer knowledge (good, poor) and four media exposure variables.

- Do you read newspapers?
- Do you listen to the radio?
- Do you read books and magazines? (solid reading)
- Do you attend lectures?

Table 16.10. Cancer Knowledge Data

Read Newspapers	Listen to Radio	Solid Reading	Attend Lectures	Cancer Knowledge Good	Cancer Knowledge Poor
Yes	Yes	Yes	Yes	23	8
Yes	Yes	Yes	No	102	67
Yes	Yes	No	Yes	8	4
Yes	Yes	No	No	35	59
Yes	No	Yes	Yes	27	18
Yes	No	Yes	No	201	177
Yes	No	No	Yes	7	6
Yes	No	No	No	75	156
No	Yes	Yes	Yes	1	3
No	Yes	Yes	No	16	16
No	Yes	No	Yes	4	3
No	Yes	No	No	13	50
No	No	Yes	Yes	3	8
No	No	Yes	No	67	83
No	No	No	Yes	2	10
No	No	No	No	84	393

Since this was a cross-sectional study, it is reasonable to treat all five variables as response variables and to investigate the patterns of dependence using loglinear models.

16.4.2 Hierarchical Loglinear Models

There are a large number of possible hierarchical models that can be considered for the cancer knowledge data of Table 16.10. The possible terms to be included in a model are the intercept μ, 5 main effects, 10 two-factor interaction terms, 10 three-factor interactions, 5 four-factor interactions, and the five-factor interaction.

The following statements read in the observed cell frequencies and fit the loglinear model of no five-factor interaction.

```
data cancer;
    input news $ radio $ reading $ lectures $ knowledg $ count;
    datalines;
Yes Yes Yes Yes  Good  23
Yes Yes Yes Yes  Poor   8
Yes Yes Yes No   Good 102
Yes Yes Yes No   Poor  67
Yes Yes No  Yes  Good   8
Yes Yes No  Yes  Poor   4
Yes Yes No  No   Good  35
Yes Yes No  No   Poor  59
Yes No  Yes Yes  Good  27
Yes No  Yes Yes  Poor  18
Yes No  Yes No   Good 201
Yes No  Yes No   Poor 177
Yes No  No  Yes  Good   7
```

```
Yes No   No   Yes  Poor    6
Yes No   No   No   Good   75
Yes No   No   No   Poor  156
No  Yes  Yes  Yes  Good    1
No  Yes  Yes  Yes  Poor    3
No  Yes  Yes  No   Good   16
No  Yes  Yes  No   Poor   16
No  Yes  No   Yes  Good    4
No  Yes  No   Yes  Poor    3
No  Yes  No   No   Good   13
No  Yes  No   No   Poor   50
No  No   Yes  Yes  Good    3
No  No   Yes  Yes  Poor    8
No  No   Yes  No   Good   67
No  No   Yes  No   Poor   83
No  No   No   Yes  Good    2
No  No   No   Yes  Poor   10
No  No   No   No   Good   84
No  No   No   No   Poor  393
;
proc catmod order=data;
   weight count;
   model news*radio*reading*lectures*knowledg=_response_
         / noresponse noiter noparm;
      loglin news|radio|reading|lectures       news|radio|reading|knowledg
             news|radio|lectures|knowledg      news|reading|lectures|knowledg
             radio|reading|lectures|knowledg;
run;
```

Although the model could be specified by listing the $5 + 10 + 10 + 5 = 30$ main effect and interaction terms, it is simpler to use the vertical bar notation to specify the five four-factor interactions.

Output 16.12 displays the population and response profiles. There is a single multinomial population and $2^5 = 32$ response profiles. The likelihood ratio goodness-of-fit statistic from this model is $G^2 = 1.02$ (see Output 16.13). This statistic, with 1 df, tests the null hypothesis of no five-factor interaction and is not significant ($p = 0.3116$). Note that it is not necessary to fit the saturated model first in order to test this hypothesis.

Output 16.12   Population and Response Profiles

```
                    Population Profiles

                Sample     Sample Size
                --------------------
                   1          1729



                    Response Profiles

Response   news    radio   reading   lectures   knowledg
----------------------------------------------------------------
    1      Yes     Yes     Yes       Yes        Good
    2      Yes     Yes     Yes       Yes        Poor
    3      Yes     Yes     Yes       No         Good
    4      Yes     Yes     Yes       No         Poor
    5      Yes     Yes     No        Yes        Good
    6      Yes     Yes     No        Yes        Poor
    7      Yes     Yes     No        No         Good
    8      Yes     Yes     No        No         Poor
    9      Yes     No      Yes       Yes        Good
   10      Yes     No      Yes       Yes        Poor
   11      Yes     No      Yes       No         Good
   12      Yes     No      Yes       No         Poor
   13      Yes     No      No        Yes        Good
   14      Yes     No      No        Yes        Poor
   15      Yes     No      No        No         Good
   16      Yes     No      No        No         Poor
   17      No      Yes     Yes       Yes        Good
   18      No      Yes     Yes       Yes        Poor
   19      No      Yes     Yes       No         Good
   20      No      Yes     Yes       No         Poor
   21      No      Yes     No        Yes        Good
   22      No      Yes     No        Yes        Poor
   23      No      Yes     No        No         Good
   24      No      Yes     No        No         Poor
   25      No      No      Yes       Yes        Good
   26      No      No      Yes       Yes        Poor
   27      No      No      Yes       No         Good
   28      No      No      Yes       No         Poor
   29      No      No      No        Yes        Good
   30      No      No      No        Yes        Poor
   31      No      No      No        No         Good
   32      No      No      No        No         Poor
```

Output 16.13   Analysis of Variance Table from Model with No Five-Factor Interaction

```
                  Maximum Likelihood Analysis of Variance

          Source                    DF    Chi-Square    Pr > ChiSq
          --------------------------------------------------------
          news                       1       47.96        <.0001
          radio                      1       49.57        <.0001
          news*radio                 1        5.60        0.0180
          reading                    1        1.66        0.1976
          news*reading               1       28.96        <.0001
          radio*reading              1        0.15        0.7024
          news*radio*reading         1        0.01        0.9099
          lectures                   1      377.03        <.0001
          news*lectures              1        2.72        0.0989
          radio*lectures             1       11.28        0.0008
          news*radio*lectures        1        1.52        0.2177
          reading*lectures           1        2.94        0.0864
          news*reading*lectures      1        0.08        0.7833
          radio*reading*lectures     1        1.78        0.1825
          news*radio*readin*lectur   1        0.39        0.5348
          knowledg                   1        5.34        0.0209
          news*knowledg              1       15.37        <.0001
          radio*knowledg             1        4.96        0.0259
          news*radio*knowledg        1        0.07        0.7868
          reading*knowledg           1        5.22        0.0224
          news*reading*knowledg      1        0.01        0.9307
          radio*reading*knowledg     1        0.27        0.6013
          news*radio*readin*knowle   1        0.19        0.6667
          lectures*knowledg          1        2.41        0.1208
          news*lectures*knowledg     1        2.26        0.1329
          radio*lectures*knowledg    1        1.79        0.1806
          news*radio*lectur*knowle   1        0.14        0.7115
          reading*lecture*knowledg   1        4.67        0.0307
          news*readi*lectur*knowle   1        0.77        0.3795
          radio*readi*lectu*knowle   1        0.32        0.5703

          Likelihood Ratio           1        1.02        0.3116
```

In Output 16.13, none of the Wald statistics for the four-factor interaction terms is larger than 0.77. Thus, the next statements fit the loglinear model with no four-way and five-way interaction terms. Again, the vertical bar notation is used to specify the 10 three-factor interactions. This is simpler than explicitly listing the 5 main effects, 10 two-factor interactions, and 10 three-factor interactions.

```
proc catmod order=data;
   weight count;
   model news*radio*reading*lectures*knowledg=_response_
       / noprofile noresponse noiter noparm;
   loglin news|radio|reading      news|radio|lectures
          news|radio|knowledg     news|reading|lectures
          news|reading|knowledg   news|lectures|knowledg
          radio|reading|lectures  radio|reading|knowledg
          radio|lectures|knowledg reading|lectures|knowledg;
run;
```

In the analysis of variance table displayed in Output 16.14, the likelihood ratio goodness-of-fit statistic tests the null hypothesis that the four-factor and five-factor interactions are jointly equal to zero. You would not reject this hypothesis ($G^2 = 3.23$, 6 df, $p = 0.7791$). You can also compare this model to the model with no five-way interaction. The value of the test statistic is $3.23 - 1.02 = 2.21$ with 5 df, which is clearly nonsignificant.

Output 16.14 Analysis of Variance Table from Model with No Four-Factor Interactions

```
                  Maximum Likelihood Analysis of Variance

         Source                DF    Chi-Square    Pr > ChiSq
         ---------------------------------------------------------
         news                   1       47.86        <.0001
         radio                  1       57.25        <.0001
         news*radio             1        6.36        0.0117
         reading                1        2.96        0.0855
         news*reading           1       28.47        <.0001
         radio*reading          1        0.04        0.8408
         news*radio*reading     1        1.26        0.2616
         lectures               1      391.77        <.0001
         news*lectures          1        2.37        0.1236
         radio*lectures         1       13.84        0.0002
         news*radio*lectures    1        1.51        0.2187
         knowledg               1        4.70        0.0302
         news*knowledg          1       16.85        <.0001
         radio*knowledg         1        5.20        0.0226
         news*radio*knowledg    1        0.00        0.9528
         reading*lectures       1        5.24        0.0221
         news*reading*lectures  1        0.01        0.9418
         reading*knowledg       1        9.29        0.0023
         news*reading*knowledg  1        2.76        0.0969
         lectures*knowledg      1        3.10        0.0782
         news*lectures*knowledg 1        3.03        0.0818
         radio*reading*lectures 1        1.50        0.2213
         radio*reading*knowledg 1        0.01        0.9434
         radio*lectures*knowledg 1       1.38        0.2409
         reading*lecture*knowledg 1      3.84        0.0500

         Likelihood Ratio       6        3.23        0.7791
```

Output 16.14 also displays Wald statistics for the 10 three-factor interactions. The one df Wald statistics for seven of these interactions are relatively small (1.5 or less). However, the statistics for the NEWS × READING × KNOWLEDG, NEWS × LECTURES × KNOWLEDG, and READING × LECTURES × KNOWLEDG interactions range from 2.76 to 3.84. The next statements fit the model that includes these three interaction terms, as well as all main effects and two-factor interactions.

```
proc catmod order=data;
   weight count;
   model news*radio*reading*lectures*knowledg=_response_
        / noprofile noresponse noiter noparm;
   loglin news|radio              radio|reading
          radio|lectures          radio|knowledg
          news|reading|knowledg   news|lectures|knowledg
          reading|lectures|knowledg;
run;
```

As shown in Output 16.15, this model provides a good fit relative to the saturated model ($G^2 = 9.50$, 13 df, $p = 0.7341$). In comparison to the previous model with no four-factor and higher interactions, the likelihood ratio test that the seven excluded three-factor interactions are jointly equal to zero is $G^2 = 9.50 - 3.23 = 6.27$ with $13 - 6 = 7$ df ($p = 0.39$). Thus, you would not reject the null hypothesis that the excluded three-factor interactions are jointly equal to zero.

Output 16.15   Analysis of Variance Table from Model with 3 Three-Factor Interactions

```
                Maximum Likelihood Analysis of Variance

         Source                     DF    Chi-Square    Pr > ChiSq
         ------------------------------------------------------------
         news                        1       53.36        <.0001
         radio                       1      100.01        <.0001
         news*radio                  1       43.62        <.0001
         reading                     1        2.88        0.0899
         radio*reading               1        0.48        0.4864
         lectures                    1      434.50        <.0001
         radio*lectures              1       10.60        0.0011
         knowledg                    1        5.22        0.0224
         radio*knowledg              1        6.28        0.0122
         news*reading                1      142.56        <.0001
         news*knowledg               1       17.99        <.0001
         reading*knowledg            1        9.17        0.0025
         news*reading*knowledg       1        3.03        0.0818
         news*lectures               1        4.29        0.0384
         lectures*knowledg           1        2.80        0.0941
         news*lectures*knowledg      1        3.09        0.0787
         reading*lectures            1        7.90        0.0049
         reading*lecture*knowledg    1        5.49        0.0191

         Likelihood Ratio           13        9.50        0.7341
```

With five exceptions, all of the Wald tests in Output 16.15 are statistically significant ($p < 0.05$). One of these nonsignificant effects is the main effect for READING. Since the main effect terms in a loglinear model fix the marginal totals, these are generally included in the model regardless of statistical significance. In addition, since READING is a lower-order effect for interactions included in the model, its removal would complicate interpretation of the retained interaction terms involving READING. This principle also motivates retention of the nonsignificant LECTURES × KNOWLEDG interaction; its removal would make it difficult to interpret the statistically significant READING × LECTURES × KNOWLEDG interaction.

Both of the nonsignificant three-factor interactions are suggestive with $p < 0.10$: NEWS × READING × KNOWLEDG ($Q_W = 3.03$, $p = 0.0818$) and NEWS × LECTURES × KNOWLEDG ($Q_W = 3.09$, $p = 0.0787$). Higher-order interactions like these are often retained to avoid oversimplification of the model. Finally, the RADIO × READING interaction has the smallest chi-square statistic of the effects in Output 16.15 ($Q_W = 0.48$, $p = 0.4864$). In addition, none of the higher-order interactions involves this effect. Therefore, the following statements fit a reduced model that excludes the RADIO × READING interaction.

```
proc catmod order=data;
    weight count;
    model news*radio*reading*lectures*knowledg=_response_
          / noprofile noresponse noiter noparm;
    loglin news|radio radio|lectures
           radio|knowledg news|reading|knowledg
           news|lectures|knowledg reading|lectures|knowledg;
run;
```

Output 16.16 indicates that this model provides a good fit to the observed cell counts
($G^2 = 9.99$, 14 df, $p = 0.7632$). The likelihood ratio statistic for testing the RADIO $\times$
READING effect is $G^2 = 9.99 - 9.50 = 0.49$ with 1 df ($p = 0.48$).

Output 16.16   Analysis of Variance Table from Reduced Hierarchical Model

```
                   Maximum Likelihood Analysis of Variance

           Source                  DF    Chi-Square    Pr > ChiSq
           ----------------------------------------------------------
           news                     1        54.21       <.0001
           radio                    1        99.63       <.0001
           news*radio               1        50.94       <.0001
           lectures                 1       434.52       <.0001
           radio*lectures           1        10.97        0.0009
           knowledg                 1         5.09        0.0240
           radio*knowledg           1         7.35        0.0067
           reading                  1         2.47        0.1160
           news*reading             1       149.31       <.0001
           news*knowledg            1        17.82       <.0001
           reading*knowledg         1         9.35        0.0022
           news*reading*knowledg    1         3.00        0.0835
           news*lectures            1         4.20        0.0404
           lectures*knowledg        1         2.76        0.0968
           news*lectures*knowledg   1         3.08        0.0792
           reading*lectures         1         8.27        0.0040
           reading*lecture*knowledg 1         5.47        0.0194

           Likelihood Ratio        14         9.99        0.7632
```

The three-factor interaction between READING, LECTURES, and KNOWLEDG is
significant at the $\alpha = 0.05$ level. Thus, the dependence between any pair of these variables
is affected by the third variable. The two other three-factor interactions included in this
model are NEWS $\times$ READING $\times$ KNOWLEDG and NEWS $\times$ LECTURES $\times$
KNOWLEDG. Both are suggestive at the $\alpha = 0.10$ level of significance. In addition, all
three two-factor interactions involving NEWS, READING, and KNOWLEDG, as well as
two of the three NEWS, LECTURES, and KNOWLEDG two-way interactions, are
significant at $\alpha = 0.05$.

Thus, the model indicates that all four media exposure variables (NEWS, RADIO,
READING, LECTURES) are associated with cancer knowledge. In addition, there are
significant associations between NEWS and RADIO, NEWS and READING, NEWS and
LECTURES, RADIO and LECTURES, and READING and LECTURES.

16.4.3   Loglinear Models with Nested Effects

The hierarchical model displayed in Output 16.16 contains 3 three-factor interaction terms. Each of these three-way interactions represents heterogeneity of the association between two of the variables across the levels of the third variable. A useful way to interpret these patterns of association more fully is to fit a model that specifically incorporates separate two-factor interactions across the levels of the third factor.

Since all three of the three-way interactions in Output 16.16 include the KNOWLEDG variable, the next statements use nested-by-value effects in the LOGLIN statement to fit a model with separate two-factor interactions for each level of KNOWLEDG.

```
proc catmod order=data;
   weight count;
   model news*radio*reading*lectures*knowledg=_response_
         / noprofile noresponse noiter noparm;
   loglin news|radio              radio|lectures
          radio|knowledg          reading|knowledg
          news|knowledg           lectures|knowledg
          news*reading(knowledg='Good')
          news*reading(knowledg='Poor')
          news*lectures(knowledg='Good')
          news*lectures(knowledg='Poor')
          reading*lectures(knowledg='Good')
          reading*lectures(knowledg='Poor');
run;
```

Output 16.17 displays the analysis of variance table for this model. Note that the likelihood ratio statistic is identical to that in Output 16.16. The two models are equivalent, even though they are parameterized differently. For example, Output 16.16 includes the NEWS × READING and NEWS × READING × KNOWLEDG interactions. These two effects are replaced in Output 16.17 by two NEWS × READING interactions, one for subjects with good cancer knowledge and one for subjects with poor cancer knowledge. Since the two models are equivalent, the hierarchical principle is actually still maintained, but the structure of the model is modified through nesting.

Output 16.17   Analysis of Variance Table from Model with Nested Effects

```
              Maximum Likelihood Analysis of Variance

         Source                    DF    Chi-Square    Pr > ChiSq
         --------------------------------------------------------
         news                       1       54.21        <.0001
         radio                      1       99.63        <.0001
         news*radio                 1       50.94        <.0001
         lectures                   1      434.52        <.0001
         radio*lectures             1       10.97        0.0009
         knowledg                   1        5.09        0.0240
         radio*knowledg             1        7.35        0.0067
         reading                    1        2.47        0.1160
         reading*knowledg           1        9.35        0.0022
         news*knowledg              1       17.82        <.0001
         lectures*knowledg          1        2.76        0.0968
         news*readin(knowle=Good)   1       44.17        <.0001
         news*readin(knowle=Poor)   1      128.92        <.0001
         news*lectur(knowle=Good)   1        6.05        0.0139
         news*lectur(knowle=Poor)   1        0.06        0.8001
         readi*lectu(knowle=Good)   1        0.15        0.6962
         readi*lectu(knowle=Poor)   1       12.91        0.0003

         Likelihood Ratio          14        9.99        0.7632
```

The nested effects for NEWS × LECTURES in individuals with poor cancer knowledge
and for READING × LECTURES in individuals with good cancer knowledge are clearly
nonsignificant. The next CATMOD invocation excludes these two effects.

```
proc catmod order=data;
   weight count;
   model news*radio*reading*lectures*knowledg=_response_
         / noprofile noresponse noiter;
   loglin news|radio            radio|lectures
          radio|knowledg        reading|knowledg
          news|knowledg         lectures|knowledg
          news*reading(knowledg='Good')
          news*reading(knowledg='Poor')
          news*lectures(knowledg='Good')
          reading*lectures(knowledg='Poor');
run;
```

The analysis of variance table displayed in Output 16.18 indicates that this model provides
a good fit to the observed data ($G^2 = 10.20$, 16 df, $p = 0.8557$). In addition, it has two
fewer parameters than the final model in Section 16.4.2 (Output 16.16). The READING
main effect and the LECTURES × KNOWLEDG interaction are the only nonsignificant
effects at the $\alpha = 0.05$ level of significance; both of these terms are retained in the model
to preserve hierarchy.

Output 16.18   Analysis of Variance Table from Reduced Model with Nested Effects

```
                  Maximum Likelihood Analysis of Variance

          Source                    DF   Chi-Square   Pr > ChiSq
          -----------------------------------------------------------
          news                       1       74.74      <.0001
          radio                      1      100.03      <.0001
          news*radio                 1       51.18      <.0001
          lectures                   1      444.55      <.0001
          radio*lectures             1       11.32       0.0008
          knowledg                   1        5.07       0.0244
          radio*knowledg             1        7.32       0.0068
          reading                    1        3.11       0.0778
          reading*knowledg           1       12.06       0.0005
          news*knowledg              1       28.40      <.0001
          lectures*knowledg          1        2.96       0.0856
          news*readin(knowle=Good)   1       45.47      <.0001
          news*readin(knowle=Poor)   1      131.33      <.0001
          news*lectur(knowle=Good)   1        6.75       0.0094
          readi*lectu(knowle=Poor)   1       15.71      <.0001

          Likelihood Ratio          16       10.20       0.8557
```

You interpret the results of this model in the same manner as was described in Section 16.3.5. Output 16.19 displays the parameter estimates, from which you can compute estimated odds ratios.

Output 16.19   Parameter Estimates from Reduced Model with Nested Effects

```
                   Analysis of Maximum Likelihood Estimates

                                                 Standard      Chi-
          Effect                    Parameter  Estimate   Error   Square   Pr > ChiSq
          ----------------------------------------------------------------------------
          news                          1       0.4443    0.0514   74.74     <.0001
          radio                         2      -0.4979    0.0498  100.03     <.0001
          news*radio                    3       0.2308    0.0323   51.18     <.0001
          lectures                      4      -1.2210    0.0579  444.55     <.0001
          radio*lectures                5       0.1597    0.0475   11.32      0.0008
          knowledg                      6      -0.1305    0.0580    5.07      0.0244
          radio*knowledg                7       0.0808    0.0299    7.32      0.0068
          reading                       8       0.0732    0.0415    3.11      0.0778
          reading*knowledg              9       0.1441    0.0415   12.06      0.0005
          news*knowledg                10       0.2564    0.0481   28.40     <.0001
          lectures*knowledg            11       0.0960    0.0559    2.96      0.0856
          news*readin(knowle=Good)     12       0.3017    0.0447   45.47     <.0001
          news*readin(knowle=Poor)     13       0.3995    0.0349  131.33     <.0001
          news*lectur(knowle=Good)     14       0.2296    0.0884    6.75      0.0094
          readi*lectu(knowle=Poor)     15       0.2716    0.0685   15.71     <.0001
```

For example, the model includes two parameters pertaining to the NEWS × READING association. The estimate for individuals with good cancer knowledge is 0.3017, from which the odds of reading newspapers are estimated to be $\exp(4 \times 0.3017) = 3.3$ times higher in individuals who do solid reading than in individuals who do not. The corresponding parameter estimate in individuals with poor cancer knowledge is 0.3995, with an associated estimated odds ratio of $\exp(4 \times 0.3995) = 4.9$. Similarly, in individuals with good cancer knowledge, the odds of reading newspapers are estimated to be $\exp(4 \times 0.2296) = 2.5$ times higher in those who attend lectures than in individuals who do not, and in individuals with poor cancer knowledge, the odds of reading books and magazines are estimated to be $\exp(4 \times 0.2716) = 3.0$ times higher in those who attend lectures than in individuals who do not.

When higher-order interactions are included in a model, either directly or through the use of nested effects, you should be cautious in interpreting lower-order main effects or interactions. Since the NEWS × READING × KNOWLEDG interaction is included through the two nested effects, the NEWS × KNOWLEDG effect is difficult to interpret in light of the interaction with the READING effect. Similarly, the READING × KNOWLEDG and LECTURES × KNOWLEDG effects are difficult to interpret due to the three-factor interactions containing these effects.

You can, however, interpret the NEWS × RADIO, RADIO × LECTURES, and RADIO × KNOWLEDG effects. For example, the odds of good cancer knowledge are estimated to be $\exp(4 \times 0.0808) = 1.4$ times higher for individuals who listen to the radio than for individuals who do not listen to the radio.

16.5 Correspondence Between Logistic Models and Loglinear Models

In Section 16.3.5, loglinear models were used to investigate the patterns of association in the three-way cross-classification of quality of management, supervisor's job satisfaction, and worker's job satisfaction for a random sample of 715 workers selected from Danish industry. The final model included main effects for MANAGMNT, SUPERVIS, and WORKER, as well as the three two-factor interactions.

Now suppose that the data displayed in Table 16.9 had instead been obtained from the four subpopulations defined by the cross-classification of quality of management and supervisor job satisfaction. In this case, you would be interested in modeling the probability of worker job satisfaction as a function of management quality and supervisor's job satisfaction. In practice, for situations where all of the variables are technically response variables, interest often focuses on modeling one of the variables as a function of the remaining ones.

The following statements model the logit of the probability of low worker job satisfaction as a function of management quality and supervisor job satisfaction.

```
proc catmod data=satisfac order=data;
   weight count;
   model worker=managmnt supervis
         / noprofile noresponse noiter p=freq;
run;
```

Output 16.20 displays the resulting analysis of variance table. For comparison, examine Output 16.21 for the corresponding analysis of variance table from the loglinear model with no three-factor interaction (Output 16.9 from Section 16.3.5).

Output 16.20   Analysis of Variance Table from Logistic Model

```
              Maximum Likelihood Analysis of Variance

         Source            DF    Chi-Square    Pr > ChiSq
         ----------------------------------------------------
         Intercept          1       25.96        <.0001
         managmnt           1       19.57        <.0001
         supervis           1        5.33        0.0210

         Likelihood Ratio   1        0.06        0.7989
```

Output 16.21   Analysis of Variance Table from Loglinear Model

```
              Maximum Likelihood Analysis of Variance

       Source              DF    Chi-Square    Pr > ChiSq
       -----------------------------------------------------
       managmnt             1       38.37        <.0001
       supervis             1        8.32        0.0039
       managmnt*supervis    1       67.06        <.0001
       worker               1       25.96        <.0001
       managmnt*worker      1       19.57        <.0001
       supervis*worker      1        5.33        0.0210

       Likelihood Ratio     1        0.06        0.7989
```

The likelihood ratio goodness of fit statistics for the two models are identical. In addition, the logistic model Wald chi-square statistics for the MANAGMNT and SUPERVIS main effects are identical to the loglinear model Wald statistics for the MANAGMNT × WORKER and SUPERVIS × WORKER interactions, and the logistic model INTERCEPT chi-square statistic is identical to the loglinear model Wald statistic for the WORKER main effect.

Output 16.22 displays the parameter estimates from the logistic model. The log odds of low worker satisfaction are estimated to be $\exp(2 \times 0.3739) = 2.1$ times higher when the quality of management is bad than when the management quality is good and $\exp(2 \times 0.1924) = 1.47$ times higher when the supervisor's job satisfaction is low than when the supervisor's job satisfaction is high. These estimates are the same as those computed in Section 16.3.5.

Output 16.22   Parameter Estimates from Logistic Model

```
              Analysis of Maximum Likelihood Estimates

                                      Standard     Chi-
     Effect       Parameter  Estimate   Error     Square   Pr > ChiSq
     --------------------------------------------------------------------
     Intercept        1      -0.4131    0.0811     25.96      <.0001
     managmnt         2       0.3739    0.0845     19.57      <.0001
     supervis         3       0.1924    0.0833      5.33      0.0210
```

Finally, Output 16.23 displays the observed and predicted response functions and frequencies from the logistic model. The predicted cell counts in each of the four logistic model subpopulations are the same as the predicted loglinear model frequencies in the $2 \times 2 \times 2$ contingency table, as displayed in Output 16.24 (Output 16.11 from Section 16.3.5).

Output 16.23   Predicted Response Functions and Cell Frequencies from Logistic Model

```
           Maximum Likelihood Predicted Values for Response Functions

                            ------Observed------    ------Predicted-----
                   Function          Standard                Standard
managmnt  supervis  Number  Function    Error    Function     Error    Residual
-----------------------------------------------------------------------------------
Bad       Low         1     0.168821  0.145612   0.153223   0.132027   0.015598
Bad       High        1    -0.27193   0.234648  -0.23151    0.17239   -0.04042
Good      Low         1    -0.61381   0.161628  -0.59464    0.142623  -0.01918
Good      High        1    -0.9663    0.133036  -0.97937    0.12311    0.013066
```

```
              Maximum Likelihood Predicted Values for Frequencies

                             -------Observed------    ------Predicted-----
                                       Standard                Standard
managmnt  supervis  worker  Frequency    Error    Frequency     Error    Residual
-----------------------------------------------------------------------------------
Bad       Low       Low        103     6.867544   102.2639   6.234641   0.736105
                    High        87     6.867544    87.73611  6.234641  -0.73611

Bad       High      Low         32     4.261709    32.73611  3.146866  -0.73611
                    High        42     4.261709    41.26389  3.146866   0.736105

Good      Low       Low         59     6.187064    59.73611  5.490361  -0.73611
                    High       109     6.187064   108.2639   5.490361   0.736105

Good      High      Low         78     7.516766    77.26389  6.915053   0.736105
                    High       205     7.516766   205.7361   6.915053  -0.73611
```

Output 16.24   Predicted Cell Counts from Loglinear Model

```
            Maximum Likelihood Predicted Values for Response Functions

                  ------Observed------     ------Predicted-----
     Function               Standard                 Standard
      Number     Function      Error     Function       Error    Residual
    -------------------------------------------------------------------------
        1        -0.68828    0.120776    -0.69904    0.113356    0.010757
        2        -0.8571     0.127954    -0.85226    0.1261     -0.00484
        3        -1.85727    0.190074    -1.83812    0.172903   -0.01916
        4        -1.58534    0.169374    -1.60661    0.148406    0.021266
        5        -1.24547    0.14774     -1.23666    0.142911   -0.00881
        6        -0.63166    0.118543    -0.64202    0.111524    0.01036
        7        -0.9663     0.133036    -0.97937    0.12311     0.013066




               Maximum Likelihood Predicted Values for Frequencies

                              ------Observed-----    -----Predicted-----
                                        Standard               Standard
  managmnt  supervis  worker  Frequency    Error  Frequency      Error  Residual
  ------------------------------------------------------------------------------
  Bad       Low       Low          103  9.389475  102.2639    8.904231  0.736105
  Bad       Low       High          87  8.741509   87.73611   8.283433 -0.73611
  Bad       High      Low           32  5.528818   32.73611   4.783695 -0.73611
  Bad       High      High          42  6.287517   41.26389   5.525364  0.736105
  Good      Low       Low           59  7.357409   59.73611   6.811457 -0.73611
  Good      Low       High         109  9.611619  108.2639    9.138839  0.736105
  Good      High      Low           78  8.336121   77.26389   7.782186  0.736105
  Good      High      High         205 12.0923    205.7361   11.75535  -0.73611
```

In summary, the cross-classification of the explanatory variables is fixed for logistic models and the effects of the factors are specified explicitly in the model statement. The loglinear model counterpart has the effects of factors specified through interactions with the response. In addition, the cross-classification of the explanatory variables is incorporated as a further component of the structure of the model.

The general result is that you can always rewrite a logistic analysis with one response variable as a loglinear model. First, move the explanatory variables to the left-hand side of the MODEL statement and use _RESPONSE_ as the only effect on the right-hand side of the MODEL statement. In addition, use a LOGLIN statement that includes all the main effects and interactions of the explanatory variables, as well as each effect from the logistic analysis crossed with the response variable.

Appendix A: Equivalence of the Loglinear and Poisson Regression Models

Suppose $\mathbf{n} = (n_1, n_2, \ldots, n_s)'$ denotes a vector of independent Poisson variables and let $\boldsymbol{\mu} = (\mu_1, \mu_2, \ldots, \mu_s)'$ denote the corresponding vector of expected values. Suppose variation among the elements of $\boldsymbol{\mu}$ can be described with the loglinear model

$$\boldsymbol{\mu} = \exp(\mathbf{X}_p \boldsymbol{\beta}_p)$$

where $\mathbf{X}_p = [\mathbf{1}, \mathbf{X}]$ is an $(s \times (t+1))$ matrix of known coefficients with full rank $(t+1) \le s$ and $\boldsymbol{\beta}_p = [\beta_0, \boldsymbol{\beta}']'$ is a $(t+1)$ vector of unknown coefficients. The likelihood function for $\mathbf{n}$ is

$$\Phi(\mathbf{n}|\boldsymbol{\mu}) = \prod_{i=1}^{s} \mu_i^{n_i} \{\exp(-\mu_i)\}/n_i! = \{\mu_+^{n_+}\{\exp(-\mu_+)\}/n_+!\} \bullet \{n_+!/\prod_{i=1}^{s} n_i!\}\{\prod_{i=1}^{s} \pi_i^{n_i}\}$$

where $n_+ = \sum_{i=1}^{s} n_i$, $\mu_+ = \sum_{i=1}^{s} \mu_i$, and $\pi_i = (\mu_i/\mu_+)$ for $i = 1, 2, \ldots, s$. Thus, the likelihood for $\mathbf{n}$ can be expressed as the product of a Poisson likelihood for n_+ and a multinomial likelihood $\phi(\mathbf{n}|n_+, \boldsymbol{\pi})$ with $\boldsymbol{\pi} = (\pi_1, \pi_2, \ldots, \pi_s)'$ for the conditional distribution of $\mathbf{n}$ given n_+. Since

$$\boldsymbol{\mu} = \exp(\mathbf{X}_p \boldsymbol{\beta}_p) = \exp(\beta_0) \exp(\mathbf{X}\boldsymbol{\beta})$$

it follows that

$$(\pi_1, \pi_2, \ldots, \pi_s)' = \boldsymbol{\pi} = \frac{\exp(\mathbf{X}\boldsymbol{\beta})}{\mathbf{1}_t' \exp(\mathbf{X}\boldsymbol{\beta})}$$

where $\mathbf{1}_t$ is the $(t \times 1)$ vectors of 1s. The structure shown here for $\boldsymbol{\pi}$ corresponds to the loglinear model for counts $\mathbf{n}$ with a multinomial distribution, either on the basis of conditioning independent Poisson counts on their sum or through simple random sampling (with replacement).

Since the maximization of the Poisson likelihood

$$\phi(\mathbf{n}|\boldsymbol{\mu}) = \phi(\mathbf{n}|\beta_0, \boldsymbol{\beta})$$

$$= \phi(n_+|\beta_0, \mathbf{1}_t' \exp(\mathbf{X}\boldsymbol{\beta}))\phi(\mathbf{n}|n_+, \boldsymbol{\beta})$$

relative to $\boldsymbol{\beta}_p = (\beta_0, \boldsymbol{\beta}')'$ correspondingly involves maximization of the multinomial likelihood $\phi(\mathbf{n}|n_+, \boldsymbol{\beta})$ relative to $\boldsymbol{\beta}$ as a byproduct, the maximum likelihood estimates of $\boldsymbol{\beta}$ in the Poisson regression model

$$\boldsymbol{\mu} = \exp(\beta_0) \exp(\mathbf{X}\boldsymbol{\beta})$$

are as well the maximum likelihood estimates of $\boldsymbol{\beta}$ in the multinomal loglinear model

$$\boldsymbol{\pi} = \frac{\exp(\mathbf{X}, \boldsymbol{\beta})}{(\mathbf{1}_t' \exp(\mathbf{X}\boldsymbol{\beta}))}$$

Also, the maximum likelihood estimate for the covariance matrix of $\widehat{\boldsymbol{\beta}}$ is the same in both situations. A convenient consequence of these considerations is that $\widehat{\boldsymbol{\beta}}$ for the multinomial loglinear model can have convenient computation through the use of Poisson regression via the GENMOD procedure to determine the maximum likelihood estimator

$$\widehat{\boldsymbol{\beta}_p} = (\widehat{\boldsymbol{\beta}_0}, \widehat{\boldsymbol{\beta}}')'$$

for $\boldsymbol{\beta}_p$ followed by removal of $\widehat{\boldsymbol{\beta}_0}$. In other words, the estimator $\widehat{\boldsymbol{\beta}_0}$ for the intercept in Poisson regression is ignored when this method is used to obtain $\widehat{\boldsymbol{\beta}}$ for the multinomial loglinear model. A further point of interest is that the Poisson loglinear model is strictly loglinear since

$$\log(\boldsymbol{\mu}) = \mathbf{X}_p \boldsymbol{\beta}_p$$

whereas the multinomial loglinear model corresponds to

$$\log(\pi) = \mathbf{X}\boldsymbol{\beta} - \{\log(\mathbf{1}_t' \exp(\mathbf{X}\boldsymbol{\beta})\}\mathbf{1}_t$$

and so is loglinear with a constraint to assure $\mathbf{1}_s' \boldsymbol{\pi} = \sum_{i=1}^{s} = 1$.

Chapter 17
Categorized Time-to-Event Data

Chapter Table of Contents

Chapter 17
Categorized Time-to-Event Data

17.1   Introduction

Categorical data often are generated from studies that have time from treatment or exposure until some event as their outcome. Such data are known as *time-to-event* data. The event may be death, the recurrence of some condition, or the emergence of a developmental characteristic. Often, the outcome is the actual lifetime (or waiting time), which is the response analyzed in typical survival analyses. However, due to resource constraints or the need to perform a diagnostic procedure, you sometimes can determine only the interval of time during which an event occurs. Examples include examining dental patients for caries at six-month periods, evaluating animals every four hours after their exposure to bacteria, and examining patients every six weeks for the recurrence of a medical condition for which they've been treated. Such data are often referred to as *grouped survival data* or as *categorized survival data*.

Since the study is conducted over a period of time, some subjects may leave before the study ends. This is called *withdrawal*. There may be protocol violations, subjects may join the study in progress and not complete the desired number of evaluations, or the subjects may drop out for other reasons. Thus, not only is status determined for each interval between successive evaluations, but the number of withdrawals for that interval is also determined. Most analysis strategies assume that withdrawal is independent of the condition being studied and that multiple withdrawals occur uniformly throughout the interval.

Frequently, interest lies in computing the survival rates. Section 17.2 discusses life table methods for computing these results. In addition, you generally want to compare survival rates for treatment groups and determine whether there is a treatment effect. Section 17.3 discusses the Mantel-Cox test, one strategy for addressing this question. It is similar to the log rank test used in traditional survival analysis. In addition to hypothesis testing, you may be interested in describing the variation in survival rates. Section 17.4 discusses the piecewise exponential model, one that is commonly used to model grouped survival data, as well as how to implement it using a Poisson regression strategy.

For an overview of grouped survival data analysis, refer to Deddens and Koch (1988).

17.2   Life Table Estimation of Survival Rates

Consider Table 17.1. Investigators were interested in comparing an active and control treatment to prevent the recurrence of a medical condition that had been healed. They applied a diagnostic procedure at the end of the first, second, and third years to determine whether there was a recurrence (based on Johnson and Koch 1978).

Table 17.1. Recurrences of Medical Condition

Treatment	Withdrawals			Recurrences			No Recurrence	Total
	Year 1	Year 2	Year 3	Year 1	Year 2	Year 3		
Control	9	7	6	15	13	7	17	74
Active	9	3	4	12	7	10	45	90

The survival rate, or the waiting time rate, is a key measure in the analysis of time-to-event data. It is written

$$S(y) = 1 - F(y) = \Pr\{Y \geq y\}$$

where Y denotes the continuous lifetime of a subject and $F(y) = \Pr\{Y \leq y\}$ is the cumulative probability distribution function. The exact form of $S(y)$ depends on the nature of $F(y)$, the probability distribution. The Weibull distribution and the exponential distribution are commonly used.

One way of estimating survival rates is with the *life table*, or *actuarial* method. Table 17.2 displays the life table format for the data displayed in Table 17.1. You determine the number of subjects at risk for each interval (the sum of those with no recurrence, those with recurrences, and those who withdrew). By knowing the number who survived all three intervals with no recurrence, you can determine the number with no recurrence for each interval.

Table 17.2. Life Table Format for Medical Condition Data

	Controls			
Interval	No Recurrences	Recurrences	Withdrawals	At Risk
0–1 Years	50	15	9	74
1–2 Years	30	13	7	50
2–3 Years	17	7	6	30
	Active			
Interval	No Recurrences	Recurrences	Withdrawals	At Risk
0–1 Years	69	12	9	90
1–2 Years	59	7	3	69
2–3 Years	45	10	4	59

Define n_{ijk} to be the number of patients in the ith group with the jth status for the kth time interval where $j = 0$ corresponds to no recurrence during the time interval, and $j = 1, 2$ corresponds to those with recurrence and those withdrawn during the kth interval, respectively; $i = 1, 2$ for the control and active groups and $k = 1, 2, \ldots, t$. The n_{i0k} are determined from

$$n_{i0k} = \sum_{j=1}^{2} \sum_{g=k+1}^{t} n_{ijg} + n_{i0t}$$

The life table estimates for the probability of surviving at least k intervals are computed as

$$G_{ik} = \prod_{g=1}^{k} \frac{n_{i0g} + 0.5n_{i2g}}{n_{i0g} + n_{i1g} + 0.5n_{i2g}} = \prod_{g=1}^{k} p_{ig}$$

where p_{ig} denotes the estimated conditional probability for surviving the gth interval given that survival of all preceding intervals has occurred.

The standard error of G_{ik} is estimated as

$$\text{s.e.}(G_{ik}) = G_{ik} \left\{ \sum_{g=1}^{k} \frac{(1 - p_{ig})}{(n_{i0g} + n_{i1g} + 0.5n_{i2g})p_{ig}} \right\}^{1/2}$$

$$= G_{ik} \left\{ \sum_{g=1}^{k} \frac{(1 - p_{ig})}{(n_{i0g} + 0.5n_{i2g})} \right\}^{1/2}$$

where $(n_{i0g} + n_{i1g} + 0.5n_{i2g})$ is the effective number at risk during the gth interval. Since

$$p_{ig} = \frac{n_{i0g} + 0.5n_{i2g}}{n_{i0g} + n_{i1g} + 0.5n_{i2g}}$$

then

$$1 - p_{ig} = \frac{n_{i1g}}{n_{i0g} + n_{i1g} + 0.5n_{i2g}}$$

The quantity $0.5 \times n_{i2g}$ is used in the numerator and denominator of p_{ig} since uniform withdrawals throughout the interval are assumed; the average exposure to risk for the withdrawing subjects is assumed to be one-half the interval.

For the active treatment, the life table estimates of surviving the kth interval are

$$G_{21} = \frac{69 + 0.5(9)}{69 + 12 + 0.5(9)} = 0.8596$$

$$G_{22} = 0.8596 \times \frac{59 + 0.5(3)}{59 + 7 + 0.5(3)} = 0.7705$$

$$G_{23} = 0.7705 \times \frac{45 + 0.5(4)}{45 + 10 + 0.5(4)} = 0.6353$$

Their standard errors are computed as follows:

$$\text{s.e.}(G_{21}) = 0.8596 \times \left\{ \frac{12/85.5}{69 + 0.5(9)} \right\}^{1/2} = 0.0376$$

$$\text{s.e.}(G_{22}) = 0.7705 \times \left\{ \frac{12/85.5}{69 + 0.5(9)} + \frac{7/67.5}{59 + 0.5(3)} \right\}^{1/2} = 0.0464$$

$$\text{s.e.}(G_{23}) = 0.6352 \times \left\{ \frac{12/85.5}{69 + 0.5(9)} + \frac{7/67.5}{59 + 0.5(3)} + \frac{10/57}{45 + 0.5(4)} \right\}^{1/2} = 0.0545$$

Table 17.3 contains the estimated survival rates and their standard errors for both active treatment and controls. The estimated survival rates for the active treatment are higher than for the controls for each of the intervals.

Table 17.3.   Life Table Format for Medical Condition Data

	Estimated Survival Rates	Standard Errors
Controls		
0–1 Years	0.7842	0.0493
1–2 Years	0.5650	0.0627
2–3 Years	0.4185	0.0665
Active		
0–1 Years	0.8596	0.0376
1–2 Years	0.7705	0.0463
2–3 Years	0.6353	0.0545

Section 15.3 discusses the Mantel-Cox test, which tests the null hypothesis that the survival rates are the same.

17.3   Mantel-Cox Test

You are often interested in comparing survival curves to determine which treatment had the more favorable outcome. Mantel (1966) and later Cox (1972) suggested an extension of the Mantel-Haenszel methodology that applies to survival data. You restructure the usual frequency table format of the data to a set of 2×2 tables, each with a life table format, and perform the Mantel-Haenszel computations on that set of tables.

The tables are generated by regarding treatment as the row variable, the numbers recurred and not recurred as the column variable, and the intervals as the strata. You are thus proceeding as though the time interval results are uncorrelated; methodological results for survival analysis establish that you can consider the respective time intervals to be essentially uncorrelated risk sets for survival information. It turns out that the Mantel-Cox test for grouped data is equivalent to the log rank test for comparing survival curves for ungrouped data (refer to Koch, Sen, and Amara 1985). Withdrawals are handled by either grouping them with the no recurrences or eliminating them entirely (this is the more conservative approach).

Table 17.4 contains the life table format for the study of the medical condition recurrence with the data grouped together by intervals and with the withdrawals excluded.

Table 17.4. Medical Condition Data

Years	Treatment	Recurrences	No Recurrences
0–1	Control	15	50
	Active	12	69
1–2	Control	13	30
	Active	7	59
2–3	Control	7	17
	Active	10	45

The following DATA step inputs these data, and the PROC FREQ statements specify that the MH test be computed. Recall that for sets of 2×2 tables, all scores are equivalent, so no scores need to be specified.

```
data clinical;
   input time $ treatment $ status $ count @@;
   datalines;
0-1 control recur 15 0-1 control not 50
0-1 active  recur 12 0-1 active  not 69
1-2 control recur 13 1-2 control not 30
1-2 active  recur  7 1-2 active  not 59
2-3 control recur  7 2-3 control not 17
2-3 active  recur 10 2-3 active  not 45
;
proc freq order=data;
   weight count;
   tables time*treatment*status / cmh;
run;
```

Output 17.1 contains the PROC FREQ output (the individual printed tables are not displayed). $Q_{MC} = 8.0294$ with 1 df, $p = 0.0046$. There is a significant treatment effect on survival.

Output 17.1 Results for Mantel-Cox Test

```
                Summary Statistics for treatment by status
                        Controlling for time

        Cochran-Mantel-Haenszel Statistics (Based on Table Scores)

        Statistic    Alternative Hypothesis      DF     Value     Prob
        ----------------------------------------------------------------
            1         Nonzero Correlation          1     8.0294    0.0046
            2         Row Mean Scores Differ       1     8.0294    0.0046
            3         General Association          1     8.0294    0.0046


                        Total Sample Size = 334
```

You can also apply the Mantel-Cox test when you have additional explanatory variables. Table 17.5 contains data from a study on gastrointestinal patients being treated for ulcers. Investigations conducted in three medical centers compared an active treatment to a placebo.

Table 17.5. Healing for Gastrointestinal Patients

Center	Treatment	Healed at Two Weeks	Healed at Four Weeks	Not Healed at Four Weeks	Total
1	A	15	17	2	34
1	P	15	17	7	39
2	A	17	17	10	44
2	P	12	13	15	40
3	A	7	17	16	40
3	P	3	17	18	38

Table 17.6 contains the life table format for the same data.

Table 17.6. Healing for Gastrointestinal Patients

Center	Weeks	Treatment	Number Healed	Number Not Healed	Total
1	0–2	A	15	19	34
		P	15	24	39
	2–4	A	17	2	19
		P	17	7	24
2	0–2	A	17	27	44
		P	12	28	40
	2–4	A	17	10	27
		P	13	15	28
3	0–2	A	7	33	40
		P	3	35	38
	2–4	A	17	16	33
		P	17	18	35

The following DATA step inputs these data, and the PROC FREQ statements specify that the MH test be computed. For these data, both TIME and CENTER are used as stratification variables.

```
data duodenal;
   input center time $ treatment $ status $ count @@;
   datalines;
1 0-2 A healed 15 1 0-2 A not 19
1 0-2 P healed 15 1 0-2 P not 24
1 2-4 A healed 17 1 2-4 A not  2
1 2-4 P healed 17 1 2-4 P not  7
2 0-2 A healed 17 2 0-2 A not 27
2 0-2 P healed 12 2 0-2 P not 28
2 2-4 A healed 17 2 2-4 A not 10
2 2-4 P healed 13 2 2-4 P not 15
3 0-2 A healed  7 3 0-2 A not 33
3 0-2 P healed  3 3 0-2 P not 35
3 2-4 A healed 17 3 2-4 A not 16
3 2-4 P healed 17 3 2-4 P not 18
;
```

```
proc freq;
   weight count;
   tables center*time*treatment*status / cmh;
run;
```

Output 17.2 contains the results.

Output 17.2 Results for Mantel-Cox Test

```
            Summary Statistics for treatment by status
                 Controlling for center and time

       Cochran-Mantel-Haenszel Statistics (Based on Table Scores)

       Statistic    Alternative Hypothesis    DF    Value    Prob
       ---------------------------------------------------------------
           1        Nonzero Correlation        1    4.2527   0.0392
           2        Row Mean Scores Differ     1    4.2527   0.0392
           3        General Association        1    4.2527   0.0392


                   Total Sample Size = 401
```

The null hypothesis is that within each center, the distribution of time to healing is the same for placebo and active treatment. $Q_{MC} = 4.2527$ with 1 df ($p = 0.0392$), so that there is a significant effect of active treatment on time to healing after adjusting for center.

17.4 Piecewise Exponential Models

Statistical models can extend the analysis of grouped survival data by providing a description of the pattern of event rates. They can describe this pattern over time as well as describe the variation due to the influence of treatment and other explanatory variables. One particularly useful model is the piecewise exponential model.

Consider Table 17.7, which contains information pertaining to the experience of patients undergoing treatment for duodenal ulcers (based on Johnson and Koch 1978). One of two types of surgeries was randomly assigned: vagotomy and drainage or antrectomy, or vagotomy and hemigastrectomy. The patients were evaluated at 6 months, 24 months, and 60 months. Death and recurrence are considered failure events, and reoperation and loss to follow-up are considered withdrawal events.

Table 17.7. Comparison of Two Surgeries for Duodenal Ulcer

Operation	Time (months)	Death or Recurrence	Reoperation or Lost	Satisfactory	Exposure (months)
	0–6	23	15	630	3894
V + D/A	7–24	32	20	578	10872
	25–60	45	71	462	18720
	0–6	9	5	329	2016
V + H	7–24	5	17	307	5724
	25–60	10	24	273	10440

In this study there are two treatment groups: $i = 1$ for V and D/A, $i = 2$ for V + H and three time intervals: $k = 1$ for 0–6 months, $k = 2$ for 7–24 months, and $k = 3$ for 25–60 months.

If you can make the following assumptions, then you can fit the piecewise exponential model to these data.

- The withdrawals are uniformly distributed during the time intervals in which they occur and are unrelated to treatment failures.

- The within-interval probabilities of the treatment failures are small. The time-to-failure events have independent exponential distributions.

The piecewise exponential likelihood is written

$$\Phi_{PE} = \prod_{i=1}^{2} \prod_{k=1}^{3} \lambda_{ik}^{n_{i1k}} \left\{ \exp[-\lambda_{ik} N_{ik}] \right\}$$

where n_{i1k} is the number of failures for the ith group during the kth interval, N_{ik} is the total person-months of exposure, and λ_{ik} is the hazard parameter. The piecewise exponential model assumes that there are independent exponential distributions with hazard parameters λ_{ik} for the respective time periods.

The N_{ik} are computed as

$$N_{ik} = a_k(n_{i0k} + 0.5n_{i1k} + 0.5n_{i2k})$$

where $a_k = 6, 18, 36$ is the length of the kth interval, n_{i0k} is the number of patients completing the kth interval without failure or withdrawal, and n_{i2k} denotes the number of withdrawals. The quantity n_{i1k} is the number of failures during the interval.

If you think of the number of deaths n_{i1k}, conditional on the exposures N_{ik}, as having independent Poisson distributions, then you can write a Poisson likelihood for these data.

$$\Phi_{PO} = \prod_{i=1}^{2} \prod_{k=1}^{3} (N_{ik} \lambda_{ik})^{n_{i1k}} \left\{ \frac{\exp[-N_{ik} \lambda_{ik}]}{n_{i1k}!} \right\}$$

$$= \Phi_{PE} \left\{ \prod_{i=1}^{2} \prod_{k=1}^{3} \frac{N_{ik}^{n_{i1k}}}{n_{i1k}!} \right\}$$

Since these likelihoods are proportional, whatever maximizes Φ_{PO} also maximizes Φ_{PE}. Thus, you can still assume the piecewise exponential model but obtain the estimates from Poisson regression computations, which are more accessible, regardless of whether you want to make the conditional arguments necessary to assume a Poisson distribution.

The relationship of the failure events to the explanatory variables is specified through models for the λ_{ik}. One class of models has the structure

$$\lambda_{ik} = \exp(\mathbf{x}_{ik}'\boldsymbol{\beta})$$

A useful subset of these models has the specification

$$\lambda_{ik} = \exp(\alpha + \eta_k + \mathbf{x}_i'\boldsymbol{\beta})$$

This latter model has the proportional hazards structure, where $\{\eta_k\}$ is the constant value of the hazard function within the kth interval when $\mathbf{x}_i = 0$. The parameter vector $\boldsymbol{\beta}$ relates the hazard function for the ith population to the explanatory variables $\mathbf{x}_i$.

Those readers familiar with survival analysis may recognize the general form of the proportional hazards model as

$$h(y, \mathbf{x}) = h_0(y)\{\exp(\mathbf{x}'\boldsymbol{\beta})\}$$

where y denotes continuous time and $h_0(y)$ is the hazard function for the reference population. In reference to this general form, $\exp(\alpha + \eta_k)$ corresponds to $h_0(y)$ for y in the kth interval.

17.4.1 An Application of the Proportional Hazards Piecewise Exponential Model

Since the GENMOD procedure fits Poisson regression models in the SAS System, you also use it to fit piecewise exponential models. The DATA step inputs the duodenal ulcer data and computes the variable NMONTHS as the log of MONTHS. The following PROC GENMOD statements request that the main effects model consisting of time and treatment be fit. The variable TREATMENT has the value 'vda' for V and D/A and the value 'vh' for V + H. Note that the value for 0–6 months for the variable TIME is '_0-6' so that it will sort last and thus become the reference value in the PROC GENMOD parameterization.

```
data vda;
    input treatment $ time $ failure months;
    nmonths=log(months);
    datalines;
vda  _0-6   23    3894
vda  7-24   32   10872
vda 25-60   45   18720
vh   _0-6    9    2016
vh   7-24    5    5724
vh  25-60   10   10440
;
proc genmod data=vda;
    class treatment time;
    model failure = time treatment
        / dist=poisson link=log offset=nmonths;
run;
```

Both TIME and TREAT are defined as CLASS variables. The LINK=LOG option is specified so that the model is in loglinear form, and the OFFSET=NMONTHS is specified since the quantity n_{ik} / N_{ik} is being modeled.

Information about the model specification and the sort levels of the CLASS variables are displayed in Output 17.3.

Output 17.3   Model Information

```
                     Model Information

          Data Set                  WORK.VDA
          Distribution               Poisson
          Link Function                  Log
          Dependent Variable         failure
          Offset Variable            nmonths
          Observations Used                6


                Class Level Information

     Class            Levels    Values

     treatment           2      vda vh
     time                3      25-60 7-24 _0-6
```

Statistics for assessing fit are displayed in Output 17.4. Q_P and the deviance both indicate an adequate fit, with values of 2.6730 and 2.5529, respectively, and 2 df for their approximately chi-square distributions.

Output 17.4   Goodness-of-Fit Criteria

```
            Criteria For Assessing Goodness Of Fit

     Criterion              DF        Value       Value/DF

     Deviance                2       2.5529        1.2764
     Scaled Deviance         2       2.5529        1.2764
     Pearson Chi-Square      2       2.6730        1.3365
     Scaled Pearson X2       2       2.6730        1.3365
     Log Likelihood                279.8914
```

The "Analysis of Parameter Estimates" table includes an intercept parameter, incremental effects for 7–24 months and 25–60 months, and an incremental effect for the V + D/A treatment. The 0–6 months time interval and the V + H treatment are the reference cell. See Output 17.5.

Output 17.5   Parameter Estimates

```
                          Analysis Of Parameter Estimates

                                   Standard   Wald 95% Confidence    Chi-
Parameter             DF  Estimate    Error         Limits         Square   Pr > ChiSq

Intercept             1    -5.8164   0.2556   -6.3174   -5.3153    517.66     <.0001
time        25-60     1    -1.0429   0.2223   -1.4787   -0.6071     22.00     <.0001
time        7-24      1    -0.8847   0.2414   -1.3579   -0.4116     13.43     0.0002
time        _0-6      0     0.0000   0.0000    0.0000    0.0000       .          .
treatment   vda       1     0.8071   0.2273    0.3616    1.2527     12.61     0.0004
treatment   vh        0     0.0000   0.0000    0.0000    0.0000       .          .
Scale                 0     1.0000   0.0000    1.0000    1.0000

NOTE: The scale parameter was held fixed.
```

All of these effects are significant. Table 17.8 contains the parameter interpretations.

Table 17.8.   Parameter Interpretations

GENMOD Parameter	Model Parameter	Value	Interpretation
INTERCEPT	α	-5.8164	log incidence density for V + H, 0-6 months (reference)
TIME 25-60	η_2	-1.0429	increment for 25-60 interval
TIME 7-24	η_1	-0.8847	increment for 7-25 interval
TREAT vda	β	0.8071	increment for treatment V + D/A

Log incidence density decreases with the 7–24 interval and further decreases with the 25–60 interval. The V + D/A treatment increases log incidence density. What this means is that the failure rate is highest for the first interval and lower for the other two intervals; the failure rate is higher for V + D/A than for V + H. Table 17.9 displays the estimated failure rates (incidence densities) per person-month for each group and interval.

The survival rates can be calculated as follows:

$$\Pr\{\text{survival for } k \text{ intervals}\} = \Pr\{\text{survival for } k - 1 \text{ intervals}\} \times e^{-\lambda_{ik} a_k}$$

where a_k is the length of the kth interval, $k = 1, 2, \ldots, t$. Table 17.9 contains the survival estimates for each interval for each treatment group.

Table 17.9.   Model-Estimated Failure Rates

Group	Interval	Failure Rate Formula	Estimated Failure Rate	Estimated Survival Rate
V + H	0–6	$e^{\hat{\alpha}}$	0.002978	0.9823
V + H	7–24	$e^{\hat{\alpha}+\hat{\eta}_1}$	0.001230	0.9608
V + H	25–60	$e^{\hat{\alpha}+\hat{\eta}_2}$	0.001050	0.9252
V + D/A	0–6	$e^{\hat{\alpha}+\hat{\beta}}$	0.006676	0.9607
V + D/A	7–24	$e^{\hat{\alpha}+\hat{\beta}+\hat{\eta}_1}$	0.002756	0.9142
V + D/A	25–60	$e^{\hat{\alpha}+\hat{\beta}+\hat{\eta}_2}$	0.002353	0.8399

17.4.2   Using PROC LOGISTIC to Fit the Piecewise Exponential Model

When the incidence rates $\{\lambda_{ik}\}$ are small (less than 0.05) and the exposures N_{ik} are very large, then you can approximate Poisson regression with logistic regression (Vine et al. 1990). Thus, you can take advantage of the features of the LOGISTIC procedure, such as its model-building facilities, to fit models such as the piecewise exponential model. You can facilitate the approximation by rescaling the exposure factor by multiplying it by a number such as 10,000; the only adjustment you need to make after parameter estimation is to add the log of the multiplier you choose to the resulting intercept estimate.

The following SAS statements fit a piecewise exponential model to the duodenal ulcer data using the LOGISTIC procedure. In the DATA step, the variable SMONTHS is the exposure in months multiplied by a factor of 100,000.

The events/trials syntax is employed in the MODEL statement, with FAILURE in the numerator and SMONTHS in the denominator. The SELECTION=FORWARD option specifies forward model selection, and INCLUDE=2 specifies that the first two variables listed in the MODEL statement, TIME and TREATMENT, be forced into the first model so that a score test is produced for the contribution of the remaining variables to the model (interactions). This serves as a goodness-of-fit test.

```
data vda;
   input treatment time $ failure months;
   smonths=100000*months;
   datalines;
1   _0-6    23    3894
1   7-24    32    10872
1   25-60   45    18720
0   _0-6    9     2016
0   7-24    5     5724
0   25-60   10    10440
;
proc logistic;
   class time/param=ref;
   model failure/smonths = time treatment time*treatment /
                           scale=none include=2 selection=forward;
run;
```

Output 17.6 contains the resulting statistics for explanatory variable contribution and the score statistic (Residual Chi-Square) for the contribution of variables not in the model. Q_S has a value of 2.6730 and is nonsignificant with 2 df and $p = 0.2628$. The model fits adequately and the proportional hazards assumption is reasonable (no time $\times$ treatment interaction).

Output 17.6   Score Statistic

```
                Testing Global Null Hypothesis: BETA=0

        Test                    Chi-Square      DF      Pr > ChiSq

        Likelihood Ratio         34.7700         3        <.0001
        Score                    39.0554         3        <.0001
        Wald                     36.2836         3        <.0001


                      Residual Chi-Square Test

              Chi-Square        DF        Pr > ChiSq

                2.6730           2          0.2628
```

Output 17.7 displays the goodness-of-fit statistics, which are adequate.

Output 17.7   Goodness-of-Fit Statistics

```
            Deviance and Pearson Goodness-of-Fit Statistics

        Criterion        DF        Value      Value/DF      Pr > ChiSq

        Deviance          2        2.5529      1.2764         0.2790
        Pearson           2        2.6730      1.3365         0.2628

              Number of events/trials observations: 6
```

The results are very similar to what was obtained with the PROC GENMOD analysis.

The resulting parameter estimates are displayed in Output 17.8. The parameter estimates for the time and treatment effects are very close to the estimates resulting from the PROC GENMOD analysis in the previous section, and if you add log(100000) to the intercept, -17.3293, you obtain -5.8164, which is identical to the intercept estimate obtained from PROC GENMOD. This approximation is usually very good.

Output 17.8   Parameter Estimates

```
                Analysis of Maximum Likelihood Estimates

                                       Standard        Wald
        Parameter           DF   Estimate    Error   Chi-Square   Pr > ChiSq

        Intercept            1    -17.3293   0.2556   4595.1888      <.0001
        time       25-60     1     -1.0430   0.2223     22.0023      <.0001
        time       7-24      1     -0.8848   0.2414     13.4324      0.0002
        treatment            1      0.8071   0.2273     12.6068      0.0004
```

At this point, you would proceed to produce survival rates and survival estimates as computed in the previous section.

References

Agresti, A. (1988). Logit models for repeated ordered categorical response data, *Proceedings of the 13th Annual SAS Users Group International Conference*, Cary, NC: SAS Institute Inc., 997–1005.

Agresti, A. (1989). A survey of models for repeated ordered categorical response data, *Statistics in Medicine*, 8, 1209–1224.

Agresti, A. (1990). *Categorical Data Analysis*, New York: John Wiley & Sons, Inc.

Agresti, A. (1992). A survey of exact inference for contingency tables, *Statistical Science*, 7 (1), 131–177.

Agresti, A. (1996). *An Introduction to Categorical Data Analysis*, New York: John Wiley & Sons, Inc.

Agresti, A., Wackerly, D., and Boyett, J. M. (1979). Exact conditional tests for cross-classifications: Approximation of attained significance levels, *Psychometrika*, 44, 75–83.

Albert, A. and Anderson, J. A. (1984). On the existence of maximum likelihood estimates in logistic regression models, *Biometrika*, 71, 1–10.

Albert, P., and McShane, L. (1995). A generalized estimating equations approach for spatially correlated binary data: Applications to the analysis of neuroimaging data, *Biometrics*, 51, 627–638.

Andersen, E. B. (1991). *The Statistical Analysis of Categorical Data, Second Edition*, Berlin: Springer-Verlag.

Armitage, P. (1955). Tests for linear trends in proportions and frequencies, *Biometrics*, 11, 375–386.

Baglivo, J., Olivier, D., and Pagano, M. (1988). Methods for the analysis of contingency tables with large and small cell counts, *Journal of the American Statistical Society*, 83, 1006–1013.

Barnhart, H. and Williamson, J. (1998). Goodness-of-fit tests for GEE modeling with binary responses, *Biometrics*, 54, 720–729.

Bauman, K. E., Koch, G. G., and Lentz, M. (1989). Parent characteristics, perceived health risk, and smokeless tobacco use among white adolescent males, *NI Monographs 8*, 43–48.

Benard, A. and van Elteren, P. (1953). A generalization of the method of m rankings, *Proceedings Koninklijke Nederlands Akademie van Wetenschappen (A)*, 56, 358–369.

Berkson, J. (1951). Why I prefer logits to probits, *Biometrics*, 7, 327–339.

Berry, D. A. (1987). Logarithmic transformations in ANOVA, *Biometrics*, 43, 439–456.

Birch, M. W. (1963). Maximum likelihood in three-way contingency tables, *Journal of the Royal Statistical Society, Series B*, 25, 220–233.

Bishop, Y. M. M., Fienberg, S. E., and Holland, P. W. (1975). *Discrete Multivariate Analysis*, Cambridge, MA: MIT Press.

Bock, R. D. (1975). *Multivariate Statistical Methods in Behavioral Research*, New York: McGraw-Hill, Inc.

Bock, R. D. and Jones, L.V. (1968). *The Measurement and Prediction of Judgement and Choice*, San Francisco: Holden-Day.

Boos, D. (1992). On generalized score tests, *The American Statistician*, 46, 327–333.

Bowker, A. H. (1948). Bowker's test for symmetry, *Journal of the American Statistical Association*, 43, 572–574.

Breslow, N. E. and Day, N. E. (1980). *Statistical Methods in Cancer Research, Volume 1: The Analysis of Case-Control Studies*, Lyon, International Agency for Research on Cancer.

Bruce, R. A., Kusumi, F., and Hosmer, D. (1973). Maximal oxygen intake and nomographic assessment of functional aerobic impairment in cardiovascular disease, *American Heart Journal,* 65, 546–562.

Carey, V., Zeger, S. L., and Diggle, P. (1993). Modelling multivariate binary data with alternating logistic regressions, *Biometrika*, 80, 517–526.

Carr, G. J., Hafner, K. B., and Koch, G. G. (1989). Analysis of rank measures of association for ordinal data from longitudinal studies, *Journal of the American Statistical Association*, 84, 797–804.

Cartwright, H. V., Lindahl, R. L., and Bawden, J. W. (1968). Clinical findings on the effectiveness of stannous fluoride and acid phosphate fluoride as caries reducing agents in children, *Journal of Dentistry for Children,* 35, 36–40.

Cawson M. J., Anderson, A. B. M., Turnbull, A. C., and Lampe, L. (1974). Cortisol, cortisone, and 11-deoxycortisol levels in human umbilical and maternal plasma in relation to the onset of labour, *Journal of Obstetrics and Gynaecology of the British Commonwealth,* 81, 737–745.

Charnes, A., Frome, E. L., and Yu, P. L. (1976). The equivalence of generalized least squares and maximim likelihood estimates in the exponential family, *Journal of the American Statistical Association*, 71, 169–172.

Cochran, W. G. (1940). The analysis of variance when experimental errors follow the Poisson or binomial laws, *Annals of Mathematical Statistics*, 11, 335–347.

Cochran, W. G. (1950). The comparison of percentages in matched samples, *Biometrika*, 37, 256–266.

Cochran, W. G. (1954). Some methods of strengthening the common χ^2 tests, *Biometrics*, 10, 417–451.

Cohen, J. (1960). A coefficient of agreement for nominal data, *Educational and Psychological Measurement*, 20, 37–46.

Collett, D. (1991). *Modelling Binary Data*, London: Chapman and Hall.

Cook, R. D. and Weisberg, S. (1982). *Residuals and Influence in Regression*, London: Chapman and Hall.

Cornoni-Huntley, J., Brock, D. B., Ostfeld, A., Taylor, J. O., and Wallace, R. B. (1986). *Established Populations for Epidemiologic Studies of the Elderly, Resource Data Book*, Bethesda, MD: National Institutes of Health (NIH Pub. No. 86–2443).

Cox, D. R. (1970). *Analysis of Binary Data*, London: Chapman and Hall.

Cox, D. R. (1972). Regression models and life tables, *Journal of the Royal Statistical Society, Series B*, 34, 187–220.

Cox, M. A. A. and Plackett, R. L. (1980). Small samples in contingency tables, *Biometrika* 67, 1–13.

Davis, C. S. (1992). Analysis of incomplete categorical repeated measures, *Proceedings of the 17th Annual SAS Users Group International Conference*, Cary, NC: SAS Institute Inc., 1374–1379.

Dean, C. B. (1998). Overdispersion, in *Encyclopedia of Biostatistics, Volume 4*, eds. P. Armitage and T. Colton, New York: John Wiley & Sons, Inc., 3226–3232.

Deddens, J. A. and Koch, G. G. (1988). Survival analysis, grouped data, in *Encyclopedia of Statistical Sciences*, 9, eds. Kotz, S. and Johnson, N. L, New York: John Wiley & Sons, Inc., 129–134.

Deming, W. E. and Stephan, F. F. (1940). On a least squares adjustment of a sample frequency table when the expected marginal totals are known, *Annals of Mathematical Statistics*, 11, 427–444.

Derr, Robert E. (2000). Performing exact logistic regression with the SAS system, *Proceedings of the Twenty-Fifth Annual SAS Users Group International Conference*, Cary, NC: SAS Institute Inc.

Diggle, P. (1992). Discussion of paper by K. Y. Liang, S. L. Zeger, and B. Qaqish, *Journal of the Royal Statistical Society, Series B*, 45, 28–29.

Diggle, P. J., Liang, K. Y., and Zeger, S. L. (1994). *Analysis of Longitudinal Data*, Oxford: Clarendon Press.

Dobson, A. J. (1990). *An Introduction to Generalized Linear Models*, London: Chapman and Hall.

Draper, N. R. and Smith, H. (1981). *Applied Regression Analysis*, New York: John Wiley & Sons, Inc.

Durbin, J. (1951). Incomplete blocks in ranking experiments, *British Journal of Mathematical and Statistical Psychology*, 4, 85–90.

Dyke, G. V. and Patterson, H. D. (1952). Analysis of factorial arrangements when the data are proportions, *Biometrics*, 8, 1–12.

Finney, D. J. (1978). *Statistical Methods in Biological Assay, Third Edition*, New York: Macmillan Publishing Company, Inc.

Firth, D. (1992). Discussion of paper by K. Y. Liang, S. L. Zeger, and B. Qaqish, *Journal of the Royal Statistical Society, Series B*, 45, 24–26.

Fisher, L. D. and van Belle, G. (1993). *Biostatistics: A Methodology for the Health Sciences,* New York: John Wiley & Sons, Inc.

Fleiss, J. L. (1975). Measuring agreement between two judges on the presence or absence of a trait, *Biometrics*, 31, 651–659.

Fleiss, J. L. (1981). *Statistical Methods for Rates and Proportions,* New York: John Wiley & Sons, Inc.

Fleiss, J. L. (1986). *The Design and Analysis of Clinical Experiments*, New York: John Wiley & Sons, Inc.

Fleiss, J. L. and Cohen, J. (1973). The equivalence of weighted kappa and the intraclass correlation coefficient as measures of reliability, *Educational and Psychological Measurement*, 33, 613–619.

Friedman, M. (1937). The use of ranks to avoid the assumption of normality implicit in the analysis of variance, *Journal of the American Statistical Association,* 32, 675–701.

Frome, E. L., Kutner, M. H., and Beauchamp, J. J. (1973). Regression analysis of Poisson-distributed data, *Journal of the American Statistician Association*, 68, 935–950.

Frome, E. L. (1981). Poisson regression analysis, *American Statistician*, 35, 262–263.

Gail, M. (1978). The analysis of heterogeneity for indirect standardized mortality ratios, *Journal of the Royal Statistical Society A*, 141, Part 2, 224–234.

Gansky, S. A., Koch. G. G., and Wilson, J. (1994). Statistical evaluation of relationships between analgesic dose and ordered ratings of pain relief over an eight-hour period, *Journal of Biopharmaceutical Statistics*, 4, 233–265.

Gart, J. J. (1971). The comparison of proportions: A review of significance tests, confidence intervals and adjustments for stratification, *Review of the International Statistical Institute*, 39 (2), 148–169.

Goodman, L. A. (1968). The analysis of cross-classified data: Independence, quasi-independence, and interactions in contingency tables with or without missing entries, *Journal of the American Statistical Association*, 63, 1091–1131.

Goodman, L. A. (1970). The multivariate analysis of qualitative data: Interaction among multiple classifications, *Journal of the American Statistical Association*, 65, 226–256.

Govindarajulu, Z. (1988). *Statistical Methods in Bioassay*, Basel: Karger.

Graubard, B. I. and Korn, E. L. (1987). Choice of column scores for testing independence in ordered $2 \times k$ contingency tables, *Biometrics*, 43, 471–476.

Grizzle, J. E., Starmer, C. F., and Koch, G. G. (1969). Analysis of categorical data by linear models, *Biometrics*, 25, 489–504.

Harrell, F. E. (1989). Analysis of repeated measurements, Unpublished course notes.

Hendricks, S. A., Wassell, J. T., Collins, J. W., and Sedlak, S. L. (1996). Power determination for geographically clustered data using generalized estimating equations, *Statistics in Medicine*, 15, 1951–1960.

Higgins, J. E. and Koch, G. G. (1977). Variable selection and generalized chi-square analysis of categorical data applied to a large cross-sectional occupational health survey, *International Statistical Review*, 45, 51–62.

Hirji, K. F., Mehta, C. R., and Patel, N. R. (1987). Computing distributions for exact logistic regression, *Journal of the American Statistical Association*, 82, 1110–1117.

Hochberg, Y., Stutts, J. C., and Reinfurt, D. W. (1977). Observed shoulder belt usage of drivers in North Carolina: A follow-up, *University of North Carolina Highway Safety Research Center Report*, May 1977.

Hodges, J. L. and Lehmann, E. L. (1962). Rank methods for combination of independent experiments in analysis of variance, *Annals of Mathematical Statistics*, 33, 482–497.

Hollander, M. and Wolfe, D. A. (1973). *Nonparametric Statistical Methods*, New York: John Wiley & Sons, Inc.

Hosmer, D. W. and Lemeshow, S. (1989). *Applied Logistic Regression*, New York: John Wiley & Sons, Inc.

Imrey, P. B., Koch, G. G., and Stokes, M. E. (1982). Categorical data analysis: some reflections on the log linear model and logistic regression, Part II, *International Statistical Review*, 50, 35–64.

Johnson, W. D. and Koch, G. G. (1978). Linear models analysis of competing risks for grouped survival times, *International Statistical Review*, 46, 21–51.

Karim, M. R. and Zeger, S. L. (1988). GEE: A SAS macro for longitudinal data analysis, Technical Report No. 674, Department of Biostatistics, The Johns Hopkins University.

Kendall, M. G. and Stuart, A. (1961). *The Advanced Theory of Statistics, Volume 2: Inference and Relationship*, London: Charles Griffin and Company.

Koch, G. G., Amara, I. A., Davis, G. W., and Gillings, D. B. (1982). A review of some statistical methods for covariance analysis of categorical data, *Biometrics,* 38, 563–595.

Koch, G. G., Amara, I. A., and Singer, J. M. (1985). A two-stage procedure for the analysis of ordinal categorical data, in *Statistics in Biomedical, Public Health and Environmental Sciences*, ed. P. K. Sen, Amsterdam: North-Holland, 357–387.

Koch, G. G., Atkinson, S. S., and Stokes, M. E. (1986). Poisson regression, in *Encyclopedia of Statistical Sciences*, 7, eds. Kotz, S. and Johnson, N. L., New York: John Wiley & Sons, Inc., 32–41.

Koch, G. G., Carr, G. J., Amara, I. A., Stokes, M. E., and Uryniak, T. J. (1990). Categorical data analysis, in *Statistical Methodology in the Pharmaceutical Sciences*, ed. D. A. Berry, New York: Marcel Dekker Inc., 391–475.

Koch, G. G. and Edwards, S. (1985). Logistic regression, in *Encyclopedia of Statistics*, 5, New York: John Wiley & Sons, Inc, 128-133.

Koch, G. G. and Edwards, S. (1988). Clinical efficacy trials with categorical data, in *Biopharmaceutical Statistics for Drug Development,* ed. K. E. Peace, New York: Marcel Dekker, 403–451.

Koch, G. G., Gillings, D. B., and Stokes, M. E. (1980). Biostatistical implications of design, sampling, and measurement to health science data, in *Annual Review of Public Health*, 1, 163–225.

Koch, G. G., Imrey, P. B, Singer, J. M., Atkinson, S. S., and Stokes, M. E. (1985). Analysis of Categorical Data, Montreal, Canada: Les Presses de L'Université de Montreal.

Koch, G. G., Landis, J. R., Freeman, J. L., Freeman, D. H., and Lehnen, R. G. (1977). A general methodology for the analysis of experiments with repeated measurement of categorical data, *Biometrics*, 33, 133–158.

Koch, G. G. and Reinfurt, D. W. (1971). The analysis of categorical data from mixed models, *Biometrics*, 27, 157–173.

Koch, G. G. and Sen, P. K. (1968). Some aspects of the statistical analysis of the mixed model, *Biometrics,* 24, 27–48.

Koch, G. G., Sen, P. K., and Amara, I. A. (1985). Log-rank scores, statistics, and tests, in *Encyclopedia of Statistics*, 5, New York: John Wiley & Sons, Inc, 136–141.

Koch, G. G., Singer, J. M., Stokes, M. E., Carr, G. J., Cohen, S. B., and Forthofer, R. N. (1989). Some aspects of weighted least squares analysis for longitudinal categorical data, in *Statistical Models for Longitudinal Studies of Health,* ed. J. H. Dywer, Oxford: Oxford University Press, Inc., 215-258.

Koch, G. G. and Stokes, M. E. (1979). Annotated computer applications of weighted least squares methods for illustrative analyses of examples involving health survey data, Technical Report prepared for the U.S. National Center for Health Statistics.

Kruskal, W. H. and Wallis, W. A. (1952). Use of ranks in one-criterion variance analysis, *Journal of the American Statistical Association,* 47, 583–621.

Kuritz, S. J., Landis, J. R., and Koch, G. G. (1988). A general overview of Mantel-Haenszel methods: Applications and recent developments, *Annual Review of Public Health*, 1988, 123–160.

Lafata, J. E., Koch, G. G., and Weissert, W. G. (1994). Estimating activity limitation in the noninstitutionalized population: A method for small areas, *American Journal of Public Health*, 84, 1813–1817.

Landis, J. R., Heyman, E. R., and Koch, G. G. (1978). Average partial association in three-way contingency tables: A review and discussion of alternative tests, *International Statistical Review*, 46, 237–254.

Landis, J. R. and Koch, G. G. (1977). The measurement of observer agreement for categorical data, *Biometrics*, 33, 159–174.

Landis, J. R. and Koch, G. G. (1979). The analysis of categorical data on longitudinal studies of behavioral development, in *Longitudinal Research in the Study of Behavior and Development*, ed. J. R. Nesselroade and P. B. Bates, New York: Academic Press, Inc., 231–261.

Landis, J. R., Miller, M. E., Davis, C. S., and Koch, G. G. (1988). Some general methods for the analysis of categorical data in longitudinal studies, *Statistics in Medicine*, 7, 233–261.

Landis, J. R., Sharp, T. J., Kuritz, S. J., and Koch, G. G. (1998). Mantel-Haenszel methods, in *Encyclopedia of Biostatistics, Volume 3*, eds. P. Armitage and T. Colton, New York: John Wiley & Sons, Inc., 2378–2391.

Landis, J. R., Stanish, W. M., and Koch, G. G. (1976). A computer program for the generalized chi-square analysis of categorical data using weighted least squares, (GENCAT), *Computer Programs in Biomedicine*, 6, 196–231.

LaVange, L. M., Keyes, L. L., Koch, G. G., and Margolis, P. E. (1994). Application of sample survey methods for modelling ratios to incidence densities, *Statistics in Medicine*, 13, 343–355.

LaVange, L. M., Koch, G. G., and Schwartz, T. A. (2000). Applying sample survey methods to clinical trials data, to appear in *Statistics in Medicine*, 2001.

Lehmann, E. L. (1975). *Nonparametrics: Statistical Methods Based on Ranks*, San Francisco: Holden-Day.

Lemeshow, S. and Hosmer, D. (1989). *Applied Logistic Regression Analysis*, New York: John Wiley & Sons, Inc.

Liang, K. Y. and Zeger, S. L. (1986). Longitudinal data analysis using generalized linear models, *Biometrika*, 73, 13–22.

Liang, K. Y., Zeger, S. L., and Qaqish, B. (1992). Multivariate regression analyses for categorical data (with discussion), *Journal of the Royal Statistical Society, Series B*, 54, 3–40.

Lipsitz, S. R., Fitzmaurice, G. M., Orav, E. J., and Laird, N. M. (1994). Performance of generalized estimating equations in practical situations, *Biometrics*, 50, 270–278.

Lipsitz, S. R., Kim, K., and Zhao, L. (1994). Analysis of repeated categorical data using generalized estimating equations, *Statistics in Medicine*, 13, 1149–1163.

Lipsitz, S. R., Laird, N. M., and Harrington, D. P. (1991). Generalized estimating equations for correlated binary data: Using the odds ratio as a measure of association, *Biometrika*, 78, 153–160.

Loomis, D., Dufort, V., Kleckner, R. C., and Savitz, D. A. (1999). Fatal occupational injuries among electric power company workers, *American Journal of Industrial Medicine*, 35, 302–309.

Lund, A. K., Williams, A. F., and Zador, P. (1986). High school driver education: Further evaluation of the De Kalb County study, *Accident Analysis and Prevention*, 18, 349–357.

Luta, G., Koch, G. G., Cascio, W. E., and Smith, W. T. (1998). An application of methods for clustered binary responses to a cardiovascular study with small sample size, *Journal of Biopharmaceutical Statistics*, 8, 87–102.

Mack, T. M., Pike, M. C., Henderson, B. E., Pfeffer, R. I., Gerkins, V. R., Arthur, M., and Brown, S. E. (1976). Estrogens and endometrial cancer in a retirement community, *New England Journal of Medicine*, 294, 23, 1262–1267.

Macknin, M. L., Mathew, S., and Medendorp, S. V. (1990). Effect of inhaling heated vapor on symptoms of the common cold, *Journal of the American Medical Association*, 264, 989–991.

MacMillan, J., Becker, C., Koch, G. G., Stokes, M. E., and Vandivire, H. M. (1981). An application of weighted least squares methods to the analysis of measurement process components of variability in an observational study, *American Statistical Association Proceedings of Survey Research Methods*, 680–685.

Madansky, A. (1963). Test of homogeneity for correlated samples, *Journal of the American Statistical Association*, 58, 97–119.

Mann, H. B. and Whitney, D. R. (1947). On a test of whether one of two random variables is stochastically larger than the other, *Annals of Mathematical Statistics*, 18, 50–60.

Mantel, N. (1963). Chi-square tests with one degree of freedom: Extensions of the Mantel-Haenszel procedure, *Journal of the American Statistical Association*, 58, 690–700.

Mantel, N. (1966). Evaluation of survival data and two new rank order statistics arising in its consideration, *Cancer Chemotherapy Report*, 50, 163–170.

Mantel, N. and Fleiss, J. (1980). Minimum expected cell size requirements for the Mantel-Haenszel one-degree of freedom chi-square test and a related rapid procedure, *American Journal of Epidemiology*, 112, 129–143.

Mantel, N. and Haenszel, W. (1959). Statistical aspects of the analysis of data from retrospective studies of disease, *Journal of the National Cancer Institute*, 22, 719–748.

McCullagh, P. (1980). Regression models for ordinal data (with discussion), *Journal of the Royal Statistical Society, Series B*, 42, 109–142.

McCullagh, P. and Nelder, J. A. (1989). *Generalized Linear Models, Second Edition*, London: Chapman and Hall.

McNemar, Q. (1947). Note on the sampling error of the difference between correlated proportions or percentages, *Psychometrika*, 12, 153–157.

Mehta, C. R. and Patel, N. R. (1983). A network algorithm for performing Fisher's exact test in r by c contingency tables, *Journal of the American Statistical Association*, 427–434.

Mehta, C. R. and Patel, N. R. (1995). Exact logistic regression: Theory and examples, *Statistics in Medicine*, 13, 2143–2160.

Mehta, C. R., Patel, N. R., and Tsiatis, A. A. (1984). Exact significance testing to establish treatment equivalence with ordered categorical data, *Biometrics*, 40, 427–434.

Miller, M. E., Davis, C. S., and Landis, J. R. (1993). The analysis of longitudinal polytomous data: Generalized estimating equations and connections with weighted least squares, *Biometrics*, 49, 1033–1044.

Moradi, T., Nyrén O., Bergström, R., Gridley, G., Linet, M., Wolk, A., Dosemeci, M. and Adami, H. (1998). Risk for endometrial cancer in relation to occupational physical activity: A nationwide cohort study in Sweden, *International Journal of Cancer*, 76, 665–670.

Moulton, L. H. and Zeger, S. L. (1989). Analyzing repeated measures on generalized linear models via the bootstrap, *Biometrics*, 45, 381–394.

Nelder, J. A. and Wedderburn, R. W. M. (1972). Generalized linear models, *Journal of the Royal Statistical Society, Series A*, 135, 370–384.

Nemeroff, C. B., Bissette, G., Prange, A. J., Loosen, P. Y., Barlow, F. S., and Lipton, M. A. (1977). Neurotensin: Central nervous system effects of a hypothalamic peptide, *Brain Research*, 128, 485–496.

Neyman, J. (1949). Contributions to the theory of the χ^2 test, in *Proceedings of the Berkeley Symposium of Mathematical Statistics and Probability*, Berkeley: University of California Press, 239–273.

Odeh, R. E., Owen, D. B., Birnbaum, Z. W., and Fisher, L. D. (1977). *Pocket Book of Statistical Tables,* New York: Marcel Dekker, Inc.

Ogilvie, J. C. (1965). Paired comparison models with tests for interaction, *Biometrics*, 21, 651–654.

Pagano, M. and Halvorsen, K. T. (1981). An algorithm for finding the exact significance levels of r × c contingency tables, *Journal of the American Statistical Society*, 76, 931–934.

Pickles, A. (1998). Generalized estimating equations, in *Encyclopedia of Biostatistics, Volume 2*, eds. P. Armitage and T. Colton, New York: John Wiley & Sons, Inc., 1626–1637.

Pirie, W. (1983). Jonckheere tests for ordered alternatives, in *Encyclopedia of Statistical Sciences, Volume 4*, eds. S. Kotz and N. L. Johnson, New York: John Wiley & Sons, Inc., 315–318.

Pitman, E. J. G. (1948). *Lecture Notes on Nonparametric Statistics,* New York: Columbia University.

Pregibon, D. (1981). Logistic regression diagnostics, *Annals of Statistics*, 9, 705–724.

Preisser, J. S. and Koch, G. G. (1997). Categorical data analysis in public health, *Annual Review of Public Health*, 18, 51–82.

Preisser, J. S. and Quaqish, B. F. (1996). Deletion diagnostics for generalised estimating equations, *Biometrika*, 83, 3, 551–562.

Prentice, R. L. (1988). Correlated binary regression with covariate specific to each binary observation, *Biometrics*, 44, 1033–1048.

Royall, R. M. (1986). Model robust confidence intervals using maximum likelihood estimators,*International Statistical Review*, 54, 221–226.

Quade, D. (1967). Rank analysis of covariance, *Journal of the American Statistical Association*, 62, 1187–1200.

Quade, D. (1982). Nonparametric analysis of covariance by matching, *Biometrics*, 38, 597–611.

Rao, C. R. (1961). Asymptotic efficiency and limiting information, *Proceedings of the 4th Berkeley Symposium on Mathematical Statistics and Probability*, 1, 531–545.

Rao, C. R. (1962). Efficient estimates and optimum inference procedures in large samples (with discussion), *Journal of the Royal Statistical Society, Series B*, 24, 46–72.

Read, C. B. (1983). Fieller's theorem, in *Encyclopedia of Statistical Sciences*, 3, eds. Kotz, S. and Johnson, N. L., New York: John Wiley & Sons, Inc., 86–88.

Roberts, G., Martyn, A. L., Dobson, A. J., and McCarthy, W. H. (1981). Tumour thickness and histological type in malignant melanoma in New South Wales, Australia, *Pathology*, 13, 763–770.

Robins, J., Breslow, N., and Greenland, S. (1986). Estimators of the Mantel-Haenszel variance consistent in both sparse data and large-strata limiting models, *Biometrics*, 42, 311–323.

Rotnitzky, A. and Jewell, N. P. (1990). Hypothesis testing of regression parameters in semiparametric generalized linear models for cluster correlated data, *Biometrika*, 77, 485–497.

SAS Institute Inc. (1999). *SAS/STAT User's Guide, Version 8*, Cary, NC: SAS Institute Inc.

Schechter, P. J., Horwitz, D., and Henkin, R. I. (1973). Sodium chloride preference in essential hypertension, *Journal of the American Medical Association*, 225, 1311–1315.

Semenya, K. A. and Koch, G. G. (1979). Linear models analysis for rank functions of ordinal categorical data, *Proceedings of the Statistical Computing Section of the American Statistical Association*, 271–276.

Semenya, K. A. and Koch, G. G. (1980). Compound function and linear model methods for the multivariate analysis of ordinal categorical data, *Institute of Statistics Mimeo Series No. 1323*, Chapel Hill: University of North Carolina.

Shah, B. V., Holt, M. M., and Folsom, R. E. (1977). Inference about regression models from sample survey data, *Bulletin of the International Statistical Institute*, 47, 43–57.

Shahpar, C. and Guohua, L. (1999). Homicide mortality in the United States, 1935-1994: Age, period, and cohort effects, *American Journal of Epidemiology*, 150, 1213–1222.

Silvapulle, M. J. (1981). On the existence of maximum likelihood estimators for the binomial response models, *Journal of the Royal Statistics Society*, 43, 310–313.

Simpson, E. H. (1951). The interpretation of interaction in contingency tables, *Journal of the Royal Statistical Society*, B13, 238–241.

Stanish, W. M. (1986). Categorical data analysis strategies using SAS software, in *Computer Science and Statistics: Proceedings of the 17th Symposium on the Interface*, ed. D. M. Allen, New York: Elsevier Science Publishing Company.

Stanish, W. M., Gillings, D. B., and Koch, G. G., (1978). An application of multivariate ratio methods for the analysis of a longitudinal clinical trial with missing data, *Biometrics*, 34, 305–317.

Stanish, W. M. and Koch, G. G. (1984). The use of CATMOD for repeated measurement analysis of categorical data, *Proceedings of the 9th Annual SAS Users Group International Conference*, Cary, NC: SAS Institute Inc., 761–770.

Stewart, J. R. (1975). An analysis of automobile accidents to determine which variables are most strongly associated with driver injury: Relationships between driver injury and vehicle model year, *University of North Carolina Highway Safety Research Center Technical Report.*

Stock, J. R., Weaver, J. K., Ray, H. W., Brink, J. R., and Sadof, M. G. (1983). *Evaluation of Safe Performance Secondary School Driver Education Curriculum Demonstration Project*, Washington, D.C.: U.S. Department of Transportation, National Highway Traffic Safety Administration.

Stokes, M. E. (1986). An application of categorical data analysis to a large environmental data set with repeated measurements and missing values, *Institute of Statistics Mimeo Series No. 1807T*, Chapel Hill: University of North Carolina.

Stram, D. O., Wei, L. J., and Ware, J. H. (1988). Analysis of repeated ordered categorical outcomes with possibly missing observations and time-dependent covariates, *Journal of the American Statistical Association*, 83, 631–637.

Tardif, S. (1980). On the asymptotic distribution of a class of aligned rank order test statistics in randomized block designs, *Canadian Journal of Statistics*, 8, 7–25.

Tardif, S. (1981). On the almost sure convergence of the permutation distribution for aligned rank test statistics in randomized block designs, *Annals of Statistics*, 9, 190–193.

Tardif, S. (1985). On the asymptotic efficiency of aligned-rank tests in randomized block designs, *Canadian Journal of Statistics*, 13, 217–232.

Thomas, D. G. (1971). Algorithm AS-36: Exact confidence limits for the odds ratio in a 2×2 table, *Applied Statistics*, 20, 105–110.

Tritchler, D. (1984). An algorithm for exact logistic regression, *Journal of the American Statistical Association*, 79, 709–711.

Tsutakawa, R. K. (1982). Statistical methods in bioassay, in *Encyclopedia of Statistical Sciences*, 1, eds. Kotz, S. and Johnson, N. L., New York: John Wiley & Sons, Inc., 237–243.

Tudor, G., Koch, G. G., and Catellier, D. (2000). Statistical methods for crossover designs in bioenvironmental and public health studies, in *Handbook of Statistics: Bioenvironmental and Public Health Statistics*, eds. Sen, P. K. and Rao, C. R., Elsevier Science Publishers.

van Elteren, P. H. (1960). On the combination of independent two-sample tests of Wilcoxon, *Bulletin of the International Statistical Institute*, 37, 351–361.

Vine, M. F., Schoenbach, V., Hulka, B. S., Koch, G. G., and Samsa, G. (1990). Atypical metaplasia as a risk factor for bronchogenic carcinoma, *American Journal of Epidemiology*, 131, 781–793.

Wald, A. (1943). Tests of statistical hypotheses concerning general parameters when the number of observations is large, *Transactions of the American Mathematical Society*, 54, 426–482.

Ware, J. H., Lipsitz, S., and Speizer, F. E. (1988). Issues in the analysis of repeated categorical outcomes, *Statistics in Medicine*, 7, 95–107.

Wei, L. J. and Stram, D. O. (1988). Analyzing repeated measurements with possibly missing observations by modelling marginal distributions, *Statistics in Medicine*, 7, 139–148.

Wilcoxon, F. (1945). Individual comparison by ranking methods, *Biometrics*, 1, 80–83.

Yule, G. U. (1903). Notes on the theory of association of attributes in statistics, *Biometrika*, 2, 121–134.

Zeger, S. L. (1988). Commentary, *Statistics in Medicine*, 7, 161–168.

Zeger, S. L. and Liang, K. Y. (1986). Longitudinal data analysis for discrete and continuous outcomes, *Biometrics*, 42, 121–130.

Zeger, S. L., Liang, K. Y., and Albert, P. S. (1988). Models for longitudinal data: A generalized estimating equation approach, *Biometrics*, 44, 1049–1060.

Zerbe, G. O. (1978). On Fieller's theorem and the general linear model, *The American Statistician*, 32 (3).

Zhao, L. P. and Prentice, R. L. (1990). Correlated binary regression using a quadratic exponential model, *Biometrika*, 77, 642–648.

Index

Books from SAS Institute's
Books by Users Press

support.sas.com/pubs

Multiple Comparisons and Multiple Tests Using SAS®
Text and Workbook Set
(books in this set also sold separately)
by **Peter H. Westfall, Randall D. Tobias,**
Dror Rom, Russell D. Wolfinger,
and **Yosef Hochberg**

Multiple-Plot Displays: Simplified with Macros
by **Perry Watts**

Multivariate Data Reduction and Discrimination with
SAS® Software
by **Ravindra Khattree**
and **Dayanand N. Naik**

Output Delivery System: The Basics
by **Lauren E. Haworth**

Painless Windows: A Handbook for SAS® Users, Third Edition
by **Jodie Gilmore**
(updated to include Version 8 and SAS 9.1 features)

PROC TABULATE by Example
by **Lauren E. Haworth**

Professional SAS® Programming Shortcuts
by **Rick Aster**

Quick Results with SAS/GRAPH® Software
by **Arthur L. Carpenter**
and **Charles E. Shipp**

Quick Results with the Output Delivery System
by **Sunil K. Gupta**

Quick Start to Data Analysis with SAS®
by **Frank C. Dilorio**
and **Kenneth A. Hardy**

Reading External Data Files Using SAS®: Examples Handbook
by **Michele M. Burlew**

Regression and ANOVA: An Integrated Approach Using
SAS® Software
by **Keith E. Muller**
and **Bethel A. Fetterman**

SAS®Applications Programming: A Gentle Introduction
by **Frank C. Dilorio**

SAS® for Forecasting Time Series, Second Edition
by **John C. Brocklebank**
and **David A. Dickey**

SAS® for Linear Models, Fourth Edition
by **Ramon C. Littell, Walter W. Stroup,**
and **Rudolf J. Freund**

SAS® for Monte Carlo Studies: A Guide for Quantitative
Researchers
by **Xitao Fan, Ákos Felsővályi, Stephen A. Sivo,**
and **Sean C. Keenan**

SAS® Functions by Example
by **Ron Cody**

SAS® Macro Programming Made Easy
by **Michele M. Burlew**

SAS® Programming by Example
by **Ron Cody**
and **Ray Pass**

SAS® Programming for Researchers and Social Scientists,
Second Edition
by **Paul E. Spector**

SAS® Survival Analysis Techniques for Medical Research,
Second Edition
by **Alan B. Cantor**

SAS® System for Elementary Statistical Analysis,
Second Edition
by **Sandra D. Schlotzhauer**
and **Ramon C. Littell**

SAS® System for Mixed Models
by **Ramon C. Littell, George A. Milliken, Walter W. Stroup,**
and **Russell D. Wolfinger**

SAS® System for Regression, Third Edition
by **Rudolf J. Freund**
and **Ramon C. Littell**

SAS® System for Statistical Graphics, First Edition
by **Michael Friendly**

The SAS® Workbook and Solutions Set
(books in this set also sold separately)
by **Ron Cody**

Selecting Statistical Techniques for Social Science Data:
A Guide for SAS® Users
by **Frank M. Andrews, Laura Klem, Patrick M. O'Malley,**
Willard L. Rodgers, Kathleen B. Welch,
and **Terrence N. Davidson**

Statistical Quality Control Using the SAS® System
by **Dennis W. King**

A Step-by-Step Approach to Using the SAS® System
for Factor Analysis and Structural Equation Modeling
by **Larry Hatcher**

A Step-by-Step Approach to Using the SAS® System
for Univariate and Multivariate Statistics, Second Edition
by **Norm O'Rourke, Larry Hatcher,**
and **Edward J. Stepanski**

Step-by-Step Basic Statistics Using SAS®: Student Guide
and Exercises
*(*books in this set also sold separately*)*
by **Larry Hatcher**

Survival Analysis Using the SAS® System:
A Practical Guide
by **Paul D. Allison**

Tuning SAS® Applications in the OS/390 and z/OS
Environments, Second Edition
by **Michael A. Raithel**

Univariate and Multivariate General Linear Models:
Theory and Applications Using SAS® Software
by **Neil H. Timm**
and **Tammy A. Mieczkowski**

Using SAS® in Financial Research
by **Ekkehart Boehmer, John Paul Broussard,**
and **Juha-Pekka Kallunki**

Using the SAS® Windowing Environment: A Quick Tutorial
by **Larry Hatcher**

Visualizing Categorical Data
by **Michael Friendly**

Web Development with SAS® by Example
by **Frederick Pratter**

Your Guide to Survey Research Using the SAS® System
by **Archer Gravely**

JMP® Books

JMP® for Basic Univariate and Multivariate Statistics: A Step-by-
Step Guide
by **Ann Lehman, Norm O'Rourke, Larry Hatcher,**
and **Edward J. Stepanski**

JMP® Start Statistics, Third Edition
by **John Sall, Ann Lehman,**
and **Lee Creighton**

Regression Using JMP®
by **Rudolf J. Freund, Ramon C. LIttell,**
and **Lee Creighton**

WILEY SERIES IN PROBABILITY AND STATISTICS

ESTABLISHED BY WALTER A. SHEWHART AND SAMUEL S. WILKS

Editors: *David J. Balding, Peter Bloomfield, Noel A. C. Cressie, Nicholas I. Fisher, Iain M. Johnstone, J. B. Kadane, Louise M. Ryan, David W. Scott, Adrian F. M. Smith, Jozef L. Teugels*
Editors Emeriti: *Vic Barnett, J. Stuart Hunter, David G. Kendall*

The *Wiley Series in Probability and Statistics* is well established and authoritative. It covers many topics of current research interest in both pure and applied statistics and probability theory. Written by leading statisticians and institutions, the titles span both state-of-the-art developments in the field and classical methods.

Reflecting the wide range of current research in statistics, the series encompasses applied, methodological and theoretical statistics, ranging from applications and new techniques made possible by advances in computerized practice to rigorous treatment of theoretical approaches.

This series provides essential and invaluable reading for all statisticians, whether in academia, industry, government, or research.

*Now available in a lower priced paperback edition in the Wiley Classics Library.

BRUNNER, DOMHOF, and LANGER · Nonparametric Analysis of Longitudinal Data in Factorial Experiments
BUCKLEW · Large Deviation Techniques in Decision, Simulation, and Estimation
CAIROLI and DALANG · Sequential Stochastic Optimization
CHAN · Time Series: Applications to Finance
CHATTERJEE and HADI · Sensitivity Analysis in Linear Regression
CHATTERJEE and PRICE · Regression Analysis by Example, *Third Edition*
CHERNICK · Bootstrap Methods: A Practitioner's Guide
CHERNICK and FRIIS · Introductory Biostatistics for the Health Sciences
CHILÈS and DELFINER · Geostatistics: Modeling Spatial Uncertainty
CHOW and LIU · Design and Analysis of Clinical Trials: Concepts and Methodologies
CLARKE and DISNEY · Probability and Random Processes: A First Course with Applications, *Second Edition*
*COCHRAN and COX · Experimental Designs, *Second Edition*
CONGDON · Bayesian Statistical Modelling
CONOVER · Practical Nonparametric Statistics, *Second Edition*
COOK · Regression Graphics
COOK and WEISBERG · Applied Regression Including Computing and Graphics
COOK and WEISBERG · An Introduction to Regression Graphics
CORNELL · Experiments with Mixtures, Designs, Models, and the Analysis of Mixture Data, *Third Edition*
COVER and THOMAS · Elements of Information Theory
COX · A Handbook of Introductory Statistical Methods
*COX · Planning of Experiments
CRESSIE · Statistics for Spatial Data, *Revised Edition*
CSÖRGŐ and HORVÁTH · Limit Theorems in Change Point Analysis
DANIEL · Applications of Statistics to Industrial Experimentation
DANIEL · Biostatistics: A Foundation for Analysis in the Health Sciences, *Sixth Edition*
*DANIEL · Fitting Equations to Data: Computer Analysis of Multifactor Data, *Second Edition*
DASU and JOHNSON · Exploratory Data Mining and Data Cleaning
DAVID · Order Statistics, *Second Edition*
*DEGROOT, FIENBERG, and KADANE · Statistics and the Law
DEL CASTILLO · Statistical Process Adjustment for Quality Control
DETTE and STUDDEN · The Theory of Canonical Moments with Applications in Statistics, Probability, and Analysis
DEY and MUKERJEE · Fractional Factorial Plans
DILLON and GOLDSTEIN · Multivariate Analysis: Methods and Applications
DODGE · Alternative Methods of Regression
*DODGE and ROMIG · Sampling Inspection Tables, *Second Edition*
*DOOB · Stochastic Processes
DOWDY and WEARDEN · Statistics for Research, *Second Edition*
DRAPER and SMITH · Applied Regression Analysis, *Third Edition*
DRYDEN and MARDIA · Statistical Shape Analysis
DUDEWICZ and MISHRA · Modern Mathematical Statistics
DUNN and CLARK · Applied Statistics: Analysis of Variance and Regression, *Second Edition*
DUNN and CLARK · Basic Statistics: A Primer for the Biomedical Sciences, *Third Edition*
DUPUIS and ELLIS · A Weak Convergence Approach to the Theory of Large Deviations
*ELANDT-JOHNSON and JOHNSON · Survival Models and Data Analysis
ENDERS · Applied Econometric Time Series
ETHIER and KURTZ · Markov Processes: Characterization and Convergence
EVANS, HASTINGS, and PEACOCK · Statistical Distributions, *Third Edition*
FELLER · An Introduction to Probability Theory and Its Applications, Volume I, *Third Edition,* Revised; Volume II, *Second Edition*
FISHER and VAN BELLE · Biostatistics: A Methodology for the Health Sciences
*FLEISS · The Design and Analysis of Clinical Experiments
FLEISS · Statistical Methods for Rates and Proportions, *Second Edition*
FLEMING and HARRINGTON · Counting Processes and Survival Analysis
FULLER · Introduction to Statistical Time Series, *Second Edition*
FULLER · Measurement Error Models
GALLANT · Nonlinear Statistical Models
GHOSH, MUKHOPADHYAY, and SEN · Sequential Estimation
GIFI · Nonlinear Multivariate Analysis
GLASSERMAN and YAO · Monotone Structure in Discrete-Event Systems
GNANADESIKAN · Methods for Statistical Data Analysis of Multivariate Observations, *Second Edition*
GOLDSTEIN and LEWIS · Assessment: Problems, Development, and Statistical Issues
GREENWOOD and NIKULIN · A Guide to Chi-Squared Testing
GROSS and HARRIS · Fundamentals of Queueing Theory, *Third Edition*
*HAHN and SHAPIRO · Statistical Models in Engineering
HAHN and MEEKER · Statistical Intervals: A Guide for Practitioners
HALD · A History of Probability and Statistics and their Applications Before 1750

*Now available in a lower priced paperback edition in the Wiley Classics Library.

*Now available in a lower priced paperback edition in the Wiley Classics Library.

*Now available in a lower priced paperback edition in the Wiley Classics Library.

SCHIMEK · Smoothing and Regression: Approaches, Computation, and Application
SCHOTT · Matrix Analysis for Statistics
SCHUSS · Theory and Applications of Stochastic Differential Equations
SCOTT · Multivariate Density Estimation: Theory, Practice, and Visualization
*SEARLE · Linear Models
SEARLE · Linear Models for Unbalanced Data
SEARLE · Matrix Algebra Useful for Statistics
SEARLE, CASELLA, and McCULLOCH · Variance Components
SEARLE and WILLETT · Matrix Algebra for Applied Economics
SEBER and LEE · Linear Regression Analysis, *Second Edition*
SEBER · Multivariate Observations
SEBER and WILD · Nonlinear Regression
SENNOTT · Stochastic Dynamic Programming and the Control of Queueing Systems
*SERFLING · Approximation Theorems of Mathematical Statistics
SHAFER and VOVK · Probability and Finance: It's Only a Game!
SMALL and McLEISH · Hilbert Space Methods in Probability and Statistical Inference
SRIVASTAVA · Methods of Multivariate Statistics
STAPLETON · Linear Statistical Models
STAUDTE and SHEATHER · Robust Estimation and Testing
STOYAN, KENDALL, and MECKE · Stochastic Geometry and Its Applications, *Second Edition*
STOYAN and STOYAN · Fractals, Random Shapes and Point Fields: Methods of Geometrical Statistics
STYAN · The Collected Papers of T. W. Anderson: 1943–1985
SUTTON, ABRAMS, JONES, SHELDON, and SONG · Methods for Meta-Analysis in Medical Research
TANAKA · Time Series Analysis: Nonstationary and Noninvertible Distribution Theory
THOMPSON · Empirical Model Building
THOMPSON · Sampling, *Second Edition*
THOMPSON · Simulation: A Modeler's Approach
THOMPSON and SEBER · Adaptive Sampling
THOMPSON, WILLIAMS, and FINDLAY · Models for Investors in Real World Markets
TIAO, BISGAARD, HILL, PEÑA, and STIGLER (editors) · Box on Quality and Discovery: with Design, Control, and Robustness
TIERNEY · LISP-STAT: An Object-Oriented Environment for Statistical Computing and Dynamic Graphics
TSAY · Analysis of Financial Time Series
UPTON and FINGLETON · Spatial Data Analysis by Example, Volume II: Categorical and Directional Data
VAN BELLE · Statistical Rules of Thumb
VIDAKOVIC · Statistical Modeling by Wavelets
WEISBERG · Applied Linear Regression, *Second Edition*
WELSH · Aspects of Statistical Inference
WESTFALL and YOUNG · Resampling-Based Multiple Testing: Examples and Methods for p-Value Adjustment
WHITTAKER · Graphical Models in Applied Multivariate Statistics
WINKER · Optimization Heuristics in Economics: Applications of Threshold Accepting
WONNACOTT and WONNACOTT · Econometrics, *Second Edition*
WOODING · Planning Pharmaceutical Clinical Trials: Basic Statistical Principles
WOOLSON and CLARKE · Statistical Methods for the Analysis of Biomedical Data, *Second Edition*
WU and HAMADA · Experiments: Planning, Analysis, and Parameter Design Optimization
YANG · The Construction Theory of Denumerable Markov Processes
*ZELLNER · An Introduction to Bayesian Inference in Econometrics
ZHOU, OBUCHOWSKI, and McCLISH · Statistical Methods in Diagnostic Medicine

*Now available in a lower priced paperback edition in the Wiley Classics Library.